Letts study aids

A-Level Mathematics

Course Companion

Duncan Graham BSc, BSc(Hons), MSc

Lecturer in Mathematics, Rolle College, Exmouth

Christine Graham BEd(Hons)

Allan Whitcombe BSc(Hons) MSc

Advisory Teacher Mathematician,
the City of Birmingham Education Authority

Charles Letts & Co Ltd
London, Edinburgh & New York

First published 1984
by Charles Letts & Co Ltd
Diary House, Borough Road, London SE1 1DW

Editor: Alan Taman
Design: Ben Sands
Illustrations: Tek Art

ISBN 0 85097 601 4

Printed and bound by
Charles Letts (Scotland) Ltd

Preface

The key aim of this book is to assist students in their study and revision of A-level, AO- level and Scottish Higher Mathematics and O-level Statistics, and it has been written for students who are studying their mathematics at school, college, evening class or at home.

Most students of A-level mathematics study single-subject mathematics leading to one A-level. Typical combinations are:

pure mathematics + mechanics;
pure mathematics + statistics;
or pure mathematics + mechanics + statistics.

If you are one of these students this book has been designed with you specifically in mind.

If you are one of the smaller number of students studying double-subject mathematics leading to two A-levels this book will cover at least a half of the work you have to do. Subject combinations are as above but with wider coverage and greater depth.

This book should act as a course companion from the beginning and throughout any A-level mathematics course (or its equivalent) and could be used as a study-aid, work-book and revision aid. It is not intended to provide a complete guide to the subject since the book is designed to complement a textbook rather than duplicate it.

This course companion has been produced after analysing the most recent appropriate mathematics syllabuses of all the Exam Boards in England, Wales and Scotland. Most of the mathematics required by these syllabuses has been divided into 86 Study Units, and Table I (p. xii) clearly shows which of these topics are included in each syllabus. A blank column is left for completion by the student whose examination is not listed. Table II (p. xiv) analyses the methods of examination.

The largest section of the book is Part II, the Study Units. Each Study Unit contains essential information, worked examples and question practice from recent examination papers relating to that unit. The essential information omits proofs and other similar material which may be found in most standard texts. The material has been presented in a clear and concise fashion which should be easily accessible to the student.

As well as the Study Units there is Part I, entitled 'Working for A-level', which includes a complete guide on how to use the book, together with hints on study, revision and examination techniques.

The majority of Study Aids/Revision Guides claim to offer a complete coverage of A-level syllabuses, but seem to have few, if any, advantages over a traditional textbook. In this Letts Course Companion the authors have made a genuine attempt to break the mould of the standard presentation of such material with their fresh, unique presentation. They are convinced that all students, including those who feel unsuccessful with conventional texts, will gain substantial assistance with their study of A-level mathematics and its examination from the use of this book.

Acknowledgements

We wish to thank the following people for their help in producing this book: Rod Parsons for constructive advice and criticism on the entire manuscript; Chris Aylott and Dave Crocker for extensive assistance in solutions to past examination questions; Norma Whitcombe for general help and encouragement, and other members of our families whose patience and understanding have been appreciated; and the staff of Charles Letts and Co. Ltd.

We are most grateful to the following Examination Boards for permission to use questions from their recent examination papers: the Associated Examining Board, the University of Cambridge, the Joint Matriculation Board, the University of London, the University of Oxford, Oxford and Cambridge, the Southern Universities Joint Board, the Welsh Joint Education Committee and the Scottish Examinations Board.

None of the above boards can accept responsibility for the answers to examination questions. In particular, the University of London University Entrance and School Examinations Council accepts no responsibility whatsoever for the accuracy or method of working in the answers given.

Contents

Part I
Working for A Level

DIFFERENCES BETWEEN O– AND A–LEVEL STUDY

At the beginning of your A-level course you may be anxious to know how it will differ from the O-level course you have probably just completed. This anxiety may have arisen because you have heard about other students failing or under-achieving because they have not successfully adjusted to their more advanced course and its demands.

Mathematics is different

In many other subjects studied at this level, students find it disconcerting that it is possible to be asked almost the same question at O level and A level but that different answers are required. Obviously the difference must be in the standard expected and students often have difficulty in assessing the correct level for their work. Mathematics is different: this same question/different answer situation cannot arise in an A-level mathematics course.

Different context

In mathematics a few of the topics studied at O level will be reconsidered at A level to a greater level of understanding but, in general, you will be studying mostly new topics and a much wider syllabus. Most of the mathematics you studied at O level is 'Pure Mathematics' but the majority of students at A level follow a course which includes both 'Pure Mathematics' and 'Applied Mathematics'; the 'Applied Mathematics' is usually mechanics and/or statistics. It will be assumed that you are able to use the mathematical background and expertise developed during an O-level course as a working part of mathematical knowledge. You will be expected to work hard on your own to acquire the new basic concepts to which you will be introduced and to develop a thorough understanding of the fundamental principles so that you can apply them in a variety of situations. Much more than at O level, you will need to be able to prove standard results underlying principles used. There is not always a single 'correct' method to solve a problem but you will learn to appreciate also that there is often an 'elegant' solution which may be preferred.

Different study skills

The major difficulty that most students experience is in adjusting to the transition from the more formal, probably highly structured pattern of most O-level teaching to the more independent approach usually encountered at A level. At O level you probably relied on your teacher to organise your work and on your class notes, and perhaps a single textbook, to provide your information. At A level you will be expected to organise your own work to a much greater extent, to use a wider range of sources of information to supplement your course notes, to maintain the necessary self-discipline to work industriously throughout your course and to have the essential motivation and determination to succeed.

Motivation

It is important to appreciate that, at this level, you have deliberately chosen to follow this course and to be aware of your own reasons for doing so. You may have chosen to study A-level mathematics because mathematics is your 'best' subject, your most 'interesting' subject, needed for another course you wish to follow, a requirement for the job you want to do ... but, whatever the reason for your choice, it must be strong enough to sustain your motivation to study. Good motivation is usually associated with an interest in a subject and you will find that you will be able to increase your own motivation if you can broaden your mathematical interests. One way to do this is to try to learn more about mathematics in general, about its history, its great scholars, its applications, its relationships with other subjects. You can do this by reading around the subject (a selection of general interest books are given in the booklist on page ix) and looking out for 'mathematical' features in television and radio programmes, etc. Besides acquiring a more extensive knowledge of mathematical ideas and methods, your mathematical maturity will increase and you will begin to appreciate the beauty and immense power of mathematics. The abilities which you will acquire, enabling you to analyse information and organise structures, will be found useful in many aspects of life.

GUIDE TO USING THIS BOOK

How this book can help you

The main aim of this book is to help you throughout your study of a mathematics course at A level (or its equivalent). It can help you because it:

- is based on a detailed analysis of the most recent appropriate mathematics syllabuses of all the Examination Boards in England, Wales and Scotland;
- gives you detailed information about the course you are following and the examination you will be taking;
- identifies which topics you need to study for your syllabus;
- provides quick and easy access to material on these topics in separate Study Units
- summarises concisely, but comprehensively, the subject content of each topic;
- reminds you of correct notation and units;
- identifies key terms and their meanings;
- demonstrates techniques in problem solving in worked examples;
- guides you through solutions to problems;
- gives you a selection of past examination questions for you to work through to check your progress and practise your skills;
- provides answers for checking;
- advises you about study skills, revision strategy and examination technique.

Identifying your course

To use this book most effectively as a course companion you need to identify the course you are studying. The answers to the following questions will help you to do so:

What is the name of your Examination Board?
What type of examination are you taking?
What is the reference number of your syllabus?
Which of the following sections are you studying: Pure Mathematics, Mechanics, Statistics?
Which options (if any) are you taking?

Your study units

Most of the mathematics required by the syllabuses analysed in this book has been divided into 86 Study Units which have been grouped into three sections: Pure Mathematics, Mechanics, and Statistics. For ease of reference, Table I, the analysis of the examination syllabuses, on pages xii-xiii is presented in these three sections and it clearly shows which of the topics are included in each syllabus. To find out which Study Units are of interest to you in each section you are studying simply look along the top of the table to find your Examination Board, then look down the column for your syllabus. (You may find this easier if you place a ruler alongside your column.) Each box in the column relates to a Study Unit in this book. If a box:

- is blank □ — ignore that Unit because you do not need it;
- has a dot ▣ — study that Unit because it is on your syllabus;
- has a letter ▣ — look for the letter(s) in the table footnotes to find out what to do about that Unit

If the syllabus you are studying does not appear in Table I, you should either ask your teacher or write to your Examination Board (address on page ix) for the details. You can then enter your syllabus in the blank end column of the appropriate part of the table.

Although Table I has been prepared very carefully from the most recent syllabuses, syllabuses sometimes change slightly from year to year. If in doubt, ask your teacher to check the column for your syllabus or you can check the syllabus yourself.

How the study units are presented

Each Study Unit is presented on a double-page spread in Part II of this book. On the left-hand page you will find all the notes on the topic, together with simple examples called Illustrations. On the right-hand page, you will find relevant Worked and Guided Examples and questions from past examination papers (answers to which are given at the end of the book).

The diagram on pages x and xi summarises all the significant features of the layout of a Study Unit. Look carefully at these two pages to understand how the information, examples and questions in a typical Study Unit are arranged.

Ways of using this book

The authors anticipate that students will use this course companion in three ways: as a study aid, as a workbook and as a revision aid.

If you use it as a study aid, you can use the Study Units to reinforce, or teach yourself, the essential features of each topic introduced in your course and the section on 'Study Skills' to improve your approach to study itself.

If you use it as a workbook, you can use the carefully graded types of questions, i.e. Illustrations, Worked Examples, Guided Examples and Exercises, in each Unit to help you to improve understanding of each topic, check progress and develop problem-solving technique.

If you use it as a revision aid, you can use the syllabus analysis and the Study Units as the basis of your revision programme and the sections on Revision Strategy and Examination Techniques to improve your approach to revision and examinations.

You may use this book in one or more of these ways or, alternatively, you may have your own ideas on how to integrate this book into your course of study. If you have any such ideas, the authors would be pleased to hear about them (write care of the publishers) so that they can be taken into account in future editions.

STUDY SKILLS

Unfortunately study skills are rarely taught to students but if you can develop good study skills your learning will be more efficient and more effective, which means that you will learn faster and better. The important thing is to have the right attitude to studying, in the end you must want to study to be successful.

Your study plan

Whether you are studying at school, college, evening class or at home, there will be a much greater onus on you at this level to organise and plan your own work. You will have to spend a considerable amount of time working on your own, possibly without a great deal of guidance, so a personal study plan is essential if you are to make the best use of your own time. If you are attending lectures or lessons at college or school, you will already have a weekly timetable for these. Your personal study plan will be an extension of this timetable and you will find it useful if it covers not only study periods but other activities which feature regularly in your week. It will help you to spread your work sensibly throughout the week on each of the subjects you are studying and to meet deadlines for assignments set on your course(s).

Although your own weekly plan must suit your particular needs and pattern of study, it is important to be realistic about deciding how much time you have to spend on it. It is better to spread your working time as evenly as possible throughout your week, organising a study period for each of five or six days during the week rather than concentrating it all on one or two days. To be most productive your study periods should be from one to three hours long with planned short breaks of 5–10 minutes duration every hour to help you to maintain concentration. These breaks will be most effective if you can do something different, such as leaving your work room, making a drink, etc., rather than just sitting at your desk. Each study period should be followed by a longer break for recreation and relaxation and these are as much a part of your study plan as the study periods.

When to study

When you study is a matter of personal preference and the time you have available. Some students are 'early birds', doing their best studying in the morning, some are 'late birds', studying best at night. You should find out, if you do not know already, when you do your 'best work' and, if it is possible, fit this into your study plan.

Where to study

You will find studying easier if you have a room, or part of a room, which you can identify as your own study place. Not only will this have a positive psychological effect on you, since you will associate the place with study and find it easier to settle down to work, but also on your family and friends, since they will associate it with your study and, hopefully, not disturb you unnecessarily. It is important that you find it comfortable and attractive: it should not be too warm but well ventilated and away from distractions. A desk or table, on which you can spread your books, is ideal for working on and the light should be good, the best form of lighting being a reading lamp. Keeping all your books, files, paper, calculator and other essential equipment readily available in this one place will save you a great deal of valuable time during your study periods.

Your study periods

Mathematics is a subject which you cannot study by simply reading: you need to 'do mathematics' to understand it. Private study periods will usually be spent either working on notes or on problems but to make them as productive as possible vary the topic and/or activity and set yourself definitive objectives. By starting each session with a brief review of the previous session on the same topic, you will find that it acts as a 'mental warming up' time and reinforces the previous session's study.

Making notes

As soon as possible after each lesson/lecture, you should write up your lesson/lecture notes whilst the topic is still fresh in your mind. The actual process of making notes will help you to understand the topic you are studying and encourage you to concentrate on what you are learning. Remember that your notes are for your use and will eventually form the basis of your revision plan, so the way you organise them must be concise and presented in a form which is best for you.

Every time you work through your lesson/lecture notes, your textbook(s) and this book, you should be jotting down notes and trying to work through the mathematics for yourself. This book will provide ideal resource material for your notes and you should use it also to vary your approach to note-making. Making card-index files, small notebooks or charts on themes such as key terms, formulas, etc. are excellent note-making activities which you will find extremely useful during your revision time. Using the alternative approaches supplied by other textbooks is a worthwhile activity and will help to broaden your mathematical knowledge.

Understanding the topic you are studying is essential but if, after making reasonable efforts to do so, you still cannot understand something, do not worry about it: make a note of it and ask about it later. Do not allow worrying about one topic to ruin the rest of your work schedule for a study session.

Problem solving

Problem solving is one of the most important aspects of your mathematics course so you will need to spend a large part of your private study time tackling problems, either for homework assignments or on your own account to improve your understanding of a topic and your ability to answer questions.

It is a good idea, if possible, to attempt problems set for homework assignments more than a day before they are due to be handed in. This will give you the opportunity, if needed, to sort out any difficulties you may encounter, either by consulting this book, your textbook, your teacher or simply by giving yourself time to 'sleep on it'. Persevere with your problem solving: do not give in too easily, but on the other hand do not waste time on a problem that is not yielding to your strategy and on which you do not have any alternative strategies to try. Leaving a problem and coming back to it can often be fruitful, the break giving you new insight into its solution.

You will find that working through the notes provided in the appropriate Study Unit in this book will remind you of the basic principles, notation, formulae, units, etc. needed in your problem solving sessions. Working through the examples in each Study Unit, stage by stage, will help to develop your problem solving technique. As you work through each section of a Study Unit, cover the solution to each Illustration (marked ⒤), try to answer the question yourself and then compare your working with that given. When you feel that you understand a topic, try to answer the Worked and Guided Examples, before referring to the given solution for guidance. If you are really stuck, either look at the first few lines of the explanation to give yourself a start and then try to complete the solution yourself or cover the working in the book with a piece of paper and only look at each line of the solution after attempting that part of it for yourself. Only rarely will you find that you can do none of a problem, so using the Worked and Guided Examples in this way will help you to locate your weaknesses in a particular topic and to build up your confidence in your problem-solving abilities. Finally try the questions from the past examination papers, those marked with * being the easiest, coming from Alternative Ordinary level papers. There is no need to restrict yourself to questions from your own Examination Board; try to answer all the questions on the Unit you have studied. You will find answers to these questions at the end of the book.

Learning

Mathematics is a very demanding discipline. To understand some of the basic ideas and concepts will involve you in some hard thinking and for the first time in your intellectual development you may be aware of consciously directing your critical powers to discover the nature of any misunderstandings which you may have.

There are a large number of identities, formulae and equations in mathematics and most probably your Examination Board will provide a book of formulae for you to use. There is no reason why many of these formulae should not be committed to memory. If you make an attempt to memorise new results when you first see them it is not such a daunting task as it may appear and the effort pays handsome dividends when you do not constantly have to refer to a formulae book when solving problems. However, if you are not confident in your memorising abilities, regular use of the formulae book will help you to get to know where in the book a formula appears so that you can check.

At A-level you may also be asked to prove results and standard bookwork: make sure that you understand the 'bookwork' and can produce these proofs when necessary.

REVISION STRATEGY

Revision should be an on-going process which starts very early in your course. The amount of knowledge to be accumulated and the variety of skills and techniques to be developed are large and they are best assimilated gradually and consolidated as you go along. Regular revision is really a part of the learning process but, of necessity, becomes more concentrated as the examination approaches.

Regular reviews

Research has shown that, although factual recall declines rapidly after the first session, the use of regular reviews keeps the level of recall much nearer to its original very high level. It is important, therefore, that the strategy of regular reviewing should be a continuous and integral aspect of a planned study programme. If you review each section of work one session, one week, one month, or three months after you have originally studied it, then you will have revised each topic at least four times before you reach your final intensive revision programme.

Final revision programme

At the start of your final revision programme (say 3 months before your actual examination) you must get organised and the best way to do this is to devise a revision timetable. The strategy involved is basically the same as that used throughout the study of your A-level course. Plan your time carefully, give yourself definite objectives for each session, revise actively, test yourself regularly, make notes, practise problem solving. Use revision sessions to study topics you have worked on before as revision is simply the process of reminding you of topics and techniques previously understood. You will appreciate how well-organised notes will help you during your revision. Write out important definitions, proofs, formulae and equations, checking them against your notes or this book. Rework previously solved problems without looking at your previous solution, then attempt questions that you have not looked at before. Make special revision notes for quick reference on cards to keep in your pocket and charts to hang on the wall of your study room. Practise your examination technique.

Examination practice

Familiarise yourself with the mode of the examination you are going to sit by studying Table II on p. xiv and analysing examples of recent examination papers for your Board. Booklets of past examination papers can be bought directly from your Board (address on p. ix) and if you are a student at school or college your teacher should have a stock of these papers. A study of these papers will enable you to calculate the amount of time you will have available to solve each question and to grow accustomed to the style of paper you are going to sit eventually.

During your ordinary study periods you will have attempted many questions but seldom given yourself strict time restrictions. In examinations the timing of your answers to questions is vitally important. Practise answering examination questions in mock examination conditions, allowing yourself only the normal available examination time and the equipment you are permitted to take into the examination room. To obtain 'mock examination' practice save one or two complete examination papers so that you can use them as final test papers 'against the clock'.

EXAMINATION TECHNIQUE

Examination nerves are common and understandable but if you have worked conscientiously and followed a sensible course of study and revision, then they should not be too serious. However, you may not do yourself justice if you have a poor examination technique. The following hints should help you to tackle the examination with greater confidence.

Before the day

Before the actual day of your examination make sure you know:

- the date, day, time and place of each paper of your examination;
- how to get to your examination centre if it is not well known to you;
- your candidate number;
- your examination centre number;
- the telephone number of your examination centre.

Prepare any equipment you will need for your particular examination:

- pens which are comfortable to use and ink,
- sharp pencils, a pencil sharpener and rubber,
- drawing instruments such as a ruler, compasses, protractor, set squares,
- calculator (if allowed) and spare batteries (check that you know how to replace them quickly),
- an accurate watch or small clock.

On the day

Before the examination Check that you have all the equipment you will need before setting off for your examination centre with plenty of time to spare. If you are delayed, contact your examination centre (have the telephone number with you) to explain what has happened. Arrive at the examination room early; a late start to an examination cannot be a good start.

Just before the start Listen carefully to the invigilator. There may be some changes or special instructions which you were not expecting or some errors in the paper. Fill in any details such as candidate number and examination centre number when the invigilator instructs you to do so.

Reading the instructions When the invigilator says that you may begin, read the instructions on your examination paper very carefully. Make sure that it is the correct examination paper! Although you will be familiar with the style and format of past papers for your examination, these can change without notice. Note these in particular:

- the number of sections and questions you have to do;
- how much time you have to do them in;
- which questions (if any) are compulsory;
- what choice of questions (if any) do you have;
- how to present your answers.

Planning your time Quickly calculate the length of time you should spend on each question. You will have practised doing this for past papers but make sure that you use the instructions on your actual examination paper, not the ones you are expecting. Try to allow about 10 minutes at the end for checking your paper.

Choosing the questions Read through the whole examination paper carefully, checking that you have read every page. If you have a choice of questions:

- cross the ones you can't do;
- tick those you can definitely do;
- choose the correct number to do;
- mark the order in which you are going to attempt them, attempting your best question(s) first.

Try to answer full questions if you can but you can sometimes pass an examination by answering a lot of part questions. Indeed, questions are often structured—the first part being easier to answer than later parts. Some Examination Boards list the marks to be awarded for each question or part question. This information will help you to decide which questions or part questions to do.

Answering the questions Before you attempt to answer a question, read it all again carefully, jotting down points such as formulae and information relating to that question. These hints should help you when writing an answer.

1. Make sure that your writing is legible.
2. Draw a large clearly labelled diagram if appropriate.
3. Present your solution in a neat, logical and concise way.
4. Show all your working; many marks are given for working, not answers.
5. Solve the problem which has been set and not the one you think is being posed.
6. Do not do things you are not asked for; for example, do not do proofs unless specifically requested.
7. State any principles, results, formulae, etc. used and indicate your reasons for using them.
8. Check any formulae you use with the formula sheet, if provided.
9. Use and state the correct units, e.g. $\mathrm{m\ s^{-2}}$, N s, etc.
10. Always do a rough estimate of any calculation to check that your answer is sensible.
11. When using a calculator, make sure that each calculation is shown clearly in your answer.
12. Give your final answer to the required degree of accuracy.
13. In questions saying 'hence or otherwise', try 'hence' first since it is usually easier and uses the suggestion given in the first part of the question.
14. If you get 'stuck', re-read the question carefully to check that you have not missed any important information or hints given in the question itself.
15. When you have completed your solution, re-read the question to check that you have answered all parts.

Examination discipline It is important that you try to keep to the times you have allocated to answering a question or section and that you answer the correct number of questions. If you answer less than the number required you are limiting the number of marks available to you.

In short-answer papers or sections, which are often compulsory, if you cannot see how to solve a problem fairly quickly, leave it and return to it later if you have time. A fresh look at a question often helps.

In longer-question papers or sections do not overrun your time allocation on any question by more than a minute or so. Do not be lured into thinking 'just a few more minutes and I'll have the answer'. In most examinations, the first parts to many questions are easier than the later parts so it is usually easier to gain more marks by attempting all the questions required than by completing a question.

Different Examination Boards have different policies regarding candidates who have answered too many questions. This can vary from year to year so check on your board's policy. If you answer too many questions and your board:

1. Marks all your questions and ignores your worst marks— hand in all your answers.
2. Ignores your last questions—cross out the questions you feel you have done badly, leaving only the correct number to be marked.

At the end Before handing in your examination paper check that:

- any 'front sheet' is completed according to the instructions;
- every loose page is clearly marked with your examination number, etc;
- every answer is numbered correctly;
- pages are numbered clearly and in order.

BOOKLIST

Pure Mathematics

Essential Mathematics for A-level. D. C. Taylor and I. S. Atkinson. Thomas Nelson and Sons Ltd.

Advanced Level Pure Mathematics. C. J. Tranter. The English Universities Press Ltd.

Advanced Level Pure Mathematics. S. L. Green. University Tutorial Press.

Pure Mathematics. E. D. Hodge and B. G. J. Wood. Blackie and Son Ltd.

Pure Mathematics for Advanced Level. B. D. Bunday and H. Mulholland. Butterworths.

Mechanics

Applied Mathematics. E. D. Hodge and B. G. J. Wood. Blackie and Son Ltd.

Applied Mathematics I and II. L. Bostock and S. Chandler. Stanley Thornes (Publishers) Ltd.

Applied Mathematics for Advanced Level. H. Mulholland and J. H. G. Phillips. Butterworths.

Advanced Level Applied Mathematics. C. G. Lambe. The English Universities Press Ltd.

A-Level Applied Mathematics A. J. Francis. Bell & Hyman.

Statistics

Statistics and Probability. S. E. Hodge and M. L. Seed. Blackie and Son Ltd.

Modern Applied Mathematics. J. C. Turner. Hodder and Stoughton.

Fundamentals of Statistics. H. Mulholland and C. R. Jones. Butterworths.

Advanced General Statistics. B. C. Erricker. Hodder and Stoughton.

Advanced Level Statistics. A. Francis. Stanley Thornes (Publishers) Ltd.

General and background reading

A Path to Modern Mathematics. W. W. Sawyer. Pelican.

Mathematician's Delight. W. W. Sawyer. Pelican.

Prelude to Mathematics. W. W. Sawyer. Pelican.

Men of Mathematics 1, 2. E. T. Bell. Pelican.

How to Lie with Statistics. D. Huff. Pelican.

Mathematics in Western Culture. M. Kline. Pelican.

The World of Mathematics. (4 volumes.) J. R. Newman. Allen and Unwin Ltd.

The Gentle Art of Mathematics. D. Pedoe. Pelican.

The Scientific American Book of Mathematical Puzzles and Diversions. M. Gardner. Penguin.

The Mathematical Experience. P. J. Davis and R. Hersh. Penguin.

EXAMINATION BOARDS AND ABBREVIATIONS

(A) *AEB* — Associated Examining Board
Wellington House, Aldershot, Hampshire GU11 1BQ

(C) *Cambridge UCLES* — University of Cambridge Local Examinations Syndicate
Syndicate Buildings, 17 Harvey Road, Cambridge CB1 2EU

(J) *JMB* — Joint Matriculation Board, Manchester M15 6EU
(For JMB Publications contact John Sherratt & Son Ltd,
78 Park Road, Altrincham, Cheshire)

(L) *London* — University Entrance & Schools Examination Council
University of London, 66-72 Gower Street, London WC1H 0PJ
Publications Office: 52 Gordon Square, London WC1H 0PJ

(OLE) *Oxford* — Oxford Local Examinations
Delegacy of Local Examinations, Ewert Place, Summertown, Oxford OX2 7BX

(O & C) *Oxford & Cambridge* — Oxford & Cambridge Schools Examination Board
10 Trumpington Street, Cambridge & Elsfield Way, Oxford

(S) *SUJB* — Southern Universities Joint Board for School Examinations
Cotham Road, Bristol BS6 6DD

(W) *WJEC* — Welsh Joint Education Committee
245 Western Avenue, Cardiff CF5 2YX

(H) *SEB* — Scottish Examination Board
Ironmills Road, Dalkeith, Midlothian EH22 1BR
(For SEB Publications contact Robert Gibson & Sons Ltd,
17 Fitzroy Place, Glasgow G3 7SF)

(N) *NIGCE* — Northern Ireland General Certificate of Education Examinations Board
Beechill House, 42 Beechill Road, Belfast BT8 4RS

LAYOUT OF A STUDY UNIT

Look at the items ①–⑭ in order, to find out the things you need to know about the organization of a Study Unit.

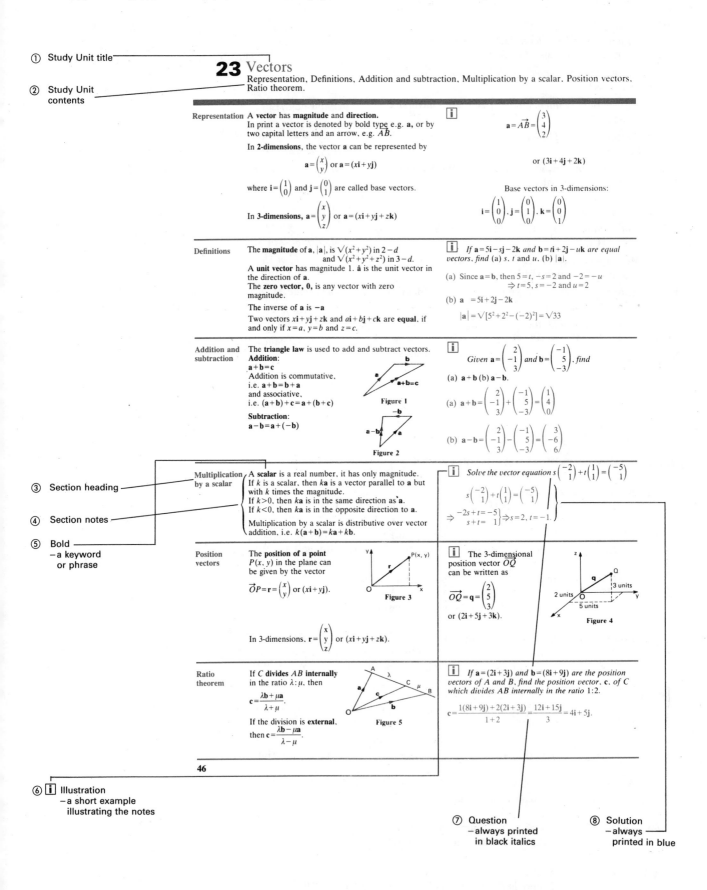

① Study Unit title

② Study Unit contents

23 Vectors
Representation, Definitions, Addition and subtraction, Multiplication by a scalar. Position vectors. Ratio theorem.

Representation A **vector** has **magnitude** and **direction**.
In print a vector is denoted by bold type e.g. **a**, or by two capital letters and an arrow, e.g. $\vec{AB}$.

In **2-dimensions**, the vector **a** can be represented by

$$\mathbf{a} = \begin{pmatrix} x \\ y \end{pmatrix} \text{ or } \mathbf{a} = (x\mathbf{i} + y\mathbf{j})$$

where $\mathbf{i} = \begin{pmatrix} 1 \\ 0 \end{pmatrix}$ and $\mathbf{j} = \begin{pmatrix} 0 \\ 1 \end{pmatrix}$ are called base vectors.

In **3-dimensions**, $\mathbf{a} = \begin{pmatrix} x \\ y \\ z \end{pmatrix}$ or $\mathbf{a} = (x\mathbf{i} + y\mathbf{j} + z\mathbf{k})$

ⓘ

$$\mathbf{a} = \vec{AB} = \begin{pmatrix} 3 \\ 4 \\ 2 \end{pmatrix}$$

or $(3\mathbf{i} + 4\mathbf{j} + 2\mathbf{k})$

Base vectors in 3-dimensions:

$$\mathbf{i} = \begin{pmatrix} 1 \\ 0 \\ 0 \end{pmatrix}, \mathbf{j} = \begin{pmatrix} 0 \\ 1 \\ 0 \end{pmatrix}, \mathbf{k} = \begin{pmatrix} 0 \\ 0 \\ 1 \end{pmatrix}$$

Definitions The **magnitude** of **a**, $|\mathbf{a}|$, is $\sqrt{(x^2 + y^2)}$ in $2-d$ and $\sqrt{(x^2 + y^2 + z^2)}$ in $3-d$.
A **unit vector** has magnitude 1. $\hat{\mathbf{a}}$ is the unit vector in the direction of **a**.
The **zero vector, 0**, is any vector with zero magnitude.
The inverse of **a** is $-\mathbf{a}$
Two vectors $x\mathbf{i} + y\mathbf{j} + z\mathbf{k}$ and $a\mathbf{i} + b\mathbf{j} + c\mathbf{k}$ are **equal**, if and only if $x = a$, $y = b$ and $z = c$.

ⓘ *If $\mathbf{a} = 5\mathbf{i} - s\mathbf{j} - 2\mathbf{k}$ and $\mathbf{b} = t\mathbf{i} + 2\mathbf{j} - u\mathbf{k}$ are equal vectors, find (a) s, t and u. (b) $|\mathbf{a}|$.*

(a) Since $\mathbf{a} = \mathbf{b}$, then $5 = t$, $-s = 2$ and $-2 = -u$
$\Rightarrow t = 5$, $s = -2$ and $u = 2$

(b) $\mathbf{a} = 5\mathbf{i} + 2\mathbf{j} - 2\mathbf{k}$
$|\mathbf{a}| = \sqrt{[5^2 + 2^2 - (-2)^2]} = \sqrt{33}$

Addition and subtraction The **triangle law** is used to add and subtract vectors.
Addition:
$\mathbf{a} + \mathbf{b} = \mathbf{c}$
Addition is commutative,
i.e. $\mathbf{a} + \mathbf{b} = \mathbf{b} + \mathbf{a}$
and associative,
i.e. $(\mathbf{a} + \mathbf{b}) + \mathbf{c} = \mathbf{a} + (\mathbf{b} + \mathbf{c})$

Figure 1

Subtraction:
$\mathbf{a} - \mathbf{b} = \mathbf{a} + (-\mathbf{b})$

Figure 2

ⓘ *Given $\mathbf{a} = \begin{pmatrix} 2 \\ -1 \\ 3 \end{pmatrix}$ and $\mathbf{b} = \begin{pmatrix} -1 \\ 5 \\ -3 \end{pmatrix}$, find*

(a) $\mathbf{a} + \mathbf{b}$ (b) $\mathbf{a} - \mathbf{b}$.

(a) $\mathbf{a} + \mathbf{b} = \begin{pmatrix} 2 \\ -1 \\ 3 \end{pmatrix} + \begin{pmatrix} -1 \\ 5 \\ -3 \end{pmatrix} = \begin{pmatrix} 1 \\ 4 \\ 0 \end{pmatrix}$

(b) $\mathbf{a} - \mathbf{b} = \begin{pmatrix} 2 \\ -1 \\ 3 \end{pmatrix} - \begin{pmatrix} -1 \\ 5 \\ -3 \end{pmatrix} = \begin{pmatrix} 3 \\ -6 \\ 6 \end{pmatrix}$

③ Section heading

④ Section notes

⑤ Bold
  – a keyword or phrase

Multiplication by a scalar A **scalar** is a real number, it has only magnitude.
If k is a scalar, then $k\mathbf{a}$ is a vector parallel to **a** but with k times the magnitude.
If $k > 0$, then $k\mathbf{a}$ is in the same direction as **a**.
If $k < 0$, then $k\mathbf{a}$ is in the opposite direction to **a**.
Multiplication by a scalar is distributive over vector addition, i.e. $k(\mathbf{a} + \mathbf{b}) = k\mathbf{a} + k\mathbf{b}$.

ⓘ *Solve the vector equation $s\begin{pmatrix} -2 \\ 1 \end{pmatrix} + t\begin{pmatrix} 1 \\ 1 \end{pmatrix} = \begin{pmatrix} -5 \\ 1 \end{pmatrix}$*

$s\begin{pmatrix} -2 \\ 1 \end{pmatrix} + t\begin{pmatrix} 1 \\ 1 \end{pmatrix} = \begin{pmatrix} -5 \\ 1 \end{pmatrix}$

$\Rightarrow \begin{matrix} -2s + t = -5 \\ s + t = 1 \end{matrix} \Rightarrow s = 2, t = -1.$

Position vectors The **position** of a point $P(x, y)$ in the plane can be given by the vector
$$\vec{OP} = \mathbf{r} = \begin{pmatrix} x \\ y \end{pmatrix} \text{ or } (x\mathbf{i} + y\mathbf{j}).$$

Figure 3

In 3-dimensions, $\mathbf{r} = \begin{pmatrix} x \\ y \\ z \end{pmatrix}$ or $(x\mathbf{i} + y\mathbf{j} + z\mathbf{k})$.

ⓘ The 3-dimensional position vector $\vec{OQ}$ can be written as
$$\vec{OQ} = \mathbf{q} = \begin{pmatrix} 2 \\ 5 \\ 3 \end{pmatrix}$$
or $(2\mathbf{i} + 5\mathbf{j} + 3\mathbf{k})$.

Figure 4

Ratio theorem If C divides AB **internally** in the ratio $\lambda : \mu$, then
$$\mathbf{c} = \frac{\lambda\mathbf{b} + \mu\mathbf{a}}{\lambda + \mu}.$$
If the division is **external**, then $\mathbf{c} = \frac{\lambda\mathbf{b} - \mu\mathbf{a}}{\lambda - \mu}$.

Figure 5

ⓘ *If $\mathbf{a} = (2\mathbf{i} + 3\mathbf{j})$ and $\mathbf{b} = (8\mathbf{i} + 9\mathbf{j})$ are the position vectors of A and B, find the position vector, c, of C which divides AB internally in the ratio 1:2.*

$\mathbf{c} = \frac{1(8\mathbf{i} + 9\mathbf{j}) + 2(2\mathbf{i} + 3\mathbf{j})}{1 + 2} = \frac{12\mathbf{i} + 15\mathbf{j}}{3} = 4\mathbf{i} + 5\mathbf{j}.$

46

⑥ ⓘ Illustration
  – a short example illustrating the notes

⑦ Question
  – always printed in black italics

⑧ Solution
  – always printed in blue

⑨ **WE** Worked example
– shows you how to
solve the question
and how to set out
your working

⑪ **EX** Exercise
– past examination
questions for you
to practice answering

Vectors
Worked example, Guided example and Exam questions

 WE *In the diagram, $ST = 2TQ$, $\overrightarrow{PQ} = a$, $\overrightarrow{SR} = 2a$ and $\overrightarrow{SP} = b$.*

(a) Find in terms of a and b:

(i) $\overrightarrow{SQ}$

(ii) $\overrightarrow{TQ}$

(iii) $\overrightarrow{RQ}$

(iv) $\overrightarrow{PT}$

(v) $\overrightarrow{TR}$

(b) What do your answers to (iv) and (v) tell you about the the points P, T, R?

(a) (i) $\overrightarrow{SQ} = \overrightarrow{SP} + \overrightarrow{PQ} = b + a$ (or $a + b$ by commutativity)

(ii) $\overrightarrow{TQ} = \frac{1}{3} \overrightarrow{SQ} = \frac{1}{3}(a + b)$

(iii) $\overrightarrow{RQ} = \overrightarrow{RS} + \overrightarrow{SQ} = -2a + (a + b)$
$= b - a$

(iv) $\overrightarrow{PT} = \overrightarrow{PS} + \overrightarrow{ST}$
$= -b + \frac{2}{3} \overrightarrow{SQ}$ (since $\overrightarrow{ST} = \frac{2}{3} \overrightarrow{SQ}$, i.e. $\overrightarrow{ST} = 2\overrightarrow{TQ}$)
$= -b + \frac{2}{3}(a + b) = \frac{2}{3} a - \frac{1}{3} b$
$= \frac{1}{3}(2a - b)$

(v) $\overrightarrow{TR} = \overrightarrow{TS} + \overrightarrow{SR}$
$= -\frac{2}{3}(a + b) + 2a$
$= \frac{4}{3} a - \frac{2}{3} b = \frac{2}{3}(2a - b)$.

(b) Since $\overrightarrow{PT} = \frac{1}{3}(2a - b)$ and $\overrightarrow{TR} = \frac{2}{3}(2a - b)$, $\overrightarrow{PT}$ and $\overrightarrow{TR}$
are both multiples of the same vector $(2a - b)$. Hence PT and
TR are parallel and T is common to both lines, so, P, T, R lie
on the same line, i.e. they are co-linear.

 GE *(a) $\overrightarrow{OS}$ and $\overrightarrow{OT}$ represent the vectors $\lambda i + \mu j$ and $\mu i + \lambda j$ where λ
and μ are scalars and i and j are unit vectors in two mutually
perpendicular directions Ox and Oy. Show that $|\overrightarrow{OS}| = |\overrightarrow{OT}|$.
Given that OS and OT are two adjacent sides of a rhombus
OSUT, find the vectors represented by the diagonals OU and
ST.*

*(b) $\overrightarrow{PQ}$ and $\overrightarrow{PR}$ are represented by the sides PQ and PR of the
triangle PQR. Show that*

$$\overrightarrow{PQ} + \overrightarrow{PR} = 2\overrightarrow{PS}$$

*where S is the midpoint of QR.
Hence, or otherwise, find the position of the point O within the
triangle PQR such that*

$$\overrightarrow{OP} + \overrightarrow{OQ} + \overrightarrow{OR} = 0$$

(a) Use the definition of the magnitude of a vector to show
that OS = OT.
Sketch the rhombus OSUT. Use, $\overrightarrow{OU} = \overrightarrow{OS} + \overrightarrow{SU}$ and
$\overrightarrow{ST} = \overrightarrow{SO} + \overrightarrow{OT}$ to find the required vectors.

(b) Use the ratio theorem to express $\overrightarrow{PS}$ in terms of $\overrightarrow{PQ}$ and
$\overrightarrow{PR}$. Hence, required result.
Consider a point O on RT (where T is the midpoint of PQ).
Write down $\overrightarrow{OP} + \overrightarrow{OQ}$ using the result just established. Hence
show that

$$\overrightarrow{OP} + \overrightarrow{OQ} + \overrightarrow{OR} = 0,$$

where O is the point which divides RT in a certain ratio. State
what this point O is called.

⑩ **GE** Guided example
– suggests a method
for solving the question
(the answer(s) are given
in the book)

EX 1 The vector p has magnitude 7 units and bearing $052°$, and
the vector q has magnitude 12 units and bearing $163°$. Draw
a diagram (which need not be to scale) showing p, q and the
resultant $p + q$. Calculate, correct to one decimal place, the
magnitude of $p + q$.
*(L)

⑫ * Alternative
O-level
question

2 From an origin O the points A, B, C have position vectors
a, b, $2b$ respectively. The points O, A, B are not collinear.
The midpoint of AB is M, and the point of trisection of AC
nearer to A is T. Draw a diagram to show O, A, B, C, M, T.
Find, in terms a and b, the position vectors of M and T. Use
your results to prove that O, M, T are collinear, and find
the ratio in which M divides OT.
*(L)

3 Given that $OA = a$, $OB = b$, $OP = \frac{1}{4}OA$ and that Q is the
midpoint of AB, express AB and PQ in terms of a and b. PQ
is produced to meet OB produced at R, so that $QR = nPQ$
and $BR = kb$. Express QR: (i) in terms of n, a and b;
(ii) in terms of k, a and b. Hence find the value of n and of k.
*(C)

4 The position vectors of three points A, B and C relative to
an origin O are p, $3q - p$, and $9q - 5p$ respectively. Show
that the points A, B and C lie on the same straight line, and
state the ratio $AB:BC$. Given that $OBCD$ is a
parallelogram and that E is the point such that $DB = \frac{1}{4}DE$,
find the position vectors of D and E relative to O.
*(C)

⑬ No symbol
– A-level
question

5 The points A, B and C have position vectors a, b and c
respectively referred to an origin O.
(a) Given that the point X lies on AB produced so that
$AB:BX = 2:1$, find x, the position vector of X, in terms of
a and b.
(b) If Y lies on BC, between B and C so that $BY:YC = 1:3$,
find y, the position vector of Y, in terms of b and c.
(c) Given that Z is the mid-point of AC, show that X, Y
and Z are collinear.
(d) Calculate $XY:YZ$.
(L)

⑭ (L) – board which
set the
question

6 O, A and B are three non-collinear points; the position
vectors of A and B with respect to O are a and b
respectively. M is the mid-point of OB, T is the point of
trisection of AB nearer to B, $AMTX$ is a parallelogram and
OX cuts AB at Y. Find, in terms of a and b, the position
vectors of:
    (a) M;   (b) T;   (c) X;   (d) Y.
(O & C)

7 The vertices A, B and C of a triangle have position vectors
a, b and c respectively relative to an origin O. The point P is
on BC such that $BP:PC = 3:1$; the point Q is on CA such
that $CQ:QA = 2:3$; the point R is on BA produced such
that $BR:AR = 2:1$. The position vectors of P, Q and R are
p, q and r respectively. Show that q can be expressed in
terms of p and r and hence or otherwise show that P, Q and
R are collinear. State the ratio of the lengths of the line
segments PQ and QR.
(J)

8 The points P and Q have position vectors p and q
respectively relative to an origin O, which does not lie on
PQ. Three points R, S, T have respective position vectors
$r = \frac{1}{4}p + \frac{3}{4}q$, $s = 2p - q$, $t = p + 3q$. Show in one diagram
the positions of O, P, Q, R, S and T.
(J)

Table I — Analysis of Examination Syllabuses Pure Mathematics

Syllabus	AEB AO 186*	AEB AO 187	AEB AO 188	AEB A 648	AEB A 636	AEB A 646	Camb AO 8175	Camb A A(9200)	Camb A B(9202)	JMB AO PMM	JMB AO PMS	JMB AO PM	JMB A P1+Mel	JMB A P1+Aml	JMB A P1+S	Lon AO 870	Lon AO 871	Lon AO 872	Lon A B(371)	Lon A 420	Oxf AO 8853	Oxf A 9850	O&C AO 8645*	O&C A 9650	SUJB AO 8177	SUJB A 9203	Wel AO 0132	Wel A 0023	Scot H
1 Polynomials	●	●	●	●	●	●	●	●	●	●	●	●	●	●	●	●	●	●	●	●	B	●	A	●	●	●	●	●	●
2 Rational Functions	●	●	●	●	●	●	●	●	●	●	●	●	●	●	●	●	●	●	●	●	B	●	●	●	●	●	●	●	
3 Quadratics	●	●	●	●	●	●	●	●	●	●	●	●	●	●	●	●	●	●	●	●	B	●	●	●	●	●	ObA	●	●
4 Sequences and Series	●			●	●	●	BP	●	●	●	●	●	●	●	●	●	●	●	●	●	B	●		●	●	●	●	●	
5 Sumation of Series	●			●	●	●	BP	●	●	●	●	●	●	●	●	●	●	●	●	●	B	●		●	●	●	●	●	
6 Permutations & Combinations							●	●	●			●		●				●		●			C						
7 Binomial Theorem	●	●	●	●	●	●	●	●	●	●	●	●		●		●	●		●		B	●	●	●			●	●	●
8 Inequations	●	●	●	●	●	●	●	●	●	●	●	●	●	●	●	●	●	●	●	●	B	●	●	●	2i	●		●	
9 Indices and Logarithms	●	●	●	●	●	●	BP	●	●	●	●	●	●	●	●	●	●		●	●	B	●	A	●		●	●		●
10 Exponential & Logarithmic Functions	●			●	●	●		●	●			●	●	●	●		●	●		●		●		●		●	ObA	●	●
11 Co-ordinates and Graphs	●	●	●	●	●	●	●	●	●	●	●	●	●	●	●	●	●	●	●	●	C	●	A	●	●	●	ObA	●	●
12 The Straight Line	●	●	●		●	●	●	●	●	●	●	●	●	●	●	●	●	●	●	●	C	O1	A	●	●	●	ObA	●	●
13 The Circle	●	●	●		●	●			●			●	●	●	●			●	●	●	C	O1	A	●		●	ObA	●	●
14 Conic Sections					●	●		●				●	●	●	●			●	●	●		O1		●		●	ObA	●	
15 Loci					●		BP	●				●	●	●				●	●	●		A				●		●	
16 Polar Co-ordinates				●					●			●	●	●				●	●	●		●		●		●		A2	
17 Experimental Laws	●	●	●	O1	●		BP					●	●	●				●	●						●		ObB	●	
18 Trigonometrical Functions	●	●	●	●	●	●	●	●	●	●	●	●	●	●	●	●	●	●	●	●	C	●	A	●	●	●	●	●	●
19 Trigonometrical Graphs	●	●	●	●	●	●	●	●	●	●	●	●	●	●	●	●	●	●	●	●	C	●	A	●	●	●	●	●	●
20 Trigonometrical Identities	●	●	●	●	●	●	BP	●	●	●	●	●	●				●	●	●	●	C	●	A	●	●	●	●	●	●
21 Plane Triangles	●	●	●	●	●	●		●	●			●	●	●	●	●	●	●	●	●	C	●	A	●	2i	●		●	●
22 3-d Figures	●	●	●	●	●	●	●	●	●			●	●	●	●	●	●	●	●	●	C	●	A	●		●		●	●
23 Vectors	●	●	●	●	●		BM BV	●	●							●	●	●	●	●	D	O3	B	●	2ii	21	ObB	A2	●
24 Vectors and Geometry	●	●	●	●			BV		●			●	●	●	●		●		●	●		O3	C	●			ObB	A2	●
25 Complex Numbers				●				●	●			●	●	●	●			●	●			●		●	2i	●		A2	
26 Complex Numbers and Graphs				●					●			●	●	●	●			●	●			●		●		●		A2	
27 Differentiation	●	●	●	●	●	●	●	●	●	●	●	●	●	●	●	●	●	●	●	●	A	●	A	●	●	●	●	●	●
28 Methods of Differentiation	●			●	●	●						●	●	●	●			●	●			●		●		●	●	●	
29 Application of Differentiation	●	●	●	●	●	●		●	●	●	●	●	●	●	●	●	●	●	●	●	A	●	A	●	●	●	ObA	●	●
30 Changes	●	●	●		●	●	●					●	●	●	●	●	●	●	●	●	A	●	A	●		●	●	●	
31 Special Points	●	●	●		●	●	●	●	●	●	●	●	●	●	●	●	●	●	●	●	A	●	A	●		●	●	●	●
32 Curve Sketching	●			●	●		●	●	●			●	●	●	●			●	●			●		●		●	●	●	●
33 Integration	●	●	●	●	●		●	●	●	●	●	●	●	●	●	●	●	●	●	●	A	●	A	●	●	●	●	●	●
34 Methods of Integration	●			●	●			●	●			●	●	●	●			●	●			●		●		●	●		
35 Applications of Integration	●	●	●	●	●		●	●	●	●	●	●	●	●				●	●	●	A	●	A	●		●	●	●	●
36 Differential Equations				●	●	●		●	●			●	●	●	●			●	●			●		●				A2	
37 Numerical Solution of Equations				O1				●				●		●				●	●			O2		●			●		
38 Numerical Integration				●	●	●			●			●		●				●	●			O2	C	●					
39 Functions				●			●		●	●	●	●	●	●	●	●	●	●	●			●		●	●	●		●	●
40 Matrices				●			BV		●							●	●	●			B	O4	C	●	●		ObA	●	●

BM Section B, mechanics
BV Section B, vectors
O1 *Option 1 etc.*   ObA *Paper Ob section A etc.*   A2 *Paper A2*   21 *Paper 2 section 1*
2i *Paper 2 section i etc.*   BP *Section B, Pure mathematics*   A *Section A etc.*   * *requires the study of probability, Study Unit 72.*

Table I, continued — Mechanics

Syllabus	AEB AO 187	AEB A 648	AEB A 636	Cambridge AO 8175	Cambridge A A(9200)	Cambridge A B(9202)	JMB AO PMM	JMB A P1+Mal	JMB A P1+Aml	London AO 871	London A B(371)	Oxford AO 8853	Oxford A 9850	Oxf & Camb AO 8645	Oxf & Camb A 9650	SUJB AO 8177	SUJB A 9203	Welsh AO 0132	Welsh A 0023
41 Force Diagrams	●	O2	●	BM	●	Oa	●	●	●	●	●	D	2M	B	●	2ii	21	ObB	A2
42 1-D Kinematics	●	O2	●	BM	●	Oa	●	●	●	●	●	D	2M	B	●	2ii	21	ObB	A2
43 Graphs in Kinematics	●	O2	●	BM	●	Oa	●	●	●	●	●	D	2M	B	●	2ii	21	ObB	A2
44 Relative Motion	●	O2		BM	●	Oa	●			●	●	D	2M		●	2ii	21		A2
45 1-D Particle Dynamics	●	O2	●	BM	●	Oa	●	●	●	●	●	D	2M	B	●	2ii	21	ObB	A2
46 Connected Particles	●	O2	●		●	Oa		●	●	●	●		2M		●		21	ObB	A2
47 Work and Energy	●	O2	●	BM	●	Oa	●	●	●	●	●		2M	B	●	2ii	21	ObB	A2
48 Power	●	O2	●	BM	●	Oa	●	●	●	●	●		2M		●	2ii	21	ObB	A2
49 Impulse and Momentum	●	O2	●	BM	●	Oa	●	●	●	●	●		2M	B	●	2ii	21	ObB	A2
50 Impact	●	O2	●	BM	●	Oa	●	●	●		●		2M		●		21		A2
51 Projectiles	●	O2			●	Oa	●	●	●	●	●		2M	B	●	2ii	21	ObB	A2
52 Motion in a Horizontal Circle	●	O2	●		●	Oa	●	●	●		●		2M		●		21		A2
53 Motion in a Vertical Circle								●	●				2M						A2
54 Variable Forces		O2	●		●	Oa		●	●		●		2M	B	●		21		A2
55 Simple Harmonic Motion		O2	●					●	●		●		2M		●		21		A2
56 Vectors in Dynamics	●	O2		BM		Oa		●	●		●		2M		●		21	ObB	A2
57 Coplanar Concurrent Forces	●	O2	●	BM	●	Oa	●	●	●	●	●	D	2M	B	●	2ii	21	ObB	A2
58 Moments	●	O2	●	BM	●	Oa	●	●	●	●	●	D	2M	B	●	2ii	21	ObB	A2
59 Equilibrium	●	O2	●	BM	●	Oa	●	●	●	●	●	D	2M	B	●	2ii	21	ObB	A2
60 Three Force Problems	●	O2	●	BM	●	Oa	●	●	●	●	●	D	2M	B	●	2ii	21	ObB	A2
61 Friction	●	O2	●	BM	●	Oa	●	●	●	●	●		2M	B	●		21	ObB	A2
62 Bodies in Contact	●	O2	●	BM	●	Oa	●	●	●	●	●	D	2M	B	●		21	ObB	A2
63 Equivalent Systems of Forces	●	O2	●					●											
64 Centre of Mass	●		●		●			●	●		●		2M		●	2ii	21	ObB	A2
65 Suspending and Toppling	●		●					●			●		2M		●		21		A2

O2 Option 2 Oa Option a 2M Paper 2 Mechanics 21 Paper 2 Section 1 A2 paper A2
BM Section B Mechanics B Section B etc. 2ii Paper 2 Section ii ObB Paper Ob Section B

Table I, continued — Statistics

The examination boards are as above

Syllabus	AEB O 063	AEB AO 185	AEB AO 188	AEB A 648	AEB A 646	Camb AO 8175	Camb AO 8180	Camb A A(9200)	Camb A B(9202)	JMB AO PMS	JMB A P1+AMl	JMB A P1+S	London AO 872	London A 420	Oxford AO 8853	Oxford A 9850	Oxf&Camb AO 8655	Oxf&Camb A 9650	SUJB O 4045	SUJB AO 8177	SUJB A 9203	Welsh O 0133	Welsh OA 0132	Welsh A 0023	Scottish O Statistics
66 Pictorial Representation	●		●	O3	●	BS	●	●		●						3S	●	●	●		22	●	ObB		●
67 Frequency Distributions	●		●	O3	●	BS	●	●						●	D	3S	●	●	●	2iii	22	●	ObC	A3	●
68 Mode and Means	●		●	O3	●	BS	●	●						●	D	3S	●	●	●	2iii	22	●	ObC	A3	●
69 Median and Quantiles	●		●	O3	●	BS	●	●		●				●	D	3S	●	●	●	2iii	22	●	ObC	A3	●
70 Measures of Dispersion	●		●	O3	●	BS	●	●	●	●				●	D	3S	●	●	●	2iii	22	●	ObC	A3	●
71 Index Numbers & Moving Averages	●					BS	●										●		●	2iii		●			●
72 Probability	●	O1	●	O3	●	BS	●	●	●	●	●	●	●	●	D	3S	●	●	●	2iii	22	●	ObC	A3	●
73 Discrete Probability Distributions		O1	●	O3		BS			Ob	●	●	●	●	●	D	3S	●	●		2iii	22		ObC	A3	●
74 Continuous Probability Distributions			●						Ob		●	●				3S	●				22		ObC	A3	
75 Binomial Distributions		●	●	O3	●	BS	●	●	Ob	●	●			●	D	3S	●	●		2iii	22	●	ObC	A3	
76 Poisson Distribution		O1		O3	●			●	Ob		●			●		3S	●				22			A3	
77 Normal Distribution			●	O3	●			●	Ob		●	●		●		3S	●				22		ObC	A3	●
78 Uses of the Normal Distribution			●	O3				●	Ob		●	●		●		3S	●						ObC	A3	●
79 Sampling			●	O3	●			●	Ob	●		●		●		3S	●				22			A3	
80 Estimation						BS		●	Ob	●		●		●		3S	●				22			A3	
81 Hypothesis Testing			●					●	Ob		●			●		3S	●				22				
82 Linear Regression	●	O1	●		●			●	●			●		●		3S	●	●	●	2iii	22	●		A3	●
83 Correlation	●	●	●	O3	●	BS	●	●						●		3S	●	●	●	2iii	22	●			●
84 χ^2					●		●									3S	●	●			22				
85 Contingency Tables					●											3S	●	●							
86 Special Graph Papers			●	O3	●																				

O1 Option 1 etc. OB Option b 3S Paper 3 Statistics 22 Paper 2 Section 2 A3 Paper A3
BS Section B Statistics C Section C etc. 2iii Paper 2 Section iii ObC Paper Ob Section C

Table II — Analysis of the Methods of Examination

Board	Syllabus	Level	No. of Papers	Duration	Style	% marks	Formula sheet	Calculator
AEB	063	O	3	P1 1 hr	P1 — objective test containing 25 multiple choice items. All questions may be attempted.	25%		
				P2 1 hr	P2 — 20 short-answer questions. All questions may be attempted.	25%		
				P3 2 hr	P3 — attempt at most 4 out of 6 questions.	50%		Yes
AEB	t85	AO	2	P1 2½ hr	P1 — on Part 1 of syllabus.			
					Section A, at most 6 out of 9 questions	25%		Yes
					Section B, at most 3 out of 4 questions	25%		Yes
				P2 2½ hr	P2 — Q1 on Part 1 of syllabus obligatory,	25%		Yes
					+ at most 3 questions out of 4 from Option 1, OR at most 3 questions out of 4 from Option 2	25%		Yes
AEB	186	AO	2	P1 2½hr	P1 — Pure mathematics Section A, attempt at most 6 out of 9 questions	25%		Yes
					Section B, attempt at most 3 out of 4 questions	25%		Yes
				P2 2½hr	P2 — Further pure mathematics, as P1.	50%		Yes
AEB	187	AO	2	P1 2½ hr	P1 — Pure Maths: Section A, Answer at most 6 out of 9 questions.	25%		Yes
					Section B, Answer at most 3 out of 4 questions.	25%		Yes
				P2 2½hr	P2 — Theoretical Mechanics: Section A, Answer at most 6 out of 9 questions.	25%		Yes
					Section B, Answer at most 3 out of 4 questions.	25%		Yes
AEB	188	AO	2	P1 2½ hr	P1 — Pure Maths: Section A, Answer at most 6 out of 9 questions.	25%		Yes
					Section B, Answer at most 3 out of 4 questions.	25%		Yes
				P2 2½ hr	P2 — Statistics: Section A, Answer at most 6 out of 9 questions.	25%		Yes
					Section B, Answer at most 3 out of 4 questions.	25%		Yes
AEB	648	A	2	P5 3 hr	P5 — 15 questions on common core syllabus. All may be attempted.	50%	Yes	Yes
				P6 3 hr	P6 — Section A, 3 questions set on common core syllabus. Section B, 6 questions set on each of 3 options. Candidates to attempt 7 questions, at least one from Section A, and restricting their choice in Section B to at most 2 options.	50%	Yes	Yes
AEB	636	A	2	P1 3 hr	P1 — Pure Maths: Attempt at most 14 out of 18 questions.	Not Specified	Yes	Yes
				P3 3 hr	P3 — Applied Maths: Attempt at most 7 out of 10 questions.		Yes	Yes
AEB	646	A	2	P1 3 hr	P1 — Pure Maths: Attempt at most 14 out of 18 questions.	Not Specified	Yes	Yes
				P9 3 hr	P9 — Statistics: Attempt at most 6 out of 10 questions.		Yes	Yes
C	8175	AO	2	P1 2½ hr	P1 Section A, Pure Maths (10 questions)		Yes	Yes
				P2 2½ hr	P2 Section B, Pure Maths (6 questions), Mechanics (6 questions), Statistics (6 questions), Vectors and Algebraic Structure (6 questions). Answer 8 questions from Section A and not more than 5 questions from Section B in each paper.	Not Specified		
C	8180	AO	1 + Projects	P1 3 hr	P1 — Section 1 on Section A of syllabus. 8 questions. All may be attempted.	48%		Yes
					Section 2 on Section B of syllabus. 6 questions on 6 sub-sections of syllabus. 3 out of 6 questions may be attempted.	32%		Yes
				Projects	Assessment on 2 short projects.	20%		
C	A(9200)	A	2	P1 3 hr	P1 — Pure Maths: Section A, About 10 short questions. Attempt all.	26%	Yes	Yes
					Section B, 8 long questions. Answer 4 questions.	24%		
				P2 3 hr	P2 — Applied Maths: Section A, 10 mechanics questions. Section B, 10 statistics questions. Answer any 7 questions.	50%	Yes	Yes
C	B(9202)	A	2	P1 3 hr	P1 Section A, 7 questions.		Yes	Yes
				P2 3 hr	P2 Section B, 15 questions (3 on each option). Answer 5 questions from Section A and 3 from Section B; not all from same option.	Not Specified		

Table II — Analysis of the Methods of Examination — *continued*

Board	Syllabus	Level	No. of Papers	Duration	Style	% marks	Formula sheet	Calculator
JMB	PMM	AO	2	P1 2½ hr	P1 — Pure Maths; Answer 7 out of 8 questions.	52.5%		Yes
				P2 2½ hr	P2 — Theoretical Mechanics: Answer 6 out of 7 questions.	47.5%		Yes
JMB	PMS	AO	2	P1 2½hr	P1 — Pure Maths: Answer 7 out of 8 questions.	52%		Yes
				P2 2 hr	P2 — Statistics: Answer 5 out of 7 questions.	48%		
JMB	PM	AO	2	P1 2½ hr	P1 — Pure Maths: Answer 7 out of 8 questions.	52.5%		Yes
				P2 2 hr	P2 — Pure Maths: Answer 6 out of 7 questions.	47.5%		
JMB	P1 + Me1	A	2	P1 3 hr	P1 — Pure Maths 1: No restriction placed on the number of questions which may be attempted.	Not Specified	Yes	Yes
				P2 3 hr	P2 — Mechanics 1: No restriction placed on the number of questions which may be attempted.			
JMB	P1 + AM1	A	2	P1 3 hr	P1 — Pure Maths 1: No restriction placed on the number of questions which may be attempted.	Not Specified	Yes	Yes
				P2 3 hr	P2 — Applied Maths 1: No restriction placed on the number of questions which may be attempted.			
JMB	P1 + S	A	2	P1 3 hr	P1 — Pure Maths: No restriction placed on the number of questions which may be attempted.	Not Specified	Yes	Yes
				P2 3 hr	P2 — Statistics: No restriction placed on the number of questions which may be attempted.			
L	870	AO	2	P1 2½ hr	P1 — Pure Maths: Section A, 8 short questions – all may be attempted. Section B, 6 questions – 4 to be attempted.	20% / 30%		Yes
				P2 2½ hr	P2 — Pure Maths: Section A, 8 short question – all may be attempted. Section B, 6 questions – 4 to be attempted.	20% / 30%		Yes
L	871	AO	2	P1 2½ hr	P1 — Pure Maths: Section A, 8 short questions – all may be attempted. Section B, 6 questions – 4 to be attempted.	20% / 30%		Yes
				P2 2½ hr	P2 — Theoretical Mechanics: Section A, 8 short questions – all may be attempted. Section B, 6 questions – 4 to be attempted.	20% / 30%		Yes
L	872	AO	2	P1 2½ hr	P1 — Pure Maths: Section A, 8 short questions – all may be attempted. Section B, 6 questions – 4 to be attempted.	20% / 30%		Yes
				P2 2½ hr	P2 — Probability: Section A, 8 short questions – all may be attempted. Section B, 6 questions – 4 to be attempted.	20% / 30%		Yes
L	B(371)	A	3	P1 1 hr	P1 — Pure Maths (Core Syllabus): Multiple choice paper.	20%	No	No
				P2 2½ hr	P2 — Pure Maths (Core Syllabus): About 15 questions set.	40%	Yes	Yes
				P3 2½ hr	P3 — Applied Maths: Answer 6 out of 8 questions.	40%	Yes	Yes
L	420	A	3	P1 1 hr	P1 — Pure Maths (Core Syllabus): Multiple choice paper.	20%	No	No
				P2 2½ hr	P2 — Pure Maths (Core Syllabus): About 15 questions set.	40%	Yes	Yes
				P3 2½ hr	P3 — Statistics: Answer 6 out of 8 questions.	40%	Yes	Yes
O	8853	AO	2	P1 2½ hr	P1 ⎫ Each paper contains 13 questions, 3 on Section A, B and C and 4 on Section D. Candidates are required to answer 6 questions in each paper; not more than 2 from any section and include at least 1 from Section A.	50%		Yes
				P2 2½ hr	P2 ⎬	50%		
O	9850	A	2	P1 3 hr	P1 — Pure Maths: Section A, Short questions on the basic syllabus. Section B, Answer 5 out of 11 questions. 6 of the questions set cover the basic syllabus and 5 (1 on each) set on the options.	18·75% / 31·25%	Yes	Yes
				P2 3 hr	P2 — Applied Maths: Candidates must answer 8 questions. These may be selected from *either* the 11 questions set on Paper 2 (Mechanics) *or* the 11 questions set on Paper 3 (Statistics) *or* from Paper 2 and 3 together.	50%	Yes	Yes

Table II — Analysis of the Methods of Examination — *continued*

Board	Syllabus	Level	No. of Papers	Duration	Style	% marks	Formula sheet	Calculator
O & C	8645	AO	2	P1 2 hr P2 2 hr	P1 ⎤ Each paper contains 14 questions as follows: P2 ⎦ Section A, 6 questions Section B, 4 questions. Section C, 4 questions. The candidate must answer not more than 7 questions.	Not Specified		Yes
O & C	8655	AO	1	P1 3 hr				Yes
O & C	9650	A	2	P1 3 hr P2 3 hr	Both P1 and P2 will have this structure: Part 1 (Short questions) — Section A, 6 questions on Pure Maths. Section B, 3 questions on Mechanics. 3 questions on Statistics. Part 2 (Long questions) — Section A, 4 questions on Pure Maths. Section B, 2 questions on Mechanics. 2 questions on Statistics. For each paper, candidates should attempt not more than 6 questions from Part 1, including not more than 4 questions from each of Sections A and B; and not more than 4 questions from Part 2, including not more than 3 questions from each of Sections A and B.	Not Specified		Yes
S	4045	O	2	P1 1½ hr P2 2 hr	P1 — About 18 questions. All questions may be attempted. No choice. P2 — Candidates may attempt 5 out of 7 questions.	50% 50%		Yes Yes
S	8177	AO	2	P1 2½ hr P2 2½ hr	P1 — Pure Maths: Section A, 10 short questions. Section B, 7 questions. Candidates are asked to answer all questions in Section A and not more than 5 questions from Section B. P2 — This is divided into 3 sections: (i) Further Pure Maths. (ii) Mechanics. (iii) Statistics. 9 questions will be set on each section. Candidates should answer 7 questions chosen from at most 2 sections of the paper.	Not Specified		Yes
S	9203	A	2	P1 3 hr P2 3 hr	P1 — Pure Maths: Candidates may answer 7 out of 10 questions. P2 — Section 1 (Mechanics) 10 questions. Section 2 (Statistics), 10 questions. Candidates may attempt 7 questions including not more than 3 from Section 2.	Not Specified		Yes
W	0132	AO	2	P(Oa) 2½ hr P(Ob) 2 hr	POa — Pure Maths: 10 questions set on Part 1 of the syllabus. POb — Section A, Further Pure Maths. Section B, Mechanics. Section C, Statistics. Each section contains 7 questions. Candidates are required to answer 5 questions from any 1 section.	 No Not Specified No	Yes Yes
W	0133	O	1	P1 2½ hr	P1 — Section A, 4 short questions. No choice. Section B, Candidates may attempt 4 out of 6 questions.	100%		
W	0023	A	2	PA1 3 hr PA2 3 hr or PA3 3 hr	PA1 — Pure Maths: Questions set on Section A1 of syllabus. PA2 — Complex Numbers, Vectors and Mechanics: Section A2 of syllabus. PA3 — Statistics: Questions set on Section A3 of syllabus.	50% ⎤ ⎬ 50% ⎦	Yes Yes	Yes Yes
Scottish	Statistics	O	2	P1 1 hr P2 1½ hr	P1 — Objective test containing 30 items. P2 — No details.	Approx. 39% Approx. 61%		Yes Yes
Scottish	Mathematics	H	2	P1 1¾ hr P2 2½ hr	P1 — Contains 40 questions. No choice. P2 — "Traditional" questions. No choice.	Approx. 45% Approx. 55%	Yes	Yes

Glossary of Symbols

LHS	left hand side (of an equation)
RHS	right hand side (of an equation)
$\angle$ or $\frown$	angle
$\perp$	is perpendicular to
$\parallel$	is parallel to
$\pm$	positive or negative, plus or minus
$=$	is equal to
$\equiv$	is identically equal to
$\approx$	is approximately equal to
$\neq$	is not equal to
$<$	is less than
$\leqslant$	is less than or equal to
$>$	is greater than
$\geqslant$	is greater than or equal to
$\sqrt{\ }$	the positive square root
$\Rightarrow$	implies that
$\Leftarrow$	is implied by
$\Leftrightarrow$	implies and is implied by (if and only if)
$\{a, b, c,...\}$	the set with elements $a, b, c, ...$
$\in$	is an element of
$:$	such that
n( )	the number of elments in the set ( )
$\emptyset$	the empty set
S or S'	the complement of the set S
$\cup$	union
$\cap$	intersection
$\subset$	is a subset of
$\leftrightarrow$	corresponds one-to-one with
$\propto$	varies directly as
$\triangle$	the discriminant of a quadratic equation
$|\ |$	the modulus of a number
δx or $\triangle x$	a small increment of x
$\dfrac{dy}{dx}$	the derivative of y with respect to x
$\int y\, dx$	the indefinite integral of y with respect to x
$\int_{b}^{a} y\, dx$	the definite integral of y with respect to x
f	a function
f^{-1}	the inverse function of f
$f(x)$	the function value for x
$(fg)(x)$ or fg or $f \circ g$	the composition of the functions f and g
$\to$	is mappped onto (for mappings) approaches, tends to (for limits)
$x \xrightarrow{f} y$	x is mapped into y under the function f
$f{:}x \to y$	f is the function under which x is mapped into y
Σ	the sum of (exact limits may be given)
$n!$	n factorial

$\dbinom{n}{r}$	binomial coefficient
$\mathbb{N}$	the set of natural numbers
$\mathbb{Z}$	the set of integers
$\mathbb{Q}$	the set of rational numbers
$\mathbb{R}$	the set of real numbers
$\mathbb{C}$	the set of complex numbers
∞	infinity
$\ln x$	the natural logarithm of x
$\lg x$	the common logarithm of x
M^{-1}	the inverse of the square matrix M
$\det M$	the determinant of the square matrix M
i	the square root of -1
$|z|$	the modulus of the complex number z
$\arg z$	the argument of the complex number z
z^*	the conjugate of the complex number z
$\dot{x}$	the derivative of x with respect ot t
g	the acceleration due to gravity (taken as 10 m/s^2 unless otherwise stated)
$\mathbf{a.b}$	the scalar product of the vectors $\mathbf{a}$ and $\mathbf{b}$
$\hat{\mathbf{r}}$	the unit vector in the direction of the vector $\mathbf{r}$
$|\mathbf{r}|$, r	the magnitude of the vector $\mathbf{r}$
$\mathbf{i, j, k}$	unit vectors in the mutually perpendicular directions Ox, Oy, Oz
E_1, E_2, etc.	events
$E_1 \cup E_2$	union of the events E_1 and E_2
$E_1 \cap E_2$	intersection of the events E_1 and E_2
$P(E)$	probability of the event E
$\bar{E}$	complement of the event E
$P(E_1|E_2)$	conditional probability of the event E_1 given that the event E_2 has occurred
X, Y, Z etc.	random variables
x, y, z etc.	values taken by random variables
p_1, p_2, etc.	probabilities of individual outcomes $x_1, x_2, \ldots$ in the distribution of a discrete random variable X
$f(x)$	probability density function of a continuous random variable X
x_1, x_2, etc.	observations
f_1, f_2, etc.	frequencies with which observations x_1, x_2, etc. occur
n	sample size
x	sample mean
$F(x)$	distribution function, $P(X<x)$ of the random variable X
$p(x)$	probability density function $P(X=x)$ of the discrete random variable X
$E[X]$	expected value of the random variable X
$Var[X]$	variance of the random variable X
μ	population mean
σ^2	population variance
σ	population standard deviation
$B(n, p)$	binomial distribution with parameter n and p
$Po(\mu)$	Poisson distribution with parameters

$N(\mu, \sigma^2)$	normal distribution with parameters and		π	population proportion
$\varphi(z)$	probability density function of the standardised normal random variable Z ($N(0, 1)$)		ϱ	population product moment correlation coefficient
$\Phi(z)$	distribution function for $N(0, 1)$		$\leftarrow x \rightarrow$	a dimension (x units)
s^2	sample variance		$- x \rightarrow$	a displacement (x units)
r	sample proportion		$\rightarrow\!-$	a velocity
r	sample product moment correlation coefficient		$\rightarrow\!\!-$	an acceleration ($a = \dot{v}$)
r_s	Spearman's rank correlation coefficient for a sample		$\longrightarrow\!\!\triangleright$	an impulse
			$\longrightarrow\!\!\blacktriangleright$	a force ($F = \dot{I}$)

The Greek Alphabet

Letters		Name			
A	α	alpha	N	ν	nu
ß	β	beta	Ξ	ξ	xi
Γ	γ	gamma	O	o	omicron
Δ	δ	delta	Π	π	pi
E	ε	epsilon	P	ϱ	rho
Z	ζ	zeta	Σ	σ	sigma
H	η	eta	T	τ	tau
Θ	θ	theta	Y	υ	upsilon
I	ι	iota	Φ	φ	phi
K	κ	kappa	X	χ	chi
Λ	λ	lambda	Ψ	ψ	psi
M	μ	mu	Ω	ω	omega

Part II
The study units

1 Polynomials

Definitions, Operations, Remainder theorem, Factor theorem, Special factors.

Definitions

A **polynomial** in x, a variable, is an expression of the form

$$c_0x^n + c_1x^{n-1} + c_2x^{n-2} + \ldots + c_{n-1}x + c_n$$

where n is a positive integer
and $c_0, c_1, c_2, \ldots, c_{n-1}, c_n$ are constants.

The **degree** of the polynomial is n, the highest power of x.

The **constant** term is c_n.

For brevity we often write $f(x)$ for

$$c_0x^n + c_1x^{n-1} + c_2x^{n-2} + \ldots + c_{n-1}x + c_n.$$

$\boxed{i}$ $2x^7 + 3x^5 - x^4 + 6x + 4$

is a polynomial of degree 7, with constant term 4.

Polynomials of low degree have special names.

Degree	Name	Example
1	linear	$3x - 2$
2	quadratic	$4x^2 + 3x + 8$
3	cubic	$8x^3 - 7x + 2$
4	quartic	$3x^4 - 2x^2$
5	quintic	$4x^5 + 1$

Operations

Addition
Add corresponding terms (powers of x).

$\boxed{i}$

$$
\begin{aligned}
f(x) &\equiv 3x^4 - 5x^3 \qquad\quad + x - 4 \\
g(x) &\equiv \qquad\quad 4x^3 - 3x^2 + 4x + 3 \\
\hline
f(x) + g(x) &\equiv 3x^4 - x^3 - 3x^2 + 5x - 1
\end{aligned}
$$

Subtraction
Subtract corresponding terms.

$\boxed{i}$

$$
\begin{aligned}
f(x) &\equiv 3x^4 - 5x^3 \qquad\quad + x - 4 \\
g(x) &\equiv \qquad\quad 4x^3 - 3x^2 + 4x + 3 \\
\hline
f(x) - g(x) &\equiv 3x^4 - 9x^3 + 3x^2 - 3x - 7
\end{aligned}
$$

Multiplication
This can be set out like a 'long multiplication'.
Leave spaces for 'missing terms'.

$\boxed{i}$

$$
\begin{aligned}
f(x) & & 3x^3 & \quad -2x + 4 \\
g(x) & & & x^2 \qquad - 3 \\
\hline
f(x) \times x^2 & & 3x^5 - 2x^3 + 4x^2 \\
f(x) \times -3 & & \quad -9x^3 \qquad + 6x - 12 \\
\hline
f(x) \times g(x) & & 3x^5 - 11x^3 + 4x^2 + 6x - 12
\end{aligned}
$$

Division
This can be set out like a 'long division'.
Leave spaces for 'missing terms'.

$\boxed{i}$

$$
\begin{array}{r}
x^2 + 4x + 11 \\
x^2 - 4x + 3 \overline{\smash{\big)}\, x^4 \qquad - 2x^2 + 3x - 6} \\
\underline{x^4 - 4x^3 + 3x^2} \\
4x^3 - 5x^2 + 3x \\
\underline{4x^3 - 16x^2 + 12x} \\
11x^2 - 9x - 6 \\
\underline{11x^2 - 44x + 33} \\
35x - 39
\end{array}
$$

$$\therefore \quad \frac{x^4 - 2x^2 + 3x - 6}{x^2 - 4x + 3} = x^2 + 4x + 11 + \frac{35x - 39}{x^2 - 4x + 3}$$

Remainder theorem

If a polynomial $f(x)$ is **divided** by $(x - a)$, then the **remainder** is $f(a)$.

$f(a)$ is the value of $f(x)$ when $x = a$.

$\boxed{i}$ *Find the remainder when $3x^5 - x^2 + 1$ is divided by $(x + 2)$.*

$$
\begin{aligned}
f(x) &= 3x^5 - x^2 + 1 \\
f(-2) &= 3(-2)^5 - (-2)^2 + 1 = -96 - 4 + 1 = -99
\end{aligned}
$$

Factor theorem

If $(x - a)$ is a **factor** of $f(x)$, then $f(a) = 0$.
Conversely if $f(a) = 0$, then $(x - a)$ is a factor of $f(x)$.

This may be used to find the factors of a polynomial. Factors of the constant term are usually tested first.

If it is suspected that $(x - a)$ is a repeated factor:
(a) 'take out' the factor, either by inspection or long division to give
$$f(x) \equiv (x - a)g(x),$$
(b) test $(x - a)$ as a factor of $g(x)$.

$\boxed{i}$ *Find the factors of $f(x) = x^3 - 5x^2 + 2x + 8$.*

Try the factors of 8, i.e. ±1, ±2, ±4, ±8.

    $a = 1$. $f(1) = (1)^3 - 5(1)^2 + 2(1) + 8 = 6 \neq 0$.
So $(x - 1)$ is not a factor.
Try $a = -1$. $f(-1) = (-1)^3 - 5(-1)^2 + 2(-1) + 8 = 0$.
So $(x + 1)$ is a factor.
Try $a = 2$. $f(2) = (2)^3 - 5(2)^2 + 2(2) + 8 = 0$.
So $(x - 2)$ is a factor.
Try $a = 4$. $f(4) = (4)^3 - 5(4)^2 + 2(4) + 8 = 0$.
So $(x - 4)$ is a factor.

Special factors

$$a^2 - b^2 = (a - b)(a + b)$$
$$a^2 \pm 2ab + b^2 = (a \pm b)^2$$

$$a^3 - b^3 = (a - b)(a^2 + ab + b^2)$$
$$a^3 + b^3 = (a + b)(a^2 - ab + b^2)$$

Polynomials
Worked examples, Guided examples and Exam questions

 The expression $2x^3+ax^2+bx+2$ is exactly divisible by $(x+2)$ and leaves a remainder of 12 on division by $(x-2)$. Calculate the values of a and b and factorise the expression completely.

Let $f(x) = 2x^3+ax^2+bx+2$
$(x+2)$ is a factor of $f(x)$, so $f(-2)=0$
$f(-2) = 2(-2)^3+a(-2)^2+b(-2)+2 = -16+4a-2b+2 = 0$

    i.e. $4a-2b=14$ [1]

Division by $(x-2)$ leaves a remainder of 12, so $f(2)=12$
$f(2) = 2(2)^3+a(2)^2+b(2)+2 = 16+4a+2b+2 = 12$

    i.e. $4a+2b=-6$ [2]

[1]+[2] gives $8a=8$, $a=1$ and [2]−[1] gives $4b=20$, $b=-5$
$\therefore f(x) = 2x^3+x^2-5x+2$
Since $(x+2)$ is a factor of $f(x)$
divide $f(x)$ by $(x+2)$ to obtain a quotient $2x^2-3x+1$
$\therefore f(x) = (x+2)(2x^2-3x+1)$
       $= (x+2)(2x+1)(x-1)$, by factorisation
$\therefore f(x) = (x+2)(2x-1)(x-1)$

 When the polynomial $P(x)$ is divided by $(x-1)$ the remainder is 7, and when divided by $(x-3)$ the remainder is 13. Find, by writing
$$P(x) \equiv (x-1)(x-3)Q(x)+ax+b$$
the remainder when $P(x)$ is divided by $(x-1)(x-3)$. If also $P(x)$ is a cubic in which the coefficient of x^3 is unity and $P(2)=6$, determine $Q(x)$.

Division by $(x-1)$ leaves a remainder of 7, so $P(1)=7$
    i.e. $a+b=7$ [1]
Division by $(x-3)$ leaves a remainder of 13, so $P(3)=13$
    i.e. $3a+b=13$ [2]
[2]−[1] gives $2a=6$, $a=3$ and from [1] $b=4$
$\therefore P(x) \equiv (x-1)(x-3)Q(x)+3x+4$

$$\frac{P(x)}{(x-1)(x-3)} \equiv Q(x)+\frac{3x+4}{(x-1)(x-3)}$$

Thus the remainder on division by $(x-1)(x-3)$ is $3x+4$
If $P(x)$ is a cubic with the coefficient of x^3 unity then
$Q(x) \equiv (x+c)$
    i.e. $P(x) = (x-1)(x-3)(x+c)+3x+4$
and $P(2)=6$, $(2-1)(2-3)(2+c)+3(2)+4=6$
                  $-(2+c)+10=6$
             i.e. $c=2$

$\therefore Q(x) = x+2$

 The quadratic polynomial $P(x)$ leaves a remainder of 3 on division by $(x-1)$, a remainder of 12 on division by $(x-2)$ and no remainder on division by $(x+2)$. Find $P(x)$ and solve $P(x)=0$

Let $P(x) \equiv ax^2+bx+c$ [1]

Let $P(1)=3$, $P(2)=12$, $P(-2)=0$, obtaining three equations in a, b and c.
The found values of a, b and c are substituted in [1] which may then be factorised and $P(x)=0$ easily solved.

---

 1 The expression $2x^3+ax^2+bx+6$ is exactly divisible by $(x-2)$ and on division by $(x+1)$ gives a remainder of -12. Calculate the values of a and b and factorise the expression completely.   *(A)*

2 $f(x) \equiv x^2+ax+b$
When $f(x)$ is divided by $(x-2)$, the remainder is 8. When $f(x)$ is divided by $(x+3)$, the remainder is 18. Find the values of the constants a and b.   *(L)*

3 When (x^4+kx^2+4x+2) is divided by $(x+3)$, the remainder is 8. Find the value of k.   *(L)*

4 If $f(x)$ denotes the polynomial $2x^3-3x^2-8x-3$, find the remainders when $f(x)$ is divided by (i) $x-1$, (ii) $x+3$, and (iii) $2x+1$. Deduce which of $(x-1)$, $(x+3)$, and $(2x+1)$ is a factor of $f(x)$, and find its remaining factors.   *(O & C)*

5 The cubic polynomial $6x^3-7x^2-ax-b$ has a remainder of 72 when divided by $(x-2)$ and is exactly divisible by $(x+1)$. Calculate a and b. Show that $(2x-1)$ is also a factor of the polynomial and obtain the third factor.   *(S)*

6 One of the factors of the cubic polynomial $2x^3+3x^2+ax+b$ is $(x-2)$. When the polynomial is divided by $(x-1)$, a remainder of -5 results. Find a and b, and hence factorise the polynomial completely.   *(S)*

7 State the Remainder Theorem.
When the cubic polynomial x^3+ax^2-3x+4 is divided by $x-3$, the remainder obtained is twice the remainder obtained when the polynomial is divided by $x-2$. Calculate a.   *(S)*

8 (a) Given that $x+1$ and $x-2$ are factors of $6x^4-x^3+ax^2-6x+b$, find the value of a and b.
(b) Given that $[x^3-3x^2-4x+16 \equiv (x-2)(x+3)(x-c)+Px+Q]$, find the value of P, of Q, and of c.   *(C)*

9 Given that $2x-1$ is a factor of $8x^3+4x^2+kx+15$, find the value of k and then factorise the expression fully when k has this value.   *(H)*

10 When $f(x)$, where $f(x) \equiv x^4-2x^3+ax^2+bx+c$, is divided by $x-2$ the remainder is -24. When $f(x)$ is divided by $x+4$ the remainder is 240. Given that $x+1$ is a factor of $f(x)$, show that $x-1$ is also a factor.   *(A)*

11 (a) Given that $f(x) = x^3+kx^2-2x+1$ and that when $f(x)$ is divided by $(x-k)$ the remainder is k, find the possible values of k.
(b) When the polynomial $p(x)$ is divided by $(x-1)$ the remainder is 5 and when $p(x)$ is divided by $(x-2)$ the remainder is 7. Given that $p(x)$ may be written in the form $(x-1)(x-2)q(x)+Ax+B$, where $q(x)$ is a polynomial and A and B are numbers, find the remainder when $p(x)$ is divided by $(x-1)(x-2)$.   *(C)*

12 When the polynomial $P(x)$ is divided by $x-2$ the remainder is 4, and when $P(x)$ is divided by $x-3$ the remainder is 7. Find, by writing $P(x) \equiv (x-2)(x-3)Q(x)+ax+b$, the remainder when $P(x)$ is divided by $(x-2)(x-3)$. If also $P(x)$ is a cubic in which the coefficient of x^3 is unity, and $P(1)=1$, determine $Q(x)$.   *(J)*

13 Given that $f(x) \equiv 3-7x+5x^2-x^3$, show that $3-x$ is a factor of $f(x)$. Factorise $f(x)$ completely and hence state the set of values of x for which $f(x) \le 0$.   *(L)*

14 (i) Three of the factors of $x^4+ax^3+bx^2+x+c$ are x, $x+1$ and $x-1$. Find a, b and c.
(ii) Write down an expression of the form x^3+px^2+qx+r which gives a remainder 4 when divided by x, $x-2$ or $x+3$.   *(O & C)*

2 Rational Functions
Definitions, Partial fractions.

Definitions

A **rational function** is a function of the form $\dfrac{P(x)}{Q(x)}$,

where $P(x)$ and $Q(x)$ are polynomials in x.
If the degree of $P(x) <$ the degree of $Q(x)$, then
$\dfrac{P(x)}{Q(x)}$ is a **proper fraction**.
If the degree of $P(x) \geqslant$ the degree of $Q(x)$, then
$\dfrac{P(x)}{Q(x)}$ is an **improper fraction**.

> **i** $\dfrac{x+1}{x^2-1}$ and $\dfrac{2x^5-x^2+1}{x^3+x-1}$
>
> are rational functions of x
>
> degree 1→ $\dfrac{x+1}{x^2-1}$ is a proper fraction.
> degree 2→
>
> degree 5→ $\dfrac{2x^5-x^2+1}{x^3+x-1}$ is an improper fraction.
> degree 3→

Partial functions

A rational function which may be expressed as a sum of separate fractions is said to be resolved into its partial fractions.

> **i** $\underbrace{\dfrac{9x+9}{(x+2)^2(x-1)}}_{\text{rational function}} \equiv \underbrace{\dfrac{3}{(x+2)^2} - \dfrac{2}{(x+2)} + \dfrac{2}{(x-1)}}_{\text{partial fractions}}$

1. If the rational function is a proper fraction.
The method of obtaining the partial fractions depends on the type of factors in the denominator.
Three types are considered at this stage

(a) For every **linear factor** $(ax + b)$ of the denominator there will be a corresponding partial fraction $\dfrac{A}{(ax+b)}$.

(b) For every **repeated linear factor** $(cx + d)^2$ of the denominator there will be two corresponding partial fractions $\dfrac{B}{(cx+d)} + \dfrac{C}{(cx+d)^2}$.

(c) For every **quadratic factor** $(ex^2 + fx + g)$ of the denominator there will be a corresponding partial fraction $\dfrac{Dx+E}{(ex^2+fx+g)}$.

To find the constants $A, B, C, \ldots$
(a) Form an identity between the original fraction and the sum of the partial fractions.
(b) Write all the fractions using a common denominator.
(c) Compare the two numerators by
  (i) comparing 'convenient coefficients' on each side of the identity.
and/or (ii) substituting values of x which reduce individual factors to zero.

Always check your answer by reversing the process, i.e. by combining the partial fractions to form a single fraction.

> **i**   *Find the partial fractions of* $\dfrac{x^3+2x^2-x-3}{(x+1)^2(x^2-2)}$.
>
> The repeated linear factor $(x + 1)^2$ has partial fractions
>
> $$\frac{A}{(x+1)} + \frac{B}{(x+1)^2}.$$
>
> The quadratic factor $(x^2 - 2)$ has partial fraction
>
> $$\frac{Cx+D}{(x^2-2)}.$$
>
> So $\dfrac{x^3+2x^2-x-3}{(x+1)^2(x^2-2)} \equiv \dfrac{A}{(x+1)} + \dfrac{B}{(x+1)^2} + \dfrac{Cx+D}{(x^2-2)}$
>
> and $\equiv \dfrac{A(x+1)(x^2-2)+B(x^2-2)+(Cx+D)(x+1)^2}{(x+1)^2(x^2-2)}$
>
> i.e. $x^3+2x^2-x-3 \equiv A(x+1)(x^2-2)+B(x^2-2)$
> $\qquad\qquad\qquad\qquad\qquad + (Cx+D)(x+1)^2 \qquad (1)$
>
> Let $x = -1$, then (1) becomes
>
> $(-1)^3+2(-1)^2-(-1)-3=B[(-1)^2-2] \Rightarrow B=1$.
>
> Compare coefficients of x^3 on L.H.S. and R.H.S. of (1)
> $\qquad\qquad\qquad 1=A+C \qquad\qquad\qquad\qquad (2)$
> Compare coefficients of x^2 on L.H.S. and R.H.S. of (1)
> $\qquad\qquad 2=A+B+2C+D \Rightarrow 1=A+2C+D \quad (3)$
> Let $x = 0$, then (1) becomes
> $\qquad\qquad -3=-2A-2B+D \Rightarrow -1=-2A+D \quad (4)$
> $(3)-(4) \qquad 2=3A+2C \qquad\qquad\qquad\qquad (5)$
> $2\times(2)-(5) \quad 0=-A \Rightarrow A=0$.
> $(5) \Rightarrow \qquad C=1$
> $(4) \Rightarrow \qquad D=-1$
>
> $\therefore \dfrac{x^3+2x^2-x-3}{(x+1)^2(x^2-2)} \equiv \dfrac{1}{(x+1)^2} + \dfrac{x-1}{(x^2-2)}$
>
> Check: $\dfrac{1}{(x+1)^2} + \dfrac{x-1}{(x^2-2)} \equiv \dfrac{(x^2-2)+(x-1)(x+1)^2}{(x+1)^2(x^2-2)}$
>
> $\equiv \dfrac{x^2-2+x^3+2x^2+x-x^2-2x-1}{(x+1)^2(x^2-2)}$
>
> $\equiv \dfrac{x^3+2x^2-x-3}{(x+1)^2(x^2-2)}$

2. If the rational function is an improper fraction.
(a) Divide to obtain a quotient and a proper fraction.
(b) Resolve the proper fraction into partial fractions as before.

> **i**
> *Express* $\dfrac{x^4+3x^3+x^2-5x-5}{(x+1)^2(x^2-2)}$ *in partial fractions.*
>
> $\dfrac{x^4+3x^3+x^2-5x-5}{(x+1)^2(x^2-2)} \equiv 1 + \dfrac{x^3+2x^2-x-3}{(x+1)^2(x^2-2)}$
>
> $\equiv 1 + \dfrac{1}{(x+1)^2} + \dfrac{x-1}{(x^2-2)}$

Rational Functions
Worked examples, Exercises and Exam questions

 Resolve into partial fractions

$$\frac{4x^3+16x^2-15x+13}{(x+2)(2x-1)^2}$$

Since the degree of the numerator is equal to the degree of the denominator, the denominator is divided into the numerator.

$$(x+2)(2x-1)^2 = (x+2)(4x^2-4x+1) = 4x^3+4x^2-7x+2$$

$$
\begin{array}{r}
1 \\
4x^3+4x^2-7x+2 \overline{\smash{\big)}4x^3+16x^2-15x+13} \\
\underline{4x^3+4x^2-7x+2} \\
12x^2-8x+11
\end{array}
$$

$$\therefore \frac{4x^3+16x^2-15x+13}{(x+2)(2x-1)^2} = 1 + \frac{12x^2-8x+11}{(x+2)(2x-1)^2}$$

Let

$$\frac{12x^2-8x+11}{(x+2)(2x-1)^2} \equiv \frac{A}{(x+2)} + \frac{B}{(2x-1)} + \frac{C}{(2x-1)^2}$$

i.e. $12x^2-8x+11 \equiv A(2x-1)^2 + B(x+2)(2x-1) + C(x+2)$

$x=-2$, $48+16+11 = 25A \Rightarrow A = 3$

$x=\frac{1}{2}$, $3-4+11 = \frac{5}{2}C \Rightarrow C = 4$

$x=0$, $11 = A-2B+2C \Rightarrow B = 0$

$$\therefore \frac{12x^2-8x+11}{(x+2)(2x-1)^2} \equiv \frac{3}{(x+2)} + \frac{4}{(2x-1)^2}$$

and

$$\frac{4x^3+16x^2-15x+13}{(x+2)(2x-1)^2} \equiv 1 + \frac{3}{(x+2)} + \frac{4}{(2x-1)^2}$$

Expressing an algebraic fraction in terms of its partial fractions is rarely of interest in itself; it is however often an important means to other ends.

Three common uses of partial fractions are given below, and briefly illustrated with Guided Examples.

SUMMATION OF SERIES

 Express $\dfrac{1}{x(x+2)}$ *in partial fractions, and hence find the sum to* n *terms of the series*

$$\frac{1}{1.3} + \frac{1}{2.4} + \frac{1}{3.5} + \dots$$

Partial fractions give $\dfrac{1}{x(x+2)} = \dfrac{x}{2} - \dfrac{1}{2(x+2)}$

The first term of the series can be represented as $\dfrac{1}{2} - \dfrac{1}{2.3}$,

the second term as $\dfrac{2}{2} - \dfrac{1}{2.4}$, and so on. Writing the terms of the series in this fashion enables the result,

$\dfrac{3}{4} - \dfrac{1}{2(n+1)} - \dfrac{1}{2(n+2)}$, to be derived.

BINOMIAL EXPANSIONS

 Express $\dfrac{2+x^2}{(2-x)^2(4+x)}$ *as partial fractions, and hence expand the function in a series of ascending powers of x as far as the term in x^3.*

Partial fractions give

$$\frac{2+x^2}{(2-x)^2(4+x)} = -\frac{1}{2(2-x)} + \frac{1}{(2-x)^2} + \frac{1}{2(4+x)}$$

Binomial expansions give,

$$-\frac{1}{4}\left(1-\frac{x}{2}\right)^{-1} = -\frac{1}{4}\left(1+\frac{x}{2}+\frac{x^2}{4}+\frac{x^3}{8}+\dots\right)$$

and similarly for the other two terms, and collecting like terms from the three expansions gives the result,

$\dfrac{1}{8} + \dfrac{3}{32}x + \dfrac{17}{128}x^2 + \dfrac{47}{512}x^3$, to the term in x^3.

INTEGRATION

 *Express* $\dfrac{1}{(x+1)(x+2)}$ *as partial fractions and hence evaluate*

$$\int_2^3 \frac{1}{(x+1)(x+2)}dx.$$

Partial fractions give

$$\frac{1}{(x+1)(x+2)} = \frac{1}{(x+1)} - \frac{1}{(x+2)}$$

Therefore the given integral can be written as

$$I = \int_2^3 \left\{ \frac{1}{(x+1)} - \frac{1}{(x+2)} \right\} dx = \left[\ln \frac{(x+1)}{(x+2)} \right]_2^3 = \ln \frac{16}{15}.$$

EX 1 Find the constants A and B in the following identity:

$$\frac{x+2}{x^2-1} \equiv \frac{A}{x-1} + \frac{B}{x+1} \tag{L}$$

Resolve into partial fractions the following:

2 $\dfrac{2x^2-x}{(2+x)(1+x^2)}$

3 $\dfrac{5x^2+6x+7}{(x-1)(x+2)^2}$

4 $\dfrac{1+2x}{(1+x)(1-2x^2)}$

5 $\dfrac{15x-6}{(1-2x)(2-x)}$

6 $\dfrac{4-3x}{(1-2x)(2+x)}$

7 $\dfrac{10-17x+14x^2}{(2+x)(1-2x)^2}$

8 $\dfrac{1}{(1-2x)(1-3x)}$

9 $\dfrac{9}{(1+2x)(1-x)^2}$

10 $\dfrac{2}{(x+7)(x+9)}$

11 $\dfrac{2x+5}{(x+2)(x+3)}$

12 $\dfrac{5x+4}{(x^2-2x-8)}$

13 $\dfrac{2x+12}{(x^2-3x+2)}$

14 $\dfrac{3x+1}{(x+1)^2}$

15 $\dfrac{2x-2}{(x+1)(x-2)^2}$

16 $\dfrac{3x^2-x-2}{(1+2x)(x+2)^2}$

17 $\dfrac{x-1}{(x+1)(x^2+1)}$

18 $\dfrac{x^2}{(x-1)(x^2+1)}$

19 $\dfrac{x^2+10x+6}{x^2+2x-8}$

20 $\dfrac{x^3}{(x+1)(x+2)}$

3 Quadratics
Quadratic equations, Quadratic functions.

Quadratic equations

The general **quadratic equation** is of the form

$$ax^2 + bx + c = 0$$

where a, b and c are constants and $a \neq 0$.
The two values of x which make $ax^2 + bx + c$ zero are called the roots of the equation.
The roots are given by

$$x = \frac{1}{2a}[-b \pm \sqrt{b^2 - 4ac}]$$

$b^2 - 4ac = \Delta$ is the **discriminant** of the equation since it discriminates between the types of roots.
If $\Delta > 0$, the roots are **real** and **different**,
$\Delta = 0$, the roots are **real** and **equal**,
$\Delta < 0$, the roots are **complex**.

Sum and product of roots
If α and β are the roots of $ax^2 + bx + c = 0$, then

$$\alpha + \beta = \frac{-b}{a} \quad \text{and} \quad \alpha\beta = \frac{c}{a}.$$

The quadratic equation may be written

$$x^2 - (\text{sum of roots})x + (\text{product of roots}) = 0$$

This can be used to obtain a new quadratic equation whose roots are functions of α and β.
Useful results in such examples are

$$\alpha^2 + \beta^2 = (\alpha + \beta)^2 - 2\alpha\beta$$

$$\alpha^3 + \beta^3 = (\alpha + \beta)^3 - 3\alpha\beta(\alpha + \beta)$$

$$\frac{1}{\alpha} + \frac{1}{\beta} = \frac{\alpha + \beta}{\alpha\beta}$$

$$\frac{1}{\alpha^2} + \frac{1}{\beta^2} = \frac{(\alpha + \beta)^2 - 2\alpha\beta}{(\alpha\beta)^2}$$

ⓘ *Find, to 2 decimal places, the roots of* $2x^2 - 3x - 4 = 0$.

Use $x = \dfrac{1}{2a}[-b \pm \sqrt{b^2 - 4ac}]$ with $a = 2$, $b = -3$, $c = -4$.

$$= \tfrac{1}{4}[3 \pm \sqrt{(-3)^2 - 4 \times 2 \times (-4)}]$$

$$= \tfrac{1}{4}[3 \pm \sqrt{41}] = \tfrac{1}{4}[3 \pm 6.403]$$

$$= 2.35 \text{ or } -0.85 \text{ (to 2 d.p.)}$$

ⓘ *Show that* $2x^2 - 3x + 4 = 0$ *has no real roots.*

Use $\Delta = b^2 - 4ac$ with $a = 2$, $b = -3$, $c = 4$.
i.e. $\Delta = (-3)^2 - 4 \times 2 \times 4 = 9 - 32 = -23 < 0$.
$\therefore 2x^2 - 3x + 4 = 0$ has no real roots.

ⓘ *If* $3x^2 - 6x + 8 = 0$ *has roots* α *and* β, *find the equation whose roots are* $\dfrac{1}{\alpha}$ *and* $\dfrac{1}{\beta}$.

$$\alpha + \beta = -\frac{(-6)}{3} = 2, \quad \alpha\beta = \frac{8}{3}$$

If A and B are the new roots,

sum of new roots is $A + B = \dfrac{1}{\alpha} + \dfrac{1}{\beta} = \dfrac{\alpha + \beta}{\alpha\beta} = \dfrac{2}{\frac{8}{3}} = \dfrac{3}{4}$

product of new roots is $AB = \dfrac{1}{\alpha} \times \dfrac{1}{\beta} = \dfrac{1}{\alpha\beta} = \dfrac{3}{8}$

The new equation is $x^2 - (A + B)x + (AB) = 0$
i.e. $x^2 - \tfrac{3}{4}x + \tfrac{3}{8} = 0$
or $8x^2 - 6x + 3 = 0$

Quadratic functions

The general **quadratic function** is $f(x) \equiv ax^2 + bx + c$.

By **completing the square** on the R.H.S.

$$f(x) \equiv a\left[\left(x + \frac{b}{2a}\right)^2 - \left(\frac{b^2 - 4ac}{4a^2}\right)\right]$$

Graph of a quadratic function
The graph of $f(x) \equiv ax^2 + bx + c$ is a **parabola**.
Its **axis of symmetry** is $x = -\dfrac{b}{2a}$.

If $a > 0$, $f(x)$ has a minimum.
If $a < 0$, $f(x)$ has a maximum.

Its maximum or minimum
value is $-\left(\dfrac{b^2 - 4ac}{4a}\right)$.

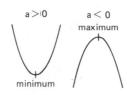

Figure 1

If $f(x) = 0$, the graph
(a) cuts the x-axis twice, if $(b^2 - 4ac) > 0$,
(b) touches the x-axis, if $(b^2 - 4ac) = 0$,
(c) does not cut the x-axis, if $(b^2 - 4ac) < 0$.

Values of a rational quadratic function
To find the range of possible values of a function of the type $y = \dfrac{ax^2 + bx + c}{px^2 + qx + r}$, when x is real
(a) rearrange the equation as a quadratic in x,
(b) use the fact that $(b^2 - 4ac) \geq 0$ for real x.

ⓘ *Sketch the graph of* $f(x) \equiv 3x^2 - 7x + 4$.
$f(x)$ is a quadratic function so its graph is a parabola.
Its axis of symmetry is $x = -\dfrac{(-7)}{2(3)} = \dfrac{7}{6}$.
Since $a = 3$, i.e. $a > 0$, $f(x)$ has a minimum.

Its value is $-\left[\dfrac{(-7)^2 - 4(3)(4)}{4(3)}\right] = -\left(\dfrac{49 - 48}{12}\right) = -\dfrac{1}{12}$.

Since $\Delta > 0$, $f(x)$ cuts the x-axis twice at

$$x = \frac{-(-7) \pm \sqrt{1}}{2(3)} = \frac{7 \pm 1}{6} = \frac{8}{6} \text{ or } \frac{6}{6}.$$

When $x = 0$, $f(x) = 4$.

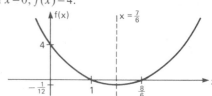

Figure 2

ⓘ *If x is real, find the possible range of values of* $y = (x^2 + x + 1)/(x + 1)$.

The given expression is $x^2 + x + 1 - y(x + 1) = 0$
$$x^2 + (1 - y)x + (1 - y) = 0$$
For real x, $b^2 - 4ac \geq 0$
i.e. $(1 - y)^2 - 4 \cdot 1 \cdot (1 - y) \geq 0$
$(1 - y)(1 - y - 4) \geq 0$
$(1 - y)(y + 3) \leq 0$ So $y \geq 1$ or $y \leq -3$.

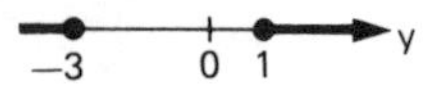

Figure 3

Quadratics
Worked examples, Guided examples and Exam questions

 Show that, for all real x, $0 < \dfrac{1}{x^2+4x+5} \leq 1$

Sketch the curve $y = \dfrac{1}{x^2+4x+5}$.

Completing the square on x^2+4x+5 gives $(x+2)^2+1$.
Since $(x+2)^2 \geq 0$, $(x+2)^2+1 \geq 1$ and so

$$1 \geq \frac{1}{x^2+4x+5}.$$

Since $\dfrac{1}{x^2}$ tends to zero through

positive values for $|x|$ large, $\dfrac{1}{x^2} > 0$.

Similarly $\dfrac{1}{(x+2)^2+1} > 0$

$\therefore 0 < \dfrac{1}{x^2+4x+5} \leq 1$

To sketch $y = \dfrac{1}{x^2+4x+5}$, consider $y_1 = x^2+4x+5$
$$= (x+2)^2+1$$
which is a parabola, axis of symmetry $x=-2$, vertex down and
minimum value $y=1$.

Clearly for each value of x

$y = \dfrac{1}{y_1}$ and y has axis of

symmetry $x=-2$, vertex up
and a maximum value of $y=1$.
The x-axis is an asymptote
for y.

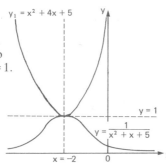

$y_1 = x^2 + 4x + 5$

$y = 1$

$y = \dfrac{1}{x^2 + x + 5}$

$x = -2$

 Given that α and β are the roots of the equation $ax^2+bx+c = 0$, where a, b and c are real and $a \neq 0$, write down the values of $\alpha+\beta$ and $\alpha\beta$ in terms of a, b and c. State the conditions that the roots α and β are equal in magnitude but opposite in sign. Hence find the value of k for which the equation

$$\frac{x^2-2x}{4x-c} = \frac{k-1}{k+1}$$

has roots equal in magnitude but opposite in sign.

If α and β are the roots of $ax^2+bx+c=0$ then
$\alpha+\beta = -\dfrac{b}{a}$, $\alpha\beta = \dfrac{c}{a}$

For α and β to be equal in magnitude but opposite in sign,
$b=0$ and a and c have opposite signs.

$\dfrac{x^2-2x}{4x-c} = \dfrac{k-1}{k+1}$

$\Rightarrow (x^2-2x)(k+1) = (4x-c)(k-1)$
$\Rightarrow x^2k-2xk+x^2-2x-4xk+4x+ck-c=0$
i.e. $x^2(k+1)+x(-6k+2)+c(k-1)=0$
For this equation to have roots equal in magnitude but
opposite in sign $-6k+2=0$, i.e. $k=\frac{1}{3}$.

 Show that $x=2$ is a root of the equation $x^3+x-10=0$. Given that the other roots are α and β, show that $\alpha+\beta=-2$ and find the value of $\alpha\beta$.

Find the equation with numerical coefficients whose roots are
(i) $\alpha+3$ and $\beta+3$;   (ii) 5, $\alpha+3$ and $\beta+3$.

Let $f(x) = x^3+x-10$.
Show $f(2)=0$, when $(x-2)$ is a factor of $f(x)$ and $x=2$ is a root
of $f(x)=0$.
Show $f(x) = (x-2)(ax^2+bx+c)$ i.e a linear factor times a
quadratic factor, when α and β will be the roots of

$ax^2+bx+c=0$ and $\alpha+\beta = -\dfrac{b}{a}$ and $\alpha\beta = \dfrac{c}{a}$

(i) Let $A = \alpha+3$, $B = \beta+3$
Find $A+B$ and AB when required equation is
$$x^2-(A+B)x+AB=0$$

(ii) The required equation will be
$(x-5)[x^2-(A+B)x+AB]=0$

1 Find the range of values of q for which the roots of the
equation $x^2+6x+q^2-7=0$ are real. If q is a positive
integer, list the values of q for which the roots of the
equation are real and of the same sign, giving reasons for
your answer. *(A)

2 (a) Given that the roots of $x^2+px+q=0$ are α and β,
form an equation whose roots are $\dfrac{1}{\alpha}$ and $\dfrac{1}{\beta}$.

(b) Given that α is a root of the equation $x^2=2x-3$ show
that (i) $\alpha^3=\alpha-6$, (ii) $\alpha^2-2\alpha^3=9$. *(C)

3 (a) Find the set of values of k for which the equation
$x^2+kx+2k-3=0$ has no real roots.
When $k=7$, the roots of the equation $x^2+kx+2k-3=0$
are α and β, where $\alpha>\beta$.
(b) Write down the values of $(\alpha+\beta)$ and $\alpha\beta$.
(c) Form an equation with integral coefficients whose
roots are $\dfrac{\alpha}{\beta}$ and $\dfrac{\beta}{\alpha}$.
(d) Prove that $\alpha-\beta=\sqrt{5}$. *(L)

4 (a) If α^2 and β^2 are the roots of $x^2-21x+4=0$ and α and
β are both positive, find:
(i) $\alpha\beta$; (ii) $\alpha+\beta$; (iii) the equation with roots $\dfrac{1}{\alpha^2}$ and $\dfrac{1}{\beta^2}$.

(b) If $\alpha+\beta=5$ and $\alpha\beta=2$, calculate $\dfrac{1}{\alpha}+\dfrac{1}{\beta}$ and hence
determine the values of m and n such that $x^2+mx+n=0$
has roots $\dfrac{1}{\alpha}$ and $\dfrac{1}{\beta}$. *(S)

5 Prove that the equation $x(x-2p)=q(x-p)$ has real roots
for all real values of p and q. If $q=3$, find a non-zero value
for p so that the roots are rational. *(H)

6 Given that α and β are the roots of the equation
$3x^2+x+2=0$,
(i) evaluate $\dfrac{1}{\alpha^2}+\dfrac{1}{\beta^2}$;
(ii) find an equation whose roots are $\dfrac{1}{\alpha^2}$ and $\dfrac{1}{\beta^2}$;
(iii) show that $27\alpha^4=11\alpha+10$. *(A)

7 (a) Find the set of real values of x for which $x^2-9x+20$ is
negative.
(b) Find the set of values of k for which x^2+kx+9 is
positive for all real values of x. *(A)

8 The real roots of the equation $x^2+6x+c=0$ differ by $2n$,
where n is real and non-zero. Show that $n^2=9-c$. Given
that the roots also have opposite signs, find the set of
possible values of n. *(J)

7

4 Sequences and Series

Definitions, Arithmetic and geometric series, Arithmetic and geometric means.

Definitions

A **sequence** is a set of terms in a defined order with a rule for obtaining each term. It is often written as:
$$u_1, u_2, u_3, \ldots, u_n, \ldots \text{ and the } r\text{th term as } u_r.$$

A **series** is formed when the terms of a sequence are added. The series formed from

$$u_1, u_2, u_3, \ldots, u_n, \ldots \text{ is}$$

$$u_1 + u_2 + u_3 + \ldots + u_n + \ldots$$

S_n is the **sum** of the **first n terms** of a series.
S_∞ is the **sum to infinity** of a series.

$\sum_{r=a}^{b} u_r$ means the sum of all terms such as u_r, where r takes all integral values from a to b inclusive. It is read as 'sigma from $r=a$ to b of u_r'.

$\boxed{i}$ 6, 24, 60 120, ... is an infinite sequence with $u_r = r(r+1)(r+2)$.

1, 4, 9, ..., n^2 is a finite sequence of n terms. The series formed from 6, 24, 60, 120, ... is

$$6 + 24 + 60 + 120 + \ldots$$

$\boxed{i}$
$$\sum_{r=1}^{n} r = 1 + 2 + 3 + \ldots + n$$

$$\sum_{r=3}^{7} r^2 = 3^2 + 4^2 + 5^2 + 6^2 + 7^2$$

Arithmetic and geometric series

An **arithmetic series** is of the form
$$a + (a+d) + (a+2d) + (a+3d) + \ldots$$

a is the **first term**, d is the **common difference**. To find d, subtract any term from the next term.

$$u_n = [a + (n-1)d] \quad S_n = \sum_{r=1}^{n} [a + (n-1)d]$$
$$= \tfrac{1}{2}n[2a + (n-1)d]$$
$$= \tfrac{1}{2}n(a+l)$$
where l is the last term.

The 'sum formula' for an arithmetic series can be used to find a formula for its nth term, $u_n = S_n - S_{n-1}$.

A **geometric series** is of the form

$$a + aR + aR^2 + aR^3 + \ldots$$

a is the **first term**, R is the **common ratio**. To find R, divide any term into the next term.

$$u_n = aR^{n-1} \qquad S_n = \sum_{r=1}^{n} aR^{n-1}$$

For $R \neq 1$, $S_n = \dfrac{a(1-R^n)}{(1-R)}$. For $R=1$, $S_n = na$.

If $|R| < 1$, a geometric series may be **summed to infinity**, and $S_\infty = \dfrac{a}{1-R}$.

For the solution of problems, the stated formulae are often used to obtain simultaneous equations in a and d, or a and R, which can then be solved.

$\boxed{i}$ *Three terms of an arithmetic series have sum 21 and product 315. Find the numbers.*

Here it is easier to let the three terms be
$$(a-d), a, (a+d).$$
Sum of terms: $(a-d) + a + (a+d) = 21$
$$3a = 21 \Rightarrow a = 7.$$
Product of terms: $(a-d)a(a+d) = 315$
$$a(a^2 - d^2) = 315$$
But $a=7$, so $7(7^2 - d^2) = 315 \Rightarrow d = \pm 2$.
So the numbers are 5, 7, 9.

$\boxed{i}$ *Three terms of a geometric series have product 343 and sum $\dfrac{49}{2}$. Find the numbers.*

Here it is easier to let the terms be $\dfrac{a}{R}, a, aR$.

Product of terms: $\dfrac{a}{R} \cdot a \cdot aR = 343$
$$a^3 = 343 \Rightarrow a = 7.$$
Sum of terms: $\dfrac{a}{R} + a + aR = \dfrac{49}{2}$
i.e. $\dfrac{a}{R}(1 + R + R^2) = \dfrac{49}{2}$
But $a=7$, so $2 + 2R + 2R^2 = 7R$
i.e. $2R^2 - 5R + 2 = 0 \Rightarrow R = \tfrac{1}{2}$ or 2.

So the numbers are $\dfrac{7}{2}$, 7, 14.

Arithmetic and geometric means

a, b, c form an arithmetic series if $b = \tfrac{1}{2}(a+c)$.
b is the **arithmetic mean** of a and c.
x, y, z form a geometric series if $y = \sqrt{(xz)}$.
y is the **geometric mean** of x and z.
For the n values $x_1, x_2, x_3, \ldots, x_n$,

the arithmetic mean is $\dfrac{1}{n}(x_1 + x_2 + x_3 + \ldots + x_n)$,

the geometric mean is $\sqrt[n]{(x_1 \cdot x_2 \cdot x_3 \ldots x_n)}$.

Given any two numbers it is possible to insert any number of arithmetic (or geometric) means between the two numbers to make the resulting terms form an arithmetic (or geometric) series.

$\boxed{i}$ *Insert*
(a) five arithmetic means between 6 and 30,
(b) four geometric means between 243 and 1.

(a) 6, the five arithmetic means and 30 form an arithmetic series.
1st term: $6 = a$; 7th term: $30 = a + 6d \Rightarrow d = 4$.
So the arithmetic means are 10, 14, 18, 22, 26 hence 6, 10, 14, 18, 22, 26, 30 form an arithmetic series.

(b) 243, the four geometric means and 1 form a geometric series.
1st term: $243 = a$; 6th term: $1 = aR^5 \Rightarrow R = \tfrac{1}{3}$.
So the geometric means are 81, 27, 9, 3 hence 243, 81, 27, 9, 3, 1 form a geometric series.

Sequences and Series
Worked examples, Guided example and Exam questions

 The first three terms of a geometric progression are $k-3$, $2k-4$, $4k-3$, in that order. Find the value of k and the sum of the first eight terms of the progression.

Since $k-3$, $2k-4$, $4k-3$ are three consecutive terms of a geometric progression

$$\frac{2k-4}{k-3}=\frac{4k-3}{2k-4}$$

i.e. $(2k-4)^2=(k-3)(4k-3)$
$4k^2-16k+16=4k^2-15k+9$
$k=7$

The first three terms of the progression are 4, 10, 25

i.e. $a=4$, $r=\dfrac{5}{2}$

∴ the sum to eight terms is $S_8=\dfrac{4\left(1-\left(\dfrac{5}{2}\right)^8\right)}{\left(1-\dfrac{5}{2}\right)}$

$= 4066\cdot 3438$

 In a certain arithmetic progression, the sum of the first and fifth terms is 18 and the fifth term is 6 more than the third term. Show that the sum of the first ten terms of the progression is 165.

Let the first six terms of the arithmetic progression be
a, $a+d$, $a+2d$, $a+3d$, $a+4d$, $a+5d$
Given $a+(a+4d)=18$
$(a+4d)=(a+2d)+6$
i.e. $2a+4d=18$
and $2d=6$
∴ $d=3$ and $a=3$
∴ the arithmetic progression has first term 3 and common difference 3.

So the sum of the first ten terms is

$$S_{10}=\frac{10}{2}\times(2\times3+9\times3)$$

$= 165$ as required

 The three real, distinct and non-zero numbers a, b, c are such that a, b, c are in arithmetic progression and a, c, b are in geometric progression. Find the numerical value of the common ratio of the geometric progression.
Hence find an expression, in terms of a, for the sum to infinity of the geometric series whose first terms are a, c, b.

Since a, b, c are in arithmetic progression

$b-a=c-b$ [1]

Since a, c, b are in geometric progression

$\dfrac{c}{a}=\dfrac{b}{c}$ [2]

Eliminate b between [1] and [2] and solve the resulting quadratic in c (c will be found in terms of a). Using these values of c in [2] the common ratio of the corresponding geometric progression can be found. One value of r is less than 1, hence the sum to infinity of the geometric progression can be found.

 1 Three consecutive terms in a geometric progression are c, $c+4$ and $c+6$, in that order. Determine the value of c and the value of the common ratio of the progression. *(S)

2 Show that there are 18 integers which are multiples of 17 and which lie between 200 and 500. Find the sum of all these integers. *(L)

3 The nth term of an arithmetic progression is denoted by u_n, and the sum of the first n terms is denoted by S_n.
(a) In a certain arithmetic progression, $u_5+u_{16}=44$ and $S_{18}=3S_{10}$. Calculate the value of the first term and of the common difference.
(b) In another arithmetic progression, $u_1=1$. Given that u_7, u_{11} and u_{17} are in geometric progression, find the value of each. *(C)

4 (i) A man invests £100 at the beginning of each year for ten years. The rate of compound interest is 9% per annum. Calculate the total value of the investment at the end of the ten full years.
(ii) Write down the sum of the arithmetic progression, $1+2+3+\ldots+n$. Let
$$S_n=\frac{1+2+3+\ldots+n}{n^2}.$$
Find a value of n such that $S_n-\frac{1}{2}<10^{-6}$. *(OLE)

5 Find the common ratio of the geometric sequence
$$\sin 2\alpha,\ -\sin 2\alpha\cos 2\alpha,\ \sin 2\alpha\cos^2 2\alpha,\ \ldots.$$
Prove that for $0<\alpha<\dfrac{\pi}{2}$ the series
$$\sin 2\alpha-\sin 2\alpha\cos 2\alpha+\sin 2\alpha\cos^2 2\alpha+\ldots.$$
has a sum to infinity and show that the sum to infinity is $\tan\alpha$. (H)

6 Find the sum to infinity of the geometrical progression
$$1+x/(x+1)+x^2/(x+1)^2+\ldots$$
and determine the set of values of x for which the result holds. (O & C)

7 Find the set of values of θ, $(-\pi<\theta\leqslant\pi)$, for which the series
$$1+2\cos^2\theta+4\cos^4\theta+8\cos^6\theta+\ldots+2^r\cos^{2r}\theta+\ldots.$$
has a sum to infinity.
Show that, for this set of values of θ, the sum to infinity of the series is $-\sec 2\theta$. (J)

8 It is given that $\dfrac{1}{b+c}$, $\dfrac{1}{c+a}$, $\dfrac{1}{a+b}$ are three consecutive terms of an arithmetic series. Show that a^2, b^2 and c^2 are also three consecutive terms of an arithmetic series. (J)

9 The sum of the first twenty terms of an arithmetic progression is 45, and the sum of the first forty terms is 290. Find the first term and the common difference. Find the number of terms in the progression which are less than 100. (J)

10 One sequence of alternating terms of the series
$$1+2+3+4+5+8+\ldots$$
forms an arithmetic progression, while the other sequence of alternating terms forms a geometric progression. Sum the first 10 terms of each progression and hence find the sum of the first 20 terms of the series. (L)

11 The first term of an arithmetic series is $(3p+5)$ where p is a positive integer. The last term is $(17p+17)$ and the common difference is 2.
Find, in terms of p (i) the number of terms; (ii) the sum of the series. Show that the sum of the series is divisible by 14, only when p is odd. (A)

5 Summation of Series

Mathematical induction, Arithmetic and geometric series, Some special results.

Mathematical induction

Mathematical induction is a method of proving a given (or suspected) result for positive integers. This method is often used to prove the formula for the sum to n terms of a series.

To prove by induction

(a) Assume that the stated result is true for some positive integral value of n, say k, and show that it is true for the next integral value, i.e. $k+1$.

(b) Show that the result is true for the first value, i.e. for $k=1$.

(c) Conclude that, by the principle of mathematical induction, the result holds for all positive integers k.

This concludes the proof because:
since the result is true for $k=1$,
then by (a) it has been shown to be true for $k=2$,
since it is true for $k=2$,
then by (a) it is true for $k=3$, ... and so on.
So it is true for all positive integers k.

ⓘ *Show by induction that*

$$S_n=\sum_{r=1}^{n}[a+(r-1)d]=\tfrac{1}{2}n[2a+(n-1)d]$$

i.e. the sum to n terms of an arithmetic series.

Assume the result is true when $n=k$,
i.e. $S_k=\tfrac{1}{2}k[2a+(k-1)d]$
Add the next term, the $(k+1)$th, giving

$$\begin{aligned}S_{k+1}&=\tfrac{1}{2}k[2a+(k-1)d]+[a+(k+1-1)d]\\&=\tfrac{1}{2}[2ak+k^2d-kd+2a+2kd]\\&=\tfrac{1}{2}[2a(k+1)+kd(k+1)]\\&=\tfrac{1}{2}(k+1)[2a+kd]\end{aligned}$$

But this is S_n with n replaced by $(k+1)$.
∴ if the result is true for k, it is true for $(k+1)$.
Put $k=1$, $S_1=\tfrac{1}{2}[2a+(1-1)d]=a$,
i.e. the first term is a, which is true.

∴ by the principle of mathematical induction the formula is true for all n.

Arithmetic and geometric series

If the terms of a series can be recognised as an arithmetic series or geometric series then the appropriate formula for its sum can be used.

For an **arithmetic series**

$$S_n=\tfrac{1}{2}n[2a+(n-1)d]$$

For a **geometric series**

$$S_n=\frac{a(1-R^n)}{(1-R)}$$

$$S_\infty=\frac{a}{1-R}\ \text{if}-1<R<1.$$

ⓘ *Find the sum to* 21 *terms of the series*

$$(-20)+(-18)+(-16)+\dots$$

This is an arithmetic series with $a=-20$ and $d=2$.
Use $S_n=\tfrac{1}{2}n[2a+(n-1)d]$
$$\begin{aligned}S_{21}&=\tfrac{1}{2}\times21[2\times(-20)+(21-1)(2)]\\&=\tfrac{1}{2}\times21[-40+40]\\&=0\end{aligned}$$

ⓘ *Find the sum to infinity of the series*

$$1+\tfrac{1}{2}+\tfrac{1}{4}+\dots$$

This is a geometric series with $a=1$ and $R=\tfrac{1}{2}$.
Since $-1<R<1$,

$$\text{use } S_\infty=\frac{a}{1-R}=\frac{1}{1-\tfrac{1}{2}}=2.$$

Some special results

The following special results can be useful when summing series.

1. Distributive property of Σ
i.e. $\Sigma[af(x)+bg(x)+ch(x)+\dots]$
$=a\Sigma f(x)+b\Sigma g(x)+c\Sigma h(x)+\dots$

2. Natural number series

(a) $1+2+3+4+\dots+n=\sum_{r=1}^{n}r=\tfrac{1}{2}n(n+1)$

(b) $1^2+2^2+3^2+4^2+\dots+n^2=\sum_{r=1}^{n}r^2=\tfrac{1}{6}n(n+1)(2n+1)$

(c) $1^3+2^3+3^3+4^3+\dots+n^3=\sum_{r=1}^{n}r^3=\tfrac{1}{4}n^2(n+1)^2$

Note: $\sum_{r=1}^{n}r^3=\left(\sum_{r=1}^{n}r\right)^2.$

3. Sum of a constant

$$\sum_{r=1}^{n}a=\frac{\overbrace{a+a+a+\dots}^{n\text{ times}}}{}=an$$

ⓘ *Find the sum to n terms of a series whose general term is* $4r^2+3r+1$.

We require $S_n=\sum_{r=1}^{n}(4r^2+3r+1)$

$$=\sum_{r=1}^{n}4r^2+\sum_{r=1}^{n}3r+\sum_{r=1}^{n}1$$

$$=4\sum_{r=1}^{n}r^2+3\sum_{r=1}^{n}r+\sum_{r=1}^{n}1$$

$$=4[\tfrac{1}{6}n(n+1)(2n+1)]+3[\tfrac{1}{2}n(n+1)]+n$$

$$=\tfrac{1}{6}n(8n^2+21n+19)$$

Summation of Series
Worked example, Guided example and Exam questions

WE *(a) Prove by mathematical induction that*

$$\frac{1}{1.2}+\frac{1}{2.3}+\ldots+\frac{1}{n(n+1)}=\frac{n}{n+1}$$

for all positive integers n.

(b) Obtain, and simplify, an expression for the sum of each of the following series:

$$1.2+3.4+5.6+\ldots+(2n-1).2n$$
$$2-3+4-5+6-7+\ldots+2n-(2n+1)$$

(a) Assume

$$S_k=\frac{1}{1.2}+\frac{1}{2.3}+\ldots+\frac{1}{k(k+1)}=\frac{k}{k+1}$$

Then

$$S_{k+1}=\left[\frac{1}{1.2}+\frac{1}{2.3}+\ldots+\frac{1}{k(k+1)}\right]+\frac{1}{(k+1)(k+2)}$$

$$=\frac{k}{(k+1)}+\frac{1}{(k+1)(k+2)}$$

$$=\frac{k(k+2)+1}{(k+1)(k+2)}=\frac{(k+1)^2}{(k+1)(k+2)}=\frac{k+1}{k+2}$$

(*N.B.* The result for S_{k+1} is that for S_k with k replaced by $k+1$.)

$\therefore$ assuming S_k is true, S_{k+1} is true.

$S_1=\dfrac{1}{1.2}=\dfrac{1}{1+1}=\dfrac{1}{2}$, which is true.

$\therefore$ by mathematical induction the result is true for all positive integral n.

(b) Let $S_n=1.2+3.4+5.6\ldots+(2n-1)\times 2n$

Let the rth term be $u_r=(2r-1)\times 2r$

We require

$$\sum_{r=1}^{n}u_r=\sum_{r=1}^{n}(2r-1)\times 2r$$

$$=4\sum_{r=1}^{n}r^2-2\sum_{1}^{n}r$$

$$=4\times\frac{n}{6}(n+1)(2n+1)-2\times\frac{n}{2}(n+1)$$

$$=n(n+1)\left[\frac{2}{3}(2n+1)-1\right]$$

$$=\frac{n(n+1)}{3}[4n+2-3]=\frac{n(n+1)(4n-1)}{3}$$

Let $S=(2-3)+(4-5)+\ldots+[2n-(2n+1)]$

This series contains $2n$ terms which taken in n pairs, each pair having sum -1, has sum $(-1)\times n=-n$.

GE *Use induction to prove that the sum to n terms of the series*

$$\frac{5}{2\times3\times4}+\frac{7}{3\times4\times5}+\frac{9}{4\times5\times6}+\ldots$$

is $\dfrac{(2n+3)(2n+5)}{2(n+2)(n+3)}-\dfrac{5}{4}$. *Find the sum to infinity of the series.*

Assume

$$S_k=\frac{5}{2\times3\times4}+\frac{7}{3\times4\times5}+\ldots+\frac{(2k+3)}{(k+1)(k+2)(k+3)}$$

$$=\frac{(2k+3)(2k+5)}{2(k+2)(k+3)}-\frac{5}{4}$$

Then show

$$S_{k+1}=\frac{(2k+5)(2k+7)}{2(k+3)(k+4)}-\frac{5}{4}$$

$\therefore$ assuming S_k is true, S_{k+1} is true.

$$S_1=\frac{5}{2\times3\times4}=\frac{5\times7}{2\times3\times4}-\frac{5}{4}=\frac{5\times7-2\times15}{2\times3\times4}$$

which is true

$\therefore$ by mathematical induction the result is true for all positive integral n

write S_n as $\dfrac{\left(2+\dfrac{3}{n}\right)\left(2+\dfrac{5}{n}\right)}{2\left(1+\dfrac{2}{n}\right)\left(1+\dfrac{3}{n}\right)}-\dfrac{5}{4}$

Show $\displaystyle\lim_{n\to\infty}S_n=\frac{3}{4}$

EX **1** (a) Find the sum of n terms of the following series:
(i) $1.1+2.3+3.5+\ldots$; (ii) $2x+4x^2+8x^3+\ldots$, $x<\frac{1}{2}$.
In each case give your answer in as simple a form as possible.

(b) Show that the nth term of the series
$1+(1+2)+(1+2+2^2)+(1+2+2^2+2^3)+\ldots$ is 2^n-1.
Hence find (i) the sum of n terms of the series; (ii) the value of n if $T_{n+1}-T_n=64$, where T_n denotes the nth term.
*(W)

2 Prove by induction, or otherwise, that

(a) $\displaystyle\sum_{r=1}^{n}r(r+1)=\frac{n}{3}(n+1)(n+2)$,

(b) $\displaystyle\sum_{r=1}^{n}r(r+1)(r+2)=\frac{n}{4}(n+1)(n+2)(n+3)$.

Show that $r^3\equiv r(r+1)(r+2)-3r(r+1)+r$ and hence prove that

$$\sum_{r=1}^{n}r^3=\frac{n^2}{4}(n+1)^2.$$

Evaluate $\displaystyle\sum_{r=1}^{20}r(r+3)(r+6)$

(L)

3 Prove by induction that, for all positive integers n,
$1^3+2^3+\ldots+n^3=\frac{1}{4}n^2(n+1)^2$.
Deduce that
$(n+1)^3+(n+2)^3+\ldots+(2n)^3=\frac{1}{4}n^2(3n+1)(5n+3)$.
(J)

4 Prove by induction that, for every positive integer n,
$(1\times4)+(2\times5)+(3\times6)+\ldots+n(n+3)=\frac{1}{3}n(n+1)(n+5)$.
(J)

5 By induction, or otherwise, prove the following results:

(i) $\displaystyle\sum_{r=1}^{n}\cos(2r-1)x=\frac{\sin 2nx}{2\sin x}$;

(ii) $\displaystyle\sum_{r=1}^{n}\frac{(r+4)}{2^r r(r+1)(r+2)}=\frac{1}{2}-\frac{1}{2^n(n+1)(n+2)}$.
(O & C)

6 *Use induction* to prove that the sum of the first $2n$ terms of the series $1^2-3^2+5^2-7^2+\ldots$ is $-8n^2$.
Write down the sum to $(2n+1)$ terms. (O & C)

7 If $x^3=x+1$, prove, by induction or otherwise, that
$x^{3n}=a_nx+b_n+c_nx^{-1}$, where $a_1=1$, $b_1=1$, $c_1=0$, and
$a_n=a_{n-1}+b_{n-1}$, $b_n=a_{n-1}+b_{n-1}+c_{n-1}$, $c_n=a_{n-1}+c_{n-1}$,
for $n=2, 3,\ldots$. (OLE)

6 Permutations and Combinations

Factorial notation, Permutations, Permutations with identical items, Combinations.

Factorial notation

$n! = n(n-1)(n-2)(n-3)\ldots 2 \cdot 1$
i.e. the product of all integers from n to 1 inclusive.
$n!$ is read as '**factorial n**' or '**n factorial**'.
We define $0! = 1$.

ⓘ $5! = 5 \times 4 \times 3 \times 2 \times 1$
$= 120$

Permutations

A **permutation** is an arrangement of items.

ⓘ The permutations of A, B, C
ABC, ACB, BAC, BCA, CAB, CBA.

The number of permutations of n different items taken r at a time is written as ^nP_r or $_nP_r$.

$$^nP_r = \frac{n!}{(n-r)!}$$

ⓘ *Find the number of permutations of the letters A, B, C taken 2 at a time.*

$$^3P_2 = \frac{3!}{(3-2)!} = 3 \times 2 = 6$$

If $r = n$, i.e. n items are taken n at a time,

$$^nP_n = \frac{n!}{0!} = n! \text{ since } 0! = 1.$$

ⓘ *In how many ways can the letters of MATHS be arranged?*

$$^5P_5 = 5! = 120 \text{ ways}$$

Permutations with identical items

(a) **One set of identical items**
The number of permutations of n items taken n at a time, when p of the items are identical and the rest are all different, is

$$\frac{n!}{p!}$$

ⓘ *In how many ways can the letters of MIME be arranged?*

There are 4 letters, so $n = 4$,
2 M's, so $p = 2$.

Number of permutations $= \dfrac{4!}{2!} = 4 \times 3 = 12$.

(b) **Two sets of identical items**
The number of permutations of n items taken n at a time,
when p items are identical and of one kind,
$\quad q$ items are identical and of a second kind,
the rest are all different, is

$$\frac{n!}{p! \, q!}$$

ⓘ *Find the number of permutations of the letters of PARALLEL.*

There are 8 letters, so $n = 8$,
2 As, so $p = 2$; 3 Ls, so $q = 3$.

Number of permutations $= \dfrac{8!}{2! \, 3!} = 3360$.

(c) **Repeated items**
The number of permutations of n different items taken r at a time, when each item may be used repeatedly, is n^r.

ⓘ *Find the number of 3 letter codes.*

The number of permutations, with repetition, is $26^3 = 17576$.

Combinations

A **combination** is a selection of items when arrangement is not important. Different permutations of the same items count as one combination.

ⓘ There is only one combination of the letters ABC.

The number of combinations of n different items taken r at a time is written as $\binom{n}{r}$ or ^nC_r.

$$\binom{n}{r} = {}^nC_r = \frac{n!}{r!(n-r)!}$$

Some useful results are:

$$\binom{n}{r} = {}^nC_r = \frac{n(n-1)(n-2)\ldots}{r(r-1)(r-2)\ldots} \begin{array}{l} \leftarrow r \text{ factors starting at } n \\ \leftarrow r \text{ factors starting at } r \end{array}$$
$$^nC_r = {}^nC_{n-r}$$

The total number of selections from n items when each can be either included or excluded is $2^n - 1$, if at least one item is to be taken.

ⓘ *A sub-committee of six, including a chairperson, is to be chosen from a main committee of twelve. If the chairperson is to be a specified member of the main committee, in how many ways can the sub-committee be chosen?*

As the specified member of the main committee has to be included, we require the number of combinations of 5 from 11, i.e.

$$\binom{11}{5} = \frac{11 \times 10 \times 9 \times 8 \times 7}{5 \times 4 \times 3 \times 2 \times 1} = 462.$$

Permutations and Combinations
Worked examples, Guided example and Exam questions

 A committee of four is chosen from five teachers and three sixth-formers. In how many ways can this be done so that the committee contains:

(a) at least one teacher; (b) at least one teacher and one sixth-former?

(a) Tabulating the possibilities for the committee gives

No. of teachers	No. of ways	No. of sixth-formers	No. of ways
1	$^5C_1$	3	$^3C_3$
2	$^5C_2$	2	$^3C_2$
3	$^5C_3$	1	$^3C_1$
4	$^5C_4$	0	$^3C_0$

$\therefore$ No. of committees

$= {}^5C_1 \times {}^3C_3 + {}^5C_2 \times {}^3C_2 + {}^5C_3 \times {}^3C_1 + {}^5C_4 \times {}^3C_0$

$= 5 \times 1 + \dfrac{5 \times 4}{1 \times 2} \times 3 + \dfrac{5 \times 4}{1 \times 2} \times 3 + 5 \times 1$

$= 5 + 30 + 30 + 5 = 70$

(b) Tabulating the possibilities for the committee would give the first three rows of the table above.

$\therefore$ No. of committees $= 5 + 30 + 30 = 65$.

 How many different arrangements of letters can be made by using all the letters of the word MINIMUM? In how many of these are the vowels separated?

There are 7 letters, including 3 Ms and 2 Is.

$\therefore$ the number of possible arrangements $= \dfrac{7!}{3!2!} = 420$

Treat the three vowels I, I, U as one letter, then the number of arrangements with the vowels together is $3 \times \dfrac{5!}{3!} = 60$

The factor 3 is introduced to allow for the three possible arrangements of the vowels among themselves. Thus the number of arrangements with the vowels separated is $420 - 60 = 360$

 A batch code on the side of a food container is composed of six dots, one blue, two white and three black. Find the number of different codes possible.
It is decided to use only five dots chosen from six where one is blue, two are white and three are black. Find the number of different codes possible in this case.

The number of codes possible is equal to the number of permutations of six objects of which two are alike and of one kind and three are alike and of another kind. In the second case, consider the possible selection of five dots, which is best presented in a tabular fashion.

Blue	White	Black
1	2	2
1	1	3
0	2	3

For each selection find the possible number of permutations of five objects, e.g. the first selection gives $\dfrac{5!}{2!2!}$

Hence the total number of codes that can be formed with five dots can be found.

EX

1 At an athletics meeting, eight lanes are marked on the running track. Find the number of ways in which the runners in a heat can be allocated to these lanes when there are:
  (i) 8 competitors in a heat;
  (ii) 5 competitors in a heat and no restriction on which lanes are used;
  (iii) 10 competitors in a heat, so that two competitors have to share lane 3 and two have to share lane 8, but no account is taken of the way in which these lanes are shared;
  (iv) two teams, each of four competitors, on the starting line, and members of the same team may not be in adjacent lanes.
  (A)

2 A concert pianist has prepared seven different pieces of music for a recital, three of which are modern and four are classical. Calculate the number of different orders in which she can play the seven pieces when:
  (a) there are no restrictions;
  (b) the recital must start and end with classical pieces;
  (c) classical and modern pieces must alternate throughout the recital.
  (A)

3 Calculate the number of different 7-letter arrangements which can be made with the letters of the word MAXIMUM. In how many of these do the 4 consonants all appear next to one another?
  (C)

4 Four visitors Dan, May, Nan and Tom arrive in a town which has five hotels. In how many ways can they disperse themselves among the five hotels:
  (a) if four hotels are used to accommodate them;
  (b) if three hotels are used to accommodate them in such a way that May and Nan stay in the same hotel?
  (L)

5 The result (home-win, score-draw, goalless-draw or away-win) is forecast for each of ten football matches. In how many different ways can the results of these ten matches be forecast to give exactly seven correct results?
  (L)

6 From eight persons, including Mr and Mrs Smith, a committee of four persons is to be chosen. Mr Smith will not join the committee without his wife, but his wife will join the committee without her husband. In how many ways can the committee be formed?
  (L)

7 Find the number of different permutations of the 8 letters of the word *SYLLABUS*. Find the number of different selections of 5 letters which can be made from the letters of the word *SYLLABUS*.
  (L)

8 Five people, of whom three are women and two are men, are to form a queue. Find how many different arrangements there are (i) if no two people of the same sex are to stand next to each other, (ii) if the first and last people in the queue are both to be men.
  (C)

9 [In this question, answers may be left in factorial form.] State the number of possible arrangements of the nine letters of the word *FACETIOUS*. Determine in how many of these arrangements: (i) the order in which the vowels occur is the same as in the original word; (ii) the vowels, in any order, are separated from each other by a consonant.
  (C)

7 Binomial Theorem

Positive integral index, Any rational index, Applications, Approximations.

Positive integral index

The **Binomial Theorem**, for all x and positive integral values of n, is given by

$$(a+x)^n = a^n + na^{n-1}x + \frac{n(n-1)}{2!}a^{n-2}x^2 + \frac{n(n-1)(n-2)}{3!}a^{n-3}x^3 + \ldots + x^n$$

Note:
There are $(n+1)$ terms in this finite expression.
The degree of each term is n, i.e. the sum of the powers of a and x in each term is n.
The general term is the $(r+1)$ term and is

$$\frac{n(n-1)(n-2)\ldots(n-r+1)}{r!}a^{n-r}x^r$$

The coefficient of x^r in the general term is sometimes written as $\binom{n}{r}$ or ^nC_r.

The coefficients in the expansion of $(a+x)^n$ form a pattern which is useful for small values of n.
It is shown in **Pascal's triangle.**

n	Pascal's triangle	expansion
1	1 1	$(a+x)^1 = 1a + 1x$
2	1 2 1	$(a+x)^2 = 1a^2 + 2ax + 1x^2$
3	1 3 3 1	$(a+x)^3 = 1a^3 + 3a^2x + 3ax^2 + 1x^3$
4	1 4 6 4 1	$(a+x)^4 = 1a^4 + 4a^3x + 6a^2x^2 + 4ax^3 + 1x^4$
5	1 5 10 10 5 1	$(a+x)^5 = 1a^5 + 5a^4x + 10a^3x^2 + 10a^2x^3 + 5ax^4 + 1x^5$

Any rational index

If n is any rational value, positive or negative, and $-1 < x < 1$, then the Binomial Theorem is

$$(1+x)^n = 1 + nx + \frac{n(n-1)}{2!}x^2 + \frac{n(n-1)(n-2)}{3!}x^3 + \ldots$$

Note:
The R.H.S. is an infinite series since $n(n-1)(n-2)$ . . . will never become zero.
The general term is $\dfrac{n(n-1)(n-2)\ldots(n-r+1)}{r!}x^r$.

The first term must be 1. If it is not, it must be rewritten: $(a+x)^n = a^n\left(1 + \dfrac{x}{a}\right)^n$

and the expansion can be applied provided that $-1 < \dfrac{x}{a} < 1$.

Important special cases are:
$(1+x)^{-1} = 1 - x + x^2 - x^3 + x^4 - \ldots$ $(1+x)^{-2} = 1 - 2x + 3x^2 - 4x^3 + 5x^4 - \ldots$
$(1-x)^{-1} = 1 + x + x^2 + x^3 + x^4 + \ldots$ $(1-x)^{-2} = 1 + 2x + 3x^2 + 4x^3 + 5x^4 + \ldots$
These are only true if $-1 < x < 1$.

Applications

In any form of the Binomial Theorem, the 'x term' may be replaced by a negative value or a power of x or a group of terms.
The range of values of x over which the expansion is valid must be found using

$$-1 < \text{'}x\text{ term'} < 1.$$

ℹ️ *Find the first four terms of $(1-2x)^{-\frac{3}{2}}$.*

$(1-2x)^{-\frac{3}{2}} = 1 + (-\frac{3}{2})(-2x) + \dfrac{(-\frac{3}{2})(-\frac{5}{2})}{2.1}(-2x)^2$

$\qquad\qquad + \dfrac{(-\frac{3}{2})(-\frac{5}{2})(-\frac{7}{2})}{3.2.1}(-2x)^3 + \ldots$

$\qquad = 1 + 3x - \frac{15}{2}x^2 + \frac{35}{2}x^3 - \ldots$

This expansion is valid for $-1 < -2x < 1$,
i.e. $\frac{1}{2} > x > -\frac{1}{2}$.

Approximations

The Binomial Theorem, in the form $(1+x)^n$, may be used to obtain approximations and to evaluate roots of numbers to any degree of accuracy.

ℹ️ *Evaluate $\sqrt{25.1}$ to four decimal places.*

$\sqrt{25.1} = (25+0.1)^{\frac{1}{2}} = 5(1+0.004)^{\frac{1}{2}}$

$\qquad = 5\left[1 + (\frac{1}{2})(0.004) + \dfrac{(\frac{1}{2})(-\frac{1}{2})}{2!}(0.004)^2 + \ldots\right]$

$\qquad = 5(1 + 0.002 - 0.000002 + \ldots)$

$\qquad \approx 5(1.001998) = 5.00999$

So $\sqrt{25.1} \approx 5.0100$ (to four decimal places)

Binomial Theorem
Worked examples and Exam questions

 (i) Find the expansion of $(1+2x+x^2)^4$ in ascending powers of x up to and including the term in x^2.

(ii) Calculate the value of the term independent of x in the expansion of
$$\left(x-\frac{3}{x^2}\right)^{15}.$$

(i) $(1+2x+x^2)^4 = [1+(2x+x^2)]^4$

$$= 1+4(2x+x^2)+\frac{4\times3}{1\times2}(2x+x^2)^2 + \text{terms} > x^2$$

$$= 1+8x+4x^2+24x^2+\text{terms} > x^2$$

$$= 1+8x+28x^2+\text{terms} > x^2$$

(ii) The general term in the expansion of $\left(x-\frac{3}{x^2}\right)^{15}$ is

$$^{15}C_r(x)^{15-r}\left(-\frac{3}{x^2}\right)^r$$

for the term independent of x the index of r must be zero
i.e. $15-r+(-2r)=0$ i.e. $r=5$
$\therefore$ the 6th term is the one independent of x and has value

$$^{15}C_5(-3)^5 = \frac{15\times14\times13\times12\times11}{1\times2\times3\times4\times5}\times(-3)^5$$

$$= -729\ 729$$

 Obtain the first four terms in the expansion of $\left(1-\frac{x}{2}\right)^6$. Use these terms to find the value of $(0\cdot99)^6$, giving your answer to four decimal places.

$$\left(1-\frac{x}{2}\right)^6 = 1+6\left(-\frac{x}{2}\right)+\frac{6\times5}{1\times2}\left(-\frac{x}{2}\right)^2$$
$$+\frac{6\times5\times4}{1\times2\times3}\left(-\frac{x}{2}\right)^3 \text{ to 4 terms}$$

$$= 1-3x+\frac{15}{4}x^2-\frac{5}{2}x^3 \text{ to 4 terms}$$

Let $x=0\cdot02$

then $1-\dfrac{x}{2}=1-\dfrac{0\cdot02}{2}=1-0\cdot01=0\cdot99$

$\therefore (0\cdot99)^6 = \left(1-\dfrac{0\cdot02}{2}\right)^6$

$$\approx 1-3\times(0\cdot02)+\frac{15}{4}\times(0\cdot02)^2-\frac{5}{2}\times(0\cdot02)^3$$

$$= 1-0\cdot06+\frac{15}{4}\times0\cdot0004-\frac{5}{2}\times0\cdot000\,008$$

$$= 0\cdot94+0\cdot0015-0\cdot000\,02$$

$$= 0\cdot9415-0\cdot000\,02$$

$$= 0\cdot9415 \text{ to four decimal places.}$$

---

 1 Expand $\left(2-\dfrac{x}{2}\right)^5$ in ascending powers of x. Use the first four terms of the expansion to find an approximate value for $(1\cdot99)^5$. *(C)*

2 (a) Find the first four terms in the expansion of $(1-4x)^{\frac14}$ in ascending powers of x and state the range of validity of the expansion. Use the expansion to determine the value of $\sqrt{15\cdot36}$, correct to 5 decimal places.

(b) The coefficient of x^2 in the expansion of $(1+ax)^{\frac14}$ is -6. Calculate the two possible values of a. *(S)*

3 Using the binomial theorem, or otherwise, find an expression, in descending powers of x and with whole number coefficients, for $(2x-3)^4-(2x+3)^4$. *(L)*

4 Obtain the binomial expansion of $(1+2x)^{-3}$ in ascending powers of x as far as the term in x^3. *(L)*

5 Show that
$$\frac{1}{\left(1-\dfrac{x}{3}\right)}+\frac{1}{\left(1+\dfrac{x}{3}\right)}=\frac{18}{(9-x^2)}.$$
Hence, or otherwise, express $\dfrac{18}{(9-x^2)}$ as a series in ascending powers of x up to and including the term in x^4, assuming that the value of x is such as to make the expansions valid. *(A)*

6 Show that $\dfrac{\sqrt{(1-x)}-2\sqrt{(1-2x)}}{\sqrt{(1-3x+2x^2)}}$ can be expressed in the form $(1-2x)^{-\frac12}-2(1-x)^{-\frac12}$.
Using the binomial expansion write $(1-2x)^{-\frac12}-2(1-x)^{-\frac12}$ as a series in ascending powers of x up to and including terms in x^4. Hence show that the first non-vanishing term in the expansion of
$$1-\tfrac34x^2-\frac{15}{8}x^3+\frac{\sqrt{(1-x)}-2\sqrt{(1-2x)}}{\sqrt{(1-3x+2x^2)}}$$
as a series in ascending powers of x is kx^4, where k is a constant. Find the value of k. *(O)*

7 Find, in ascending powers of x, the first three terms in the expansion of $(2-3x)^8$. Use the expansion to find the value of $(1\cdot997)^8$ correct to the nearest whole number. *(C)*

8 Expand $(2-x)^5$ in ascending powers of x up to and including the term in x^2. Use these terms to find the value of $2\cdot01^5$, giving your answer to three decimal places. *(S)*

9 Write down the first four terms of the binomial series expansion in ascending powers of x of the function $(1-x)^{-\frac14}$, and hence show that
$$\sqrt{10}=3\left\{1+\frac{1}{20}+\frac{1\times3}{20\times40}+\frac{1\times3\times5}{20\times40\times60}+\dots\right\}.$$
(OLE)

10 Find the coefficient of x^n, where $n\geqslant2$, in the expansion, in ascending powers of x, of $(a+bx+cx^2)e^x$. Given that, for $n\geqslant0$, the coefficient of x^n is $\dfrac{(n+1)(n+2)}{n!}$, find a, b and c. *(J)*

11 Write down the expansion in ascending powers of x up to the term in x^2 of (i) $(1+x)^{\frac12}$, (ii) $(1-x)^{-\frac12}$, and simplify the coefficients. Hence, or otherwise, expand
$$\sqrt{\left(\frac{1+x}{1-x}\right)}$$
in ascending powers of x up to the term in x^2. By using $x=1/10$ obtain an estimate, to three decimal places, for $\sqrt{11}$. *(J)*

12 Find the first four terms of the expansion, in ascending powers of x, of: (i) $(1-x)^{-1}$; (ii) $(1-x)^{-2}$.
Hence show that, if $p=1-q$; (iii) $p+pq+pq^2+pq^3+\dots=1$; (iv) $p+2pq+3pq^2+4pq^3+\dots=1/p$. *(A)*

15

8 Inequations
Rules for manipulating inequations, Solving inequations, Modulus.

Rules for manipulating inequations

If a, b, c, d and k are numbers such that $a>b$ and $c>d$, then:

(a) $a\pm k>b\pm k$

(b) $ak>bk$ for $k>0$
     $ak<bk$ for $k<0$.

(c) $a+c>b+d$

A similar set of results arise for $<$.

Note: We cannot make any deductions about:
$a-c$ and $b-d$ or ac and bd or $a\div c$ and $b\div d$.

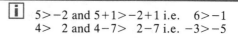 5>−2 and 5+1>−2+1 i.e. 6>−1
   4> 2 and 4−7> 2−7 i.e. −3>−5

   4>−6 and 2×4> 2×−6 i.e. 8>−12
   3>−1 and −1×3<−1×−1 i.e. −3<1

4>2 and 3>−1, so 4+3>2+−1 i.e. 7>1

Solving inequations

The **solution of an inequation** is a range (or ranges) of values of the variable.

1. Linear inequations in one unknown
These can be solved using the rules of inequations. The solution set can be illustrated on a number line. Note the symbols used:

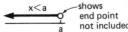

| Figure 1 | Figure 2 |

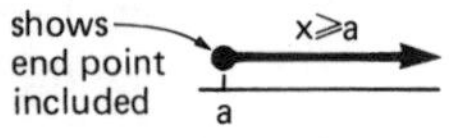

i *Find the solution set of $8-x\geqslant 5x-4$.*

$8-x\geqslant 5x-4\Rightarrow 8+4\geqslant 5x+x\Rightarrow 12\geqslant 6x$
$\Rightarrow 2\geqslant x$

This is illustrated as:

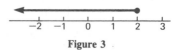

Figure 3

2. Linear inequations in two unknowns
These are best solved graphically.
The corresponding equality gives the boundary line. This is drawn as:

(a) a continuous line if the inequation is $\geqslant$ or $\leqslant$,
(b) a dotted line if the inequation is $>$ or $<$.

A convenient point is chosen to identify on which side of the line the inequation applies. The solution set of the inequation is usually left unshaded.

i *Solve $y\geqslant 0$, $x+y\leqslant 2$ and $y-2x<2$.*

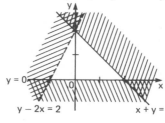

Draw the lines
$y=0$ (continuous)
$x+y=2$
$y-2x=2$ (dotted)
Test the point $(0, 1)$.
The unshaded region gives the solution set.

Figure 4

3. Quadratic inequations
(a) solution using a number line
If the inequation is written as $f(x)>0$ or $f(x)<0$, then the end points of the inequation can be found by solving the corresponding quadratic equation. A convenient point is tested to identify the range.

(b) graphical solution
The graph of the corresponding quadratic equality (a parabola) is sketched. By inspection the range corresponding to the inequation can be seen.

4. Other inequations
These are usually solved graphically.

i *Find the solution set of $x^2-2x-8\leqslant 0$.*

(a) $x^2-2x-8\leqslant 0\Rightarrow (x-4)(x+2)\leqslant 0$
End points are $x=4$, $x=-2$.
Test $x=0$: $(-4)(2)=-8\leqslant 0$ (true)
So $-2\leqslant x\leqslant 4$.

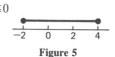

Figure 5

(b) Sketch of $y=x^2-2x-8$, i.e. of $y=(x-4)(x+2)$

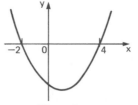

$y\leqslant 0$ for
$-2\leqslant x\leqslant 4$
which is the solution set of the inequation.

Figure 6

Modulus

The **modulus function** $f(x)=|x|$ is defined as

$$|x|=x \text{ for } x\geqslant 0,$$
$$|x|=-x \text{ for } x<0.$$

So $|x|$ is the numerical value of x.
$|x|$ is read as 'mod x'.
The graphs of $y=x$ and $y=|x|$ are:

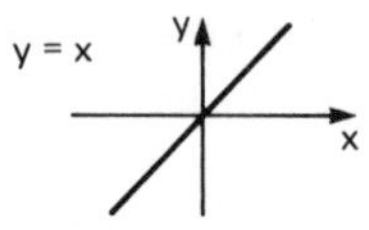

| Figure 7 | Figure 8 |

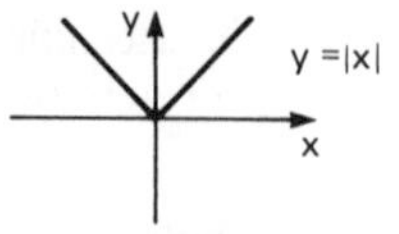

For $x<0$, $y=|x|$ is the reflection of $y=x$ in x-axis.

i $|x-4|\leqslant 2$ means $x-4\leqslant 2\Rightarrow x\leqslant 6$
   or $-(x-4)\leqslant 2$ i.e. $x-4\geqslant -2\Rightarrow x\geqslant 2$ So $2\leqslant x\leqslant 6$.

i *Solve $|\sin x|\leqslant \frac{1}{2}$ for $-\pi\leqslant x\leqslant \pi$.*

$|\sin x|\leqslant \frac{1}{2}$ means $\sin x\leqslant \frac{1}{2}$
or $-\sin x\leqslant \frac{1}{2}$ i.e. $\sin x\geqslant -\frac{1}{2}$

From the graph we see:
   $-\pi\leqslant x\leqslant -5\pi/6$
or $-\pi/6\leqslant x\leqslant \pi/6$
or $5\pi/6\leqslant x\leqslant \pi$

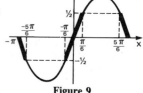

Figure 9

Inequations
Worked examples, Guided example and Exam questions

WE *Obtain the three sets of values of x for which*

(a) $2x > \dfrac{1}{x}$

(b) $\dfrac{1}{x-1} > \dfrac{x}{3-x}$

(c) $2|x-1| > |x+1|$

(a) To preserve the order of the inequality, multiply both sides by x^2 to give $2x^3 > x$

i.e. $2x^3 - x > 0$

or $x(2x^2 - 1) > 0$

giving $x(\sqrt{2}x - 1)(\sqrt{2}x + 1) > 0$

The values $x = -\dfrac{1}{\sqrt{2}}$, $x = 0$ and $x = \dfrac{1}{\sqrt{2}}$ divide the domain

into four parts. The sign of the factors in each part is best investigated using a tabular display.

	$x < -\dfrac{1}{\sqrt{2}}$	$-\dfrac{1}{\sqrt{2}} < x < 0$	$0 < x < \dfrac{1}{\sqrt{2}}$	$\dfrac{1}{\sqrt{2}} < x$
x	$-$	$-$	$+$	$+$
$(\sqrt{2}x - 1)$	$-$	$-$	$-$	$+$
$(\sqrt{2}x + 1)$	$-$	$+$	$+$	$+$
$x(\sqrt{2}x - 1)(\sqrt{2}x + 1)$	$-$	$+$	$-$	$+$

$$-\dfrac{1}{\sqrt{2}} < x < 0 \qquad\qquad \dfrac{1}{\sqrt{2}} < x$$

$\therefore$ Solution set is $\left\{ x: -\dfrac{1}{\sqrt{2}} < x < 0 \text{ or } \dfrac{1}{\sqrt{2}} < x \right\}$

(b) $\dfrac{1}{x-1} > \dfrac{x}{3-x}$

To preserve the order of the inequality multiply both sides by $(x-1)^2$ and $(3-x)^2$, giving $(x-1)(3-x)^2 > x(3-x)(x-1)^2$

i.e. $(x-1)(3-x)[(3-x) - x(x-1)] > 0$

$\qquad\qquad (x-1)(3-x)(3-x^2) > 0$

or $\qquad (x-1)(3-x)(\sqrt{3}-x)(\sqrt{3}+x) > 0$

The four values $x = -\sqrt{3}$, $x = 1$, $x = \sqrt{3}$, $x = 3$ divide the domain into five parts, the sign of each factor in each part is best investigated in a tabular form.

	$x < -\sqrt{3}$	$-\sqrt{3} < x < 1$	$1 < x < \sqrt{3}$	$\sqrt{3} < x < 3$	$3 < x$
$(\sqrt{3}+x)$	$-$	$+$	$+$	$+$	$+$
$(x-1)$	$-$	$-$	$+$	$+$	$+$
$(\sqrt{3}-x)$	$+$	$+$	$+$	$-$	$-$
$(3-x)$	$+$	$+$	$+$	$+$	$-$
Product	$+$	$-$	$+$	$-$	$+$

$$x < -\sqrt{3} \qquad\qquad 1 < x < \sqrt{3} \qquad\qquad 3 < x$$

$\therefore$ the solution set is

$\{x: x < -\sqrt{3} \text{ or } 1 < x < \sqrt{3} \text{ or } 3 < x\}$

(c) $2|x-1| > |x+1|$

$2(x-1) > (x+1)$	or	$-2(x-1) > (x+1)$
$2x - 2 > x + 1$	or	$-2x + 2 > x + 1$
$x > 3$	or	$1 > 3x$
		$\frac{1}{3} > x$

$\therefore$ the solution set is $\{x: x < \frac{1}{3} \text{ or } 3 < x\}$

GE *Find the set of values of x for which*

(a) $x^2 - 5x + 6 \geqslant 2$

(b) $\dfrac{1}{x^2 - 5x + 6} \leqslant \dfrac{1}{2}$

(a) Rearrange the inequality to give $x^2 - 5x + 4 \geqslant 0$ and solve directly.

(b) A sketch of $y = x^2 - 5x + 6$ is helpful.

Write $y = (x-3)(x-2) = \left(x - \dfrac{5}{2}\right)^2 - \dfrac{1}{4}$

from which we deduce that the graph has zeros at $x = 2$, $x = 3$, is vertex down with axis of symmetry $x = \dfrac{5}{2}$ and minimum value $-\dfrac{1}{4}$. The graph indicates that there are three sets of values of x to be considered, two where y is positive and one where y is negative.

Now consider $\dfrac{1}{y} \leqslant \dfrac{1}{2}$ for each of these sets of values of x,

taking care with the direction of the inequality when rearranging, deduce the values of x within these ranges for which the inequality holds.

EX

1 (a) Find the range of values of x for which $4x^2 - 12x + 5 < 0$.

(b) Find the ranges of values of x such that $x > \dfrac{2}{x-1}$.

$\qquad\qquad\qquad *(A)$

2 Find in each case the set of real values of x for which:

(i) $3(x+1) \geqslant x - 1$; (ii) $\dfrac{3}{x-1} \geqslant \dfrac{1}{x+1}$.

$\qquad\qquad\qquad (A)$

3 For what real values of x is $\left|\dfrac{1}{1+2x}\right| = 1$?

Solve the inequality $\left|\dfrac{1}{1+2x}\right| = 1$?

$\qquad\qquad\qquad (O \& C)$

4 Find the sets of real values of x for which:

(i) $(2-3x)(1+x) < 0$; (ii) $\dfrac{x}{(2-3x)(1+x)} > 0$;

(iii) $\dfrac{2}{x} - 3x < 1$.

$\qquad\qquad\qquad (A)$

5 Solve the inequality $x^2 - |x| - 12 < 0$. $\qquad (J)$

6 Find the set of values of x for which $\dfrac{x(x+2)}{x-3} < x + 1$.

$\qquad\qquad\qquad (C)$

7 Sketch the region in the xy plane within which all three of the following inequalities are satisfied: (i) $y < x + 1$; (ii) $y > (x-1)^2$; (iii) $xy < 2$. Determine the area of this region. $\qquad (J)$

9 Indices and Logarithms

Index notation, Basic rules of indices, Logarithms, Basic rules of logarithms, Solving exponential equations.

Index notation	a^n means $\underbrace{a \times a \times a \times a \times \ldots \times a}_{n \text{ factors}}$	ⓘ $8^5 = \underbrace{8 \times 8 \times 8 \times 8 \times 8}_{5 \text{ factors}}$

n is the **index** (plural **indices**). a is the **base**. An index is also called a **power** or an **exponent**.

5 is the index. 8 is the base. 8^5 is read as '8 to the power 5'.

Basic rules of indices

When m and n are positive rational numbers:

ⓘ The basic laws of indices can be illustrated as follows:

multiplication: $\quad a^m \times a^n = a^{m+n}$

$$a^3 \times a^2 = a^{3+2} = a^5$$

division: $\quad a^m \div a^n = a^{m-n}$

$$a^7 \div a^3 = a^{7-3} = a^4$$

raising to a power: $\quad (a^m)^n = a^{mn}$

$$(a^3)^2 = a^{3 \times 2} = a^6$$

zero index: $\quad a^0 = 1$

$$7^0 = 1$$

negative index: $\quad a^{-m} = \dfrac{1}{a^m}$

$$5^{-2} = \dfrac{1}{5^2} = \dfrac{1}{25}$$

fractional indices: $a^{\frac{1}{n}} = \sqrt[n]{a}$

$$a^{\frac{m}{n}} = (\sqrt[n]{a})^m$$

$$8^{\frac{1}{3}} = \sqrt[3]{8} = 2$$

$$\sqrt[3]{a^6} = a^{\frac{6}{3}} = a^2$$

Logarithms

If $N = a^x$, then we define x as the **logarithm of N to the base a**,
i.e. if $N = a^x$, then $x = \log_a N$.
This can be used to convert from 'index form' to 'logarithmic form' and vice versa.
Logarithms to the **base e**, written **ln** x or **log$_e$** x, are called **natural** or **Naperian** logarithms.
Logarithms to the base 10, written lg x or $\log_{10} x$, are called **common** logarithms.

ⓘ If $8 = 2^3$, then 3 is the logarithm of 8 to the base 2
i.e. $8 = 2^3 \Rightarrow \log_2 8 = 3$

If $7 = e^x$, then $x = \log_e 7$.

$100 = 10^2$, so $\log_{10} 100 = 2$.

Basic rules of logarithms

ⓘ

multiplication: $\quad \log_a(p \times q) = \log_a p + \log_a q$

$$\lg 3 + \lg 2 = \lg(3 \times 2) = \lg 6$$

division: $\quad \log_a(p \div q) = \log_a p - \log_a q$

$$\lg 8 - \lg 4 = \lg(8 \div 4) = \lg 2$$

raising to a power: $\quad \log_a p^n = n \log_a p$

$$\log_a 8 = \log_a(2^3) = 3\log_a 2$$

logarithm of unity: $\quad \log_a 1 = 0$

$$\log_{10} 1 = 0$$

logarithm of the base: $\quad \log_a a = 1$

$$\log_e e = 1$$

To change a logarithm from one base to another use

$$\log_b N = \frac{\log_a N}{\log_a b}$$

$$\log_2 10 = \frac{\log_{10} 10}{\log_{10} 2} = \frac{1}{\log_{10} 2}$$

In particular, $\log_e N = \dfrac{\log_{10} N}{\log_{10} e}$ and $\log_b a = \dfrac{1}{\log_a b}$.

Solving exponential equations

Exponential equations, i.e. equations in which the variable is an index, can often be solved by
either (a) taking logs of both sides of the equation and using the basic rules of logarithms,
or (b) using a substitution of the form $y = a^x$ to obtain an equation in y (usually quadratic) which can then be solved.

ⓘ *Solve (a)* $5^x = 4$ *(b)* $2^{2x+1} - 5(2^x) + 2 = 0$.

(a) Taking logs gives $\lg 5^x = \lg 4$
$$x \lg 5 = \lg 4$$
$$x = \lg 4 \div \lg 5 = 0.8614$$

(b) The equation is $2(2^x)^2 - 5(2^x) + 2 = 0$
Using $y = 2^x$, this is $2y^2 - 5y + 2 = 0$
$$(2y - 1)(y - 2) = 0$$
So $2^x = \frac{1}{2} = 2^{-1} \Leftrightarrow x = -1$ or $2^x = 2^1 \Leftrightarrow x = 1$

Indices and Logarithms
Worked examples and Exam questions

 (a) *Evaluate*: (i) $\left(\dfrac{81}{256}\right)^{\frac{3}{4}}$, (ii) $\left(\dfrac{25}{49}\right)^{-\frac{1}{2}}$,

(b) *Simplify*: $\dfrac{27^{n+2} - 6.3^{3n+3}}{3^n.9^{n+2}}$

(c) *Simplify*: $\dfrac{x}{y^{\frac{1}{2}} + x^{\frac{1}{4}}} + \dfrac{x}{y^{\frac{1}{2}} - x^{\frac{1}{4}}}$.

(a)
(i) $\left(\dfrac{81}{256}\right)^{\frac{3}{4}} = \dfrac{(81)^{\frac{3}{4}}}{(256)^{\frac{3}{4}}} = \dfrac{(81^{\frac{1}{4}})^3}{(256^{\frac{1}{4}})^3} = \dfrac{(3)^3}{(4)^3} = \dfrac{27}{64}$.

(ii) $\left(\dfrac{25}{49}\right)^{-\frac{1}{2}} = \left(\dfrac{49}{25}\right)^{\frac{1}{2}} = \dfrac{(49)^{\frac{1}{2}}}{(25)^{\frac{1}{2}}} = \dfrac{7}{5} = 1\frac{2}{5}$.

(b) $\dfrac{27^{n+2} - 6.3^{3n+3}}{3^n.9^{n+2}} = \dfrac{(3^3)^{n+2} - 2.3^1.3^{3n+3}}{3^n.(3^2)^{n+2}}$

$= \dfrac{3^{3(n+2)} - 2.3^{3n+4}}{3^n.3^{2(n+2)}}$

$= \dfrac{3^{3n}.3^6 - 2.3^{3n}.3^4}{3^{3n}.3^4}$

$= \dfrac{3^{3n}(3^6 - 2.3^4)}{3^{3n}.3^4}$

$= \dfrac{3^4(3^2 - 2)}{3^4}$

$= 7$.

(c) $\dfrac{x}{y^{\frac{1}{2}} + x^{\frac{1}{4}}} + \dfrac{x}{y^{\frac{1}{2}} - x^{\frac{1}{4}}}$

$= \dfrac{x(y^{\frac{1}{2}} - x^{\frac{1}{4}}) + x(y^{\frac{1}{2}} + x^{\frac{1}{4}})}{(y^{\frac{1}{2}} + x^{\frac{1}{4}})(y^{\frac{1}{2}} - x^{\frac{1}{4}})}$

$= \dfrac{xy^{\frac{1}{2}} - x^{\frac{3}{4}} + xy^{\frac{1}{2}} + x^{\frac{3}{4}}}{y - x}$

$= \dfrac{2x\sqrt{y}}{y - x}$.

 Solve the following equations:
(a) $\log_x 3 + \log_x 27 = 2$;   (b) $\log_3 x + 3\log_x 3 = 4$.

(a) $\log_x 3 + \log_x 27 = 2$

∴ $\log_x(3 \times 27) = 2$

i.e. $\log_x 81 = 2$

so, $81 = x^2$

⇒ $x = 9$

(b) $\log_3 x + 3\log_x 3 = 4$,
can be rewritten as

$\log_3 x + 3\dfrac{1}{\log_3 x} = 4$

i.e. $(\log_3 x)^2 - 4\log_3 x + 3 = 0$.

Let $\log_3 x = y$, then we need to solve,

$y^2 - 4y + 3 = 0$,

i.e. $(y - 3)(y - 1) = 0$

⇒ $y = 3$ or 1.

Hence, $\log_3 x = 3 \Rightarrow x = 3^3 = 27$,

or, $\log_3 x = 1 \Rightarrow x = 3^1 = 3$.

So, the solutions of the equations are 3 and 27.

EX **1** (a) Given that $x^y = z$, find:
(i) z when $x = 9$ and $y = \frac{1}{2}$;
(ii) z when $x = 64$ and $y = -\frac{1}{3}$;
(iii) x when $y = -\frac{1}{2}$ and $z = 4$;
(iv) y when $z = 2$ and $x = \frac{1}{2}$.

(b) Express $3^{2y} - 3^{y+1} - 3^y + 3$ in terms of z, where $z = 3^y$.
Hence solve the equation $3^{2y} - 3^{y+1} - 3^y + 3 = 0$.

(c) Without using tables evaluate $\log 6 + \log 4 + \log 20 - \log 3 - \log 16$, where all logarithms are to the base 10.   *(W)

2 (a) Simplify: $\dfrac{(x^{\frac{2}{3}} + x^{\frac{1}{2}})(x^{\frac{1}{2}} - x^{-\frac{1}{2}})}{(x^{\frac{2}{3}} - x^{\frac{1}{2}})^2}$

(b) Without using a calculator find the logarithm of 8 to the base a: (i) when $a = 64$; (ii) when $a = \frac{1}{4}$.   *(A)

3 Without the use of tables, slide rules or calculators:
(i) evaluate $8^{-\frac{2}{3}} + 81^{\frac{1}{4}}$;
(ii) find the value of x given that $\log_5 x + \log_2 8 = 0$. *(L)

4 (a) Express in its simplest form, $\log_2 64 - \log_2 16$
(b) Given that $\log_x u + \log_x v = p$ and $\log_x u - \log_x v = q$, prove that $u = x^{\frac{1}{2}(p+q)}$ and find a similar expression for v.   *(H)

5 (a) Solve the equation $\log_5 x = 16 \log_x 5$.
(b) Find the values of y which satisfy the equation:

$$(8^y)^y \cdot \dfrac{1}{32^y} = 4.$$ *(S)

6 (i) If $2^{2x} = 51 \times 3^{3x}$, prove that $x \log \dfrac{4}{27} = \log 51$ and hence find x.
(ii) If y varies directly as $x^{-\frac{3}{2}}$ and $x = 4$ when $y = 64$, find the value of y when $x = 16$.   (O & C)

7 (a) Given that $\log_9 x = p$ and $\log_{\sqrt{3}} y = q$, express xy and $\dfrac{x^2}{y}$ as powers of 3.
(b) Solve for x the equation $e^{2x} + e^x - 6 = 0$.   (A)

8 (a) Simplify (i) $20 \times 8^{2n} - 5 \times 4^{3n+1}$; (ii) $(\log_2 5) \times (\log_5 8)$.
(b) Find x and y given that $e^x + 3e^y = 3$ and $e^{2x} - 9e^{2y} = 6$, expressing each answer as a logarithm to base e.   (A)

9 If $2 \log_y x + 2 \log_x y = 5$, show that $\log_y x$ is either $\frac{1}{2}$ or 2. Hence find all pairs of values of x and y which satisfy simultaneously the equation above and the equation $xy = 27$.   (J)

10 (a) Prove that, if $x = \log_{10}(a - by) - \log_{10} a$, where a and b are constants, then $y = (a/b)(1 - 10^x)$.
Find the value of y when $a = 4$, $b = 2$ and $x = -2.065$.
(b) Without using tables find the value of

$$\dfrac{\log 0.8 - \log 32 + \log 8}{\log 0.7 + \log 7 - \log 49}$$

given that $\log 2 = 0.30103$. All logarithms are to the base 10.   (OLE)

11 (a) Find the real value of k such that $10^x = e^{kx}$ for all x.
(b) Find $g(x)$ such that $x^x = e^{xg(x)}$ for all x.
Hence, or otherwise, find the derivatives of 10^x and of x^x with respect to x.   (A)

12 Given that $\log_2 x + 2 \log_4 y = 4$, show that $xy = 16$.
Hence solve for x and y the simultaneous equations:
$$\log_{10}(x + y) = 1$$
$$\log_2 x + 2 \log_4 y = 4.$$ (A)

10 Exponential and Logarithmic Functions

Graphs, Exponential series, Logarithmic series.

Graphs

Exponential functions are functions in which the variable is in the index, or exponent, e.g. a^x.
e^x is called **the exponential function**.
e is the number such that the gradient of $y = e^x$ at $(0, 1)$ is 1.

The **logarithmic function** $y = \log_e x$ is the inverse of the exponential function $y = e^x$.

The graph shows the typical exponential and logarithmic curve shapes.

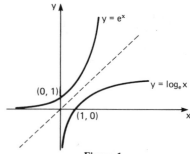

Figure 1

Exponential series

e^x can be expressed as an **infinite series**.

$$e^x = 1 + \frac{x}{1!} + \frac{x^2}{2!} + \frac{x^3}{3!} + \ldots + \frac{x^n}{n!} + \ldots$$

The series is valid for all values of x.

The value of e can be calculated to a required degree of accuracy by substituting $x = 1$ in the series for e^x.

Other exponential functions such as e^{kx} and e^{x+k}, where k is a constant, can be expressed as infinite series.
To write e^{kx} as a series, replace x by kx in the expansion, so

$$e^{kx} = 1 + \frac{(kx)}{1!} + \frac{(kx)^2}{2!} + \frac{(kx)^3}{3!} + \ldots \frac{(kx)^n}{n!} + \ldots$$

To write e^{x+k} as a series, use $e^{x+k} = e^k . e^x$, so

$$e^{x+k} = e^k \left(1 + \frac{x}{1!} + \frac{x^2}{2!} + \frac{x^3}{3!} + \ldots + \frac{x^n}{n!} + \ldots \right)$$

These series are valid for all values of x too.

ⓘ *Calculate e to six significant figures by using the series for e^x as far as the 10th term.*

$$e = e^1 = 1 + \frac{1}{1!} + \frac{1^2}{2!} + \frac{1^3}{3!} + \ldots + \frac{1^9}{9!}$$

Evaluating the terms using a calculator.
$e = 2.71828$ (to six significant figures)

ⓘ *Expand e^{3x} and e^{x+2} as far as the terms in x^3.*

$$e^{3x} = 1 + \frac{(3x)}{1!} + \frac{(3x)^2}{2!} + \frac{(3x)^3}{3!} + \ldots$$

$$= 1 + 3x + \frac{9x^2}{2} + \frac{9x^3}{2} + \ldots$$

$$e^{x+2} = e^2 . e^x$$

$$= e^2 \left(1 + x + \frac{x^2}{2!} + \frac{x^3}{3!} + \ldots \right)$$

Logarithmic series

There is no simple series for **ln**x, but

$$\ln(1+x) = x - \frac{x^2}{2} + \frac{x^3}{3} - \frac{x^4}{4} + \ldots$$

This series converges provided $-1 < x \leqslant 1$.

To write $\ln(1 + kx)$ as a series, replace x by kx in the expansion and find the new range of validity.

To express logarithms of products and quotients as series, rewrite them using the basic rules of logarithms first.

Approximate values of logarithms can be calculated using logarithmic series.

The series for $\ln\left(\frac{1+x}{1-x}\right)$ is often used for these

approximations because it converges rapidly.

ⓘ *Expand ln $(1-x)$ as far as the third term and give the range over which it is valid.*

$$\ln(1-x) = \ln[1+(-x)] = (-x) - \frac{(-x)^2}{2} + \frac{(-x)^3}{3} - \ldots$$

$$= -x - \frac{x^2}{2} - \frac{x^3}{3} - \ldots$$

This series is valid for $-1 < (-x) \leqslant 1$
i.e. $-1 \leqslant x < 1$.

ⓘ *Expand ln $\{(1+x)/(1-x)\}$ as far as the third term.*

$$\ln\{(1+x)/(1-x)\} = \ln(1+x) - \ln(1-x)$$

$$= \left(x - \frac{x^2}{2} + \frac{x^3}{3} - \frac{x^4}{4} + \ldots \right) - \left(-x - \frac{x^2}{2} - \frac{x^3}{3} - \frac{x^4}{4} - \ldots \right)$$

$$= 2 \left(x + \frac{x^3}{3} + \frac{x^5}{5} + \ldots \right)$$

The two series are valid if $-1 < x \leqslant 1$ and $-1 \leqslant x < 1$. So for both to be valid $-1 < x < 1$.

ⓘ *Find the value of ln 1.5. to 4 decimal places.*

We use $\ln\left(\frac{1+x}{1-x}\right) = \ln 1.5$ i.e. $x = 0.2$.

So $\ln 1.5 = \ln\left(\frac{1+0.2}{1-0.2}\right) = 2\left[(0.2) + \frac{(0.2)^3}{3} + \frac{(0.2)^5}{5} + \ldots \right]$

$$= 2(0.202731) = 0.4055 (4 d.p.)$$

Exponential and Logarithmic Functions
Worked examples, Guided example and Exam questions

 (a) *Expand e^{1-x^3} as a series of ascending powers of x as far as the term in x^9 and find the general term.*

(b) *Find the first four terms and the term in x^n in the expansion of $(1+x)e^x$.*

(a) $e^{1-x^3} = e^1 \cdot e^{-x^3}$

$$= e\left[1 + (-x^3) + \frac{(-x^3)^2}{2!} + \frac{(-x^3)^3}{3!} + \dots + \frac{(-x^3)^n}{n!} + \dots\right]$$

$$= e\left[1 - x^3 + \frac{1}{2}x^6 - \frac{1}{6}x^9 + \dots + \frac{(-1)^n}{n!}x^{3n} + \dots\right]$$

Hence, $e^{1-x^3} = e(1 - x^3 + \frac{1}{2}x^6 - \frac{1}{6}x^9)$ as far as terms in x^9, and

the general term is $\frac{(-1)^n e}{n!}x^{3n}$.

(b) $(1+x)e^x = (1+x)\left(1 + \frac{x}{1!} + \frac{x^2}{2!} + \frac{x^3}{3!} + \dots + \frac{x^n}{n!} + \dots\right)$

$$= 1 + \frac{x}{1!} + \frac{x^2}{2!} + \frac{x^3}{3!} + \dots + \frac{x^n}{n!} + \dots$$
$$+ x + \frac{x^2}{1!} + \frac{x^3}{2!} + \frac{x^4}{3!} + \dots + \frac{x^n}{(n-1)!} + \dots$$

$$= 1 + \left(1 + \frac{1}{1!}\right)x + \left(\frac{1}{1!} + \frac{1}{2!}\right)x^2 + \left(\frac{1}{2!} + \frac{1}{3!}\right)x^3 +$$

$$\dots + \left(\frac{1}{(n-1)!} + \frac{1}{n!}\right)x^n + \dots$$

$$= 1 + \frac{2}{1!}x + \left(\frac{2}{2!} + \frac{1}{2!}\right)x^2 + \left(\frac{3}{3!} + \frac{1}{3!}\right)x^3 + \dots$$

$$+ \left(\frac{n}{n!} + \frac{1}{n!}\right)x^n + \dots$$

$$= 1 + \frac{2}{1!}x + \frac{3}{2!}x^2 + \frac{4}{3!}x^3 + \dots + \frac{(n+1)}{n!}x^n + \dots$$

 Express $\ln\sqrt{\dfrac{1+x}{1-x}}$ as a series of terms in ascending powers of x up to and including the term in x^5.

Use your series to obtain an approximate value for $\ln\left(\dfrac{\sqrt{11}}{3}\right)$ giving your answer correct to six decimal places.

$\ln\sqrt{\dfrac{1+x}{1-x}} = \ln(1+x)^{\frac{1}{2}} - \ln(1-x)^{\frac{1}{2}}$

$$= \frac{1}{2}[\ln(1+x) - \ln(1-x)]$$

$$= \frac{1}{2}\left[\left(x - \frac{x^2}{2} + \frac{x^3}{3} - \frac{x^4}{4} + \frac{x^5}{5}\right) - \right.$$
$$\left.\left(-x - \frac{x^2}{2} - \frac{x^3}{3} - \frac{x^4}{4} - \frac{x^5}{5}\right)\right]$$

$$= x + \frac{x^3}{3} + \frac{x^5}{5} + \dots$$

If $\sqrt{\dfrac{1+x}{1-x}} = \dfrac{\sqrt{11}}{3}$, then $x = \dfrac{1}{10}$.

Putting $x = \dfrac{1}{10}$ in the series obtained,

$\ln\left(\dfrac{\sqrt{11}}{3}\right) = \dfrac{1}{10} + \dfrac{\frac{1}{3}}{1000} + \dfrac{\frac{1}{5}}{100\,000}$

$$= 0\cdot1 + 0\cdot000\,333 + 0\cdot000\,002$$

So $\ln\left(\dfrac{\sqrt{11}}{3}\right) = 0\cdot100\,335$, correct to six decimal places.

 Expand $e^{-x}\ln(1+2x)$ in ascending powers of x as far as the term in x^4. Write down the range of values of x for which this series is valid.

Expand e^{-x} as far as the term in x^3.
Expand $\ln(1+2x)$ as far as the term in x^4.
(Think why there is no need to expand e^{-x} as far as the term in x^4.)
Multiply the two series obtained together, discarding any terms which are of order x^5 or bigger.
Hence you have obtained the required series. Write down the ranges of validity for (i) e^{-x} and (ii) $\ln(1+2x)$.
Hence write down the range of values of x over which *both* series are valid.

EX 1 Expand $\dfrac{e^{2x}}{1+2x}$ as a series of ascending powers of x as far as the term in x^3 and give the set of values of x for which the expansion is valid.

(L)

2 Expand $\dfrac{e^x + e^{-x}}{2}$ as a series of ascending powers of x.

Deduce the general term in the expansion of $\dfrac{e^{\sqrt{x}} + e^{-\sqrt{x}}}{2}$.

(L)

3 (a) Write down the first 4 terms and the term in x^n in the expansion of e^x. Prove that the coefficient of x^n in the expansion of $(x^2+x)e^x$ is

$$\frac{n}{(n-1)!} \text{ when } n \geqslant 2.$$

Deduce that $(x^2+x)e^x - x$ is always positive when x is positive.

(b) Express $\dfrac{\sqrt{1-x}}{1+x}$ in ascending powers of x up to and including the term in x^2. By substituting $x = \dfrac{1}{9}$, show that

$$\sqrt{2} \approx \frac{2755}{1944}.$$

(S)

4 Expand, in ascending powers of x as far as the term in x^3:
(a) $\log_e(1-2x)$; (b) $\log_e(1-3x)$.

Hence expand $\log_e\dfrac{(1-2x)^3}{(1-3x)^2}$ as far as the term in x^3 and find the coefficient of x^n in this series. Write down the range of values of x for which this series is valid.

(A)

5 (a) Expand $e^{\frac{1}{2}x}\log_e(1+x)$ in ascending powers of x as far as the term in x^4 and hence show that, for certain values of x to be stated,

$$e^{\frac{1}{2}x}\log_e(1+x) + e^{-\frac{1}{2}x}\log_e(1-x) = ax^4 + \dots$$

and give the value of a.

(b) Expand $(1+x)^{-\frac{1}{2}}$ in ascending powers of x, giving the first four terms and the general term. Hence, without using tables or calculator, obtain the value of $1/\sqrt{101}$, correct to 6 decimal places, showing all working.

(S)

6 Express $\log_e\left(\dfrac{1+2x}{1-2x}\right)$ as a series of terms in ascending powers of x up to and including the term in x^5. Use your series to obtain an approximate value for $\log_e(5/3)$, giving your answer to 5 decimal places.

(A)

11 Coordinates and Graphs

Rectangular cartesian coordinates, Definitions, Drawing graphs, Solutions of equations

---

Rectangular cartesian coordinates

Rectangular cartesian coordinates determine the position of a point in the plane by reference to:
    a fixed point O (**the origin**),
    a pair of perpendicular lines (**axes**) through 0.
Any point P can be described by an ordered pair of numbers (x, y). The **x-coordinate** (or **abscissa**) is given first, the **y-coordinate** (or **ordinate**) second.

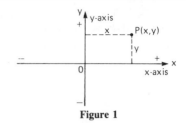

Figure 1

Definitions

Let A, B, C be (x_1, y_1), (x_2, y_2), (x_3, y_3) respectively.

The **length** of AB is $\sqrt{(x_1-x_2)^2+(y_1-y_2)^2}$.

The **gradient** of AB is $\dfrac{y_2-y_1}{x_2-x_1}$.

The **midpoint** of AB is $(\frac{1}{2}[x_1+x_2], \frac{1}{2}[y_1+y_2])$

The point P which divides AB in the ratio $\lambda:\mu$ is
$$\left(\frac{\lambda x_2+\mu x_1}{\lambda+\mu}, \frac{\lambda y_2+\mu y_1}{\lambda+\mu}\right).$$
Area of $\triangle ABC$ is
$\frac{1}{2}[x_1(y_2-y_3)+x_2(y_3-y_1)+x_3(y_1-y_2)]$

ℹ️ A, B, C are $(2, 3)$, $(4, 7)$, $(7, 3)$ respectively.

The length of AB is $\sqrt{(2-4)^2+(3-7)^2}=\sqrt{20}=2\sqrt{5}$

The gradient of BC is $\dfrac{7-4}{3-7}=\dfrac{3}{-4}=-\dfrac{3}{4}$

The midpoint of AC is $(\frac{1}{2}[2+7], \frac{1}{2}[3+3])=(4\frac{1}{2}, 3)$

The point P which divides BC in the ratio 1:2 is
$$\left(\frac{1\times7+2\times4}{1+2}, \frac{1\times3+2\times7}{1+2}\right)=(5, 5\frac{2}{3})$$

Area of ABC is
$\frac{1}{2}[2(7-3)+4(3-3)+7(3-7)]=10$ units^2

Drawing graphs

To draw the graph of $y=f(x)$

(a) Ascertain the range of values of x and calculate, to two decimal places, the corresponding values of y. The results are usually displayed in a table.

(b) Draw axes Ox and Oy intersecting at right angles on graph paper. The position of the axes on the paper is determined by the ranges of values of x and y.

(c) Choose suitable scales and sensible units (and sub-units) for the axes. The graph should use as much of the graph paper as possible. Different scales may be used on the axes but they distort familiar shapes.

(d) Mark the ordered pairs with ⊙ or ×.

(e) Draw the curve faintly first to obtain a general impression and then use a heavier line.

When points are required from the graph:

(a) if the point(s) lie between calculated, or given, values, the process is known as interpolation,

(b) if the graph has to be extended to find the point(s), the process is known as extrapolation.

ℹ️ Draw the graph of $y=4x^3-4x^2-x+6$, plotting points for values of $x=-2, -1.5, -1, -0.5, \ldots, +2$.

A table of values gives

x	-2	-1.5	-1	-0.5	0	$+0.5$	$+1$	$+1.5$	$+2$
y	-20	0	$+9$	$+10$	$+6$	0	-5	-6	0

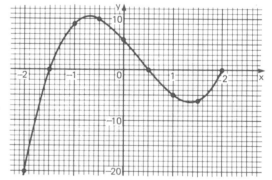

Figure 2

Solutions of equations

1. To solve $f(x)=0$

(a) Draw the graph of $y=f(x)$.

(b) Find the point(s) where the curve cuts the x-axis $(y=0)$.

2. To solve $f(x)=a$

(a) Draw the curve $y=f(x)$ and the line $y=a$.

(b) Find the point(s) where the curve cuts the line.

3. To solve $f(x)=0$, rewritten as $F(x)=G(x)$

(a) Draw the curves $y=F(x)$ and $y=G(x)$ on the same graph paper using the same axes and scales.

(b) Find any point(s) of intersection of the curves.

This 'two graph method' is often used when $F(x)$ is an algebraic function and $G(x)$ is non-algebraic. Non-algebraic functions, e.g. sin x, log x, are called transcendental functions

ℹ️ The solution of $4x^3-4x^2-11x+6=0$ can be found at the points where the curve $y=4x^3-4x^2-11x+6$ cuts the x-axis $(y=0)$. From the above graph these are where $x=-1.5$, $x=+0.5$, $x=+2$.

The solution of $4x^3-4x^2-11x+10=0$
    i.e. $4x^3-4x^2-11x+6=-4$
is found from the graph of $y=4x^3-4x^2-11x+6$ by drawing the line $y=-4$. This line cuts the curve where $x\approx-1.65$, $x\approx+0.9$, $x\approx+1.75$.

The solution of $4x^3-4x^2-8x+6=0$
    rewritten as $4x^3-4x^2-11x+6=-3x$
is found from the graphs of $y=4x^3-4x^2-11x+6$ and $y=-3x$.
If $y=-3x$ is drawn on the above graph, it cuts the curve where $x\approx-1.35$, $x\approx+0.7$, $x\approx+1.65$.

Co-ordinates and Graphs
Worked example, Guided example and Exam questions

WE *A, B, C are the points (2, 3), (4, 7), (7, 3) respectively.*
Find (i) the length of AB;
(ii) the co-ordinates of the mid-point of AC;
(iii) the co-ordinates of the point of trisection of BC;
(iv) the angle BAC;
(v) the area of triangle ABC.

(i) Length of AB $= \sqrt{(2-4)^2 + (3-7)^2}$
$= \sqrt{4+16}$
$= 2\sqrt{5}$ units.

(ii) If N is the mid-point of AC, then N has co-ordinates
$\left(\dfrac{2+7}{2}, \dfrac{3+3}{2}\right) = \left(\dfrac{9}{2}, 3\right)$

(iii) If P and Q are the points of trisection as indicated then P has co-ordinates
$\left(\dfrac{1\times7+2\times4}{3}, \dfrac{1\times3+1\times7}{3}\right) = \left(5, 5\dfrac{2}{3}\right)$
and Q has co-ordinates
$\left(\dfrac{2\times7+1\times4}{3}, \dfrac{2\times3+1\times7}{3}\right) = \left(6, 4\dfrac{1}{3}\right)$

(iv) Gradient of AB = angle BAC $= \dfrac{7-3}{4-2} = 2$.

$\therefore \angle BAC = 63° \, 26'$

(v) Area ABC $= \dfrac{1}{2}[2(3-7) + 7(7-3) + 4(3-3)]$

$= \dfrac{1}{2}(-8+28) = 10$ units^2

GE *Draw the graph of $y = \dfrac{2}{x}$, plotting values of x at ½-unit intervals*
from $x = \frac{1}{2}$ to $x = 4$, using the same scale on both axes. With the
same scales and axes draw the graph of $y = x(3-x)$ from $x = 0$ to
$x = 4$ for ½-unit intervals. Read off the values of x at their
intersections. Of what equation in x are they the roots?

Table of values for $y = \dfrac{2}{x}$

x	$\frac{1}{2}$	1	$1\frac{1}{2}$	2	$2\frac{1}{2}$	3	$3\frac{1}{2}$	4
y	4	2	1·33	1	0·80	0·67	0·57	0·50

Table of values for $y = x(3-x)$

x	0	$\frac{1}{2}$	1	$1\frac{1}{2}$	2	$2\frac{1}{2}$	3	$3\frac{1}{2}$	4
y	0	1·25	2	2·25	2	1·25	0	−1·75	−4

Draw a graph of these two functions.
The values of x at the points of intersection are approximately
1 and 2·75.

They are the roots of the equation $\dfrac{2}{x} = x(3-x)$,

i.e. $x^3 - 3x^2 + 2 = 0$

EX 1 Three points have co-ordinates P (−2, −3), Q (2, 0),
R (8, −8): (i) Prove that $P\hat{Q}R = 90°$; (ii) Calculate the area
of △PQR; (iii) Calculate the length of PR and hence, or
otherwise, find the perpendicular distance of Q from PR.
(C)

2 The midpoint of the line joining the points A (3, 0) and
B (5, 6) is M. A point C (t, 4t) is such that CM is
perpendicular
to AB. Calculate: (i) the value of t; (ii) the area of the
triangle ABC.
(S)

3 Draw the graph $y = \dfrac{4}{x}$, plotting values of x at ½-unit

intervals from $x = \frac{1}{2}$ to $x = 4$, taking 2 cm to represent a unit
on both axes. With the same scales and axes, again plotting
values of x at ½-unit intervals, draw the graph $y = x(4-x)$
from $x = 0$ to $x = 4$. Read off the values of x at their
intersections. Of what equation in x are they roots?
(O & C)

4 Three points have co-ordinates $A(-2, -1)$, $B(6, 9)$ and
$C(2 -3)$. The line through the midpoint of AB parallel to
AC meets the x-axis at X and the y-axis at Y. Calculate the
co-ordinates of X and Y. Hence deduce the area of △XOY,
where O is the origin.*
(C)

5 The points A and B have co-ordinates (−3, −1) and (7, 4)
respectively. Find the coordinates of the point C which
divides AB internally in the ratio 2:3. Find also the
equation of the line through B perpendicular to AB.
(L)

6 A, B, C and D are the points (−1, 3, 4), (2, 9, −2),
(−2, 4, 3) and (9, 9, −40) respectively.
(a) P and Q divide AB internally and externally in the ratio
2:1. Find the co-ordinates of P and Q.
(b) Find the ratio AP:AQ.
(c) M is the mid-point of AD. Find the co-ordinates of M.
(d) Prove that CP is perpendicular to MQ.
(H)

7 Given the three points A (4, 0), B (0, 2) and C (−2, −2),
show that: (i) AB = BC; (ii) AB is perpendicular to BC.
A square ABCD is formed. Calculate the co-ordinates
of D. [A diagram will be found helpful, but a solution using
measurements from an accurate drawing is not acceptable.]
(C)

8 The points C(6, 6), O(0, 0) and A(4, 3) are three of the
vertices of a parallelogram COAB:
(i) Calculate the co-ordinates of the point B;

(ii) Show that the tangent of the acute angle AOC is $\dfrac{1}{7}$ and

hence write down the exact value of the sine of the
angle AOC;
(iii) Calculate the area of the parallelogram COAB.
(A)

12 The Straight Line
Equations, Perpendicular distance, Pairs of lines.

Equations

The general equation of a **straight line** is

$$ax + by + c = 0$$

where a, b and c are constants.

The line with **gradient** m and **intercept** c on the y-axis has equation

$$y = mx + c.$$

Note: $y = mx$ passes through the origin $(0, 0)$.
$y = c$ is parallel to the x-axis.
$x = k$ (k is a constant) is parallel to the y-axis.

The line which cuts the x-axis at $(a, 0)$ and the y-axis at $(0, b)$ has equation

$$\frac{x}{a} + \frac{y}{b} = 1$$

The line of gradient m, through (x_1, y_1) has equation

$$y - y_1 = m(x - x_1)$$

The line through (x_1, y_1) and (x_2, y_2) has equation

$$\frac{y - y_1}{y_2 - y_1} = \frac{x - x_1}{x_2 - x_1}$$

If the perpendicular from the origin to the line is of length p and at an angle θ to the x-axis, then the equation of the line is

$$x \cos \theta + y \sin \theta = p$$

Note: $ax + by + c = 0$ and $x \cos \theta + y \sin \theta = p$ represent the same straight line if

$$\cos \theta = \pm \frac{a}{\sqrt{(a^2 + b^2)}}, \sin \theta = \pm \frac{b}{\sqrt{(a^2 + b^2)}},$$

$$p = \mp \frac{c}{\sqrt{(a^2 + b^2)}}$$

The signs are chosen so as to ensure p is positive.

Perpendicular distance

The **length of the perpendicular** from (x_1, y_1) to the straight line $ax + by + c = 0$ is

$$\pm \frac{ax_1 + by_1 + c}{\sqrt{(a^2 + b^2)}}$$

The perpendicular distance from the origin is taken as positive.
Points on the same side of a line give the same sign.
Points on opposite sides of a line give different signs.

ⓘ *Show that $(-1, 2)$ and $(3, 4)$ are on opposite sides of the line $x + 2y = 6$.*

Perpendicular distance from $(-1, 2)$ to $x + 2y - 6 = 0$

$$= \frac{(-1) + 2(2) - 6}{\sqrt{(1^2 + 2^2)}} = \frac{-3}{\sqrt{5}}$$

Perpendicular distance from $(3, 4)$ to $x + 2y - 6 = 0$

$$= \frac{(3) + 2(4) - 6}{\sqrt{(1^2 + 2^2)}} = \frac{5}{\sqrt{5}}$$

Since these distances are oppositely signed they are on opposite sides of the line.

Pairs of lines

The **angle θ between two straight lines** with gradients m_1 and m_2 ($m_1 > m_2$) is given by

$$\tan \theta = \frac{m_1 - m_2}{1 + m_1 m_2}$$

The two lines are (a) parallel if $m_1 = m_2$,
(b) perpendicular if $m_1 m_2 = -1$.

ⓘ *Find the angle between the lines $y = \sqrt{3}x + 2$ and $\sqrt{3}y = x - 4$.*

$y = \sqrt{3}x + 2 \Rightarrow m_1 = \sqrt{3}; \sqrt{3}y = x - 4 \Rightarrow m_2 = \frac{1}{\sqrt{3}}$

$$\tan \theta = \frac{m_1 - m_2}{1 + m_1 m_2} = \frac{\sqrt{3} - \frac{1}{\sqrt{3}}}{1 + \sqrt{3} \times \frac{1}{\sqrt{3}}} = \frac{3 - 1}{\sqrt{3}(1 + 1)} = \frac{1}{\sqrt{3}}$$

$$\Rightarrow \theta = \tan^{-1} \frac{1}{\sqrt{3}} = 30°$$

The **equations of the bisectors** of the angles between the lines $a_1x + b_1y + c_1 = 0$ and $a_2x + b_2y + c_2 = 0$ are given by

$$\frac{a_1x + b_1y + c_1}{\sqrt{(a_1^2 + b_1^2)}} = \pm \frac{a_2x + b_2y + c_2}{\sqrt{(a_2^2 + b_2^2)}}$$

ⓘ *Find the equations of the bisectors of the angles between $5x + 12y + 4 = 0$ and $3x - 4y + 1 = 0$.*

The equations are $\dfrac{5x + 12y + 4}{\sqrt{(5^2 + 12^2)}} = \pm \dfrac{3x - 4y + 1}{\sqrt{(3^2 + (-4)^2)}}$

$$\Rightarrow 5(5x + 12y + 4) = \pm 13(3x - 4y + 1)$$

giving $2x - 16y - 1 = 0$ and $64x + 8y + 33 = 0$.

The **equation of a line through the intersection of the lines** $a_1x + b_1y + c_1 = 0$ and $a_2x + b_2y + c_2 = 0$ is given by

$$(a_1x + b_1y + c_1) + \lambda(a_2x + b_2y + c_2) = 0$$

where λ is a constant.

λ is usually found from the given conditions.

ⓘ *Find the equation of the line through the origin and concurrent with $2x - 5y - 3 = 0$ and $3x - 4y + 2 = 0$.*

Any line concurrent with the given lines has an equation of the form
$(2x - 5y - 3) + \lambda(3x - 4y + 2) = 0$

Since $(0, 0)$ lies on the required line

$$-3 + 2\lambda = 0 \Rightarrow \lambda = \frac{3}{2}$$

So the required line is $13x - 22y = 0$.

The Straight Line
Worked examples and Exam questions

 A triangle has vertices at $A(-4, 10)$, $B(2, 2)$, $C(5, 8)$. Calculate the co-ordinates of D, the mid-point of AB. The line through D parallel to AC meets BC at E. Calculate the equation of the line DE and the co-ordinates of the point E.

D has co-ordinates $\left(\dfrac{-4+2}{2}, \dfrac{10+2}{2}\right) = (-1, 6)$

Gradient of $AC = \dfrac{10-8}{-4-5} = -\dfrac{2}{9}$

∴ equation of line through D parallel to AC is

$\dfrac{y-6}{x+1} = -\dfrac{2}{9}$

i.e. $9y - 54 = -2x - 2$
$2x + 9y = 52$ [1]
Equation of BC is

$\dfrac{y-2}{x-2} = \dfrac{8-2}{5-2} = 2$

i.e. $y - 2 = 2x - 4$
$2x - y = 2$ [2]
The co-ordinates of E are found by solving [1] and [2]
$[1] - [2]$ $10y = 50$ $y = 4$
$y = 4$ in [2], $x = 3$
∴ co-ordinates of E are $(3, 4)$.

 The points $A(-4, 6)$ and $C(-1, 1)$ are opposite vertices of a parallelogram ABCD. The sides BC, CD of the parallelogram lie along the lines

$$x + 3y - 2 = 0 \text{ and } x - y + 2 = 0$$

respectively. Calculate:

  (i) *the co-ordinates of D;*
 (ii) *the tangent of the acute angle between the diagonals of the parallelogram;*
(iii) *the length of the perpendicular from A to the side CD;*
(iv) *the area of the parallelogram.*

Gradient of BC and AD is $-\frac{1}{3}$.
Gradient of AB and CD is 1.
Equation of AD is

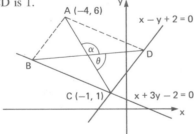

$\dfrac{y-6}{x+4} = -\dfrac{1}{3}$

i.e. $3y - 18 = -x - 4$
$x + 3y = 14$ [1]
Similarly, equation of AB is $x - y = -10$ [2]
Given equation of BC is $x + 3y = 2$ [3]
and equation of CD is $x - y = -2$ [4]
  (i) $[1] - [4]$ $4y = 16$, $y = 4$
$y = 4$ in [4] gives $x = 2$
∴ co-ordinates of D are $(2, 4)$.
 (ii) $[3] - [2]$ $4y = 12$, $y = 3$
$y = 3$ in [2] gives $x = -7$
∴ co-ordinates of B are $(-7, 3)$.

$m_1 = $ gradient $AC = \dfrac{6-1}{-4+1} = \dfrac{5}{-3} = -\dfrac{5}{3}$

$m_2 = $ gradient $BD = \dfrac{4-3}{2+7} = \dfrac{1}{9}$

$\tan \alpha = \dfrac{m_1 - m_2}{1 + m_1 m_2} = \dfrac{-\dfrac{5}{3} - \dfrac{1}{9}}{1 + \left(-\dfrac{5}{3}\right)\left(\dfrac{1}{9}\right)} = -\dfrac{16}{9} \times \dfrac{27}{22} = -\dfrac{24}{11}$

Hence $\tan \theta = \dfrac{24}{11}$.

(iii) Let p be the length of the perpendicular from A to DC then

$p = \dfrac{(x_1 - y_1 + 2)}{\sqrt{1^2 + (-1)^2}}$ where $(x_1, y_1) = (-4, 6)$

$= \dfrac{(-4 - 6 + 2)}{\sqrt{2}} = 4\sqrt{2}$ units

(iv) Length of $CD = \sqrt{(2+1)^2 + (4-1)^2}$
           $= \sqrt{9+9} = 3\sqrt{2}$ units
∴ Area of parallelogram $= p \times CD = 4\sqrt{2} \times 3\sqrt{2} = 24$ units^2

EX

1 Three points have co-ordinates $A(1, 7)$, $B(7, 5)$, and $C(0, -2)$. Find: (i) the equation of the perpendicular bisector of AB, (ii) the point of intersection of this perpendicular bisector and BC.
                             *(C)

2 The midpoint of the line joining the points $A(7, 3)$ and $B(-1, -5)$ is C. Find both values of k if the straight line joining C to the point $P(k^2, k)$ is perpendicular to AB.
                             *(A)

3 The perpendicular bisector of the line joining the points $(1, 2)$ and $(5, 4)$ meets the y-axis at the point $(0, k)$. Calculate k.
                             *(S)

4 P and Q are the points of intersection of the line

$$\dfrac{x}{a} + \dfrac{y}{b} = 1, \quad (a > 0, b > 0),$$

with the x and y axes respectively. The distance PQ is 10 and the gradient of PQ is -2. Find the value of a and b.
                             *(C)

5 Find the equation of the perpendicular bisector of the line joining the points $A(0, 8)$ and $B(4, 2)$.
                             *(S)

6 The triangle OAB has vertices $O(0, 0)$, $A(25, 0)$ and $B(18, 24)$. Prove that $AO = AB$ and find the equation of the internal bisector of the angle OAB. Given that the internal bisector of the angle AOB is $x - 2y = 0$, deduce the equation of the circle which touches the sides of the triangle OAB and has its centre inside this triangle. State the co-ordinates of the point of contact of the circle with the side OB.
                         (O & C)

7 The equations of the sides of a triangle ABC are:
$AB\ x - 2y + 11 = 0$, $BC\ y = 7$, $AC\ 2x - y + 7 = 0$.
Without using tables or calculator, find: (i) the tangent of the angle BAC; (ii) the area of the triangle ABC.
                             (A)

13 The Circle
Equations, Tangents, Intersecting circles.

Equations

The **general equation of a circle** is of the form
$$x^2+y^2+2gx+2fy+c=0$$
It is a second degree equation in which
(a) the coefficients of x^2 and y^2 are unity,
(b) there is no xy term.
   This circle has centre $(-g, -f)$
   and radius $\sqrt{(g^2+f^2-c)}$.

The circle, centre (a, b), radius r, has equation
$$(x-a)^2+(y-b)^2=r^2$$
If (a, b) is the origin, the circle equation is
$$x^2+y^2=r^2$$

The **equation of the circle on AB as diameter** where A is (x_1, y_1) and B is (x_2, y_2) is given by
$$(x-x_1)(x-x_2)+(y-y_1)(y-y_2)=0$$

To find the **equation of a circle through three points**
(a) Substitute the coordinates of each point in turn in the general equation of the circle.
(b) Solve the three simultaneous equations to find the values of g, f and c.
(c) Substitute g, f and c in the general equation.

ⓘ *Find the coordinates of the centre and the radius of the circle given by* $3x^2+3y^2+6x+12y+9=0$.

The circle equation can be rewritten as
$$x^2+y^2+2x+4y+3=0$$
Comparing this with the general equation of a circle
$$x^2+y^2+2gx+2fy+c=0$$
gives $g=1$, $f=2$, $c=3$.
So, the centre of the circle is $(-1, -2)$,
   its radius is $\sqrt{(1^2+2^2-3)}=\sqrt{2}$.

ⓘ The equation of the circle with $(0, 0)$ and $(2, 2)$ as end points of a diameter is given by
$$(x-0)(x-2)+(y-0)(y-2)=0$$
i.e. $x^2+y^2-2x-2y=0$

ⓘ *Find the equation of the circle which circumscribes the triangle with vertices* $(1, 0)$, $(2, 1)$ *and* $(0, 2)$.

Let the equation be $x^2+y^2+2gx+2fy+c=0$.
Substituting the coordinates in turn give
$1+2g+c=0$; $5+4g+2f+c=0$; $4+4f+c=0$
Solving these equations gives $g=-\frac{5}{6}$, $f=-\frac{7}{6}$, $c=\frac{2}{3}$.
Hence the equation of the circle is
$$3x^2+3y^2-5x-7y+2=0.$$

Tangents

The **equation of the tangent** at (x_1, y_1) to the circle
$$x^2+y^2+2gx+2fy+c=0$$
is $xx_1+yy_1+g(x+x_1)+f(y+y_1)+c=0$
Note the relation between these equations:
$x^2 \rightarrow xx_1$, $y^2 \rightarrow yy_1$, $2x \rightarrow (x+x_1)$, $2y \rightarrow (y+y_1)$.

To find the **condition that** $y=mx+c$ **be a tangent** to $x^2+y^2+2gx+2fy+c=0$, see [WE]

The **length of the tangent** from (x_1, y_1) to the circle
$x^2+y^2+2gx+2fy+c=0$ is
$$\sqrt{(x_1^2+y_1^2+2gx_1+2fy_1+c)}.$$

ⓘ *Find the equation of the tangent at* $(1, -2)$ *to the circle* $2x^2+2y^2-3x+4y+1=0$.

Rewrite the circle equation as
$$x^2+y^2-\tfrac{3}{2}x+2y+\tfrac{1}{2}=0$$
So the tangent equation is
$$(1)x+(-2)y-\tfrac{3}{4}(x+1)+(y-2)+\tfrac{1}{2}=0$$
i.e. $x-4y-9=0$.

ⓘ *Find the length of the tangent from* $(-5, 8)$ *to the circle* $x^2+y^2-4x-6y+3=0$.

The length of the tangent is
$$\sqrt{[(-5)^2+8^2-4(-5)-6(8)+3]}=\sqrt{64}=8$$

Intersecting circles

The **equation of any circle through the intersections of the circles** $x^2+y^2+2g_1x+2f_1y+c_1=0$ and $x^2+y^2+2g_2x+2f_2y+c_2=0$ is given by an equation of the form
$$(x^2+y^2+2g_1x+2f_1y+c_1)+\lambda(x^2+y^2+2g_2x+2f_2y+c_2)=0$$
where λ is a constant.

λ is usually found from the given conditions.

If $\lambda=-1$, the equation reduces to a straight line. This is the **common chord** of the two circles and is known as the **radical axis** of the circles.

ⓘ *Find the equation of the circle which passes through the point* $(-3, 1)$ *and the points of intersection of* $x^2+y^2-y-5=0$ *and* $x^2+y^2+2x+5y-1=0$.

A circle through the intersection of the given circles has equation of the form
$$(x^2+y^2-y-5)+\lambda(x^2+y^2+2x+5y-1)=0.$$
Since $(-3, 1)$ must lie on this circle
$[(-3)^2+1^2-1-5]+\lambda[(-3)^2+1^2+2(-3)+5(1)-1]=0$
$\Rightarrow \lambda=-\frac{1}{2}$

So the required equation is
$$(x^2+y^2-y-5)-\tfrac{1}{2}(x^2+y^2+2x+5y-1)=0$$
i.e. $x^2+y^2-2x-7y-9=0$

The Circle
Worked example, Guided example and Exam questions

WE *Find the values of c such that the line $x+y=c$ shall be a tangent to the circle $x^2+y^2-4x+2=0$. For each value of c find the co-ordinates of the point of contact. Draw a sketch of the circle and the two tangents.*

To find the points of intersection of $x+y=c$ and $x^2+y^2-4x+2=0$ put $y=c-x$ in the circle equation.
$$x^2+(c-x)^2-4x+2=0,$$
i.e. $2x^2-x(2c+4)+(2+c^2)=0$ [1]

If $x+y=c$ is a tangent to the circle then this quadratic will have equal roots,

    i.e. $\triangle=b^2-4ac=0$
    i.e. $(2c+4)^2-4\times2(2+c^2)=0$
      $4c^2+16c+16-16-8c^2=0$
          $-4c^2+16c=0,$ $\therefore c=0$ or 4

The equations of the tangents are $x+y=0$ and
                         $x+y=4$

$c=0$ in [1] gives $x=1$, when $x+y=0$ gives $y=-1$
$c=4$ in [1] gives $x=3$, when $x+y=4$ gives $y=1$
$\therefore$ the co-ordinates of the points of contact are $(1,-1)$ and $(3,1)$.

The diagram below is a sketch of the circle and the two tangents.

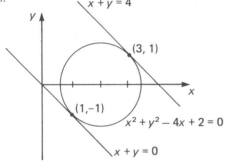

GE *Find the centre and radius of each of the circles C_1 and C_2 whose equations are $x^2+y^2-16y+32=0$, $x^2+y^2-18x+2y+32=0$ respectively and show that the circles touch externally. Find the co-ordinates of their point of contact and show that the common tangent at that point passes through the origin. The other tangents from the origin, one to each circle, are drawn. Find, correct to the nearest degree, the angle between these tangents.*

Find the centre and radius of C_1.
Find the centre and radius of C_2.
Find the distance between the centres and show that it is equal to the sum of the radii, thus showing that the circles touch externally.
Eliminate x^2 and y^2 from the equations of C_1 and C_2 to obtain $x=y$ which is substituted into C_1 giving a quadratic in x (or y) which has equal roots, thus the co-ordinates of the point of contact can be found. A sketch at this stage is useful.

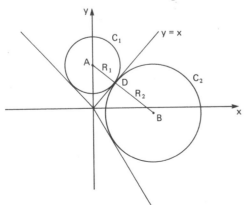

Note: The section formula could also be used to find the co-ordinates of D, the point of contact, which is the point dividing AB in the ratio $R_1:R_2$.

Let $y=mx$ be the line through the origin tangential to C_1. Applying the tangency condition gives $m=\pm1$.
So the required tangent is $y=-x$.
Similarly let $y=Mx$ be the required tangent to C_2. Apply the tangency condition to find the value of M.
The angle between the tangents will be calculated as θ, hence the required angle α is easily found.

EX 1 Obtain the equation of the circle which passes through the origin and has its centre at the point $(3,-4)$. The line $y=x-6$ meets this circle at the points P and Q. Find the co-ordinates of P and Q. Calculate the distance PQ.

                                  *(A)

2 Find the distance between the centres of the circles $x^2+y^2+6y+8=0$ and $x^2+y^2-12x-10y-60=0$ and prove that these circles touch one another. Obtain the equation of the smallest circle that passes through the centres of the given circles.

                                  *(S)

3 Prove that the circles $x^2+y^2+2x+2y=23$ and $x^2+y^2-10x-7y+31=0$ touch each other externally, and calculate the co-ordinates of the point of contact. (A graphical solution will not be acceptable.)

                              *(O & C)

4 [In this question the use of tables, a calculator or accurate drawing is forbidden.]
The triangle ABC has vertices $A(0,12)$, $B(-9,0)$, $C(16,0)$. Find the equations of the internal bisectors of the angles ABC and ACB. Hence, or otherwise, find the equation of the inscribed circle of the triangle ABC. Find also the equation of the circle passing through A, B and C.

                                  (C)

5 Find the equation of the circle which touches the line $y=x$ at the point $(4,4)$ and whose centre lies on the line which passes through the point $(-1,-3)$ and the origin. Deduce the equation of the circle which touches the line $y=x$ at the point $(4,4)$ and whose centre is on the line $3y=x$. Find the equations of the three tangents common to the two circles.

                               (O & C)

6 The points A and B have co-ordinates $(8,0)$ and $(0,6)$ respectively and O is the origin. Find the equation of:
(i) the circumcircle; (ii) the inscribed circle; of the triangle AOB.

                                    (A)

14 Conic Sections

Definitions, Summary for parabola, ellipse and rectangular hyperbola.

Definitions

A **conic section** is the locus of a point in a plane such that its distance from a fixed point (the **focus**) in the plane bears a constant ratio (the eccentricity) to its distance from a fixed straight line (the **directrix**) in the plane.

The value of e determines the type of curve obtained.

If $e = \infty$, the locus is a **pair of straight lines**.
If $e = 0$, the locus is a **circle**.
If $e = 1$, the locus is a **parabola**.
If $0 < e < 1$, the locus is an **ellipse**.
If $e > 1$, the locus is a **hyperbola**.
If $e = \sqrt{2}$, the locus is a **rectangular hyperbola**.

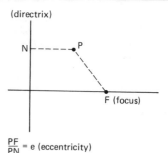

$\dfrac{PF}{PN} = e$ (eccentricity)

Figure 1

Summary for parabola, ellipse and rectangular hyperbola

The most frequently used information about the parabola, ellipse and rectangular hyperbola is summarised in the table below.

	Parabola	Ellipse	Rectangular hyperbola
	Axes chosen so that the focus is $(a, 0)$ and the directrix is $x + a = 0$.	Axes chosen so that the focus is $(ae, 0)$ and the directrix is $x = \dfrac{a}{e}$.	Both axes are asymptotes to the curve.
Equation	$y^2 = 4ax$	$\dfrac{x^2}{a^2} + \dfrac{y^2}{b^2} = 1$	$xy = c^2$
Lines of symmetry	x-axis	x-axis and y-axis	Lines $y = \pm x$
Tangent of gradient m	$y = mx + \dfrac{a}{m}$	$y = mx \pm \sqrt{(a^2m^2 + b^2)}$	$y = mx \pm 2c\sqrt{-m}$ (Note: tangents only exist for $m < 0$)
Tangent at (x_1, y_1)	$yy_1 = 2a(x + x_1)$	$\dfrac{xx_1}{a^2} + \dfrac{yy_1}{b^2} = 1$	$y_1 x + x_1 y = 2c^2$
Condition that $lx + my = n$ be a tangent	$am^2 = ln$	$a^2l^2 + b^2m^2 = n^2$	$n^2 = 4c^2lm$
Normal at (x_1, y_1)	$(y - y_1) = -\dfrac{y_1}{2a}(x - x_1)$	$b^2 \dfrac{(y - y_1)}{y_1} = a^2 \dfrac{(x - x_1)}{x_1}$	$xx_1 - yy_1 = x_1^2 - y_1^2$
Parametric equation	$x = at^2,\ y = 2at$	$x = a\cos\phi,\ y = b\sin\phi$	$x = ct,\ y = \dfrac{c}{t}$
Tangent in terms of parameter	$ty = x + at^2$	$bx\cos\phi + ay\sin\phi = ab$	$x + t^2 y - 2ct = 0$
Normal in terms of parameter	$y + tx = 2at + at^3$	$ax\sin\phi - by\cos\phi = (a^2 - b^2)\sin\phi\cos\phi$	$t^2 x - y - ct^3 + \dfrac{c}{t} = 0$

Conic Sections
Worked examples, Guided example and Exam questions

 Draw a sketch of the parabola $y^2 = 4x$. Prove that the point $(t^2, 2t)$ lies on the parabola for all values of t. The line $2x + 3y = 8$ cuts the parabola at the points A and B. $D(t^2, 2t)$, $E(1, -2)$ and $F(4, -4)$ are three other points on the parabola.

(i) Calculate the co-ordinates of the mid-point of AB.
(ii) Prove that the normal at D to the parabola has the equation $tx + y = 2t + t^3$.
(iii) Prove that the normals at E and F meet on the parabola.
(iv) Find the co-ordinates of the point C on the parabola where the tangent to the curve is parallel to the line $5y = 8x - 1$.

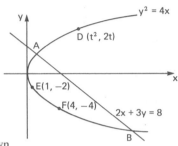

The sketch is as shown.
For the point $(t^2, 2t)$,
$4x = 4t^2 = (2t)^2 = y^2$
$\therefore y^2 = 4x$ for all t.
Solve $2x + 3y = 8$ and $y^2 = 4x$ to find the co-ordinates of A and B.
$y^2 + 6y - 16 = 0$, i.e. $(y+8)(y-2) = 0$
$y = -8$ gives $x = 16$, $(16, -8)$ are the co-ordinates of B.
$y = 2$ gives $x = 1$, $(1, 2)$ are the co-ordinates of A.

(i) M, the mid-point of AB, has co-ordinates $\left(\dfrac{17}{2}, -3\right)$.

(ii) Bookwork.

(iii) The normal at E has equation $-x + y = -3$
i.e. $x - y = 3$ [1]
The normal at F has equation $-2x + y = -12$
i.e. $2x - y = 12$ [2]
$[2] - [1]$ $x = 9$, $y = 6$.
$\therefore$ the normals at E and F meet at $Q(9, 6)$.
$y^2 = 36 = 4 \times 9 = 4x$, as required for Q to lie on the parabola.

(iv) The required tangent has gradient $\dfrac{8}{5}$.

Gradient at the required point on the parabola is
$\dfrac{dy}{dx} = \dfrac{2}{y} = \dfrac{8}{5}$, i.e. $y = \dfrac{5}{4}$ and $x = \dfrac{25}{64}$.

$\therefore C$ has co-ordinates $\left(\dfrac{25}{64}, \dfrac{5}{4}\right)$.

 Prove that the normal to the hyperbola $xy = c^2$ at the point $P(ct, c/t)$ has equation $y = t^2 x + \dfrac{c}{t} - ct^3$. If the normal at P meets the line $y = x$ at N, and O is the origin, show that $OP = PN$ provided that $t \neq 1$. The tangent to the hyperbola at P meets the line $y = x$ at T. Prove that $OT.ON = 4c^2$.

Bookwork

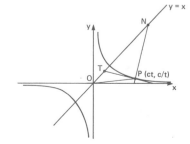

The distance OP can be found immediately (using the distance formula). The co-ordinates of N are found by solving the normal equation with the equation $y = x$.
The distance PN can now be found and the result follows.
($t \neq 1$ excludes the normal which coincides with the line $y = x$).
Write down (or derive) the equation of the tangent at P, solve this equation with the equation $y = x$ to give the co-ordinates of T.
Write down expressions for OT and ON (using the distance formula) form the product $OT.ON$ and the required result is easily obtained.

EX

1 Find, in terms of h, the equation of the tangent to the curve $y = x^2$ at the point P, whose x co-ordinate is h. This tangent intersects the x-axis at A and the y-axis at B. The midpoint of AB is Q.
(i) Find the co-ordinates of Q in terms of h.
(ii) Find the equation of the locus of Q as h varies.
(iii) Given that $h = 4$, find the co-ordinates of the point at which PQ produced meets the locus again.
$*(C)$

2 Sketch the parabola $x = t^2$, $y = 2t$.
Show that the equation of the tangent to the parabola at the point with parameter t_1, is $t_1 y = x + t_1^2$. The tangent to the parabola at the point A is parallel to the line $3y = x + 6$. Find the co-ordinates of A and the equations of the tangent and normal to the parabola at A. Given that the normal at A meets the parabola again at the point B, find the co-ordinates of B.
The tangent at A meets the x-axis at D and the line AB meets the same axis at E. Find the lengths of DE, AE and AD.
$*(W)$

3 Show that the equation of the tangent to the parabola $y^2 = 4ax$ at the point $(at^2, 2at)$ is $ty = x + at^2$. The three points P, Q, R on the curve have parameters p, q, r respectively. The tangents at P, Q meet the tangent at R at the points A and B respectively. Given that B is the mid-point of AR, show that $2q = p + r$. In this case, show that the tangent at Q is parallel to PR.
(C)

4 Derive the equations of the tangent and the normal to the ellipse $\dfrac{x^2}{a^2} + \dfrac{y^2}{b^2} = 1$ at the point P ($a \cos t$, $b \sin t$). The normal at P meets the axes at the points Q and R. Lines are drawn parallel to the axes through the points Q and R; these lines meet at the point V. Find the co-ordinates of V, and prove that, as P moves round the ellipse, the point V moves round another ellipse, and find its equation.
(OLE)

5 Find the equation of the chord joining the points $P(cp, c/p)$, $Q(cq, c/q)$ on the rectangular hyperbola $xy = c^2$. Deduce the equation of the tangent at P. If the tangents at the points P, Q on a rectangular hyperbola, centre O, intersect at T, show that OT cuts PQ at M the mid-point of PQ.
$(O \& C)$

6 The normals to the parabola $y^2 = 4x$ at the points $P(p^2, 2p)$ and $Q(q^2, 2q)$ are perpendicular and meet at R. Show that $pq = -1$. Given that the co-ordinates of r are $(p^2 + q^2 + 1, p + q)$, find the equation of the locus of R.
(A)

15 Loci

Definition, Equation of a locus, Parametric equations.

Definition

A **locus** (plural **loci**) is the set of points having a given property.

Only loci where all the points lie in a plane will be considered here.

In locus problems it is useful to sketch a diagram with points marked in roughly correct positions using the given property.

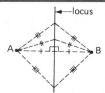 The locus of points equidistant from two given points is the perpendicular bisector of the line joining the two given points.

The locus of points equidistant from a given point is a circle, centre the given point.

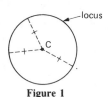

Figure 1

Equation of a locus

A locus can be expressed as an equation.

If a typical point P of the locus has coordinates (x, y), then it is usually possible to write the given condition(s) in terms of x and y (and no other variable) and so obtain the general equation for the locus.

$\boxed{i}$ *A point is equidistant from the x-axis and the point $F(0, 2)$. Find the locus of the point.*

A typical point $P(x, y)$ is shown in the sketch. P is equidistant from the x-axis and $F(0, 2)$.

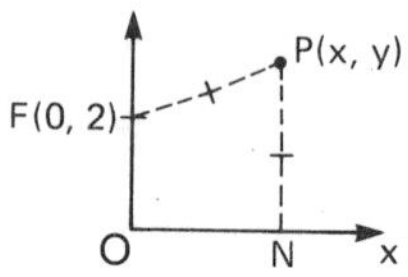

$$\therefore PF = PN$$
$$\text{But } PF = \sqrt{[(x-0)^2 + (y-2)^2]}$$
$$= \sqrt{[x^2 + (y-2)^2]}$$
$$\text{and } PN = y$$
$$\text{So } \sqrt{[x^2 + (y-2)^2]} = y$$
$$\text{i.e. } x^2 + y^2 - 4y + 4 = y^2 \Rightarrow y = \frac{x^2}{4} + 1.$$

Figure 2

This locus is a parabola, with Oy as the axis of symmetry, vertex 'down', through $(0, 1)$.

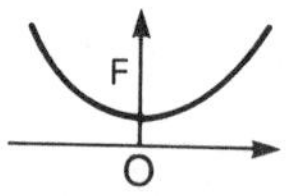

Figure 3

Parametric equations

Locus problems often involve **parametric equations**. Each point used in the problem has a **parametric value** associated with it. Different points have different parametric values. For example, $(at^2, 2at)$ and $(at_1^2, 2at_1)$ would represent two points on the parabola $y^2 = 4ax$.

The cartesian equation (in x and y) of the locus may be found by eliminating the parameters using the given conditions.

Sometimes it is not necessary to eliminate each parameter separately as fixed combinations of the parameters occur in the working. For example, $t + t_1$ and $t_1 t_1$ are frequently occurring combinations in many problems (see).

Further conditions are given to enable values for the parameter combinations to be obtained.

$\boxed{i}$ *Find the locus of the mid-point of a chord of the circle $x^2 + y^2 = a^2$ which subtends an angle of $90°$ at the centre.*

If p has parametric coordinates $(a \cos \theta, a \sin \theta)$, and Q has parameteric coordinates $(a \cos \phi, a \sin \phi)$, then the mid-point M of the chord PQ has coordinates

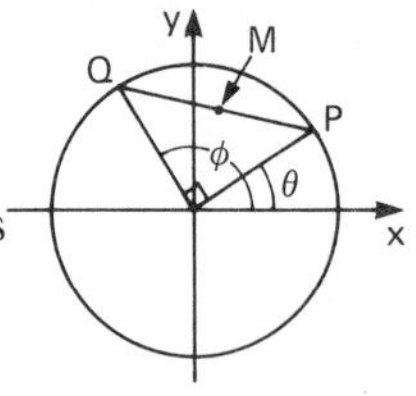

Figure 4

$$x = \frac{a \cos \theta + a \cos \phi}{2}; \quad y = \frac{a \sin \theta + a \sin \phi}{2}.$$

Since the chord PQ subtends an angle of $90°$ at the centre, $\phi - \theta = 90° \Rightarrow \phi = \theta + 90°$.

So $x = \frac{a}{2}(\cos \theta + \cos(\theta + 90°))$; $y = \frac{a}{2}(\sin \theta + \sin(\theta + 90°))$

$$\Rightarrow 2x = a(\cos \theta - \sin \theta); \quad 2y = a(\sin \theta + \cos \theta)$$

Squaring and adding gives
$$4x^2 + 4y^2 = a^2(\cos \theta - \sin \theta)^2 + a^2(\sin \theta + \cos \theta)^2$$
$$= a^2(\cos^2 \theta + \sin^2 \theta - 2\cos \theta \sin \theta$$
$$+ \sin^2 \theta + \cos^2 \theta + 2\cos \theta \sin \theta)$$
$$= a^2(1+1)$$
i.e. $4x^2 + 4y^2 = 2a^2$
$$\Rightarrow \quad x^2 + y^2 = \frac{a^2}{2}.$$

This is the required locus which is a circle, centre $(0, 0)$, radius $\frac{a\sqrt{2}}{2}$.

Loci
Worked examples and Exam questions

O is the origin and Q is a point on the circle $x^2+y^2-4x+3=0$. Find the equation of the locus of P, the point of trisection of OQ nearest to O. Identify the locus.

$x^2+y^2-4x+3=0$ represents the circle centre $(2, 0)$ and radius $\sqrt{4+0-3}=1$.

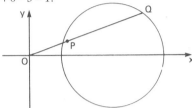

Let Q have co-ordinates (p, q), then P has co-ordinates $\left(\dfrac{p}{3}, \dfrac{q}{3}\right)$, being the point of trisection of OQ nearest to O.

Representing by (X, Y) the co-ordinates of P

$$X=\frac{p}{3}, \quad Y=\frac{q}{3}$$

i.e. $p=3X, \quad q=3Y$

As (p, q) lies on the given circle
$$9X^2+9Y^2-12X+3=0$$
As (p, q) varies the locus of P will be
$$9x^2+9y^2-12x+3=0$$

Write this equation as $x^2+y^2-\dfrac{4}{3}x+\dfrac{1}{3}=0$

representing a circle centre $\left(\dfrac{2}{3}, 0\right)$ and radius

$$\sqrt{\frac{4}{9}+0-\frac{1}{3}}=\sqrt{\frac{1}{9}}=\frac{1}{3}.$$

Find the locus of the mid-points of the chords of $xy=c^2$ which pass through $(6c, 4c)$.

Let $P(ct, c/t)$ and $Q(cT, c/T)$ be two points on $xy=c^2$.
The mid-point of PQ has co-ordinates

$$\left[\frac{ct+cT}{2}, \frac{ct+cT}{2tT}\right]$$

Equation of PQ is

$$\frac{y-c/t}{x-ct}=\frac{c/T-c/t}{cT-ct}=\frac{t-T}{tT(T-t)}=-\frac{1}{tT} \qquad [1]$$

It is required that the chord passes through the point $(6c, 4c)$, so substituting into [1] gives

$$\frac{4c-c/t}{6c-ct}=-\frac{1}{tT}$$

i.e. $4tT+6=t+T \qquad [2]$

From the mid-point co-ordinates

$$x=\frac{ct+cT}{2} \text{ and } y=\frac{ct+cT}{2tT}$$

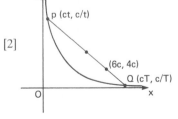

These equations together with [2] enable t and T to be eliminated.

From [2] $tT=\dfrac{1}{4}(t+T)-\dfrac{3}{2}$, and substituting into the mid-point

value of y gives $y=\dfrac{c(t+T)}{\frac{1}{2}(t+T)-3}$

Substitute for $(t+T)$ from the mid-point value of x to give

$$y=\frac{2x}{\dfrac{x}{c}-3}, \text{ i.e. } xy=2xc+3cy.$$

---

1 It it known that the point P is such that the sum of the squares of its distances from the lines $y=0$, $\sqrt{3}x+y-2\sqrt{3}=0$ and $\sqrt{3}x-y+2\sqrt{3}=0$ is 6 units. Use the fact that the perpendicular distance of the point (h, k) from the line $ax+by+c=0$ is

$$\pm\left(\frac{ah+bk+c}{\sqrt{(a^2+b^2)}}\right)$$

to prove that the locus of P is a circle touching the x-axis. (You are advised not to approximate to $\sqrt{3}$ at any stage of your working.)

(O & C)

2 A curve is defined by parametric co-ordinates $x=\sqrt{3}\cos t$, $y=\sin t$.
(i) Sketch the section of the curve for values of t in the range $0\leqslant t\leqslant\pi$.
(ii) Show that the gradient of the curve at the point $(\sqrt{3}\cos t_1, \sin t_1)$ is $-(\cot t_1)/\sqrt{3}$.
(iii) The gradient of the curve at the point P is $\frac{1}{3}$. Calculate the co-ordinates of P and find the area of the triangle whose vertices are P and the two points of intersection of the curve and the x-axis.

(S)

3 A curve is defined parametrically by the equations

$$x=t^3-6t+4, y=t-3+\frac{2}{t}.$$ Find (i) the equations of the

normals to the curve at the points where the curve meets the x-axis; (ii) the co-ordinates of their point of intersection.

(C)

4 Find the equations of the loci of points which are:
(i) equidistant from A $(1, 5)$ and B $(3, 1)$; (ii) equidistant from C $(7, -1)$ and D $(15, 5)$. Verify that A, B, C, D all lie on a
circle, and find the equation of this circle.

(O & C)

5 Sketch the curve given parametrically by $x=t^2$, $y=t^3$. Show that an equation of the normal to the curve at the point $A(4, 8)$ is $x+3y-28=0$. This normal meets the x-axis at the point N. Find the area of the region enclosed by the arc OA of the curve, the line segment AN and the x-axis.

(L)

6 Show that the equation of the normal to the parabola $y^2=4ax$ at the point $P(at^2, 2at)$ is $y+tx=2at+at^3$. If this normal meets the x-axis at Q show that the mid-point M of PQ has the co-ordinates $(a+at^2, at)$. If P is a variable point on the parabola, find the cartesian equation of the locus of M.

(A)

7 Find the equation of the tangent to the curve $ay^2=x^3$ at the point (at^2, at^3) and prove that, apart from one exceptional case, the tangent meets the curve again. Find the co-ordinates of the point of intersection. What is the exceptional case?

(O & C)

16 Polar Coordinates

Definition, Relation with cartesian coordinates, Curve sketching, Common curves.

Definition

Polar coordinates determine the position of a point in a plane by reference to:
   a **fixed point** (the **pole**),
   a **fixed line through the pole** (the **initial line**).
The point P in the diagram has polar coordinates $(r, \theta.)$
r is conventionally taken as positive.
$-\pi < \theta < \pi$ with θ positive when measured anticlockwise, θ negative when measured clockwise.

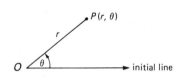

O is the pole

Figure 1

Relation with cartesian coordinates

There is a simple relationship between polar and cartesian coordinates.

If a point P has cartesian coordinates (x, y) and polar coordintes (r, θ), then to change from one system to the other use
$$x = r\cos\theta$$
$$y = r\sin\theta$$

or $r = \sqrt{(x^2 + y^2)}$ and $\tan\theta = \dfrac{y}{x}$.

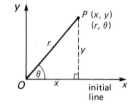

Figure 2

ℹ️ (a) Equation of a circle radius a, centre $(0, 0)$ is $x^2 + y^2 = a^2$ in cartesian coordinates.
Substituting $x = r\cos\theta$, $y = r\sin\theta$ gives
$$r^2(\cos^2\theta + \sin^2\theta) = a^2 \Rightarrow r = a.$$
So $r = a$ is the polar equation.

(b) Consider the curve whose polar equation is $r = a\cos\theta$.

Substituting $r = \sqrt{(x^2+y^2)}$ and $\cos\theta = \dfrac{x}{\sqrt{(x^2+y^2)}}$ gives

$x^2 + y^2 - ax = 0$ as cartesian equation, i.e. a circle, centre $(\frac{1}{2}a, 0)$, radius $\frac{1}{2}a$.

Curve sketching

To **sketch a curve $r = f(\theta)$**
(a) Tabulate values of r for some special values of θ, from $-180°$ to $180°$.
(b) Plot the corresponding points.

Although polar graph paper is available, a sketch on plain paper is usually required.

ℹ️ *Sketch the curve whose polar equation is $r = a\cos\theta$.*

Tabulating θ and r gives

θ	$-180°$	$-150°$	$-120°$	$-90°$	$-60°$	$-30°$	$0°$
r	$-a$	$-0.86a$	$-0.5a$	0	$0.5a$	$0.86a$	a

θ	$30°$	$60°$	$90°$	$120°$	$150°$	$180°$
r	$0.86a$	$0.5a$	0	$-0.5a$	$-0.86a$	$-a$

Since we consider $r \geq 0$, the curve will not exist for $-180° \leq \theta < -90°$ or for $90° < \theta \leq 180°$.

Plotting the remaining points and sketching the curve gives a circle, centre $(\frac{1}{2}a, 0)$, radius $\frac{1}{2}a$.

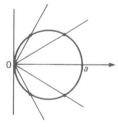

Figure 3

Common curves

Straight line
Consider a line ℓ as shown.
p is the perpendicular distance from the pole O to the line ℓ.
p is inclined at an angle α to the initial line. In $\triangle OPA$,
$$p = OP\cos(\theta - \alpha)$$
i.e. $p = r\cos(\theta - \alpha)$
is one form of the polar equation of a straight line.

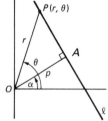

Figure 4

Cardioid
$r = a(1 + \cos\theta)$
is a cardioid as shown.
This polar equation is simpler than the cartesian form:
$$ay + a\sqrt{(x^2+y^2)} = x^2 + y^2.$$

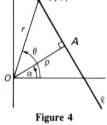

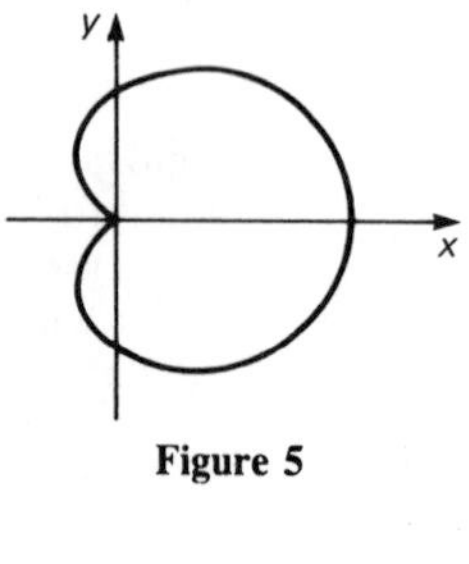

Figure 5

Circle
$r = a$ is a circle, centre the pole, radius a.
$r = a\cos\theta$ is a circle with the pole on its circumference and the initial line a diameter.

In general, if the circle has centre (c, α) in polar coordinates and radius a, then by the cosine rule in $\triangle OCP$,
$$a^2 = r^2 + c^2 - 2ra\cos(\theta - \alpha)$$
is its polar equation.

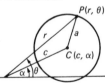

Figure 6

Parabola
$2a = r(1 - \cos\theta)$
is a parabola with the pole at the focus $(a, 0)$ as shown.
Its cartesian equation is
$$y^2 = 4a(x + a).$$

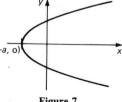

Figure 7

Polar Co-ordinates
Worked examples and Exam questions

 Draw the curve whose polar equation is $r^2 = \cos 2\theta$.

For $-180° \leqslant \theta \leqslant -135°$, $-360° \leqslant 2\theta \leqslant -270°$ and $\cos 2\theta \geqslant 0$, hence $r^2 \geqslant 0$.

For $-135° < \theta < -45°$, $-270° < 2\theta < -90°$ and $\cos 2\theta < 0$, hence $r^2 < 0$ and so the curve will not exist for this range of values of θ.

For $-45° \leqslant \theta \leqslant 45°$, $-90° \leqslant 2\theta \leqslant 90°$ and $\cos 2\theta \geqslant 0$, hence $r^2 \geqslant 0$.

For $45° < \theta < 135°$, $90° < 2\theta < 270°$ and $\cos 2\theta < 0$, hence $r^2 < 0$ and so the curve will not exist for this range of values of θ.

For $135° \leqslant \theta \leqslant 180°$, $270° \leqslant 2\theta \leqslant 360°$ and $\cos 2\theta \geqslant 0$, hence $r^2 \geqslant 0$.

The curve is symmetrical about the initial line because $\cos(-2\theta) = \cos(2\theta)$.

A table of values for $-180° \leqslant \theta \leqslant -135°$ and $-45° \leqslant \theta \leqslant 0°$ will enable the complete curve to be drawn.

θ	$-180°$	$-175°$	$-170°$	$-165°$	$-160°$	$-155°$	$-150°$	$-145°$
2θ	$-360°$	$-350°$	$-340°$	$-330°$	$-320°$	$-310°$	$-300°$	$-290°$
r	1	0.99	0.97	0.93	0.88	0.80	0.71	0.58

θ	$-140°$	$-135°$	$-45°$	$-40°$	$-35°$	$-30°$	$-25°$	$-20°$	$-15°$
2θ	$-280°$	$-270°$	$-90°$	$-80°$	$-70°$	$-60°$	$-50°$	$-40°$	$-30°$
r	0.42	0	0	0.42	0.58	0.71	0.80	0.88	0.93

θ	$-10°$	$-5°$	$0°$
2θ	$-20°$	$-10°$	$0°$
r	0.97	0.99	1

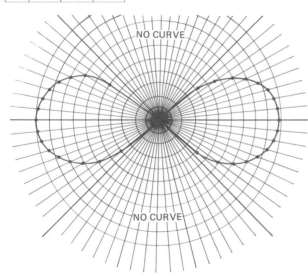

This curve is called a lemniscate.

 On the same diagram, sketch the curves given in polar co-ordinates by the equations $r = 2 + 2 \cos \theta$ and $r = 6 \cos \theta$. These curves intersect at the pole O and at the points P and Q, where P is in the first quadrant. Find the polar equation of the line PQ and of the half line OP. Find also the polar equation of the straight line through P parallel to the initial line.

$r = 2 + 2 \cos \theta$ is a cardioid with cusp at the pole.
$r = 6 \cos \theta$ is a circle centre $(3, O)$ and radius 3.

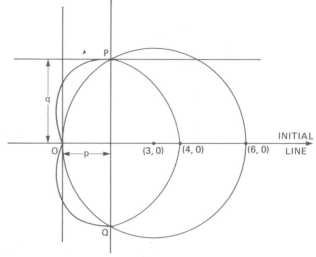

Solve $r = 2 + 2 \cos \theta$ and $r = 6 \cos \theta$ to find the co-ordinates of P and Q, i.e. $2 + 2 \cos \theta = 6 \cos \theta$ giving $\cos \theta = \frac{1}{2}$ and $\theta = \frac{\pi}{6}$ or $-\frac{\pi}{6}$ and so $r = 3$.

$\therefore$ the co-ordinates of P are $\left(3, \frac{\pi}{6}\right)$ and of Q are $\left(3, -\frac{\pi}{6}\right)$.

Let p be the perpendicular distance from O to PQ,

$p = 3 \cos \frac{\pi}{6} = \frac{3}{2}$

$\therefore$ the polar equation of PQ is $r \cos \theta = \frac{3}{2}$

The equation of the half-line OP is $\theta = \frac{\pi}{6}$

Let q be the perpendicular distance from O to the line through P parallel to the initial line.

$q = OP \sin \frac{\pi}{6} = 3 \times \frac{\sqrt{3}}{2}$

$\therefore$ the polar equation of the specified line is $r \cos \left(\theta - \frac{\pi}{2}\right) = 3\frac{\sqrt{3}}{2}$

EX 1 On the same diagram, sketch the curves given in polar co-ordinates by the equations: $r = 1 + \cos \theta$; $r = 3 \cos \theta$. These curves intersect at the pole O and the points P and Q, where P lies in the first quadrant. Find the polar equations of the line PQ and of the half-line OP. Find also the polar equation of the straight line through P parallel to the initial line. *(L)*

2 Write down the polar equations relative to the origin O of:
   (i) a circle, radius a, centre O;
   (ii) a straight line with one end at O and making an angle α with the initial line;
   (iii) a straight line parallel to the initial line and distant c from it;
   (iv) a straight line through the point with polar co-ordinates $(d, 0)$ and making an angle β with the initial line. *(O & C)*

3 Sketch on the same diagram the curves given in polar co-ordinates by the equations: $r = a(1 + \cos \theta)$; $4r \cos \theta = 3a$, where $a > 0$. These curves intersect at P and Q. Show that PQ is of length $3\sqrt{3}a/2$. Obtain the polar equation of the circle which passes through the pole O and touches PQ at the point where it intersects the half-line $\theta = 0$. *(L)*

33

17 Experimental Laws
Introduction, Linear relations, Non-linear relations.

Introduction

It is often necessary to find a relationship between two connected quantities. A table of experimental data, showing corresponding values of the two quantities, and a suggestion as to the form of the expected relationship are given. The data are usually displayed graphically and, if the expected relationship is confirmed, the graph is used to obtain the unknown constants in this relationship.

Linear relations

The simplest case is when the expected relation is a straight line of the form $y=mx+c$.

In this case simply plot y against x. If the points plotted lie approximately on a straight line then the given linear relation is approximately true.
The points rarely lie exactly on a straight line, so the line of 'best fit' is drawn 'by eye'. Consequently there is usually a small range of values in which acceptable answers will lie.

The values of m and c can be found from the graph by

either (a) finding the gradient (m) of the line and its intercept (c) on the y-axis,

or (b) substituting the coordinates of two points on the line in $y=mx+c$ and solving the resulting simultaneous equations.

Note: Do not use given values to find m and c since they may not give points on the line of 'best fit'.

ⓘ *The following data are believed to satisfy a law of the form $y=ax+b$. Find suitable values of a and b.*

x	0	1	2	3	4	5
y	5.8	8.7	12.5	15.0	18.4	20.9

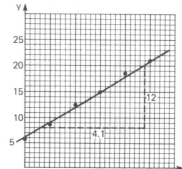

Figure 1

From the graph;

The gradient of the line gives

$$a=\frac{12}{4.1}=2.9$$

'y-intercept' gives $b=6.2$

So,

$$y=2.9x+6.2$$

Non-linear relations

If the expected relation is not in the form $y=mx+c$, then it must be transformed to a linear form before proceeding as before. The transformations for some common relationships are given below.

Relationships of the form $y=ax^n+b$ (n known)
Compare $y=ax^n+b$
with $Y=mX+c$
Plot $Y=y$ against $X=x^n$ to give a straight line.
Gradient gives a, 'Y-intercept' gives b.

Relationships of the form $\dfrac{1}{y}+\dfrac{1}{x}=\dfrac{1}{a}$

Rewrite as $\dfrac{1}{y}=-\dfrac{1}{x}+\dfrac{1}{a}$.

Compare with $Y=mX+c$.

Plot $Y=\dfrac{1}{y}$ against $X=\dfrac{1}{x}$ to give a straight line.

Gradient is -1, 'Y-intercept' gives $\dfrac{1}{a}$.

Relationships of the form $y=ax^n$ (n unknown)

Rewrite $y=ax^n$
as $\log y=n\log x+\log a$.

Compare with $Y=mX+c$
Plot $Y=\log y$ against $X=\log x$ to give a straight line.
Gradient gives n, 'Y-intercept' gives $\log a$.

Relationships of the form $y=ab^x$
Rewrite $y=ab^x$
as $\log y=x\log b+\log a$
Compare with $Y=mX+c$
Plot $Y=\log y$ against $X=x$ to give a straight line.
Gradient gives $\log b$, 'Y-intercept' gives $\log a$.

ⓘ *The data of the following table are thought to obey a law of the form $y=ax^2+b$. Find suitable values for a and b.*

x	0	1	2	3
y	-5	3	24	67

Compare $y=ax^2+b$
with $Y=mX+c$

Constructing a table for $X=x^2$ and $Y=y$ gives

$X(=x^2)$	0	1	4	9
$Y(=y)$	-5	3	24	67

The straight line represents $Y=aX+b$.
Gradient gives a, 'Y-intercept' gives b.

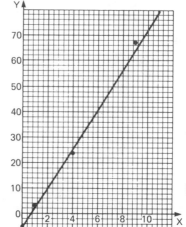

Figure 2

From the graph:

Gradient gives

$$a=\frac{70}{9}$$

'Y-intercept' gives $b=-6$

So,

$$Y=\frac{70}{9}X-6$$

i.e.

$$y=\frac{70}{9}x^2-6$$

Experimental Laws
Worked example and Exam questions

 The following corresponding values of x and y are believed to be related by the equation $y = x + ax^b$.

x	2	3	4	5	6	10
y	7·54	9·33	11·00	12·59	14·12	19·90

Draw a suitable graph to show that this may be so, and use your graph to find the probable values of a and b.

Write $y = x + ax^b$ as $y - x = ax^b$.
Taking logarithms gives $\log_{10}(y - x) = \log_{10}a + b\log_{10}x$
Let $\log_{10}(y - x) = Y$ and $\log_{10}x = X$
i.e. $Y = \log_{10}a + bX$
i.e. a straight line with slope b and y-intercept $\log_{10}a$

x	2	3	4	5	6	10
y	7·54	9·33	11·00	12·59	14·12	19·90
$y - x$	5·54	6·33	7·00	7·59	8·12	9·90
$Y = \log_{10}(y-x)$	0·74	0·80	0·85	0·88	0·91	1·00
$X = \log_{10}x$	0·30	0·48	0·60	0·70	0·78	1·00

The graph is drawn as shown.

From the graph $b = \dfrac{1 \cdot 1 - 0 \cdot 6}{1} = 0 \cdot 50$

$\log_{10}a = 0 \cdot 6$, $a = 3 \cdot 98$
$\therefore y = x + 3 \cdot 98x^{0 \cdot 50}$

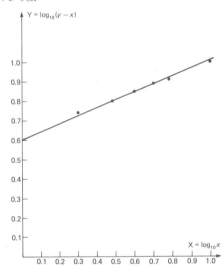

 1 It is believed that two variables x and y are related by $y = ka^x$, where k and a are constants. For the following values, plot $\log_{10}y$ against x and hence estimate k and a.

x	0·5	1	2	4
y	0·28	0·45	1·13	7·03 *(S)

2 The points $(0 \cdot 5, 0 \cdot 265)$, $(1 \cdot 6, 1 \cdot 352)$, $(2 \cdot 1, 1 \cdot 978)$ and $(3 \cdot 4, 3 \cdot 883)$ very nearly satisfy the relation $y = ax^n$, where a and n are constants. Draw a graph of $\log y$ against $\log x$, and use the points where the graph cuts the axes to determine the values of a and n. Use these values to calculate the value of y when $x = 1 \cdot 8$. *(OLE)*

3 It is believed that two variables x and y are related by $y = ax^b$, where a and b are constants. For the following values, plot $\log_{10}y$ against $\log_{10}x$ and hence estimate a and b.

x	0·6	1·2	1·8	2·4	3·0
y	0·35	1·04	1·95	3·06	4·33 *(S)

4 The variables x and y satisfy an equation of the form $y = ax^k$, where a and k are constants. Express $\log_{10}y$ in terms of $\log_{10}x$.
In an experiment values of y corresponding to values of x are as given in the table below:

x	1·6	2·0	2·8	3·9	5·0
y	5·1	7·1	11·7	19·2	28·0

By drawing a straight-line graph estimate to one decimal place the values of a and k. Use your graph to estimate:
(a) the value of x when $y = 10$;
(b) the value of y when $x = \sqrt{10}$. *(A)*

5 Measured values of x and y are given in the following table:

x	1	2	3	4	5	6	7
y	5·1	4·6	4·2	3·8	3·2	2·4	1·4

It is known that x and y are related by the equation $y^2 = a + bx$. Explain how a straight line graph may be drawn to represent the given equation and draw it for the values given. Use the graph to estimate the value of a and b. Estimate the greatest possible value of x. *(C)*

6 Given that the values of x and y in the table below are experimental values of variables that satisfy $y = ax + b$, estimate graphically the constants a and b.

x	2	4	7	8	10
y	1·8	1·2	0·5	0·3	-0·2
(L)

7 Rewrite the following equations in suitable form to display a linear relationship in each case between two of the variables, x, $\ln x$, y and $\ln y$.
(a) $\dfrac{2}{x} + \dfrac{3}{y} = \dfrac{4}{xy}$, (b) $5x = 6^y$. *(L)*

8 A relation of the form $y = ae^x + b$ is known to exist between two variables x and y. By plotting y against e^x, use the following table of experimental values of x and y to estimate the constants a and b to 1 significant figure.

x	1	2	3	4
y	24	56·7	145·6	387·2
(L)

9

v	5	10	15	20	25
R	149	175	219	280	359

The table shows corresponding values of variables R and v obtained in an experiment. By drawing a suitable linear graph, show that these pairs of values may be regarded as approximations to values satisfying a relation of the form $R = a + bv^2$, where a and b are constants. Use your graph to estimate the values of a and b, giving your answers to 2 significant figures. *(L)*

10 In each of the following cases, given experimental values of x and y, explain how straight line graphs may be drawn, using ordinary graph paper only:
(i) $y = ab^{x+1}$;
(ii) $px^2 + qy = x$, where a, b, p, q are constants.
In each case express the gradient of the line and its ordinate for $x = 0$ in terms of the constants. *(A)*

35

18 Trigonometrical Functions

Angle measure, Circular functions of the general angle, Trigonometrical ratios of any angle, Special angles, Limiting values of angles.

Angle measure

(a) **Sexagesimal measure**: 1 right angle = 90 degrees
                 1 degree = 60 minutes
                 1 minute = 60 seconds

(b) **Circular measure**: The circular measure of an angle is the number of radians it contains. A radian is the angle subtended at the centre of a circle by an arc equal in length to the radius.

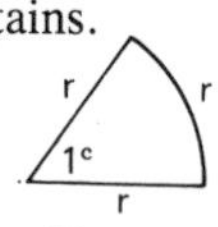

1 radian ≈ 57°17′45″

Figure 1

The **length of a circular arc** of angle θ radians is $r\theta$

The **area of a circular sector** of angle θ radians is $\frac{1}{2}r^2\theta$.

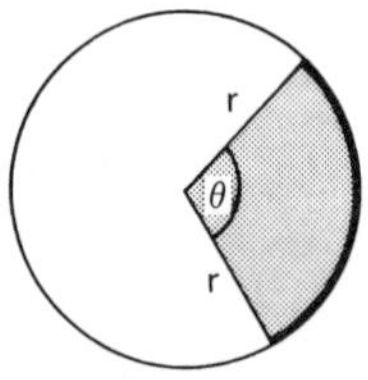

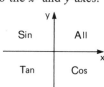

Figure 2

ℹ️ Useful common equivalences are:

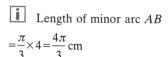

$0°=0$ radians	$30°=\frac{\pi}{6}$ $45°=\frac{\pi}{4}$
$180°=\pi$	$60°=\frac{\pi}{3}$ $90°=\frac{\pi}{2}$
$360°=2\pi$	$120°=\frac{2\pi}{3}$ $135°=\frac{3\pi}{4}$

ℹ️ Length of minor arc AB

$$=\frac{\pi}{3}\times 4=\frac{4\pi}{3}\text{ cm}$$

Area of minor sector AOB

$$=\frac{1}{2}\times 4^2\times\frac{\pi}{3}=\frac{8\pi}{3}\text{ cm}^2$$

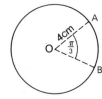

Figure 3

Circular functions of the general angle

The **general angle** is an angle of any size, positive or negative.

Let Ox, Oy be the usual perpendicular axes, with the same scales.

If P is the point (x, y), $OP = r$ (positive) and $\angle POX = \theta$ (anticlockwise positive), the definitions of the trigonometrical functions when θ is the general angle are:

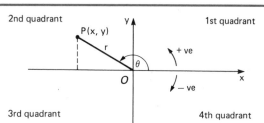

Figure 4

$$\sin\theta=\frac{y}{r},\ \cos\theta=\frac{x}{r},\ \tan\theta=\frac{y}{x},\ \operatorname{cosec}\theta=\frac{r}{y},\ \sec\theta=\frac{r}{x},\ \cot\theta=\frac{x}{y}$$

The following **identities** are obtained from the above definitions:

$$\operatorname{cosec}\theta=\frac{1}{\sin\theta},\ \sec\theta=\frac{1}{\cos\theta},\ \tan\theta=\frac{\sin\theta}{\cos\theta},\ \cot\theta=\frac{1}{\tan\theta}=\frac{\cos\theta}{\sin\theta}$$

Trigonometrical ratios of any angle

To find the trigonometrical ratios of any angle

(a) Sketch the angle in relation to the x- and y-axes.

(b) Find the sign of the ratio using the 'CAST' diagram. This shows which ratios are positive in each quadrant.

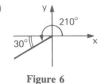

Figure 5

(c) Find the 'associated acute angle', i.e. the acute angle the radius defining the angle makes with the x-axis.

(d) Find the required ratio of this angle.

ℹ️ *Find the value of (a) sin 210° (b) cos 675°*

(a)

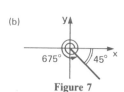

Figure 6

210° is in the 3rd quadrant. So its sine ratio is negative. Associated acute angle = 30°. So sin 210° = −sin 30° = −$\frac{1}{2}$.

(b)

Figure 7

Since 675° = 360° + 315°, 675° is in the 4th quadrant. So its cosine ratio is positive. Associated acute angle = 45°. So cos 675° = cos 45° = $1/\sqrt{2}$.

Special angles

The trigonometrical ratios of 30°, 45° and 60° can be found using these triangles.

'30°–60° triangle'   '45° triangle'

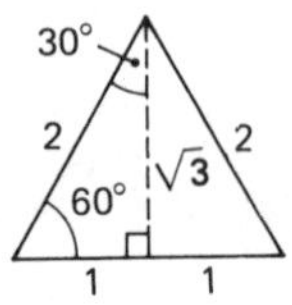

Figure 8   **Figure 9**

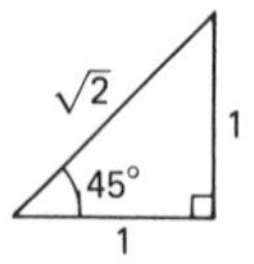

	0°	30°	45°	60°	90°
sine	0	$\frac{1}{2}$	$\frac{1}{\sqrt{2}}$	$\frac{\sqrt{3}}{2}$	1
cosine	1	$\frac{\sqrt{3}}{2}$	$\frac{1}{\sqrt{2}}$	$\frac{1}{2}$	0
tangent	0	$\frac{1}{\sqrt{3}}$	1	$\sqrt{3}$	undefined

Limiting values of angles

If θ is small and in radians, then
$\sin\theta\approx\tan\theta\approx\theta$
$\cos\theta\approx 1-\frac{1}{2}\theta^2$

ℹ️ *Solve, approximately, the equation cos θ=0.999.*

Since θ must be small, $\cos\theta\approx 1-\frac{1}{2}\theta^2$.
$\therefore 1-\frac{1}{2}\theta^2=0.999$
$\qquad\theta^2=0.002\Rightarrow\theta=0.004472$ radians

Trigonometrical Functions
Worked examples and Exam questions

 Find the number of seconds in the angle subtended at the centre of a circle of radius 5 kilometres by an arc of length 1 metre.

Let θ be required angle, in radians, and r the radius.

$\therefore$ arc length $= r\theta = 5 \times 1000 \times \theta$ metres

$\therefore 1 = 5 \times 1000 \times \theta$, i.e. $\theta = \dfrac{1}{5000}$ radians

$\qquad = \dfrac{1}{5000} \times \dfrac{180}{\pi}$ degrees

$\qquad = \dfrac{1}{5000} \times \dfrac{180}{\pi} \times 60 \times 60$ seconds

$\therefore \theta = 41 \cdot 3$ seconds.

 The development of a cone is a sector of a circle of radius r and angle $\dfrac{2\pi}{3}$ radians. Find the semi-vertical angle of the cone.

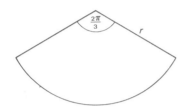 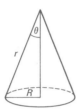

Let R be the base radius of the cone and θ its semi-vertical angle. The slant length of the cone is r and the circumference of its base $2\pi R$ is equal to the arc length of the sector,

i.e. $2\pi R = \dfrac{2\pi}{3} r$ and $R = \dfrac{r}{3}$.

$\sin \theta = \dfrac{R}{r} = \dfrac{1}{3} = 0 \cdot 3333$

$\therefore \theta = 19° 28'$

 Find the value of sin 7′, without using tables.

If θ is small and measured in radians $\sin \theta \approx \theta$.

In this case $7' = \dfrac{\pi}{180} \times \dfrac{7}{60}$ radians, $\therefore \sin 7' \approx \dfrac{\pi}{180} \times \dfrac{7}{60} = 0 \cdot 0020$

WE *Find, without using tables, the values of cos θ and tan θ, if $\sin \theta = \dfrac{5}{13}$.*

Since $\sin \theta$ is positive, θ may be in quadrants 1 or 2. If θ is in the first quadrant then

$\cos \theta = \dfrac{12}{13}$ and $\tan \theta = \dfrac{5}{12}$.

If θ is in the second quadrant then

$\cos \theta = -\dfrac{12}{13}$ and $\tan \theta = -\dfrac{5}{12}$.

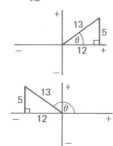

EX **1** By using suitable approximations for sin θ and cos θ obtain an approximation in radians to the positive solution of the equation $\cos \theta - \theta \sin \theta = 0 \cdot 9976$ when θ is small.
(L)

2 Express as a decimal of a right angle, to six decimal places:
(a) 44° 55′; (b) 65° 34′ 20″; (c) 134° 27′ 18″.

3 Express the following right angles in sexagesimal measure:
(a) $0 \cdot 041732$; (b) $0 \cdot 674513$; (c) $1 \cdot 397615$.

4 Change the following angles to circular measure:
(a) 45°; (b) 73° 47′; (c) 175° 45′.

5 Change the following radian measure to sexagesimal measure: (a) $\frac{1}{2}$; (b) $\dfrac{\pi}{4}$; (c) $1 \cdot 5851$.

6 Find the number of degrees subtended at the centre of a circle of radius 1 m by an arc length 30 cm.

7 Find the number of degrees subtended at the centre of a circle by an arc which is twice the length of the radius.

8 Find in radians the angle subtended at the centre of a circle of circumference 36 cm by an arc of length $7 \cdot 5$ cm.

9 Find the length of the arc of a circle of radius 2 m which subtends an angle of 37° 45′ at the centre.

10 Find the length of the arc of a circle of radius $12 \cdot 5$ cm which subtends an angle of $1 \cdot 5$ radians at the centre.

11 Find the radius of a circle in which a chord of length 10 cm subtends an angle of 85° at the centre.

12 Find the radius of a circle in which a chord of length 15 cm subtends an angle of 135° at the centre.

13 Find without using tables the value of sine, cosine and tangent of 5′.

14 Find without using tables the value of sine, cosine and tangent of 20′.

15 Find, without using tables, the value of:
(a) sin θ and cos θ, if $\tan \theta = \frac{3}{4}$;

(b) tan θ and sin θ, if $\cos \theta = \dfrac{4}{5}$;

(c) cos θ and tan θ, if $\sin \theta = \dfrac{24}{25}$.

16 Solve approximately the following equations:
(a) $\sin \theta = 0 \cdot 0095$;   (b) $\tan \theta = 0 \cdot 0099$;
(c) $\cos \theta = 0 \cdot 995$;   (d) $\cos \theta = 0 \cdot 99$.

17 By using suitable approximations for sin θ and tan θ obtain an approximate value for the positive solution of $\sin \theta - \theta \tan \theta = 0 \cdot 0016$, when θ is small.

18 By using suitable approximations for sin θ and cos θ obtain an approximate value for the positive solution of $\cos \theta - \theta \sin \theta = 0 \cdot 9925$, where θ is small.

19 The development of a cone is a sector of a circle of radius 10 cm and semi-vertical angle 135°. Find the semi-vertical angle of the cone.

20 A solid cone has a semi-vertical angle of 30° and a slant height of 12 cm, find the total surface area of the cone.

19 Trigonometrical Graphs

Graphs of the trigonometrical functions, Inverse trigonometrical functions, General solutions of trigonometrical equations, Graphical solution of trigonometrical equations.

Graphs of the trigonometrical functions

The following are the graphs of the six trigonometrical functions between $x=-360°$ and $x=360°$.

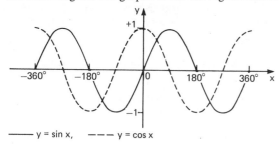

—— $y = \sin x$, – – – $y = \cos x$

Figure 1

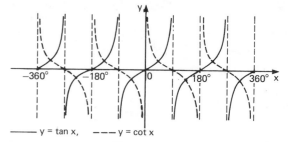

—— $y = \tan x$, – – – $y = \cot x$

Figure 2

Some features of the graph of the sine and cosine functions:

(a) Continuous functions, i.e. graphs have no breaks.

(b) Periodic functions with period 360° (2π radians), i.e. curves repeat their values every 360°.

(c) $-1 \leqslant \sin x \leqslant +1$ and $-1 \leqslant \cos x \leqslant +1$,

(d) The amplitude is 1.

Note: $\cos x$ is $\sin x$ translated through $-90°$ along Ox.

Some features of the graph of the tan function:

(a) Not continuous, being undefined when $x = \ldots 90°$, 270°, …

(b) Periodic function with period 180° (π radians).

(c) Unlimited range.

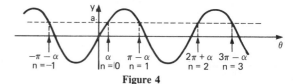

—— $y = \text{cosec } x$, – – – $y = \sec x$

Figure 3

Inverse trigonometrical functions

$\sin^{-1}a = \arcsin a$ is the angle between $-90°$ and $+90°$ satisfying $\sin \theta = a$.

$\cos^{-1}b = \arccos b$ is the angle between $0°$ and $180°$ satisfying $\cos \theta = b$.

$\tan^{-1}c = \arctan c$ is the angle between $-90°$ and $+90°$ satisfying $\tan \theta = c$.

ℹ️
$\arcsin \frac{1}{2} = 30°$
$\arcsin (-\frac{1}{2}) = -30°$

$\arccos \frac{1}{2} = 60°$
$\arccos (-\frac{1}{2}) = 120°$

$\arctan 1 = 45°$
$\arctan (-1) = -45°$

General solutions of trigonometrical equations

The **general solution of $\sin \theta = a$** is $\theta = n\pi + (-1)^n \alpha$ where $\sin \alpha = a$.

The **general solution of $\cos \theta = b$** is $\theta = 2n\pi \pm \beta$ where $\cos \beta = b$.

The **general solution of $\tan \theta = c$** is $\theta = n\pi + \gamma$ where $\tan \gamma = c$.

The diagram shows why there are many solutions to $\sin \theta = a$ and how they are generated by the formula $\theta = n\pi + (-1)^n \alpha$ by putting $n = -1, 0, 1, 2, 3$.

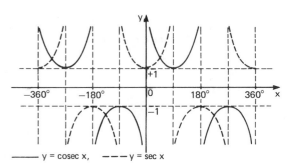

Figure 4

Graphical solution of trigonometrical equations

To find the **graphical solution of $f(x)=0$ when $f(x)=0$ can be written as $F(x)=G(x)$,** where $F(x)$ is a trigonometrical function and $G(x)$ is an algebraic function, the two graph method is used.
The curves $y=F(x)$ and $y=G(x)$ are drawn on the same sheet of graph paper using the same axes and the same scales.

At any point of intersection of the two curves the two values of y are the same and therefore at these points $F(x)=G(x)$. So the points of intersection give the required solutions.

ℹ️ *Find, by a graphical method, approximate solutions of* $\pi \sin \left(x + \frac{\pi}{6} \right) = \pi - x$ *for* $0 < x < 2\pi$.

Write the given equation as $\sin \left(x + \frac{\pi}{6} \right) = 1 - x/\pi$.

The L.H.S. is the trig. function $y = \sin (x + \pi/6)$.
The R.H.S. is the algebraic function $y = 1 - x/\pi$.
Construct a table of values for $y = \sin (x + \pi/6)$, $0 < x < 2\pi$ and draw the curve. Using the same axes and scales draw the straight line $y = 1 - x/\pi$.

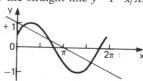

Figure 5

The points of intersection give $x \approx 3\pi/20$, $3\pi/4$, $97\pi/60$.

Trigonometrical Graphs
Worked example and Exam questions

 A taut belt passes round two pulleys of radii 6 cm and 2 cm respectively. The straight portions of the belt are direct common tangents to the pulleys and are inclined to each other at an angle of 2α radians.
If the total length of the belt is 44 cm, show that
$\pi + \alpha + \cot \alpha = 5 \cdot 5$.
Draw the graph of $y = \cot \alpha$ and $y = 5 \cdot 5 - \pi - \alpha$ and hence find α.

A diagram of the situation is shown below.

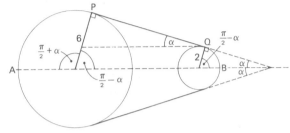

The length of the belt from A to B is 22 cm.

$$\text{arc } AP = 6 \times \left(\frac{\pi}{2} + \alpha \right) \text{ cm}$$

$$PQ = 4 \tan \left(\frac{\pi}{2} - \alpha \right) = 4 \cot \alpha \text{ cm}$$

$$\text{arc } QB = 2 \times \left(\frac{\pi}{2} - \alpha \right) \text{ cm}$$

$$\therefore 22 = 3\pi + 6\alpha + 4 \cot \alpha + \pi - 2\alpha$$
$$= 4\pi + 4\alpha + 4 \cot \alpha$$

i.e. $5 \cdot 5 = \pi + \alpha + \cot \alpha$

An appropriate table of values for α, $\cot \alpha$ and $5 \cdot 5 - \pi - \alpha$ is given below.

α	0	$\frac{\pi}{10}$	$\frac{\pi}{5}$	$\frac{3\pi}{10}$	$\frac{2\pi}{5}$	$\frac{\pi}{2}$
$\cot \alpha$	∞	3·08	1·38	0·73	0·32	0
$5 \cdot 5 - \pi - \alpha$	2·36					0·79

Using the two graph method for $y = \cot \alpha$ and $y = 5 \cdot 5 - \pi - \alpha$ gives the diagram shown below.

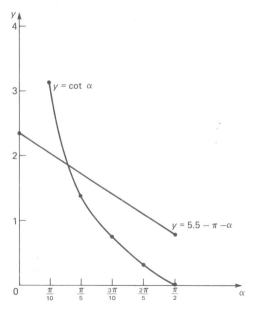

From the graph the value of α for which $\cot \alpha = 5 \cdot 5 - \pi - \alpha$ is approximately 0·50 radians.

EX

1 The chord AB of a circle, radius r, subtends an angle of 2θ radians at the centre, O, of the circle. The perimeter of the minor segment AB is k times the perimeter of the triangle AOB.
Show that $k + (k-1) \sin \theta = \theta$. By drawing graphs of the curve $y = \sin \theta$ ($0 \leqslant \theta \leqslant 1 \cdot 5$ radians) and a suitable straight line, obtain an estimate of θ in the case when $k = \frac{1}{2}$.
(S)

2 Sketch each of the following curves in the range $0 \leqslant x \leqslant \pi$:
  (i) $y = \tan 2x$;

  (ii) $y = \tan \left\{ 2 \left(x - \frac{\pi}{4} \right) \right\}$;

  (iii) $y = \tan \left(\frac{\pi}{2} - 2x \right)$.
(A)

3 Solve $\sin (\theta - 15°) = \frac{1}{2}$, for values of θ in the range $0° \leqslant \theta \leqslant 180°$.
(S)

4 Find the values of x in the range $0 < x < 360$ for which $\cos x° = \sin 57°$.
(L)

5 Sketch on separate diagrams for $0° \leqslant x \leqslant 360°$:
(i) $y = 3 \cos x$; (ii) $y = \cos 2x$; (iii) $y = \cos (x - 30°)$.
(C)

6 Draw the graph $y = \sin \frac{1}{2}x$ (x being measured in radians) from $x = 0$ to $x = 6\pi$, at intervals of $\frac{1}{2}\pi$, taking 2 cm to represent π radians on the x-axis and 1 unit on the y-axis. Plot the point P $(5\pi, \frac{1}{2})$ and join it to the origin O. Write down (in terms of π) the x co-ordinates of the points where OP (produced if necessary) meets your graph. Of what equation in x are they roots?
(O & C)

7 Given that $\sin x = \frac{12}{13}$ and that $\frac{1}{2}\pi < x < \pi$, show that $\cos x = -\frac{5}{13}$. Without using tables or a calculator, evaluate: (a) $\sin (x + \pi)$; (b) $\tan (x + 2\pi)$; (c) $\cot (\pi - x)$; (d) $\sec (x + \frac{1}{2}\pi)$.
(A)

8 Sketch the curve $y = \cos x + 2$ for the interval $0 \leqslant x \leqslant 2\pi$.
On the same diagram sketch the curve $y = \dfrac{1}{\cos x + 2}$ for the same interval.
(C)

9 Find the values of θ, lying in the interval $-\pi \leqslant \theta \leqslant \pi$, for which $\sin 2\theta = \cos \theta$. Sketch on the same axes the graphs of $y = \sin 2\theta$ and $y = \cos \theta$ in the above interval and deduce the set of values of θ in this interval for which $\sin 2\theta \geqslant \cos \theta$.
(J)

20 Trigonometrical Identities
Standard identities, Applications.

An **identity** is a relationship which is true for all values of the variable.
Basic identities:

$$\sin^2 A + \cos^2 A = 1 \qquad 1 + \tan^2 A = \sec^2 A \qquad 1 + \cot^2 A = \csc^2 A$$

Sum and difference formulae

$$\sin(A+B) = \sin A \cos B + \cos A \sin B \qquad\qquad \sin(A-B) = \sin A \cos B - \cos A \sin B$$
$$\cos(A+B) = \cos A \cos B - \sin A \sin B \qquad\qquad \cos(A-B) = \cos A \cos B + \sin A \sin B$$
$$\tan(A+B) = \frac{\tan A + \tan B}{1 - \tan A \tan B} \qquad\qquad\qquad \tan(A-B) = \frac{\tan A - \tan B}{1 + \tan A \tan B}$$

Double and triple angle formulae

$$\sin 2A = 2 \sin A \cos A \qquad\qquad\qquad \sin 3A = 3 \sin A - 4 \sin^3 A$$
$$\cos 2A = \cos^2 A - \sin^2 A = 1 - 2\sin^2 A = 2\cos^2 A - 1 \qquad \cos 3A = 4\cos^3 A - 3 \cos A$$
$$\tan 2A = \frac{2 \tan A}{1 - \tan^2 A} \qquad\qquad\qquad \tan 3A = \frac{3 \tan A - \tan^3 A}{1 - 3 \tan^2 A}$$

Half angle formulae

$$\sin A = \frac{2 \tan \frac{1}{2}A}{1 + \tan^2 \frac{1}{2}A} \qquad \cos A = \frac{1 - \tan^2 \frac{1}{2}A}{1 + \tan^2 \frac{1}{2}A} \qquad \tan A = \frac{2 \tan \frac{1}{2}A}{1 - \tan^2 \frac{1}{2}A}$$

Factor formulae

$$\sin A + \sin B = 2 \sin \tfrac{1}{2}(A+B) \cos \tfrac{1}{2}(A+B) \qquad \sin A - \sin B = 2 \cos \tfrac{1}{2}(A+B) \sin \tfrac{1}{2}(A-B)$$
$$\cos A + \cos B = 2 \cos \tfrac{1}{2}(A+B) \cos \tfrac{1}{2}(A-B) \qquad \cos A - \cos B = 2 \sin \tfrac{1}{2}(A+B) \sin \tfrac{1}{2}(B-A)$$
$$\tan A + \tan B = \frac{\sin(A+B)}{\cos A \cos B} \qquad\qquad\qquad \tan A - \tan B = \frac{\sin(B-A)}{\sin A \cos B}$$

Applications

1. To find the values of trigonometrical ratios of compound angles

$\boxed{i}$
$$\cos 75° = \cos(30° + 45°)$$
$$= \cos 30° \cos 45° - \sin 30° \sin 45°$$
$$= \frac{\sqrt 3}{2} \cdot \frac{1}{\sqrt 2} - \frac{1}{2} \cdot \frac{1}{\sqrt 2} = \frac{\sqrt 2}{4}(\sqrt 3 - 1)$$

2. To establish other trigonometrical identities
Change one side of the identity (usually the more complicated expression), step by step, to the other side using the standard identities. Do not work with both sides of the given identity simultaneously.

$\boxed{i}$ *Show that* $\sin(x+y) - \sin(x-y) \equiv 2 \cos x \sin y$.

$$\sin(x+y) - \sin(x-y)$$
$$\equiv [\sin x \cos y + \cos x \sin y] - [\sin x \cos y - \cos x \sin y]$$
$$\equiv 2 \cos x \sin y.$$

3. To solve trigonometrical equations
(a) **Equations involving the 'basic identities'**
Use the 'basic identities' to express the equation in terms of one ratio only. The equation can be solved in the same way as an algebraic equation.
Note: do not cancel a ratio from an equation or one of the solutions will be lost.

$\boxed{i}$ *Solve* $\cos \theta = 2 \sin^2\theta - 1$ for $0° \leqslant \theta \leqslant 360°$.

The equation becomes $\cos \theta = 2(1 - \cos^2\theta) - 1$
$\Rightarrow 2\cos^2\theta + \cos \theta - 1 = 0$
i.e. $(2 \cos \theta - 1)(\cos \theta + 1) = 0 \Rightarrow \cos \theta = \frac{1}{2}$ or -1.
If $\cos \theta = \frac{1}{2}$, $\theta = 60°$, $300°$. If $\cos \theta = -1$, $\theta = 180°$.

(b) **Equations involving sums, differences or products**
Combine terms to give
'an expression that will factorize' $= 0$
then solve this equation like an algebraic equation.

$\boxed{i}$ *Find the general solution of*
$\sin \theta + \sin 5\theta - \sin 3\theta = 0$.

The factor formula gives $2 \sin 3\theta \cos 2\theta - \sin 3\theta = 0$
$$\Rightarrow \sin 3\theta(2 \cos 2\theta - 1) = 0$$
If $\sin 3\theta = 0$, $3\theta = n\pi \Rightarrow \theta = \frac{1}{3}n\pi$.
If $\cos 2\theta = \frac{1}{2}$, $2\theta = 2n\pi \pm \frac{1}{3}\pi \Rightarrow \theta = n\pi \pm \frac{1}{6}\pi$.

(c) **Equations of the form** $a \cos \theta + b \sin \theta = c$
Note: for real roots $c \leqslant \sqrt{(a^2 + b^2)}$
(i) **Auxiliary angle method**
Use one of the standard forms, $R \sin(\theta \pm \alpha)$ or $R \cos(\theta \pm \alpha)$, to obtain an equivalent expression.
(ii) **Half angle method**
Use $t = \tan \frac{1}{2}\theta$, so that

$$\sin \theta = \frac{2t}{1 + t^2} \text{ and } \cos \theta = \frac{1 - t^2}{1 + t^2}.$$

Substitute these in the equation and solve the quadratic in t. Hence find θ.

$\boxed{i}$ *Solve* $12 \cos \theta + 5 \sin \theta = 3$ for $0° \leqslant \theta \leqslant 360°$.

Let $12 \cos \theta + 5 \sin \theta = R \cos \theta \cos \alpha + R \sin \theta \sin \alpha$
$= R \cos(\theta - \alpha)$

This gives $\left.\begin{array}{l} R \cos \alpha = 12 \\ R \sin \alpha = 5 \end{array}\right\} \Rightarrow \begin{array}{l} R^2 = 12^2 + 5^2 = 169 \\ \tan \alpha = \frac{5}{12} \end{array}$

So $R = 13$, $\alpha = 22°37'$ and $R \cos(\theta - \alpha) = 3$ becomes
$13 \cos(\theta - 22°37') = 3 \Rightarrow \cos(\theta - 22°37') = \frac{3}{13}$
$$= \cos 76°39'$$
$\therefore \theta - 22°37 - = 2n\pi \pm 76°39' = 76°39'$ or $(360° - 76°39')$
i.e. $\qquad \theta = 99°16'$ or $305°58'$

Trigonometrical Identities
Worked example, Guided example and Exam questions

 Given that $4 \sin x - \cos x \equiv R \sin (x - \theta)$, where $R > 0$ and $0° < \theta < 90°$, find the values of R and θ correct to one decimal place.
Hence find one value of x between $0°$ and $360°$ for which the curve $y = 4 \sin x - \cos x$ has a turning point.

$4 \sin x - \cos \equiv R \sin x \cos \theta - R \cos x \sin \theta$
  i.e. $R \cos \theta = 4$ and $R \sin \theta = 1$

$\therefore R = \sqrt{17}$ and $\tan \theta = \frac{1}{4}$, giving $R = 4 \cdot 1$ and $\theta = 14 \cdot 0°$

$\therefore 4 \sin x - \cos x = 4 \cdot 1 \sin (x - 14 \cdot 0°)$

One turning point of $y = \sin x$ is $x = 90°$, therefore
one turning point of $y = \sin (x - 14 \cdot 0°)$ is $x = 90° + 14°$
$= 104°$
$\therefore y = 4 \sin x - \cos x$ has a turning point when $x = 104°$

 Show that:

(i) $\dfrac{1 - \cos \theta}{1 + \cos \theta} = \tan^2 \dfrac{\theta}{2}$,

(ii) $\dfrac{1 - \sin \theta}{1 + \sin \theta} = \tan^2 \left(\dfrac{\pi}{4} - \dfrac{\theta}{2} \right)$,

(iii) $\dfrac{\cos \theta - \sin \theta}{\cos \theta + \sin \theta} = \tan \left(\dfrac{\pi}{4} - \theta \right)$.

Hence, or otherwise, show that $\tan 22\frac{1}{2}° = \sqrt{2} - 1$.

(i) Use the standard result $\cos \theta = \dfrac{1 - \tan^2 \dfrac{\theta}{2}}{1 + \tan^2 \dfrac{\theta}{2}}$ in LHS.

The required result is − ter a few lines of simple algebra.

(ii) Use the standard result $\sin \theta = \dfrac{2 \tan \dfrac{\theta}{2}}{1 + \tan^2 \dfrac{\theta}{2}}$ in LHS.

After some algebraic manipulation the LHS is reduced

to $\dfrac{\left(1 - \tan \dfrac{\theta}{2}\right)^2}{\left(1 + \tan \dfrac{\theta}{2}\right)^2}$. Since $\tan \dfrac{\pi}{4} = 1$ this last result may be

written as $\dfrac{\left(\tan \dfrac{\pi}{4} - \tan \dfrac{\theta}{2}\right)^2}{\left(1 + \tan \dfrac{\pi}{4} \tan \dfrac{\theta}{2}\right)^2}$ and the required result

follows.

(iii) Using the standard half-angle results for $\sin \theta$ and $\cos \theta$

the LHS is reduced to $\dfrac{1 - \tan \theta}{1 + \tan \theta}$, and since $\tan \dfrac{\pi}{4} = 1$ this

$\dfrac{\tan \dfrac{\pi}{4} - \tan \theta}{1 + \tan \dfrac{\pi}{4} \tan \theta}$

can be written as and the result follows.

Put $\theta = \dfrac{\pi}{4}$ in result (ii) to obtain

$\dfrac{1 - \sin \dfrac{\pi}{4}}{1 + \sin \dfrac{\pi}{4}} = \tan^2 \left(\dfrac{\pi}{4} - \dfrac{\pi}{8} \right) = \tan^2 \dfrac{\pi}{8} = \tan^2 22\frac{1}{2}°$

Evaluating the LHS and taking the square root leads to the required result.

---

1 (a) If $t = \tan \dfrac{\theta}{2}$, show that $\sin \theta = \dfrac{2t}{1 + t^2}$ and derive an expression for $\cos \theta$ in terms of t. Hence, or otherwise, solve the equation $3 \sin \theta + \cos \theta = 2$ for values of θ in the range $0° \leqslant \theta \leqslant 180°$.

(b) Prove the identity $\dfrac{\sin 4\theta + \sin 2\theta}{\cos 4\theta + \cos 2\theta} = \tan 3\theta$.   *(S)*

2 Prove the identity $\sec^2 A + \csc^2 A = 4 \csc^2 2A$. Find all the values of A between $0°$ and $360°$ such that $4 \csc^2 2A - \csc^2 A = 3$. For which range of values of the constant k has the equation $4 \csc^2 2A - \csc^2 A = k$ no solutions?   *(OLE)*

3 (a) Prove the identity $\tan A + \cot A \equiv 2 \csc 2A$.
  (b) It is given that $\tan B = \frac{4}{3}$ and that B is acute. Without using tables or a calculator, find the value of (i) $\cos 2B$,
  (ii) $\tan \dfrac{B}{2}$.   *(C)*

4 Find all possible values of x from $0°$ to $360°$ when:
  (i) $\sin^2 x = 0 \cdot 75$; (ii) $5 \sin x = 3 \cos x$;
  (iii) $\sec 2x = 2$; (iv) $3 \sin^2 x + 2 \cos x = 2$.   *(O & C)*

5 Given that α and β are acute angles so that $\sin \alpha = \dfrac{1}{\sqrt{10}}$ and $\sin \beta = \dfrac{1}{\sqrt{5}}$, prove, without using tables, slide rule or calculator, that $\sin (\alpha + \beta) = \dfrac{1}{\sqrt{2}}$.   *(L)*

6 Find the values of x between $0°$ and $360°$ which satisfy:
  (i) $\cos (3x - 75°) = 0 \cdot 5$; (ii) $2 - \sin x = \cos^2 x + 7 \sin^2 x$.   *(C)*

7 Show that $f(x) = 2 \sin x° + 6 \cos^2 \dfrac{x°}{2}$ can be written in the form $a + r \cos (x - \alpha)°$ where $0 \leqslant a < 360$ and $r > 0$ and state the values of a, r and α. What is the maximum value of $f(x)$?   *(H)*

8 Express $\sin x - 2 \cos x$ in the form $R \sin (x - \alpha)$, where R is positive and α is acute. Hence, or otherwise:
  (i) find the set of possible values of $\sin x - 2 \cos x$;
  (ii) solve the equation $\sin x - 2 \cos x = 1$ for $0 \leqslant x < 360°$.   *(A)*

9 Find all values of θ between $0°$ and $360°$ for which $2 \sin \theta + 8 \cos^2 \theta = 5$, giving your answers correct to the nearest $0 \cdot 1°$ where necessary.   *(C)*

10 Find all solutions in the interval $0° \leqslant \theta \leqslant 360°$ of the equations: (i) $\tan 3\theta + 1 = 0$; (ii) $2 \cos^2 \theta + 3 \sin \theta = 0$.   *(O & C)*

11 Solve the equation $\cos 3\theta + \cos \theta = 0$, giving those solutions that satisfy $0 < \theta < 360°$.   *(O & C)*

12 By expressing $\sin \theta$ and $\cos \theta$ in terms of t, where $t = \tan (\theta/2)$, in the equation $5 \sin \theta + 2 \cos \theta = 5$, form a quadratic equation in t and solve this equation to find θ, correct to the nearest $0 \cdot 1°$, in the range $40° < \theta < 50°$.   *(L)*

21 Plane Triangles

Standard notation, Solving triangles, Special triangles, Sine and cosine rules, Applications of the sine and cosine rules, Other formulae, Area of a triangle.

Standard notation	In a triangle ABC, the angles are A, B, C } side a is opposite the sides are a, b, c } angle A, etc	angles sides Figure 1 Figure 2

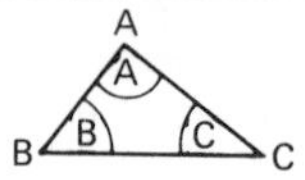

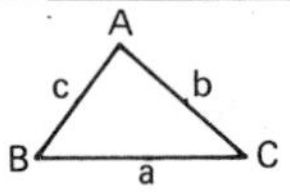

Solving triangles

Solving a triangle means finding all the unknown sides and angles in that triangle.

To solve a triangle:
(a) Sketch the triangle and mark in the given data.
(b) Use the appropriate formula(e).

Useful geometric facts about a triangle:
(a) The angle sum of a triangle is 180°, i.e. $A+B+C=180°$.
(b) The greatest side is opposite the greatest angle, the smallest side is opposite the smallest angle.

Special triangles

Right angled and **isosceles** triangles can be solved using **Pythagoras' Theorem** and/or the basic trigonometrical ratios.

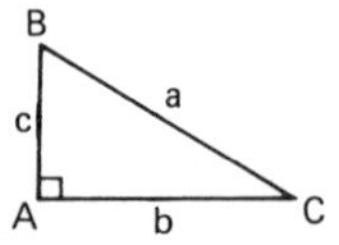

Figure 3

$a^2=b^2+c^2$ Pythagoras' Theorem

$\sin B=\dfrac{b}{a}$, $\cos B=\dfrac{c}{a}$, $\tan B=\dfrac{b}{c}$, etc.

Sine and cosine rules

Triangles without right angles can be solved using the sine and/or cosine rules.

Sine rule: $\dfrac{a}{\sin A}=\dfrac{b}{\sin B}=\dfrac{c}{\sin C}=2R$ where R is the radius of the circumcircle of the triangle.

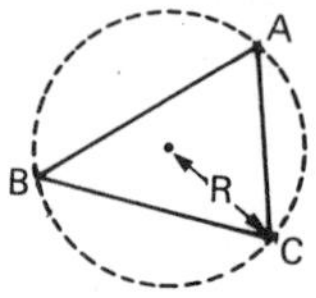

Figure 4

Cosine rule: $a^2=b^2+c^2-2bc\cos A$ $b^2=a^2+c^2-2ac\cos B$ $c^2=a^2+b^2-2ab\cos C$

or $\cos A=\dfrac{b^2+c^2-a^2}{2bc}$ or $\cos B=\dfrac{a^2+c^2-b^2}{2ac}$ or $\cos C=\dfrac{a^2+b^2-c^2}{2ab}$

Applications of the sine and cosine rules

(a) **Given three sides**, find:
the largest angle by the cosine rule,
the second angle by the cosine or sine rule,
the third angle by the 'angle sum'.
(b) **Given two sides and the included angle**, find:
the third side by the cosine rule,
the smaller angle by the sine rule,
the third angle by the 'angle sum'.
(c) **Given two sides and a non-included angle** (the ambiguous case):
try to find an angle using the sine rule (this can give two, one or no possible solutions),
find the third angle by the 'angle sum',
the third side by the sine rule.
(d) **Given one side and two angles**, find:
the third angle by the 'angle sum'
the other sides by the sine rule.
(e) **Given two or three angles only:**
The sides cannot be found.
The sine rule gives the ratios between sides.

ⓘ *Solve the triangle with sides 6, 14, 16 units.*

The largest angle is opposite to the '16 unit' side.

$\cos A=\dfrac{14^2+6^2-16^2}{2\times14\times6}$

$=-0.1429\Rightarrow A=98.2°$

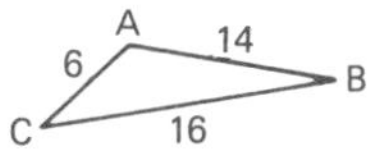

Figure 5

$\cos C=\dfrac{16^2+6^2-14^2}{2\times16\times6}=0.5\Rightarrow \hat{C}=60°$

So $\hat{B}=21.8°$ (angle sum)

ⓘ *Solve the triangle illustrated.*

Sine rule: $\dfrac{9}{\sin 50°}=\dfrac{10}{\sin A}$

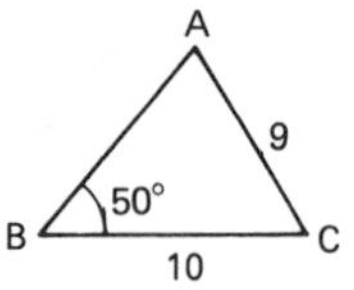

Figure 6

$\sin A=0.8512$, so $\hat{A}=58°20'$ or $121°40'$
So there are two possible triangles.
If $\hat{A}=58°20'$, $\hat{C}=71°40'$ (angle sum).
Sine rule gives $c≈11.15$ (2 d.p.)
If $\hat{A}=121°40'$, $\hat{C}=8°20'$ (angle sum)
Sine rule gives $c=1.70$ (2 d.p.)

Other formulae

Half angle formulae (used when three sides are known)

$\sin\tfrac{1}{2}A=\sqrt{\dfrac{(s-b)(s-c)}{bc}}$ $\cos\tfrac{1}{2}A=\sqrt{\dfrac{s(s-a)}{bc}}$

$\tan\tfrac{1}{2}A=\sqrt{\dfrac{(s-b)(s-c)}{s(s-a)}}$ where $s=\tfrac{1}{2}(a+b+c)$

Included angle formulae (used when two sides and an included angle are known)

$\tan\tfrac{1}{2}(B-C)=\left(\dfrac{b-c}{b+c}\right)\cot\tfrac{1}{2}A$

etc.

Area of a triangle

The **area of a triangle** is often denoted by $\triangle$.

$\triangle=\tfrac{1}{2}ab\sin C$ or $\tfrac{1}{2}ac\sin B$ or $\tfrac{1}{2}bc\sin A$

Hero's (or Heron's) formula

$\triangle=\sqrt{s(s-a)(s-b)(s-c)}$ where $s=\tfrac{1}{2}(a+b+c)$

Plane Triangles
Worked examples and Exam questions

 A triangle ABC has area 20 cm². Given that AC = 10 cm, BC = 6 cm and that ∠ACB is obtuse, calculate (i) ∠ACB, (ii) the length of AB.

(i) Area △ ABC = 20 cm².
With standard notation

$$\frac{1}{2}ab \sin C = 20$$

i.e. $\frac{1}{2} \times 10 \times 6 \times \sin C = 20$

giving $\sin C = \frac{2}{3}$

and ∠ACB = 138·2°

(ii) The cosine rule for AB gives,
$$AB^2 = 10^2 + 6^2 - 2 \times 10 \times 6 \times \cos 138·2°$$
$$= 100 + 36 + 120 \times 0·7455$$
$$= 225·46$$
∴ AB = 15·02 cm

CHECK   a = 6   s = 15·51
        b = 10   s − a = 9·51
        c = 15·02   s − b = 5·51
        2s = 31·02   s − c = 0·49
        s = 15·51 cm

$$\triangle = \sqrt{s(s-a)(s-b)(s-c)}$$
$$= \sqrt{15·51 \times 9·51 \times 5·51 \times 0·49}$$
$$= 19·96 \text{ cm}^2$$

 The perimeter of a triangle is 42 cm, one side is of length 14 cm and the area is 21√15 cm². Find the lengths of the other two sides and show that the cosine of the largest angle is ¼.

With the standard notation
2s = a + b + c = 42, i.e. s = 21
Let a = 14, then b + c = 28

Using $\triangle = \sqrt{s(s-a)(s-b)(s-c)}$
$$21\sqrt{15} = \sqrt{21(21-14)(21-b)(21-c)}$$
Squaring both sides and putting c = 28 − b gives
$$441 \times 15 = 21 \times 7 \times (21-b) \times (21-28+b)$$
$$45 = (21-b)(b-7)$$
i.e. $45 = -147 + 28b - b^2$
i.e. $b^2 - 28b + 192 = 0$
(b − 16)(b − 12) = 0, so b = 16 or 12

If b = 16 then c = 12 or if b = 12 then c = 16.
∴ the lengths of the other two sides are 12 cm and 16 cm.
Let a = 14, b = 16 and c = 12, when B̂ will be the largest angle,

and $\cos \hat{B} = \dfrac{a^2 + c^2 - b^2}{2ac} = \dfrac{196 + 144 - 256}{2 \times 14 \times 12}$

$$= \frac{84}{2 \times 14 \times 12} = \frac{1}{4}, \text{ as required.}$$

---

 1 In △ABC, BC = 8 cm, AC = 5cm and ∠ABC = 30°.
(a) Calculate the two possible values of ∠BAC, giving your answers in degrees to one decimal place.
(b) Draw a diagram to illustrate your answers.
(L)

2 In △ABC, AB = 12 cm, BC = 6√3 cm and AB̂C = 150°.
Calculate (i) AC, (ii) AĈB.*
(C)

3 (i) The sides of a triangle are 3, 7 and 8 units respectively. Prove that one of the angles is 60° and calculate the other two angles to the nearest degree. (ii) Solve the equation √3 cos θ − sin θ = 1 for 0° < θ < 360°.
(O & C)

4 In the triangle ABC, AB = 12 cm, BC = 10 cm and angle CAB = 45°. Find, to the nearest degree, the two possible values of angle BCA. Find also the corresponding lengths of the side AC.
(A)

5 In △ABC, BC = 12 cm, AB = 4 cm and angle C is acute with sin C = ⅛. Find, in radians, the two possible values of the angle A, leaving your answer in terms of π.
(L)

6 In triangle ABC, angle C = $\dfrac{\pi}{3}$.
(a) Prove that sin A = ½(√3 cos B + sin B).
(b) Given that b = 2a, where the usual notation for triangle ABC applies, find, by using the sine rule or otherwise, the size of angle B.
(H)

7 In any triangle ABC, prove, by using the sine rule or otherwise, that tan ½(B − C) = $\dfrac{b-c}{b+c}$ tan ½(B + C).

In a particular triangle the angle A is 51° and b = 3c. Find the angle B in degrees and minutes. The area of this triangle is 0·47 m². Find a to 3 significant figures.
(A)

8 In the triangle ABC, angle CAB = α°, D is a point on AB such that AD = 3DB, and angle ACD = angle BCD = 15°. Prove that cot α = 6 − √3.
(J)

9 A man walks due north. When he is at a point A he sees a pole on a bearing of 40°. After walking 200 m he is at the point B from which the bearing of the pole is 70°. Find, to the nearest metre, the distance of the pole from:
(i) the man's path, (ii) the mid-point of AB.
(A)

10 An isosceles triangle ABC, in which AB = AC and ∠A = 2θ, is inscribed in a circle of radius 5 cm. Prove that the two equal altitudes of the triangle have length 10 cos θ sin 2θ cm. If the sum of the lengths of the three altitudes is 10 cm, find the three angles of the triangle to the nearest degree.
(O & C)

11 ABC is a triangle, with sides of lengths a, b, c opposite A, B, C respectively. The point P is on the opposite side of BC to A, as shown in the diagram, and the triangle BCP is equilateral. Write down an expression for AP² in terms of a, c and the angle ABC. If the area of the triangle ABC is S, deduce that AP² = λ (a² + b² + c²) + μS, where λ and μ are numerical constants, and find λ and μ.
(J)

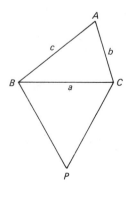

3-d problems

When solving problems in three dimensions:
(a) Draw a clear diagram and mark in the given information.
(b) Identify any right-angled triangles formed by horizontal and vertical lines since these triangles are the easiest to use.
(c) Pick out the relevant triangles, draw each one separately and mark in the given data.

Some three dimensional problems reduce simply to the solution of plane triangles (see first ⓘ).

Other problems, although truly three dimensional, still only depend on the solution of plane triangles (see second ⓘ).

Three dimensional problems often involve:
(a) **angles of elevation** or **depression**
An angle of elevation (or depression) is always measured upwards (or downwards) from the horizontal line of sight.

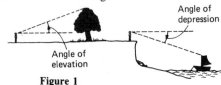

Figure 1

(b) directions given as:
(i) **compass directions**
(measured as acute angles from N or S to E or W)

Figure 2

(ii) **bearings**
(measured as clockwise angles from North and given as three-digit numbers)

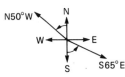

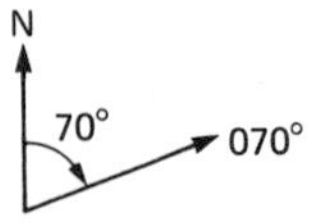

Figure 3

ⓘ *An observer notes that the angle of elevation of the top of a tower is $\alpha°$ from a point A, while at a point B, x metres nearer to the tower, the angle of elevation is $\beta°$. What is the height of the tower?*

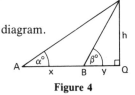

The situation is shown in the diagram. If PQ represents the tower of height h metres, then
in $\triangle APQ$: $h=(x+y)\tan\alpha$
in $\triangle BPQ$: $h=y\tan\beta$.

Figure 4

Eliminating y gives: $h=\left(x+\dfrac{h}{\tan\beta}\right).\tan\alpha$

i.e. $h=\dfrac{x\tan\beta\tan\alpha}{\tan\beta-\tan\alpha}$

ⓘ *AB is a vertical tower standing on a horizontal plane BXY. If $A\hat{X}B=\alpha°$, $A\hat{X}Y=\beta°$, $A\hat{Y}X=\gamma°$, and XY is x metres long, show that the height of AB is $x\sin\alpha\sin\gamma\operatorname{cosec}(\beta+\gamma)$.*

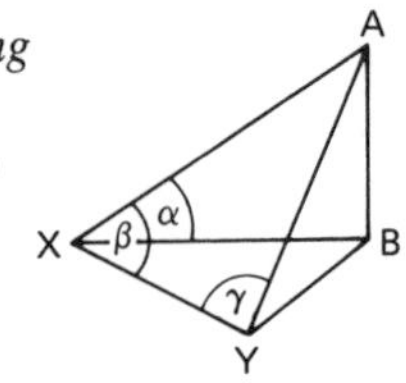

Let $AB=h$ metres

Figure 5

In $\triangle ABX$, $\dfrac{h}{XA}=\sin\alpha$

So $XA=\dfrac{h}{\sin\alpha}$

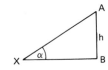

Figure 6

In $\triangle XAY$, $X\hat{A}Y=180°-(\beta+\gamma)$
Sine rule gives

$\dfrac{XA}{\sin\gamma}=\dfrac{x}{\sin(180°-(\beta+\gamma))}$

But $\sin(180°-(\beta+\gamma))=\sin(\beta+\gamma)$ and substituting for XA gives

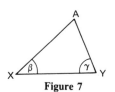

Figure 7

$\dfrac{h}{\sin\alpha\sin\gamma}=\dfrac{x}{\sin(\beta+\gamma)}\Rightarrow h=x\sin\alpha\sin\gamma\operatorname{cosec}(\beta+\gamma)$

Problems involving planes

The solution of some three dimensional problems require the use of one or more of the following facts about angles and planes.

A line perpendicular to a plane is perpendicular to every line in that plane.

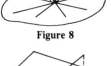

Figure 8

The angle between a line and a plane is the angle between the line and its projection in the plane.

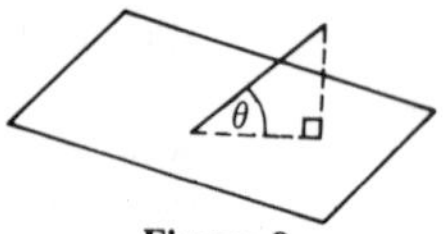

Figure 9

The angle between two planes is the angle between two lines, one in each plane, both perpendicular to the line common to the two planes.

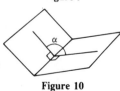

Figure 10

The line of greatest slope in a plane is a line perpendicular to the line of intersection of the plane and the horizontal plane.

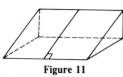

Figure 11

ⓘ *Find (a) the angle between an edge and a face, and (b) the angle between two faces of a regular tetrahedron.*

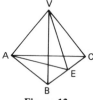

Figure 12

Each face angle is 60°.
E is the mid-point of BC and so AE lies on the projection of VA onto ABC. VE and AE are both perpendicular to BC.
So $V\hat{A}E$ is the angle for (a), and $V\hat{E}A$ is the angle for (b).

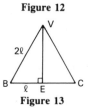

Figure 13

If $2l$ is the side of the tetrahedron, then $BE=l$ and $VE=\sqrt{3}l$ (Pythagoras') $AE=VE$ (altitudes of congruent triangles).

Figure 14

Cosine rule in $\triangle VAE$ gives:

$\cos E=\dfrac{3l^2+3l^2-4l^2}{2.\sqrt{3}l.\sqrt{3}l}=\dfrac{1}{3}$

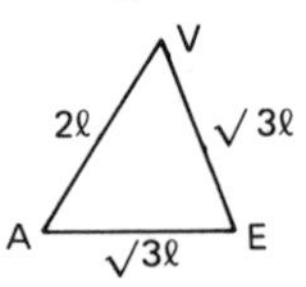

$\therefore \hat{E}=70°32'$
and $\hat{A}=\hat{V}=54°44'$

$\therefore$ the angle between an edge and a face is 54°44' and the angle between two faces is 70°32'

3-d Figures
Worked example and Exam questions

WE *ABCD is a horizontal rectangle with AB = 4 cm, AD = 3 cm. PA is a vertical line of length 9 cm. Calculate (i) the angle between PC and the plane ABCD, (ii) the angle between the planes PBD and ABCD.*

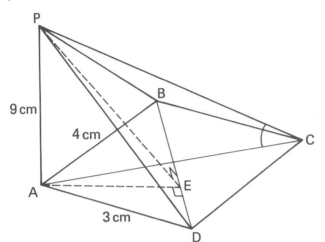

(i) ∠PCA is the required angle.
Pythagoras' Theorem in △ADC gives
AC = 5 cm.

In △PAC, $\tan \angle PCA = \dfrac{PA}{AC} = \dfrac{9}{5} = 1·8$

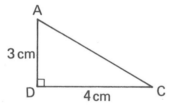

∴ ∠PCA = 60·9°

(ii) E is on BD such that PE and AE are both perpendicular to
BD. In △ABE, $AE = AB \sin \angle ABE = 4 \times \dfrac{3}{5} = \dfrac{12}{5}$ cm.

Pythagoras' Theorem in △PAD gives PD = √90 cm
Pythagoras' Theorem in △PAB gives PB = √97 cm

In △PDB $\cos \angle PBD = \dfrac{97 + 25 - 90}{2 \times 5 \times \sqrt{97}} = \dfrac{16}{5\sqrt{97}}$

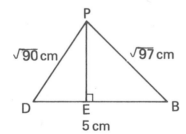

i.e. ∠PBD = 71·0°
∴ PE = PB sin ∠PBD = √97 × sin 71·0°
     = 9·85 cm

In △PEA, $\sin \angle PEA = \dfrac{9}{9·85}$, giving ∠PEA = 66·0°

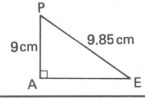

EX 1 An eagle, 200 metres above ground level, is observed from two points *x* metres apart at ground level. From one point, which is due South of the eagle, the angle of elevation is 30° and from the other point, which is due East of the eagle, the angle of elevation is 40°. Calculate the value of *x*.
*(C)

2 *PA* is a straight horizontal path; *ABC* is a straight path uphill so that *PABC* is a vertical plane; *PAB* = 150°, *AB* = 50 m and the angle of depression of *P* from *B* is 10°. What must be the length of *BC* (to the nearest metre) if the angle of depression of *P* from *C* is 16°?   *(O & C)

3 A vertical tower *AB* of height 40 metres is observed from two points *C* and *D* in the same horizontal plane as *B*, the foot of the tower. The points *B*, *C* and *D* lie in a straight line and *BC* = *CD*. Given that the angle of elevation of *A* from *D* is 60°, calculate:
 (i) the distance of *C* from the foot of the tower; and
 (ii) the angle of elevation of *A* from *C*.   *(W)

4 To find the height of a pylon a surveyor sets up his theodolite some distance from the pylon and finds the angle of elevation of the top of the pylon to be 30°. He then moves 50 m nearer to the pylon and finds the angle of elevation to be 45°. Find the height of the pylon given that the ground is horizontal and that the instrument is 1·5 m above ground level.   *(W)

5 A vertical mast, *OM*, of height 80 m, is built in a horizontal field. The angles of elevation of the top, *M*, of the mast from two points in the field, *A* and *B*, are 20° and 30°, respectively. *A* is due South of the mast, whilst *B* is on a bearing of N50°E (050°) from the mast. Calculate the distance *OA*, *OB* and *AB*. Find the area of the triangle *AOB*.   *(S)

6 *ABCD* is a tetrahedron in which ∠*BAC* = ∠*CAD* = ∠*DAB* = 60°, *AB* = *AD* = *BD* = 9 cm, *BC* = 10 cm. Calculate ∠*ACB*, the length of the perpendicular from *B* to *AC*, and the angle between the planes *ABC* and *ADC*.   *(O & C)

7 A pyramid *VABC* has *VA* = *VB* = *VC*, where *V* is the vertex, and its base forms an equilateral triangle *ABC* of side 2*a*. The height of the pyramid is also 2*a*. Find, leaving your answers in surd form if desired:
 (i) the area of the triangle *VAB*;
 (ii) the volume of the pyramid;
 (iii) the perpendicular distance from *C* to the face *VAB*;
 (iv) the cosine of the angle between the faces *VAB* and *VAC*.   (C)

8 The trapezium *ABCD* lies in a horizontal plane, with *AD* parallel to *BC*, *AB* = *p*, *BC* = *q*, *CD* = *r*, *DA* = *s*, angle *ABC* = ½π, and *q* > *s*. The points *E*, *F* are vertically below *C*, *D* respectively, and *A*, *B*, *E*, *F* lie on a plane making an angle α with the horizontal. Prove that *ABEF* is a trapezium, and find its area. Find also the inclination of *EF* to the horizontal.   (OLE)

9 A plane is inclined at angle α to the horizontal and a line *PQ* on the plane makes an acute angle β with *PR* which is a line of greatest slope on the plane. Show that the inclination θ of *PQ* to the horizontal is given by sin θ = sin α cos β. Show that the angle φ between the vertical plane through *PQ* and the vertical plane through *PR* is given by cos φ cos θ = cos α cos β.   (A)

10 Three points *O*, *A*, *B* lie on level ground. *OA* = *OB* = 5 m and *AB* = 6 m. A pole *OT* of length 6 m has its base at *O* and its other end *T* is held above the ground by two ropes *AT* and *BT* each of length 8 m. Find the angle that *OT* makes with the horizontal.   (A)

45

23 Vectors

Representation, Definitions, Addition and subtraction, Multiplication by a scalar, Position vectors, Ratio theorem.

Representation A **vector** has **magnitude** and **direction**.

In print a vector is denoted by bold type e.g. **a**, or by two capital letters and an arrow, e.g. $\vec{AB}$.

In **2-dimensions**, the vector **a** can be represented by

$$\mathbf{a} = \begin{pmatrix} x \\ y \end{pmatrix} \text{ or } \mathbf{a} = (x\mathbf{i} + y\mathbf{j})$$

where $\mathbf{i} = \begin{pmatrix} 1 \\ 0 \end{pmatrix}$ and $\mathbf{j} = \begin{pmatrix} 0 \\ 1 \end{pmatrix}$ are called base vectors.

In **3-dimensions**, $\mathbf{a} = \begin{pmatrix} x \\ y \\ z \end{pmatrix}$ or $\mathbf{a} = (x\mathbf{i} + y\mathbf{j} + z\mathbf{k})$

i
$$\mathbf{a} = \vec{AB} = \begin{pmatrix} 3 \\ 4 \\ 2 \end{pmatrix}$$

or $(3\mathbf{i} + 4\mathbf{j} + 2\mathbf{k})$

Base vectors in 3-dimensions:

$$\mathbf{i} = \begin{pmatrix} 1 \\ 0 \\ 0 \end{pmatrix}, \mathbf{j} = \begin{pmatrix} 0 \\ 1 \\ 0 \end{pmatrix}, \mathbf{k} = \begin{pmatrix} 0 \\ 0 \\ 1 \end{pmatrix}$$

Definitions The **magnitude** of **a**, $|\mathbf{a}|$, is $\sqrt{(x^2+y^2)}$ in $2-d$ and $\sqrt{(x^2+y^2+z^2)}$ in $3-d$.

A **unit vector** has magnitude 1. $\hat{\mathbf{a}}$ is the unit vector in the direction of **a**.

The **zero vector, 0,** is any vector with zero magnitude.

The inverse of **a** is $-\mathbf{a}$

Two vectors $x\mathbf{i} + y\mathbf{j} + z\mathbf{k}$ and $a\mathbf{i} + b\mathbf{j} + c\mathbf{k}$ are **equal**, if and only if $x = a$, $y = b$ and $z = c$.

i *If* $\mathbf{a} = 5\mathbf{i} - s\mathbf{j} - 2\mathbf{k}$ *and* $\mathbf{b} = t\mathbf{i} + 2\mathbf{j} - u\mathbf{k}$ *are equal vectors, find* (a) s, t *and* u, (b) $|\mathbf{a}|$.

(a) Since $\mathbf{a} = \mathbf{b}$, then $5 = t$, $-s = 2$ and $-2 = -u$
$$\Rightarrow t = 5, s = -2 \text{ and } u = 2$$

(b) $\mathbf{a} = 5\mathbf{i} + 2\mathbf{j} - 2\mathbf{k}$

$$|\mathbf{a}| = \sqrt{[5^2 + 2^2 - (-2)^2]} = \sqrt{33}$$

Addition and subtraction The **triangle law** is used to add and subtract vectors.

Addition:
$$\mathbf{a} + \mathbf{b} = \mathbf{c}$$
Addition is commutative,
i.e. $\mathbf{a} + \mathbf{b} = \mathbf{b} + \mathbf{a}$
and associative,
i.e. $(\mathbf{a} + \mathbf{b}) + \mathbf{c} = \mathbf{a} + (\mathbf{b} + \mathbf{c})$

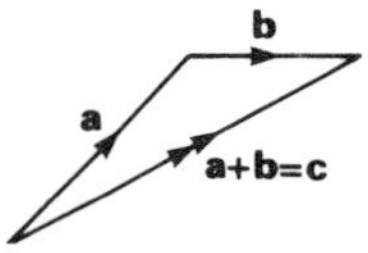

Figure 1

Subtraction:
$$\mathbf{a} - \mathbf{b} = \mathbf{a} + (-\mathbf{b})$$

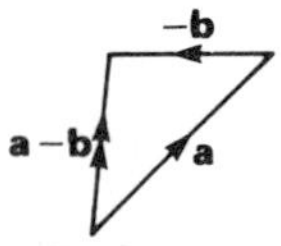

Figure 2

i *Given* $\mathbf{a} = \begin{pmatrix} 2 \\ -1 \\ 3 \end{pmatrix}$ *and* $\mathbf{b} = \begin{pmatrix} -1 \\ 5 \\ -3 \end{pmatrix}$, *find*

(a) $\mathbf{a} + \mathbf{b}$ (b) $\mathbf{a} - \mathbf{b}$.

(a) $\mathbf{a} + \mathbf{b} = \begin{pmatrix} 2 \\ -1 \\ 3 \end{pmatrix} + \begin{pmatrix} -1 \\ 5 \\ -3 \end{pmatrix} = \begin{pmatrix} 1 \\ 4 \\ 0 \end{pmatrix}$

(b) $\mathbf{a} - \mathbf{b} = \begin{pmatrix} 2 \\ -1 \\ 3 \end{pmatrix} - \begin{pmatrix} -1 \\ 5 \\ -3 \end{pmatrix} = \begin{pmatrix} 3 \\ -6 \\ 6 \end{pmatrix}$

Multiplication by a scalar A **scalar** is a real number, it has only magnitude.
If k is a scalar, then $k\mathbf{a}$ is a vector parallel to **a** but with k times the magnitude.
If $k > 0$, then $k\mathbf{a}$ is in the same direction as **a**.
If $k < 0$, then $k\mathbf{a}$ is in the opposite direction to **a**.

Multiplication by a scalar is distributive over vector addition, i.e. $k(\mathbf{a} + \mathbf{b}) = k\mathbf{a} + k\mathbf{b}$.

i *Solve the vector equation* $s\begin{pmatrix} -2 \\ 1 \end{pmatrix} + t\begin{pmatrix} 1 \\ 1 \end{pmatrix} = \begin{pmatrix} -5 \\ 1 \end{pmatrix}$.

$$s\begin{pmatrix} -2 \\ 1 \end{pmatrix} + t\begin{pmatrix} 1 \\ 1 \end{pmatrix} = \begin{pmatrix} -5 \\ 1 \end{pmatrix}$$

$$\Rightarrow \left. \begin{array}{r} -2s + t = -5 \\ s + t = 1 \end{array} \right\} \Rightarrow s = 2, t = -1.$$

Position vectors The **position of a point** $P(x, y)$ in the plane can be given by the vector

$$\vec{OP} = \mathbf{r} = \begin{pmatrix} x \\ y \end{pmatrix} \text{ or } (x\mathbf{i} + y\mathbf{j}).$$

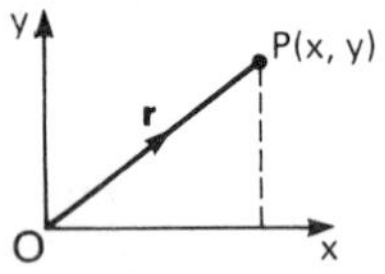

Figure 3

In 3-dimensions, $\mathbf{r} = \begin{pmatrix} x \\ y \\ z \end{pmatrix}$ or $(x\mathbf{i} + y\mathbf{j} + z\mathbf{k})$.

i The 3-dimensional position vector $\vec{OQ}$ can be written as

$$\vec{OQ} = \mathbf{q} = \begin{pmatrix} 2 \\ 5 \\ 3 \end{pmatrix}$$

or $(2\mathbf{i} + 5\mathbf{j} + 3\mathbf{k})$.

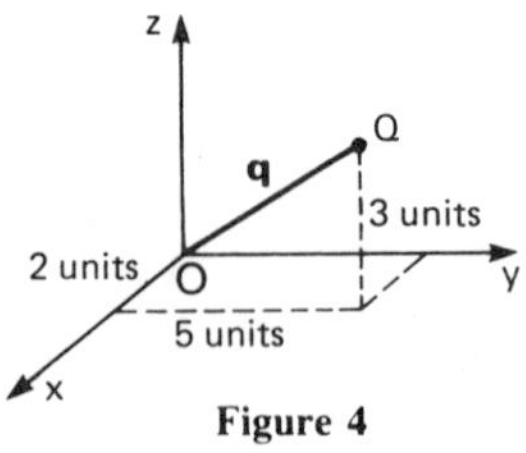

Figure 4

Ratio theorem If C divides AB **internally** in the ratio $\lambda : \mu$, then

$$\mathbf{c} = \frac{\lambda \mathbf{b} + \mu \mathbf{a}}{\lambda + \mu}.$$

If the division is **external**, then $\mathbf{c} = \frac{\lambda \mathbf{b} - \mu \mathbf{a}}{\lambda - \mu}$.

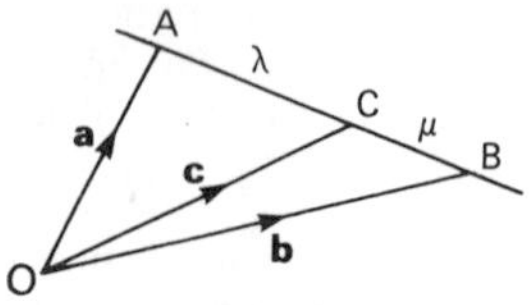

Figure 5

If $\mathbf{a} = (2\mathbf{i} + 3\mathbf{j})$ *and* $\mathbf{b} = (8\mathbf{i} + 9\mathbf{j})$ *are the position vectors of A and B, find the position vector,* $\mathbf{c}$, *of C which divides AB internally in the ratio* 1:2.

$$\mathbf{c} = \frac{1(8\mathbf{i} + 9\mathbf{j}) + 2(2\mathbf{i} + 3\mathbf{j})}{1 + 2} = \frac{12\mathbf{i} + 15\mathbf{j}}{3} = 4\mathbf{i} + 5\mathbf{j}.$$

Vectors
Worked example, Guided example and Exam questions

WE *In the diagram, ST = 2TQ, $\overrightarrow{PQ}=a$, $\overrightarrow{SR}=2a$ and $\overrightarrow{SP}=b$.*

*(a) Find in terms of **a** and **b**:*

(i) $\overrightarrow{SQ}$

(ii) $\overrightarrow{TQ}$

(iii) $\overrightarrow{RQ}$

(iv) $\overrightarrow{PT}$

(v) $\overrightarrow{TR}$

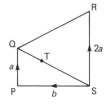

(b) What do your answers to (iv) and (v) tell you about the the points P, T, R?

(a) (i) $\overrightarrow{SQ} = \overrightarrow{SP} + \overrightarrow{PQ} = b+a$ (or $a+b$ by commutativity)

(ii) $\overrightarrow{TQ} = \frac{1}{3}\overrightarrow{SQ} = \frac{1}{3}(a+b)$

(iii) $\overrightarrow{RQ} = \overrightarrow{RS} + \overrightarrow{SQ} = -2a+(a+b)$
$= b-a$

(iv) $\overrightarrow{PT} = \overrightarrow{PS} + \overrightarrow{ST}$

$= -b+\frac{2}{3}\overrightarrow{SQ}$ (since $\overrightarrow{ST} = \frac{2}{3}\overrightarrow{SQ}$, i.e. $\overrightarrow{ST} = 2\overrightarrow{TQ}$)

$= -b+\frac{2}{3}(a+b) = \frac{2}{3}a-\frac{1}{3}b$

$= \frac{1}{3}(2a-b)$

(v) $\overrightarrow{TR} = \overrightarrow{TS} + \overrightarrow{SR}$

$= -\frac{2}{3}(a+b)+2a$

$= \frac{4}{3}a-\frac{2}{3}b = \frac{2}{3}(2a-b).$

(b) Since $\overrightarrow{PT} = \frac{1}{3}(2a-b)$ and $\overrightarrow{TR} = \frac{2}{3}(2a-b)$, $\overrightarrow{PT}$ and $\overrightarrow{TR}$ are both multiples of the same vector ($2a-b$). Hence PT and TR are parallel and T is common to both lines, so, P, T, R lie on the same line, i.e. they are co-linear.

GE *(a) $\overrightarrow{OS}$ and $\overrightarrow{OT}$ represent the vectors $\lambda i+\mu j$ and $\mu i+\lambda j$ where λ and μ are scalars and i and j are unit vectors in two mutually perpendicular directions Ox and Oy. Show that $|\overrightarrow{OS}|=|\overrightarrow{OT}|$. Given that OS and OT are two adjacent sides of a rhombus OSUT, find the vectors represented by the diagonals OU and ST.*

(b) $\overrightarrow{PQ}$ and $\overrightarrow{PR}$ are represented by the sides PQ and PR of the triangle PQR, Show that

$$\overrightarrow{PQ} + \overrightarrow{PR} = 2\overrightarrow{PS}$$

where S is the midpoint of QR.
Hence, or otherwise, find the position of the point O within the triangle PQR such that

$$\overrightarrow{OP} + \overrightarrow{OQ} + \overrightarrow{OR} = 0$$

(a) Use the definition of the magnitude of a vector to show that OS = OT.

Sketch the rhombus OSUT. Use, $\overrightarrow{OU} = \overrightarrow{OS} + \overrightarrow{SU}$ and $\overrightarrow{ST} = \overrightarrow{SO} + \overrightarrow{OT}$ to find the required vectors.

(b) Use the ratio theorem to express $\overrightarrow{PS}$ in terms of $\overrightarrow{PQ}$ and $\overrightarrow{PR}$. Hence, required result.

Consider a point O on RT (where T is the midpoint of PQ). Write down $\overrightarrow{OP} + \overrightarrow{OQ}$ using the result just established. Hence show that

$$\overrightarrow{OP} + \overrightarrow{OQ} + \overrightarrow{OR} = 0,$$

where O is the point which divides RT in a certain ratio. State what this point O is called.

EX

1 The vector **p** has magnitude 7 units and bearing 052°, and the vector **q** has magnitude 12 units and bearing 163°. Draw a diagram (which need not be to scale) showing **p**, **q** and the resultant **p+q**. Calculate, correct to one decimal place, the magnitude of **p+q**.

*(L)

2 From an origin O the points A, B, C have position vectors **a**, **b**, 2**b** respectively. The points O, A, B are not collinear. The midpoint of AB is M, and the point of trisection of AC nearer to A is T. Draw a diagram to show O, A, B, C, M, T. Find, in terms **a** and **b**, the position vectors of M and T. Use your results to prove that O, M, T are collinear, and find the ratio in which M divides OT.

*(L)

3 Given that **OA = a, OB = b, OP = $\frac{4}{5}$OA** and that Q is the midpoint of AB, express **AB** and **PQ** in terms of **a** and **b**. PQ is produced to meet OB produced at R, so that **QR = nPQ** and **BR = kb**. Express **QR**: (i) in terms of n, **a** and **b**; (ii) in terms of k, **a** and **b**. Hence find the value of n and of k.

*(C)

4 The position vectors of three points A, B and C relative to an origin O are **p**, 3**q−p**, and 9**q−5p** respectively. Show that the points A, B and C lie on the same straight line, and state the ratio AB:BC. Given that OBCD is a parallelogram and that E is the point such that **DB = $\frac{1}{3}$DE**, find the position vectors of D and E relative to O.

*(C)

5 The points A, B and C have position vectors **a, b** and **c** respectively referred to an origin O.
(a) Given that the point X lies on AB produced so that AB:BX=2:1, find **x**, the position vector of X, in terms of **a** and **b**.
(b) If Y lies on BC, between B and C so that BY:YC=1:3, find **y**, the position vector of Y, in terms of **b** and **c**.
(c) Given that Z is the mid-point of AC, show that X, Y and Z are collinear.
(d) Calculate XY:YZ.

(L)

6 O, A and B are three non-collinear points; the position vectors of A and B with respect to O are **a** and **b** respectively. M is the mid-point of OB, T is the point of trisection of AB nearer B, AMTX is a parallelogram and OX cuts AB at Y. Find, in terms of **a** and **b**, the position vectors of:
    (a) M;   (b) T;   (c) X;   (d) Y.
(O & C)

7 The vertices A, B and C of a triangle have position vectors **a, b** and **c** respectively relative to an origin O. The point P is on BC such that BP:PC=3:1; the point Q is on CA such that CQ:QA=2:3; the point R is on BA produced such that BR:AR=2:1. The position vectors of P, Q and R are **p, q** and **r** respectively. Show that **q** can be expressed in terms of **p** and **r** and hence or otherwise show that P, Q and R are collinear. State the ratio of the lengths of the line segments PQ and QR.

(J)

8 The points P and Q have position vectors **p** and **q** respectively relative to an origin O, which does not lie on PQ. Three points R, S, T have respective position vectors **r = $\frac{1}{4}$p+$\frac{3}{4}$q, s = 2p−q, t = p+3q**. Show in one diagram the positions of O, P, Q, R, S and T.

(J)

24 Vectors and Geometry

Equation of a straight line, Pairs of lines, Scalar (or dot) product, Applications of the scalar product, Equation of a plane.

Equation of a straight line

The equation of a straight line **parallel to vector b through a point** with position vector **a** is

$$\mathbf{r} = \mathbf{a} + t\mathbf{b}$$

where t is a real parameter.
r is the position vector of any point on the line.

The equation of a straight line **through two fixed points** with position vectors **a** and **b** is

$$\mathbf{r} = \mathbf{a} + t(\mathbf{b} - \mathbf{a})$$

where t is a real parameter.
Note: the equation is not unique.

ⓘ The equation of the straight line parallel to the vector $\mathbf{i} + 2\mathbf{j} - 5\mathbf{k}$ through the point with position vector $2\mathbf{i} - 3\mathbf{j} + \mathbf{k}$ is

$$\mathbf{r} = (2\mathbf{i} - 3\mathbf{j} + \mathbf{k}) + t(\mathbf{i} + 2\mathbf{j} - 5\mathbf{k}).$$

ⓘ The equation of the straight line through two points with position vectors $2\mathbf{i} + \mathbf{j} - \mathbf{k}$ and $3\mathbf{i} - \mathbf{j} - 3\mathbf{k}$ is

$$\mathbf{r} = (2\mathbf{i} + \mathbf{j} - \mathbf{k}) + t[(3\mathbf{i} - \mathbf{j} - 3\mathbf{k}) - (2\mathbf{i} + \mathbf{j} - \mathbf{k})]$$

i.e. $\mathbf{r} = (2\mathbf{i} + \mathbf{j} - \mathbf{k}) + t(\mathbf{i} - 2\mathbf{j} - 2\mathbf{k}).$

Pairs of lines

Two lines in space, $\mathbf{r}_1 = \mathbf{a}_1 + t\mathbf{b}_1$ and $\mathbf{r}_2 = \mathbf{a}_2 + s\mathbf{b}_2$, **intersect** if:
(a) $\mathbf{r}_1 = \mathbf{r}_2$, i.e. $\mathbf{a}_1 + t\mathbf{b}_1 = \mathbf{a}_2 + s\mathbf{b}_2$, and
(b) unique values for t and s can be found.

If unique values of t and s cannot be found, then the lines do not intersect and they are said to be **skew**.

ⓘ *Find the point of intersection of*
$\mathbf{r}_1 = (2\mathbf{i} + 3\mathbf{j}) + t(-\mathbf{i} + 2\mathbf{j})$ *and* $\mathbf{r}_2 = (-\mathbf{i} + \mathbf{j}) + s(3\mathbf{i} - 2\mathbf{j}).$

The lines intersect where $\mathbf{r}_1 = \mathbf{r}_2$,
i.e. $(2\mathbf{i} + 3\mathbf{j}) + t(-\mathbf{i} + 2\mathbf{j}) = (-\mathbf{i} + \mathbf{j}) + s(3\mathbf{i} - 2\mathbf{j})$
$\Rightarrow 2 - t = -1 + 3s$, i.e. $t + 3s = 3$
and $3 + 2t = 1 - 2s$, i.e. $t + s = -1$
giving $t = -3$ and $s = 2$.
So, the lines meet at $(2\mathbf{i} + 3\mathbf{j}) - 3(-\mathbf{i} + 2\mathbf{j}) = 5\mathbf{i} - 3\mathbf{j}$.
So, the point of intersection is $(5, -3)$.

Scalar (or dot) product

The **scalar (or dot) product** of vectors **a** and **b** is

$$\mathbf{a} \cdot \mathbf{b} = |\mathbf{a}||\mathbf{b}| \cos \theta$$

where θ is the angle between **a** and **b**.

If **a** and **b** are **parallel**, i.e. $\theta = 0$, $\mathbf{a} \cdot \mathbf{b} = |\mathbf{a}||\mathbf{b}|$.

If **a** and **b** are **perpendicular**, i.e. $\theta = \dfrac{\pi}{2}$, $\mathbf{a} \cdot \mathbf{b} = 0$.

If $\mathbf{a} = x_1\mathbf{i} + y_1\mathbf{j} + z_1\mathbf{k}$ and $\mathbf{b} = x_2\mathbf{i} + y_2\mathbf{j} + z_2\mathbf{k}$
$\mathbf{a} \cdot \mathbf{b} = (x_1\mathbf{i} + y_1\mathbf{j} + z_1\mathbf{k}) \cdot (x_2\mathbf{i} + y_2\mathbf{j} + z_2\mathbf{k})$
$= x_1x_2 + y_1y_2 + z_1z_2$
since $\mathbf{i} \cdot \mathbf{i} = \mathbf{j} \cdot \mathbf{j} = \mathbf{k} \cdot \mathbf{k} = 1$
and $\mathbf{i} \cdot \mathbf{j} = \mathbf{j} \cdot \mathbf{k} = \mathbf{k} \cdot \mathbf{i} = 0$.

ⓘ *Show that the vector $3\mathbf{i} + 2\mathbf{j} - \mathbf{k}$ is at right angles to the straight line* $\mathbf{r} = (\mathbf{i} + 7\mathbf{j} + 2\mathbf{k}) + s(2\mathbf{i} - 5\mathbf{j} - 4\mathbf{k}).$

The direction of the straight line is parallel to the direction of the vector $(2\mathbf{i} - 5\mathbf{j} - 4\mathbf{k})$.
If the vector $3\mathbf{i} + 2\mathbf{j} - \mathbf{k}$ is at right angles to the straight line, the scalar product of the vectors $2\mathbf{i} - 5\mathbf{j} - 4\mathbf{k}$ and $3\mathbf{i} + 2\mathbf{j} - \mathbf{k}$ will be zero.

$(2\mathbf{i} - 5\mathbf{j} - 4\mathbf{k}) \cdot (3\mathbf{i} + 2\mathbf{j} - \mathbf{k}) =$
$\qquad (3)(2) + (-5)(2) + (-4)(-1) = 0$
Hence, the given vector and the straight line are perpendicular.

Applications of the scalar product

The **angle θ between vectors** **a** and **b** is given by

$$\cos \theta = \frac{\mathbf{a} \cdot \mathbf{b}}{|\mathbf{a}||\mathbf{b}|}$$

The **projection of a on b**, i.e. the resolved part of **a** in the direction of **b**, is

$$OP = |\mathbf{a}| \cos \theta$$
$$= \frac{\mathbf{a} \cdot \mathbf{b}}{|\mathbf{b}|}$$

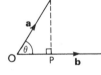

Figure 1

ⓘ *Find, for the vectors $\mathbf{a} = \mathbf{i} + 2\mathbf{j}$ and $\mathbf{b} = 3\mathbf{i} + \mathbf{j}$,*
*(a) the acute angle θ between **a** and **b**,*
*(b) the resolved part of **a** in the direction of **b**.*

(a) $\cos \theta = \dfrac{(\mathbf{i} + 2\mathbf{j}) \cdot (3\mathbf{i} + \mathbf{j})}{\sqrt{(1^2 + 2^2)} \cdot \sqrt{(3^2 + 1^2)}} = \dfrac{(1)(3) + (2)(1)}{\sqrt{5} \cdot \sqrt{10}} = \dfrac{1}{\sqrt{2}}.$

So, $\theta = 45°$.

(b) The resolved part of **a** in the direction of **b** is

$$\frac{\mathbf{a} \cdot \mathbf{b}}{|\mathbf{b}|} = \frac{3 + 2}{\sqrt{(3^2 + 1^2)}} = \frac{5}{\sqrt{10}} = \frac{\sqrt{10}}{2}.$$

Equation of a plane

The equation of a plane **through a point** with position vector **a**, **perpendicular to a vector n**, is

$$(\mathbf{r} - \mathbf{a}) \cdot \mathbf{n} = 0$$
i.e. $\mathbf{r} \cdot \mathbf{n} = \mathbf{a} \cdot \mathbf{n}$

The equation of a plane **through any three points** with position vectors **a**, **b** and **c** is

$$\mathbf{r} \cdot (d\mathbf{i} + e\mathbf{j} + f\mathbf{k}) = 1$$

where d, e and f are found by
(a) substituting **a**, **b** and **c** in turn for **r**,
(b) solving the three simultaneous equations formed.

ⓘ *Find the equation of the plane through the point with position vector $2\mathbf{i} - 3\mathbf{j} + \mathbf{k}$ and perpendicular to the vector $\mathbf{i} - \mathbf{j} - 2\mathbf{k}$.*

If **r** is the position vector of any point lying in the plane, then the equation of the plane is
$\mathbf{r} \cdot (\mathbf{i} - \mathbf{j} - 2\mathbf{k}) = (2\mathbf{i} - 3\mathbf{j} + \mathbf{k}) \cdot (\mathbf{i} - \mathbf{j} - 2\mathbf{k})$
$\qquad = (2.1) + (-3. -1) + (1. -2)$
$\qquad = 2 + 3 - 2$
i.e. $\mathbf{r} \cdot (\mathbf{i} - \mathbf{j} - 2\mathbf{k}) = 3$

Vectors and Geometry
Worked example, Guided example and Exam questions

 Relative to an origin the points A, B, C, D and E shown in the diagram have position vectors $i+11j$, $2i+8j$, $-i+7j$, $-2i+8j$ and $-4i+6j$ respectively. The lines AB and DC intersect at F.

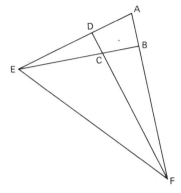

Calculate:
(i) the vector equations of the lines AB and DC.
(ii) the position vector of the point F.
Show that FD is perpendicular to EA and hence find
(iii) the position vector of the centre of the circle through E, D and F.

(i) First find $\overrightarrow{AB}$ and $\overrightarrow{DC}$.

$\overrightarrow{AB} = (2i+8j)-(i+11j) = i-3j$,

$\overrightarrow{DC} = (-i+7j)-(-2i+8j) = i-j$
Equation of line AB is $r = i+11j+s(i-3j)$ and, equation of line DC is $r = -2i+8j+t(i-j)$.

(ii) F is the point of intersection of AB and DC.
AB and DC intersect where
$$i+11j+s(i-3j) = -2i+8j+t(i-j)$$
i.e. $(3+s-t)i+(3-3s+t)j = 0$
and $3-3s+t = 0$,
$\Rightarrow 3+s-t = 0$, and $3-3s+t = 0$,
giving $s = 3$ and $t = 6$.

So, the lines intersect at the point F with position vector $r = i+11j+3(i-3j)$
i.e. $r = 4i+2j$.

$\overrightarrow{FD} = (-2i+8j)-(4i+2j) = -6i+6j$
$\overrightarrow{EA} = (i+11j)-(-4i+6j) = 5i+5j$
$\overrightarrow{FD}.\overrightarrow{EA} = (-6i+6j).(5i+5j) = -30+30 = 0$
Hence, FD and EA are perpendicular.

(iii) Since FD is perpendicular to ED, the circle passing through E, D and F must have the midpoint of EF as centre. (Angle in a semicircle is a right angle). Hence, position vector of the centre is
$$\frac{(-4i+6j)+(4i+2j)}{2} = 4j.$$

GE *Let $a = i-2j+k$, $b = 2i+j-k$. Given that $c = \lambda a+\mu b$ and that c is perpendicular to a, find the ratio of λ to μ.*

Let A, B be the points with position vectors a, b respectively with respect to an origin O. Write down in terms of a and b, a vector equation of the line l through A, in the plane of O, A and B, which is perpendicular to OA.

Find the position vector of P, the point of intersection of l and OB. JMB 9.10 section 9 question sheet

Since a and c are perpendicular, calculate $a.c = 0$.
Hence find ratio of λ to μ.
Since c is perpendicular to a and coplanar with a and b the equation of l can be written $r = a+kc$.
Write kc as $t(\alpha i+\beta j+\gamma k)$ using $c = \lambda a+\mu b$ and a and b as given and the ratio λ to μ as found. Hence, equation of l is found as $r = a+t(\alpha i+\beta j+\gamma k)$ with α, β and γ determined from your working. l meets OB where $r = b$.

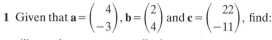 **1** Given that $\mathbf{a} = \begin{pmatrix} 4 \\ -3 \end{pmatrix}$, $\mathbf{b} = \begin{pmatrix} 2 \\ 4 \end{pmatrix}$ and $\mathbf{c} = \begin{pmatrix} 22 \\ -11 \end{pmatrix}$, find:

(i) a unit vector perpendicular to $\mathbf{a}$;
(ii) the value of the constants m and n for which $m\mathbf{a}+n\mathbf{b} = \mathbf{c}$.

Evaluate the scalar product of $\mathbf{a}$ and $\mathbf{b}$ and hence find the cosine of the angle between the direction of $\mathbf{a}$ and the direction of $\mathbf{b}$. *(C)*

2 Find the position vector of the point of intersection of the two lines
$$\mathbf{r} = \begin{pmatrix} 3 \\ 4 \end{pmatrix}+s\begin{pmatrix} 1 \\ -2 \end{pmatrix}$$
and
$$\mathbf{r} = \begin{pmatrix} 1 \\ -1 \end{pmatrix}+t\begin{pmatrix} 2 \\ 2 \end{pmatrix}.$$

Show that this position vector is perpendicular to the first line, and find its projection on the second line. Write down in vector form the equations of the lines parallel to the given lines and passing through $(0, 0)$, and find the exact value of the cosine of the angle between these two lines. *(O & C)*

3 A, B and C are the points $(0,5,5)$, $(4,1,1)$ and $(2\frac{1}{2}, 2\frac{1}{2}, 2\frac{1}{2})$ respectively.
(a) Prove that A, B and C are collinear and find the ratio in which C divides AB.
(b) O is the origin. Prove that OC bisects angle AOB.
(c) Given that $\mathbf{p}, \mathbf{q}$ and $\mathbf{r}$ are position vectors of P, Q and R relative to the origin O and that OQ bisects angle POR prove that
$$\frac{\mathbf{p}\cdot\mathbf{q}}{\mathbf{r}\cdot\mathbf{q}} = \frac{|\mathbf{p}|}{|\mathbf{r}|} \qquad (H)$$

4 Find the angle between the lines with vector equations $\mathbf{r} = \mathbf{a}+t\mathbf{c}$ and $\mathbf{r} = \mathbf{b}+s\mathbf{d}$, where
$$\mathbf{a} = \begin{pmatrix} 1 \\ 2 \\ -1 \end{pmatrix} \quad \mathbf{b} = \begin{pmatrix} 5 \\ 5 \\ -4 \end{pmatrix} \quad \mathbf{c} = \begin{pmatrix} 0 \\ 1 \\ -1 \end{pmatrix} \quad \mathbf{d} = \begin{pmatrix} 2 \\ 1 \\ -1 \end{pmatrix}.$$

Show that these lines intersect, and find the position vector of P, the point of intersection. Express in the vector form $\mathbf{r}.\mathbf{n} = k$ the equation of the plane which passes through the point P, and which is perpendicular to the line joining the two points with position vectors $\mathbf{a}$ and $\mathbf{b}$. *(OLE)*

5 The lines L_1 and L_2 are given by the equations
$$L_1:\mathbf{r} = \begin{pmatrix} 1 \\ 6 \\ 3 \end{pmatrix}+t\begin{pmatrix} 2 \\ -1 \\ 1 \end{pmatrix}, \qquad L_2:\mathbf{r} = \begin{pmatrix} 3 \\ 3 \\ 8 \end{pmatrix}+s\begin{pmatrix} 1 \\ 0 \\ 1 \end{pmatrix}.$$

(i) Calculate the angle between the directions of L_1, L_2;
(ii) Show that the lines do not intersect;
(iii) Verify that the vector $\mathbf{a} = \begin{pmatrix} 1 \\ 1 \\ -1 \end{pmatrix}$ is perpendicular to each of the lines.

The point P on L_1 is given by $t = p$; the point Q on L_2 is given by $s = q$. Write down the column vector representing $\overrightarrow{PQ}$. Hence calculate p and q so that the vectors $\overrightarrow{PQ}$ and $\mathbf{a}$ are parallel. *(J)*

6 The point A has position vector $i+4j-3k$ referred to the origin O. The line L has vector equation $\mathbf{r} = ti$. The plane Π contains the line L and the point A. Find:
(a) a vector which is normal to the plane Π;
(b) a vector equation for the plane Π;
(c) the cosine of the acute angle between OA and the line L. *(L)*

49

25 Complex Numbers

Definitions, Operations, Operations with the conjugate, Roots of equations.

Definitions

A **complex number**, z, is a number of the form
$$z = x + iy$$
where x and y are **real** numbers and $i = \sqrt{-1}$.
x is called the **real part** of z; y the **imaginary part**.

Since $i = \sqrt{-1}$
$$i^2 = -1,\ i^3 = -i,\ i^4 = 1, \ldots$$

The **modulus** of z is $|z| = \sqrt{x^2 + y^2}$.

The **argument** of z is $\arg(z) = \tan^{-1}\left(\dfrac{y}{x}\right)$

where $-\pi < \arg(z) \leq \pi$.
The conjugate of z, denoted by z^* or $\bar{z}$, is $x - iy$.
$z_1 = a + ib$ and $z_2 = c + id$ are **equal** if and only if $a = c$ and $b = d$, i.e. if the real parts are equal and the imaginary parts are equal.

$z = x + iy$ is zero if and only if $x = 0$ and $y = 0$.

[i] *For the complex number* $z = \dfrac{\sqrt{3}}{2} + \dfrac{1i}{2}$, *find*:

(a) $|z|$, (b) $\arg z$, (c) z^*.

(a) $|z| = \sqrt{\left(\dfrac{\sqrt{3}}{2}\right)^2 + \left(\dfrac{1}{2}\right)^2} = 1$

(b) $\arg z = \tan^{-1}\left(\dfrac{1/2}{\sqrt{3}/2}\right) = \tan^{-1}\left(\dfrac{1}{\sqrt{3}}\right) = \dfrac{\pi}{6}$

(c) $z^* = \dfrac{\sqrt{3}}{2} - \dfrac{1}{2}i$

[i] *Find the real values of* x *and* y *if*
$$(x-1) + i(y-2) = 0.$$

If $(x-1) + i(y-2) = 0$
then $(x-1) = 0$ and $(y-2) = 0$.
So $x = 1$ and $y = 2$.

Operations

Let $z_1 = a + ib$ and $z_2 = c + id$.

Addition: $z_1 + z_2 = (a+ib) + (c+id)$
$$= (a+c) + i(b+d)$$

Subtraction: $z_1 - z_2 = (a+ib) - (c+id)$
$$= (a-c) + i(b-d)$$

Multiplication: $z_1 z_2 = (a+ib)(c+id)$
$$= ac + i^2 bd + iad + ibc$$
$$= (ac - bd) + i(ad + bc)$$

Division: $z_1 \div z_2 = \dfrac{(a+ib)}{(c+id)} = \dfrac{(a+ib)(c-id)}{(c+id)(c-id)}$
$$= \left(\dfrac{ac+bd}{c^2+d^2}\right) + i\left(\dfrac{bc-ad}{c^2+d^2}\right)$$

[i] *If* $p = -2 + 3i$ *and* $q = 1 + 2i$, *express as complex numbers in the form* $x + iy$,

(a) $p+q$, (b) $p-q$, (c) pq, (d) $p \div q$.

(a) $p + q = (-2+3i) + (1+2i) = -1 + 5i$.

(b) $p - q = (-2+3i) - (1+2i) = -3 + i$.

(c) $pq = (-2+3i)(1+2i)$
$$= -2 + 6i^2 + 3i - 4i$$
$$= -2 - 6 - i = -8 - i.$$

(d) $p \div q = \dfrac{(-2+3i)}{(1+2i)} = \dfrac{(-2+3i)(1-2i)}{(1+2i)(1-2i)}$
$$= \left(\dfrac{-2+6}{1+4}\right) + i\left(\dfrac{3--4}{1+4}\right) = \dfrac{4}{5} + \dfrac{7}{5}i$$

Operations with the conjugate

Addition: $z + z^* = (x+iy) + (x-iy) = 2x$

Subtraction: $z - z^* = (x+iy) - (x-iy) = 2iy$

Multiplication: $zz^* = (x+iy)(x-iy) = x^2 + y^2$

Division: $\dfrac{z}{z^*} = \dfrac{(x+iy)}{(x-iy)} = \dfrac{(x+iy)(x+iy)}{(x-iy)(x+iy)}$
$$= \left(\dfrac{x^2 - y^2}{x^2 + y^2}\right) + i\left(\dfrac{2xy}{x^2 + y^2}\right)$$

[i] *If* $z = 3 + 4i$, *evaluate*:
(a) $z + z^*$ (b) $z - z^*$ (c) zz^* (d) $z \div z^*$.

(a) $z + z^* = (3+4i) + (3-4i) = 6$.

(b) $z - z^* = (3+4i) - (3-4i) = 8i$.

(c) $zz^* = (3+4i)(3-4i) = 9 + 16 = 25$.

(d) $z \div z^* = \dfrac{(3+4i)(3+4i)}{(3-4i)(3+4i)} = \left(\dfrac{-7}{25}\right) + \left(\dfrac{24}{25}\right)i$

Roots of equations

If the complex number $p + iq$ is a root of a polynomial equation with real coefficients then its conjugate, $p - iq$, is also a root.

[i] *If* $(2+3i)$ *is a root of a quadratic equation with real coefficients, find the equation.*

Since $(2+3i)$ is a root, $(2-3i)$ is the other root.
The required equation is
$$[x - (2+3i)][x - (2-3i)] = 0$$
i.e. $x^2 - [(2+3i) + (2-3i)]x + (2+3i)(2-3i) = 0$
i.e. $\qquad x^2 - 4x + 13 = 0$

Complex Numbers
Worked example, Guided example and Exam questions

(a) *Given that $z_1=2-3i$ and $z_2=3+4i$ find*

(i) z_1z_2,

(ii) $\dfrac{z_1}{z_2}$, *in the form $p+iq$ where p and q are real.*

(b) *Given that $2+3i$ is a root of the equation*
$z^3-6z^2+21z-26=0$, *find the other two roots.*

(a)

(i) $\begin{aligned}
z_1z_2 &= (2-3i)(3+4i)\\
&= (6+12)+i(-9+8)\\
&= 18-i
\end{aligned}$

(ii) $\begin{aligned}
\dfrac{z_1}{z_2} &= \dfrac{2-3i}{3+4i}\\
&= \dfrac{(2-3i)(3-4i)}{(3+4i)(3-4i)}\\
&= \dfrac{(6-12)+i(-9-8)}{9+16}\\
&= -\dfrac{6}{25}-\dfrac{7}{25}i
\end{aligned}$

(b) $z^3-6z^2+21z-26=0$

We are given that $2+3i$ is a root of this equation. Since the coefficients of the equation are real, $2-3i$ is also a root.
Hence $z-(2+3i)$ and $z-(2-3i)$ are factors of the equation.
The product of these factors is
$$[z-(2+3i)][z-(2-3i)]$$
$$= z^2-4z+13$$
Dividing the L.H.S. of the original equation by $z^2-4z+13$ gives $z-2$.
Hence $z^3-6z^2+21z-26=0$ can be written as
$$(z-2)[z-(2+3i)][z-(2-3i)]=0,$$
giving the other two required roots as $z=2$ and $z=2-3i$.

(i) *Express the square roots of $-2i$ in the form $\pm(a+ib)$ where a and b are real numbers.*

(ii) *Solve the equation*
$$z^2-3(1+i)z+5i=0$$
giving your answers in the form $a+ib$. Hence or otherwise solve the equation
$$z^2-3(1-i)z-5i=0.$$

(i) Let $(a+ib)^2=-2i$. Work out $(a+ib)^2$. Equate real and imaginary parts. Find a and b. Hence roots are $\pm(a+ib)$.

(ii) Let $z=a+ib$. Work out L.H.S. Equate real and imaginary parts. Find a and b. Hence roots are $a+ib$. Second equation is obtained from first by replacing i by $-i$. Hence the roots are $a-ib$.

---

1 Express $(6+5i)(7+2i)$ in the form $a+ib$. Write down $(6-5i)(7-2i)$ in a similar form. Hence find the prime factors of 32^2+47^2. *(J)*

2 Expand $z=(1+ic)^6$ in powers of c and find the five real finite values of c for which z is real. *(J)*

3 If $(1+i)z-iw+i=iz+(1-i)w-3i=6$, find the complex numbers z, w, expressing each in the form $a+bi$ where a, b are real. *(O & C)*

4 (a) Express $\dfrac{-1+i\sqrt{3}}{-1-i\sqrt{3}}$ in the form $a+ib$, where a and b are real numbers.

(b) Find the quadratic equation whose roots are $-3+4i$ and $-3-4i$, expressing your answer in the form $x^2+px+q=0$, where p and q are real numbers. *(C)*

5 Given that $(a+ib)^2=i$, find the real values of a and b. Hence, or otherwise, find the solutions of the equation $z^2+2z+1-i=0$ in the form $z=p+iq$.

6 Let $z=x+iy$ be any non-zero complex number.

Express $\dfrac{1}{z}$ in the form $u+iv$.

Given that $z+\dfrac{1}{z}=k$ with k real, prove that either

$y=0$ or $x^2+y^2=1$. Show
(i) that if $y=0$ then $|k|\geqslant 2$,
(ii) that if $x^2+y^2=1$ then $|k|\leqslant 2$. *(J)*

7 (a) Given that $z=x+iy$, where x and y are real numbers, find z^2 in terms of x and y. Hence, or otherwise, find both square roots of i.
(b) One root of a quadratic equation with real coefficients is $(7-24i)/5$. State the other root of this equation, and find the equation in its simplest form.
(A)

8 The roots of the quadratic equation $z^2+pz+q=0$ are $1+i$ and $4+3i$. Find the complex numbers p and q. It is given that $1+i$ is also a root of the equation $z^2+(a+2i)z+5+ib=0$, where a and b are real. Determine the values of a and b. *(J)*

9 Obtain a quadratic function: $f(z)=z^2+az+b$, where a and b are real constants such that $f(-1-2i)=0$.
(L)

10 Express the complex number $\dfrac{1-3i}{1+3i}$ in modulus-argument form. Show that, as the real number t varies, the point representing $\dfrac{1-it}{1+it}$ in the Argand plane moves round a circle. Write down the radius on centre of the circle.

11 Given that $a=1+3i$ is a root of the equation $z^2-(p+2i)z+q(1+i)=0$, and that p and q are real, determine p, q and the other root of the equation.
(J)

12 Given that $(x+iy)^2=a+ib$, where x,y,a,b are real, prove that $4x^4-4ax^2-b^2=0$. Hence, or otherwise, find the values of $(5+12i)^{1/2}$. What are the values of $(5-12i)^{1/2}$? Solve the equation $z^2-(7+4i)z+(7+11i)=0$. State the roots of $z^2-(7-4i)z+(7-11i)=0$. [Give all your answers in the form $u+iv$ where u, v are real.]
(O & C)

13 In the quadratic equation $x^2+(p+iq)x+3i=0$, p and q are real. Given that the sum of the squares of the roots is 8, find all possible pairs of values of p and q.
(J)

14 Given that ω denotes either one of the non-real roots of the equation: $z^3=1$, show that: (i) $1+\omega+\omega^2=0$; and
(ii) the other non-real root is ω^2. Show that the non-real roots of the equation
$$\left(\dfrac{1-u}{u}\right)^3=1$$
can be expressed in the form $A\omega$ and $B\omega^2$, where A and B are real numbers, and find A and B.
(J)

26 Complex Numbers and Graphs

Argand diagram, Polar form, Multiplication and division in polar form, Geometric representation of operations, Loci.

Argand diagram

Any complex number $z = x + iy$ may be represented on an **Argand diagram** by

either (a) the point $P(x, y)$,

or (b) the position vector $\overrightarrow{OP}$.

The **modulus** of z, $|z|$, is the length of OP.
The **argument** of z, $\arg z$, is the angle θ between OP and the positive real axis, where $-\pi < \theta \leqslant \pi$.

imaginary axis

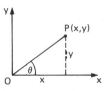

Figure 1   **real** axis

$$|z| = \sqrt{(x^2 + y^2)}$$

$$\arg z = \theta = \tan^{-1}\left(\frac{y}{x}\right)$$

Polar form (also called modulus-argument form)

The polar form of a complex number is
$z = r(\cos\theta + i\sin\theta)$,
where $r = OP$ and $\theta = x\hat{O}P$.

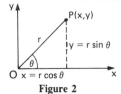

Figure 2

$|z| = r$, where $r \geqslant 0$.
$\arg z = \theta$, where $-\pi < \theta \leqslant \pi$.

$z^* = r(\cos\theta - i\sin\theta)$
$\quad = r(\cos(-\theta) + i\sin(-\theta))$

$|z^*| = r$ and $\arg z^* = -\theta$.

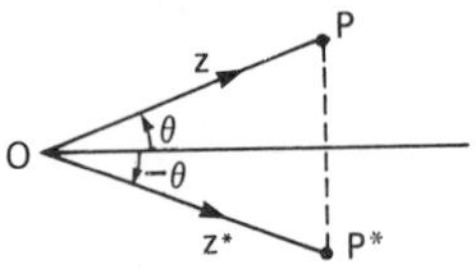

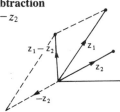

Figure 3

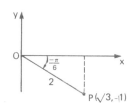
Express the complex number $\sqrt{3} - i$ in polar form and illustrate it on an Argand diagram.

Let $\sqrt{3} - i = r(\cos\theta + i\sin\theta)$.

$r = |z|$
$\quad = \sqrt{((\sqrt{3})^2 + (-1)^2)} = 2$
$\theta = \arg z$

$\quad = \tan^{-1}\left(\frac{-1}{\sqrt{3}}\right) = -\frac{\pi}{6}$

So, in polar form $\sqrt{3} - i$ is
$2(\cos(-\pi/6) + i\sin(-\pi/6))$

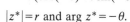

Figure 4

Multiplication and division in polar form

Let $z_1 = r_1(\cos\theta + i\sin\theta)$ and $z_2 = r_2(\cos\phi + i\sin\phi)$.

multiplication: $z_1 z_2 = r_1 r_2[\cos(\theta + \phi) + i\sin(\theta + \phi)]$

$|z_1 z_2| = |z_1||z_2|$ and $\arg(z_1 z_2) = \arg z_1 + \arg z_2$

division: $\dfrac{z_1}{z_2} = \dfrac{r_1}{r_2}[\cos(\theta - \phi) + i\sin(\theta - \phi)]$

$\left|\dfrac{z_1}{z_2}\right| = \dfrac{|z_1|}{|z_2|}$ and $\arg\left(\dfrac{z_1}{z_2}\right) = \arg z_1 - \arg z_2$

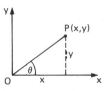
If $z_1 = 4(\cos \pi/3 + i\sin \pi/3)$ and $z_2 = 2(\cos \pi/6 + i\sin \pi/6)$, evaluate:
(a) $z_1 z_2$ and (b) $z_1 \div z_2$.

(a) $z_1 z_2 = 4(\cos \pi/3 + i\sin \pi/3) \times 2(\cos \pi/6 + i\sin \pi/6)$
$\quad = 8[\cos(\pi/3 + \pi/6) + i\sin(\pi/3 + \pi/6)]$
$\quad = 8(\cos \pi/2 + i\sin \pi/2)$

(b) $z_1 \div z_2 = \dfrac{4(\cos \pi/3 + i\sin \pi/3)}{2(\cos \pi/6 + i\sin \pi/6)}$
$\quad = 2[\cos(\pi/3 - \pi/6) + i\sin(\pi/3 - \pi/6)]$
$\quad = 2(\cos \pi/6 + i\sin \pi/6)$

Geometric representation of operations

addition $z_1 + z_2$

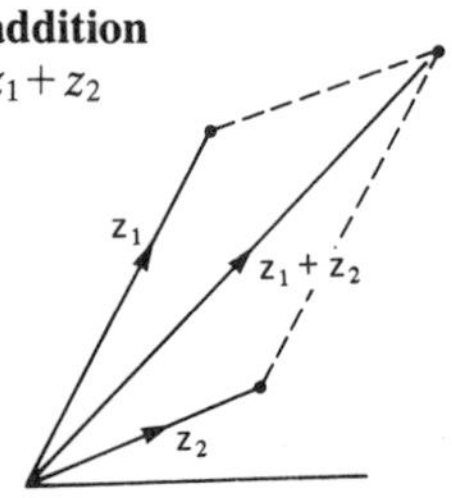

Figure 5

subtraction $z_1 - z_2$

Figure 6

multiplication $z_1 z_2$

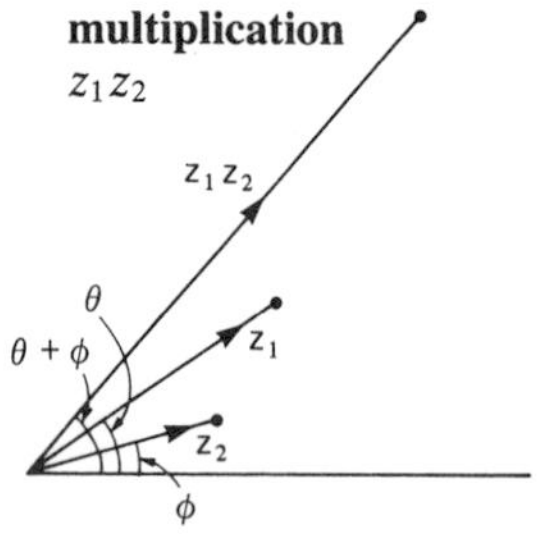

Figure 7

division $z_1 \div z_2$

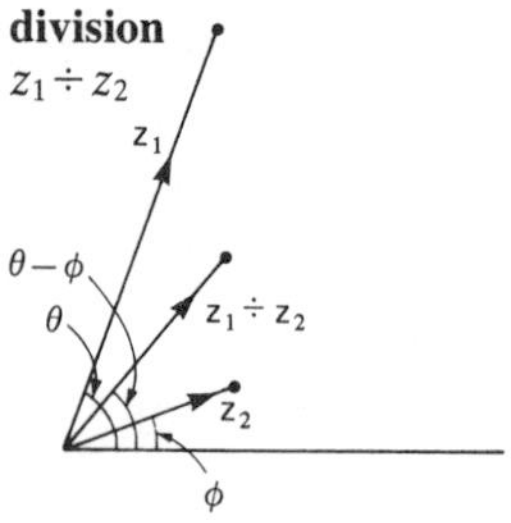

Figure 8

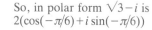

Loci

If z is a **variable complex number**, represented by the position vector $\overrightarrow{OZ}$, then the **locus** of Z under certain conditions can be sketched. Four common loci are illustrated below.

The locus of Z when
$|z| = a$
is a circle, centre O radius a.

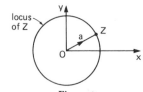

Figure 9

The locus of Z when
$|z - p| = a$,
where p is a fixed complex number, is a circle, centre P, radius a.

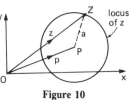

Figure 10

The locus of Z when
$\arg z = \alpha$, $(-\pi < \alpha \leqslant \pi)$
is a half line from 0, at an angle α with the real axis.

Figure 11

The locus of Z when
$\arg(z - p) = \arg q$,
where p and q are fixed complex numbers, is a half line from P, parallel to OQ.

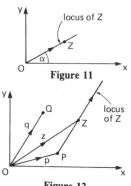

Figure 12

Complex Numbers and Graphs
Worked example, Guided example and Exam questions

 (a) Indicate on an Argand diagram the region in which z lies if $|z-2+3i| \leq 2$.

(b) If the real part of $\dfrac{z+1}{z-1}$ *is zero, show that the locus of the point representing z in the Argand plane is a circle and write down its centre and radius.*

(a) $|z-2+3i| \leq 2$ can be rewritten as $|z-(2-3i)| \leq 2$. This says that the distance between the fixed point $2-3i$ and the variable point z in the Argand plane must always be less than or equal to 2. i.e. the region in which z lies is the circular disc, centre $2-3i$ and radius 2, shown in the Argand diagram.

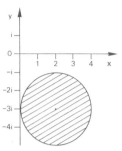

(b) Let $z = x+iy$,

then $\dfrac{z+1}{z-1} = \dfrac{x+iy+1}{x+iy-1}$

$= \dfrac{[(x+1)+iy][(x-1)-iy]}{[(x-1)+iy][(x-1)-iy]}$

$= \dfrac{(x^2-1+y^2)+i(-2y)}{(x-1)^2+y^2}$

If the real part of $\dfrac{z+1}{z-1} = 0$,

then $x^2-1+y^2 = 0$

$\Rightarrow \quad x^2+y^2 = 1$.

Hence the locus of z is a circle, centre $(0,0)$, radius 1.

Express the complex numbers $z = \sqrt{2}+i\sqrt{2}$ *and*

$w = -3+i3\sqrt{3}$ *in modulus-argument form and hence write down the modulus and argument of each of the following:*

(i) $\dfrac{1}{z}$ *(ii)* zw *(iii)* $\dfrac{z}{w}$

Show in an Argand diagram the points representing the complex numbers $\dfrac{1}{z}$, zw, $\dfrac{z}{w}$.

Let $z = r(\cos\theta + i\sin\theta) = \sqrt{2}+i\sqrt{2}$, and find r and θ. Hence z can be written in modulus-argument form. Do the same for $w = -3+i3\sqrt{3}$.

(i) $\dfrac{1}{z} = \dfrac{1}{r}(\cos\theta - i\sin\theta)$

(ii) To find zw, multiply the moduli and add the arguments.

(iii) To find $\dfrac{z}{w}$, divide the moduli and subtract the arguments.

Having written down (i), (ii) and (iii), the points representing these complex numbers can easily be shown in an Argand diagram.

---

 1 (a) The complex number $z_1 = 2i$. Find the values of a and b such that: $(a+ib)^2 = z_1$.
If these two resulting complex numbers are z_2 and z_3, express z_1, z_2 and z_3 in modulus-argument form and display all three on the same Argand diagram.
(b) The complex number $z_4 = \sqrt{3}+i$. Find $(z_4)^2$.
Express z_4 and $(z_4)^2$ in modulus-argument form and display them on the same Argand diagram. Deduce a further complex number z_5 such that: $(z_5)^2 = (z_4)^2$. *(S)

2 Express $\dfrac{1}{1+i\sqrt{3}}$ in the form $r(\cos\theta + i\sin\theta)$ where $r>0$ and $-\pi < \theta \leq \pi$.

(J)

3 Given that $z = \sqrt{3}+i$, find the modulus and argument of
(a) z^2, (b) $\dfrac{1}{z}$.
Show in an Argand diagram the points representing the complex numbers z, z^2 and $\dfrac{1}{z}$.

(L)

4 You are given that $z = \cos\theta + i\sin\theta$ $(0 < \theta < \frac{1}{2}\pi)$. Draw an Argand diagram to illustrate the relative positions of the points representing z, $z+1$, $z-1$. Hence, or otherwise,
(a) determine the modulus and argument of each of these three complex numbers;
(b) prove that the real part of $\dfrac{z-1}{z+1}$ is zero. *(O & C)*

5 Shade in an Argand diagram the region of the z-plane in which one or the other, but not both, of the following inequalities is satisfied: (i) $|z| \leq 1$, (ii) $|z-1-i| \leq 2$. Your diagram should show clearly which parts of the boundary are included.

(J)

6 (a) The points P and Q in an Argand diagram represent the complex numbers $7-i$ and $12+4i$ respectively, and O is the origin. Prove that the triangle OPQ is isosceles, and calculate to the nearest degree the size of angle OPQ.
(b) Identify the locus of the point representing the complex number z in each of the following cases:
(i) $|z-(1+i)| = 1$; (ii) $|z-1| = |z+i|$; (iii) $\arg(z-1) = \frac{1}{2}\pi$.

(C)

7 Sketch on the same Argand diagram the loci of z and w, where $|z| = |z-4i|$ and $|w+2| = 1$. State (i) the minimum value of $|z-w|$; (ii) the cartesian form of z for which $\arg z = \dfrac{\pi}{4}$.

(A)

8 If $z = 5+5i$ and $w = 7+i$, find (a) $z-w$, (b) $1/(z+w)$, expressing each in the form $x+iy$, where x and y are real. Verify that the real part of $(z-w)/(z+w)$ is zero. If, in the Argand diagram, P represents the position of $z-w$ and Q the position of $z+w$, what does the result imply about $P\hat{O}Q$? *(O & C)*

9 In an Argand diagram, the point P represents the complex number z, where $z = x+iy$. Given that $z+2 = \lambda i(z+8)$, where λ is a real parameter, find the Cartesian equation of the locus of P as λ varies.
If also $z = \mu(4+3i)$, where μ is real, prove that there is only one possible position for P. *(J)*

10 Find the modulus and argument of the complex number $\dfrac{1-3i}{1+3i}$. *Show that, as the real number t varies, the point representing* $\dfrac{1-it}{1+it}$ *in the Argand diagram moves round a circle, and write down the radius and centre of the circle.*

(OLE)

27 Differentiation
Notation, Standard results, Rules, Higher derivatives.

Notation

Differentiation is the process of finding the **derivative** of a function.
The derivative of a function is also called its **derived function** and also its **differential coefficient**.

The derivative of y with respect to x is usually written as $\dfrac{dy}{dx}$ or y' or $\dot{y}$.

The derivative of $f(x)$ with respect to x is usually written as $f'(x)$ or $\dfrac{d}{dx}[f(x)]$.

Standard results

	function	derivative
algebraic	constant	0
	x^n	nx^{n-1}
trigonometrical (x in radians)	$\sin x$	$\cos x$
	$\cos x$	$-\sin x$
	$\tan x$	$\sec^2 x$
	$\operatorname{cosec} x$	$-\operatorname{cosec} x \cot x$
	$\sec x$	$\sec x \tan x$
	$\cot x$	$-\operatorname{cosec}^2 x$

	function	derivative
inverse trigonometrical	$\sin^{-1} x$	$\dfrac{1}{\sqrt{(1-x^2)}}$
	$\cos^{-1} x$	$\dfrac{-1}{\sqrt{(1-x^2)}}$
	$\tan^{-1} x$	$\dfrac{1}{1+x^2}$
logarithmic and exponential	$\ln x$	$\dfrac{1}{x}$
	e^x	e^x

Rules

1. Sum
A sum of terms can be differentiated term by term.

If $y = x^3 + \cos x - \ln x + 4$
$$\frac{dy}{dx} = 3x^2 - \sin x - \frac{1}{x}$$

2. Product
(a) If $y = au$, where a is a constant and
$\quad\quad\quad\quad u$ is a function of x,

then $\dfrac{dy}{dx} = a\dfrac{du}{dx}$

If $y = 7x^4$
$$\frac{dy}{dx} = 7(4x^3)$$
$$= 28x^3$$

(b) If $y = uv$, where u and v are functions of x,

then $\dfrac{dy}{dx} = v.\dfrac{du}{dx} + u.\dfrac{dv}{dx}$

If $y = xe^x$

let $u = x \quad$ and $v = e^x$

then $\quad \dfrac{du}{dx} = 1 \quad$ and $\dfrac{dv}{dx} = e^x$

Using $\dfrac{dy}{dx} = v\dfrac{du}{dx} + u\dfrac{dv}{dx}$
$$= e^x.1 + x.e^x$$
$$= e^x(1+x)$$

3. Quotient
If $y = \dfrac{u}{v}$, where u and v are functions of x,

then $\dfrac{dy}{dx} = \dfrac{v.\dfrac{du}{dx} - u.\dfrac{dv}{dx}}{v^2}$

If $\quad y = \dfrac{\sin x}{x^2}$

let $\quad u = \sin x \quad$ and $\quad v = x^2$

then $\dfrac{du}{dx} = \cos x \quad$ and $\dfrac{dv}{dx} = 2x$

Using $\dfrac{dy}{dx} = \dfrac{v.\dfrac{du}{dx} - u.\dfrac{dv}{dx}}{v^2}$
$$= \frac{x^2.\cos x - \sin x.2x}{x^4}$$
$$= \frac{x.\cos x - 2\sin x}{x^3}$$

Higher derivatives

If the derivative of a function of x is differentiated with respect to x, the 2nd derivative of the function is obtained. If the 2nd derivative is differentiated, the 3rd derivative is obtained, and so on.
The 2nd, 3rd, . . . , nth derivatives of y with respect

to x are usually written as $\dfrac{d^2y}{dx^2}, \dfrac{d^3y}{dx^3}, \ldots, \dfrac{d^ny}{dx^n}$.

The usual function notation is $f'(x), f''(x), \ldots,$ $f^{(n)}(x)$.

If $y = x^6 + 4x^2 - \dfrac{3}{x}$
$$\frac{dy}{dx} = 6x^5 + 8x + \frac{3}{x^2}$$
$$\frac{d^2y}{dx^2} = 30x^4 + 8 - \frac{6}{x^3}$$
$$\frac{d^3y}{dx^3} = 120x^3 + \frac{18}{x^4}$$

Differentiation
Worked example, Guided example and Exam questions

 (a) *Differentiate with respect to x.*

(i) $4x^3 - 3x^2 + \dfrac{5}{x^2}$

(ii) $(2x^3 - 1)\sin x$

(iii) $\dfrac{7x-4}{\ln(\ln x)}$

(b) *Given that $y = x^4 - x^3 + 4x - 1$, find $\dfrac{d^2y}{dx^2}$ and state the set of values of x for which $\dfrac{d^2y}{dx^2}$ is zero.*

(a)

(i) Let $y = 4x^3 - 3x^2 + \dfrac{5}{x^2}$

i.e. $y = 4x^3 - 3x^2 + 5x^{-2}$.

Now $\dfrac{dy}{dx} = 4.\,3x - 3.\,2x + 5.\,(-2)x^{-3}$

$= 12x^2 - 6x - \dfrac{10}{x^3}$.

(ii) Let $y = (2x^3 - 1)\sin x$, a product,
so let $u = 2x^3 - 1$ and $v = \sin x$

Now $\dfrac{du}{dx} = 6x$ and $\dfrac{dv}{dx} = \cos x$

Using $\dfrac{d}{dx}(uv) = v\dfrac{du}{dx} + u\dfrac{dv}{dx}$,

$\dfrac{dy}{dx} = \sin x.\,6x^2 + (2x^3 - 1).\,\cos x$

$= 2x^3 \cos x + 6x^2 \sin x - \cos x$.

(iii) Let $y = \dfrac{7x - 4}{\ln x}$, a quotient,

so let $u = 7x - 4$ and $v = \ln x$.

Now $\dfrac{du}{dx} = 7$ and $\dfrac{dv}{dx} = \dfrac{1}{x}$.

Using $\dfrac{d}{dx}\left(\dfrac{u}{v}\right) = \dfrac{v\dfrac{du}{dx} - u\dfrac{dv}{dx}}{v^2}$

$= \dfrac{\ln x.\,7 - (7x - 4).\dfrac{1}{x}}{[\ln x]^2}$

$= \dfrac{7x(\ln x - 1) + 4}{x[\ln x]^2}$.

(b) $y = x^4 - x^3 + 4x - 1$

$\dfrac{dy}{dx} = 4x^3 - 3x^2 + 4$

$\dfrac{d^2y}{dx^2} = 12x^2 - 6x$

$= 6x(2x - 1)$.

$\dfrac{d^2y}{dx^2} = 0$ when $x = 0$ or $\tfrac{1}{2}$,

so, $\dfrac{d^2y}{dx^2}$ is zero for $\{x : x = 0 \text{ or } \tfrac{1}{2}\}$.

GE **(a)** *Differentiate with respect to x*

(i) $e^x \sin x$ (ii) $\dfrac{x^3}{1 + \sec x}$.

(b) *Given that $y = x^5 \ln x + \cos x$, find $\dfrac{dy}{dx}$.*

(a) (i) A product, so use product rule with $u = e^x$ and $v = \sin x$.
(ii) A quotient, so use quotient rule with $u = x^3$ and $v = 1 + \sec x$.
(b) $x^5 \ln x$ is a product so use product rule with $u = x^5$ and $v = \ln x$. Then differentiate $\cos x$. Add the derivatives together to get the final result.

EX **1** Differentiate with respect to x (a) $\dfrac{2}{x} + \dfrac{1}{\sqrt{x}}$; (b) $\dfrac{x^2 + 1}{x + 1}$.

*(S)

2 Differentiate with respect to x: (a) $(2x - 1)(3x + 2)$;

(b) $x^4 - 2x + \dfrac{1}{x^2}$; (c) $\dfrac{(2x - 3)}{\sqrt[4]{x}}$.

*(A)

3 Prove that $\dfrac{d}{dx}(\tan x - x) = \tan^2 x$.

*(L)

4 Differentiate with respect to x: (a) $x^{-\frac{3}{2}}$; (b) $2x^2(x + 1) + 2$;

(c) $\dfrac{3x^4 - x}{x^3}$.

*(O & C)

5 (a) Differentiate with respect to x
(i) $x^{\frac{1}{2}}(x + 1)$,
(ii) $\dfrac{(x^5 + x^2 - 2)}{x^4}$.

(b) If $x = t^3$ and $y = t^2$, express y in terms of x and hence, or otherwise, calculate the value of $\dfrac{dy}{dx}$ when $x = 8$.

*(A)

6 Given that: $y = \dfrac{\sin x - \cos x}{\sin x + \cos x}$, show that $\dfrac{dy}{dx} = 1 + y^2$.

Prove that $\dfrac{d^2y}{dx^2}$ is zero only when $y = 0$.

(J)

7 Differentiate with respect to x: (a) $x^3 \ln x$. (b) $\dfrac{1 + \cos x}{x}$.

(L)

8 Given that $y = \dfrac{x^2 - 1}{2x^2 + 1}$, find $\dfrac{dy}{dx}$ and state the set of values of x for which $\dfrac{dy}{dx}$ is positive.

Find the greatest and least values of y for $0 \le x \le 1$.

(L)

9 Given that $y = \dfrac{x^2 - 2x - 4}{x^2 - 4}$, find and simplify $\dfrac{dy}{dx}$. Find also the greatest and least values of y for $-1 \le x \le 1$.

(A)

10 (a) Given that $y = \dfrac{e^x - e^{-x}}{e^x + e^{-x}}$, show that $\dfrac{dy}{dx} = 1 - y^2$.

(b) The equation of a curve is $y = \cot x - 8\cos x$, $(0 < x < \pi)$. Find the co-ordinates of the points on the curve where $\dfrac{dy}{dx} = 0$. Sketch the curve.

(C)

28 Methods of Differentiation

Chain rule, Implicit differentiation, Parametric differentiation, Logarithmic differentiation.

Chain rule

If y is a function of u and u is a function of x, then y is called a **function of a function** of x.

This can be differentiated using the chain rule

$$\frac{dy}{dx} = \frac{dy}{du} \cdot \frac{du}{dx}$$

It is useful to remember that, by the **chain rule**,

$$\frac{d(y^2)}{dx} = 2y \cdot \frac{dy}{dx}$$

$$\frac{d(y^3)}{dx} = 3y^2 \cdot \frac{dy}{dx} \text{ and so on.}$$

ⓘ *Differentiate $y = (3x^4 - 5)^7$ with respect to x.*

y is a function (i.e. the 7th power) of $(3x^4 - 5)$, which is a function of x.
So y is a function of a function of x.

Let $y = u^7$ where $u = (3x^4 - 5)$

so $\dfrac{dy}{du} = 7u^6$ and $\dfrac{du}{dx} = 12x^3$

Using $\dfrac{dy}{dx} = \dfrac{dy}{du} \cdot \dfrac{du}{dx}$

$= 7u^6 \cdot 12x^3$

$= 84x^3(3x^4 - 5)^6$

Implicit differentiation

An **implicit function** in two variables, x and y say, is one in which neither variable can be easily expressed in terms of the other.

To differentiate such an implicit function
(a) Differentiate it term by term to give an equation in $\dfrac{dy}{dx}$, x and y.

(b) Make $\dfrac{dy}{dx}$ the subject of the equation.

To obtain the second derivative
(a) Differentiate the '$\dfrac{dy}{dx}$ equation' to obtain an equation for $\dfrac{d^2y}{dx^2}$.

(b) Substitute for $\dfrac{dy}{dx}$ if necessary.

Repeat the procedure for higher derivatives.

ⓘ *Find $\dfrac{dy}{dx}$ when $x^3 + y^3 = 3xy$*

$x^3 + y^3 = 3xy$ is an implicit function.

Differentiating term by term gives

$$3x^2 + 3y^2 \cdot \frac{dy}{dx} = 3y + \underbrace{3x \cdot \frac{dy}{dx}}_{\text{product rule}}$$

$$(y^2 - x)\frac{dy}{dx} = (y - x^2)$$

$$\frac{dy}{dx} = \frac{(y - x^2)}{(y^2 - x)}$$

Parametric differentiation

If x and y are each expressed in terms of a third variable, t say, called the **parameter**, then $x = f(t)$ and $y = g(t)$ give the **parametric form** of the equation relating x and y.

To differentiate such a parametric form of an equation
(a) Find $\dfrac{dx}{dt}$ and $\dfrac{dy}{dt}$ separately. Do not attempt to eliminate the parameter.

(b) Use $\dfrac{dy}{dx} = \dfrac{dy}{dt} \cdot \dfrac{dt}{dx}$.

To obtain the second derivative
(a) Find $\dfrac{d}{dt}\left(\dfrac{dy}{dx}\right)$.

(b) Use $\dfrac{d^2y}{dx^2} = \dfrac{d}{dx}\left(\dfrac{dy}{dx}\right) = \dfrac{d}{dt}\left(\dfrac{dy}{dx}\right) \cdot \dfrac{dt}{dx}$.

ⓘ *The parametric equation of an ellipse is $x = a \cos\theta$, $y = b \sin\theta$, where a and b are constants. Find $\dfrac{dy}{dx}$ as a function of θ.*

$x = a \cos\theta \qquad y = b \sin\theta$

$\dfrac{dx}{d\theta} = -a \sin\theta \qquad \dfrac{dy}{d\theta} = b \cos\theta$

Using $\dfrac{dy}{dx} = \dfrac{dy}{d\theta} \cdot \dfrac{d\theta}{dx}$

$= b \cos\theta \cdot \dfrac{1}{-a \sin\theta}$

$= -\dfrac{b}{a} \cot\theta$

Logarithmic differentiation

To differentiate a function in the form $y = [f(x)]^{g(x)}$
(a) Take logarithms of the given function.
(b) Differentiate the new function as usual.

This method is useful when differentiating complicated products and quotients.

ⓘ *Differentiate x^{2x} with respect to x.*

Let $y = x^{2x}$

so $\ln y = 2x \ln x$

$$\frac{1}{y}\frac{dy}{dx} = 2x \cdot \frac{1}{x} + 2\ln x$$

$$\frac{dy}{dx} = 2y(1 + \ln x)$$

$$= 2x^{2x}(1 + \ln x)$$

Methods of Differentiation
Worked example and Exam questions

 (a) *Differentiate with respect to x*
(i) $x^7 \sin 3x$

(ii) $\dfrac{\ln (5x)}{x^2}$.

(b) *Given that* $x^2 - 3xy + 2y^2 - 2x = 4$, *find the value of* $\dfrac{dy}{dx}$ *at the point* $(1, -1)$.

(c) *If* $x = \dfrac{2t}{t+2}$ *and* $y = \dfrac{3t}{t+3}$, *find the value of* $\dfrac{dy}{dx}$ *at the point* $\left(\dfrac{2}{3}, \dfrac{3}{4}\right)$.

(a)
(i) Let $y = x^7 \sin 3x$, a product,
so let $u = x^7$ and $v = \sin 3x$

Now $\dfrac{du}{dx} = 7x^6$ and $\dfrac{dv}{dx} = 3 \cos 3x$ (chain rule)

Using $\dfrac{d}{dx}(uv) = v\dfrac{du}{dx} + u\dfrac{dv}{dx}$,

$= \sin 3x.\, 7x^6 + x^7.\, 3 \cos 3x$
$= 3x^7 \cos 3x + 7x^6 \sin 3x$.

(ii) Let $y = \dfrac{\ln (5x)}{x^2}$, a quotient,

so let $u = \ln (5x)$ and $v = x^2$.

Now $\dfrac{du}{dx} = 5.\dfrac{1}{5x}$ (chain rule) and $\dfrac{dv}{dx} = 2x$.

Using $\dfrac{d}{dx}\left(\dfrac{u}{v}\right) = \dfrac{v\dfrac{du}{dx} - u\dfrac{dv}{dx}}{v^2}$

$\dfrac{dy}{dx} = \dfrac{x^2.\dfrac{1}{x} - \ln (5x).\, 2x}{x^4}$

$= \dfrac{1 - 2 \ln (5x)}{x^3}$

(b) $x^2 - 3xy + 2y^2 - 2x = 4$ is an implicit function.
Differentiating with respect to x gives

$2x - 3.\underbrace{\left(y + x\dfrac{dy}{dx}\right)}_{\substack{\text{product rule} \\ \text{and chain rule}}} + \underbrace{4y\dfrac{dy}{dx}}_{\text{chain rule}} - 2 = 0.$

Factorize to find $\dfrac{dy}{dx}$ explicitly,

i.e. $(4y - 3x)\dfrac{dy}{dx} = 2 - 2x + 3y$

$\Rightarrow \qquad \dfrac{dy}{dx} = \dfrac{2 - 2x + 3y}{4y - 3x}.$

At the point $(1, -1)$,

$\dfrac{dy}{dx} = \dfrac{2 - 2 \times 1 + 3\,(-1)}{4\,(-1) - 3}$

$= \dfrac{3}{7}.$

(c) x and y are expressed in parametric form.

$x = \dfrac{2t}{t+2},$

so, $\dfrac{dx}{dt} = \dfrac{(t+2).2 - 2t.1}{(t+2)^2} = \dfrac{4}{(t+2)^2}$ (using quotient rule).

$y = \dfrac{3t}{t+3},$

so, $\dfrac{dy}{dt} = \dfrac{(t+3).3 - 3t.1}{(t+3)^2} = \dfrac{9}{(t+3)^2}$ (using quotient rule).

Using $\dfrac{dy}{dx} = \dfrac{dy}{dt} \times \dfrac{dt}{dx}$, we have,

$\dfrac{dy}{dx} = \dfrac{9}{(t+3)^2} \times \dfrac{(t+2)^2}{4}.$

At the point $\left(\dfrac{2}{3}, \dfrac{3}{4}\right)$, $t = 1$,

so, $\dfrac{dy}{dx} = \dfrac{9}{(1+3)^2} \times \dfrac{(1+2)^2}{4}$

$= \dfrac{81}{64}.$

EX 1 Differentiate with respect to x: (i) $(2x^2 - 1)(x^3 + 4)^3$,

(ii) $\dfrac{x^4 + 1}{\sqrt{(1+x)}}$, (iii) $\left(x + \dfrac{1}{x}\right)^{-1}$,

(iv) $\sin \dfrac{x}{2} \cos^3 x$, (v) $\tan^4 2x$.

*(O & C)

2 (i) Differentiate the following with respect to x:
(a) $(x^2 + 1)^3$; (b) $\sin^4 3x$; (c) $x(2x + 1)^{\frac{1}{2}}$.

(ii) Given that $y = \dfrac{x \cos x + \sin x}{x^2}$, calculate $\dfrac{dy}{dx}$ and

simplify your answer as much as possible.

3 (a) Differentiate $f(x) = \dfrac{x^2 + 1}{\sqrt{x}}$ with respect to x.

(b) Given that $f(x) = \sin^4 x - \cos^4 x$, prove that
$f'(x) = 2\sin 2x$.

(H)

4 Differentiate the following expressions with respect to x,
giving your answers in as simple a form as possible:

(i) $x^2 \cos 3x$; (ii) $e^x \log_e x$; (iii) $(x^2 + 2)^3$; (iv) $\dfrac{3x - 1}{\sqrt{(x^2 + 1)}}$.

*(W)

5 Differentiate with respect to x: (i) $\tan^4 2x$; (ii) $\dfrac{x - 1}{2x - 3}$;
(iii) $x^2 \log_e x$.

(S)

6 Given that x and y are related by $x^3 + y^3 = 3xy$, find $\dfrac{dy}{dx}$ in

terms of x and y.

(L)

7 Given that $x = \theta - \sin \theta$, $y = 1 - \cos \theta$,

show that $\dfrac{dy}{dx} = \cot \dfrac{\theta}{2}$ and that $\dfrac{d^2y}{dx^2} + \dfrac{1}{y^2} = 0.$

(J)

8 The co-ordinates of a point on a curve are given
parametrically by $x = a(t - \sin t)$, $y = a(1 - \cos t)$. Find dy/dx
in terms of t. Deduce that the tangent at the point on the
curve where $t = T$ makes an angle $\frac{1}{2}T$ with the y-axis.
Hence, or otherwise, sketch the curve in the
neighbourhood of the origin.

(A)

29 Applications of Differentiation
Gradient, Tangent and normal, Velocity and acceleration.

Gradient, tangent and normal

The **gradient** of a curve at any point is the gradient of the **tangent** to the curve at that point.
Its value is given by the **derivative** at the point.
The **normal** to a curve at a point is perpendicular to the tangent at that point.

i Gradient of tangent $= \dfrac{dy}{dx}$.

Tangent and normal are perpendicular, so $m_1 m_2 = -1$.

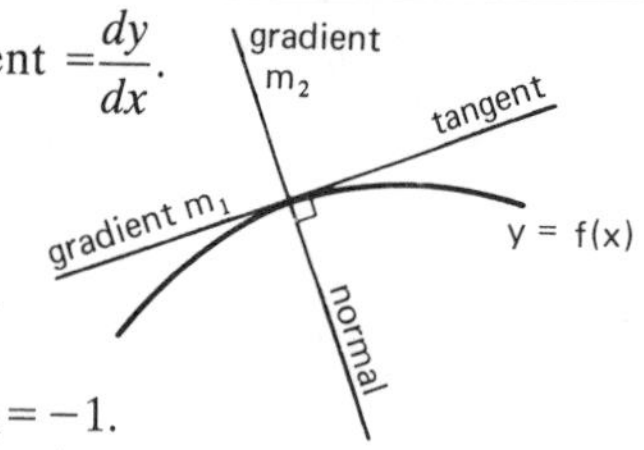

Figure 1

Consider a curve $y = f(x)$ and a point $P(x_1, y_1)$ on it.

1. To find the **gradient** m of the curve at $P(x_1, y_1)$
(a) Differentiate to find $\dfrac{dy}{dx}$.
(b) Substitute x_1 and y_1 into the derived function.

2. To find the equation of the **tangent** at $P(x_1, y_1)$
(a) Find the gradient m at $P(x_1, y_1)$.
(b) Substitute x_1, y_1 and m in the equation
$$y - y_1 = m(x - x_1).$$

3. To find the equation of the **normal** at $P(x_1, y_1)$
(a) Find the gradient m at $P(x_1, y_1)$.
(b) Since the normal is perpendicular to the tangent, the gradient of the normal is $-\dfrac{1}{m}$.
Substitute x_1, y_1 and m in the equation
$$y - y_1 = -\dfrac{1}{m}(x - x_1).$$

i *Find the gradient and the equations of the tangent and normal to $y = 5x^3 - 7x^2 + 3x + 2$ at $(1, 3)$.*

The gradient of the curve is given by
$$\frac{dy}{dx} = 15x^2 - 14x + 3$$

At $(1, 3)$, $m = 15(1)^2 - 14(1) + 3$
$= 4$

The equation of the tangent at $(1, 3)$ is
$$y - 3 = 4(x - 1)$$
i.e. $\quad y = 4x - 1$

The gradient of the tangent at $(1, 3)$ is 4, so the gradient of the normal at $(1, 3)$ is $-\frac{1}{4}$ since $(4)(-\frac{1}{4}) = -1$.

The equation of the normal at $(1, 3)$ is
$$y - 3 = -\tfrac{1}{4}(x - 1)$$
i.e. $\quad 4y + x = 13$

Velocity and acceleration

Consider a particle which is moving in a straight line such that its displacement from a fixed point is s after time t.

Velocity v is the **rate of change** of **displacement** s with respect to time t,

i.e. $\quad v = \dfrac{ds}{dt}$.

If $v = 0$, the particle is at **rest**.
if $v < 0$, the particle is moving in the **opposite direction** to that in which s is measured.

Acceleration a is the **rate of change** of **velocity** v with respect to time t,

i.e. $\quad a = \dfrac{dv}{dt}$.

But $\quad v = \dfrac{ds}{dt}$, so $\dfrac{dv}{dt} = \dfrac{d^2s}{dt^2}$.

Also $\dfrac{dv}{dt} = \dfrac{dv}{ds} \cdot \dfrac{ds}{dt} = \dfrac{dv}{ds} \cdot v = v \cdot \dfrac{dv}{ds}$.

So acceleration may be written as
$$a = \frac{dv}{dt} \text{ or } \frac{d^2s}{dt^2} \text{ or } v \cdot \frac{dv}{ds}.$$

If $a = 0$, the velocity of the particle is **constant**.
If $a > 0$, the particle is **accelerating**.
If $a < 0$, the particle is being **retarded**.

i *A particle moves in a straight line so that its distance from a fixed point O after t seconds is s metres where $s = \frac{1}{3}t^3 - \frac{3}{2}t^2 + 2t$. Show that the particle is at rest at two different times and find these times. Find the acceleration of the particle at these times and interpret the results.*

$$s = \tfrac{1}{3}t^3 - \tfrac{3}{2}t^2 + 2t$$

So velocity $v = \dfrac{ds}{dt} = t^2 - 3t + 2$

The particle is at rest when $v = 0$,
i.e. when $t^2 - 3t + 2 = 0$
or $(t - 1)(t - 2) = 0$
Therefore $t = 1$ second and $t = 2$ seconds.
So the particle is at rest at two different times.

Acceleration $a = \dfrac{dv}{dt} = \dfrac{d^2s}{dt^2} = 2t - 3$

When $t = 1$ second, $a = -1 \text{ m s}^{-2}$,
i.e. the particle is being retarded (slowing down).

When $t = 2$ seconds, $a = 1 \text{ m s}^{-2}$,
i.e. the particle is being accelerated (speeding up).

Applications of Differentiation
Worked examples, Guided example and Exam questions

 A particle P moves in a straight line such that its distance s m *from a fixed point O at time t* s, *where t ≥ 0, is given by* $s = 9t^2 - 2t^3$. *What is the velocity and acceleration of P when* $t = 3$ s? *Find also the distance of P from O when $t = 4$ s and show that it is then moving towards O.*

$s = 9t^2 - 2t^3$
The velocity of P at time t s is given by $\dfrac{ds}{dt} = 18t - 6t^2$.

At time $t = 3$ s, the velocity of P is $18 \times 3 - 6 \times 3^2 = 0$ m/s
i.e. P is stationary.
The acceleration of P at time t s is
$\dfrac{d^2s}{dt^2} = 18 - 12t$.

At $t = 3$ s, the acceleration of P is $18 - 12 \times 3 = -18$ m/s².
When $t = 4$ s, $s = 9 \times 4^2 - 2 \times 4^3 = 16$ m

and $\dfrac{ds}{dt} = 18 \times 4 - 6 \times 4^2 = -24$ m/s.

Hence at $t = 4$ s, P is 16 m from O and moving towards O.

 Find the equations of the tangent and normal to the parabola $x^2 = 4y$ *at the point* (6, 9).

$y = \tfrac{1}{4}x^2$
Differentiating with respect to x gives

$\dfrac{dy}{dx} = \tfrac{1}{2}x$.

At the point (6, 9) *the gradient of the curve is given by the*

value of $\dfrac{dy}{dx}$ *when* $x = 6$.

i.e. when $x = 6$, $\dfrac{dy}{dx} = \tfrac{1}{2} \times 6 = 3$.

So the gradient of the tangent to the curve at (6, 9) is 3 and the gradient of the normal to the curve at this point is $-\tfrac{1}{3}$.
The equation of the tangent at (6, 9) is
$\qquad y - 9 = 3(x - 6)$
i.e. $\quad y = 3x - 9$.
The equation of the normal at (6, 9) is
$\qquad y - 9 = -\tfrac{1}{3}(x - 6)$
i.e. $3y + x = 33$.

 A particle moves along the x-axis, Ox, such that its distance x m *from O at the time t s is given by* $x = t^3 - 6t^2 + 9t$. *Calculate the*
(a) *times at which the particle is stationary,*
(b) *distance of P from O at these times,*
(c) *acceleration of P at these times,*
(d) *velocity of P when its acceleration is zero.*

(a) Calculate $\dfrac{dx}{dt}$ (the velocity) and equate this to zero. Solve

the resulting equation and find the two times (values of t) at which the particle is stationary.
(b) Use the two values of t from (a) to find the two values of x.

(c) Calculate $\dfrac{d^2x}{dt^2}$ (the acceleration). Substitute the two values

of t from (a) to find the two accelerations.

(d) Put $\dfrac{d^2x}{dt^2} = 0$. Find the value of t. Use this in the expression

for $\dfrac{dx}{dt}$ to find the necessary velocity.

 1 A particle P moves in a straight line so that its velocity v m/s at time t seconds, where $t \geq 0$, is given by $v = 12t - 5t^2$. Calculate: (a) the value of t when the acceleration of P is zero; (b) the distance covered by P between the instants when $t = 0$ and $t = 2$.
(L)

2 A particle moving along a straight line has a velocity of 4 cm/s as it passes a point A on the line and t s after passing A, an acceleration of $\left(\pi \sin \dfrac{\pi t}{3} + 3t \right)$ cm/s².

Calculate when $t = 2$ (a) the velocity of the particle, (b) the distance it has travelled.
(O & C)

3 A particle moves in a straight line so that its velocity, v m/s, is given by $v = 12t - t^2$, where t is the time, in seconds, measured from the start of the motion. Find: (i) the acceleration when $t = 20$; (ii) the values of t at which the particle is stationary, (iii) the value of t when the particle is again at its starting point.
(C)

4 Find the equation of the tangent to the curve $y = 3x^3 - x$ at the point where $x = \tfrac{2}{3}$. Find also the equation of the tangent at the point where $x = \tfrac{1}{3}$.
(O & C)

5 (a) Differentiate $(3x + 1\sqrt{x})^2$ with respect to x.
(b) Find the x co-ordinates of the points on the curve

$y = \dfrac{x^2 - 1}{x}$ at which the gradient of the curve is 5.
(C)

6 Show that the normal to the curve $y = \tan x$ at the point P

whose co-ordinates are $\left(\dfrac{\pi}{4}, 1 \right)$ meets the x-axis at the

point $A\left(\dfrac{\pi + 8}{4}, 0 \right)$.
(L)

7 A curve has the equation $y = x \sin 2x$. Find the gradient of the curve at $x = \pi/3$.
(J)

8 A curve is defined by the parametric equations $x = \theta - \sin \theta$, $y = 1 - \cos \theta$, $0 < \theta < 2\pi$. Show that $dy/dx = \cot \theta/2$, and find the equation of the tangent and of the normal to the curve at the point where $\theta = \pi/2$.
(J)

9 (i) Differentiate $\tan^2 3x$ with respect to x;
(ii) If $x = te^t$ and $y = t^2 e^t$, find dy/dx in terms of t;
(iii) Determine the slope of the curve $x^3 + x^2y + y^3 = 3$ at the point (1, 1).
(O & C)

10 Using implicit differentiation, or otherwise, find an expression for the slope of the tangent at the point (x_0, y_0) on the curve $x^2 + xy + y^2 - x - y = 0$. Find:
(a) the equation of the tangent to the curve at the point $(\tfrac{2}{3}, \tfrac{2}{3})$,
(b) the co-ordinates of the points at which the tangents to the curve are parallel to the x-axis;
(c) the maximum and minimum values of y, distinguishing between them;
(d) the co-ordinates of the points at which the tangents to the curve are parallel to the y-axis.
(O & C)

30 Changes
Rates of change, Small changes.

Rates of change

Rates of change can be expressed using **differentials**.
Rates of **increase** are **positive**.
Rates of **decrease** are **negative**.
They are often, but not always, rates of change 'with respect to time'.
However, by convention
'the rate of change of a quantity Q' means
'the rate of change of Q with respect to time'

i.e. $\dfrac{dQ}{dt}$.

This use of differentials has important applications in science.

Rates of change can be related by the chain rule

$$\frac{dy}{dx} = \frac{dy}{du} \cdot \frac{du}{dx}$$

It enables us to find the rate of change of y with respect to x, if y is a function of u and the rate of change of u with respect to x is known.

This is useful when solving problems concerning rates of change of physical quantities.

ℹ️ For a sphere, volume V, radius r
$\dfrac{dV}{dt}$ is the 'rate of change of volume V with respect to time t'
$\dfrac{dV}{dr}$ is the 'rate of change of volume V with respect to radius r'
$\dfrac{dr}{dt}$ is the 'rate of change of radius r with respect to time t'

ℹ️ *An inverted right circular cone of semi-vertical angle 45° is collecting water from a tap at a steady rate of $18\pi \, cm^3 s^{-1}$ Find the rate at which the depth h of water is rising when $h=3$ cm.*

Given h and the rate of increase of the water volume V, we must write V as a function of h and find the rate of increase of h.
Volume of a cone $V = \frac{1}{3}\pi r^2 h$.
Since the semi-vertical angle
is 45°, $r = h$.
So $V = \frac{1}{3}\pi h^3$

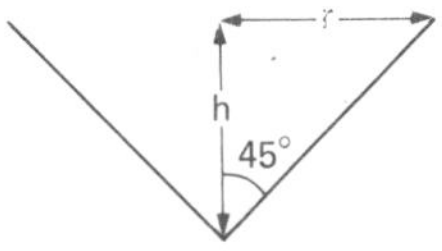

and $\dfrac{dV}{dh} = \pi h^2$

Using $\dfrac{dV}{dt} = \dfrac{dV}{dh} \cdot \dfrac{dh}{dt}$

$\qquad = \pi h^2 \cdot \dfrac{dh}{dt}$

So $\quad \dfrac{dh}{dt} = \dfrac{1}{\pi h^2} \cdot \dfrac{dV}{dt}$

Figure 1

When $h = 3$ cm and $\dfrac{dV}{dt} = 18\pi \, \text{cm}^3\,\text{s}^{-1}$,

$$\frac{dh}{dt} = \frac{1}{\pi 3^2} \cdot 18\pi \, \text{cm s}^{-1} = 2 \, \text{cm s}^{-1}$$

So the depth of water is rising at a rate of $2 \, \text{cm s}^{-1}$.

Small changes

Since $\lim\limits_{\delta x \to 0} \left(\dfrac{\delta y}{\delta x} \right) = \dfrac{dy}{dx}$

then $\qquad \dfrac{\delta y}{\delta x} \approx \dfrac{dy}{dx}$ when δx is small.

So $\qquad \delta y \approx \dfrac{dy}{dx} \delta x$

This approximation can be used to estimate the **small change** δy in y, if $\dfrac{dy}{dx}$ can be found and the small change δx in x is given.

It can also be used to estimate **percentage changes.**
If x is increased by $P\%$,

then $\delta x = \dfrac{P}{100} \times x$

and the approximate percentage increase in y is

$$\frac{\delta y}{y} \times 100\%.$$

ℹ️ *The radius r of a circle is 5 cm. Find the increase in the area A of the circle when the radius expands by 0.01 cm.*

Let the small increase in A be δA
and the small increase in r be δr.

$$A = \pi r^2$$

and $\dfrac{dA}{dr} = 2\pi r$

Using $\delta A \approx \dfrac{dA}{dr} \cdot \delta r$

$\qquad \approx 2\pi r \cdot \delta r$

When $r = 5$ cm, $\delta r = 0.01$ cm

$$\delta A \approx 2\pi(5)(0.01) \, \text{cm}^2$$
$$\approx 0.314 \, \text{cm}^2$$

So the required increase in area $\approx 0.314 \, \text{cm}^2$.

Changes
Worked example, Guided example and Exam questions

WE *A hemispherical bowl of radius 6 cm contains water which is flowing into it at a constant rate. When the height of the water is h cm the volume V of water in the bowl is given by $V = \pi(6h^2 - \frac{1}{3}h^3)$ cm^3. Find the rate at which the water level is rising when h = 3, given that the time taken to fill the bowl is 1 minute.*

We know $V = \pi(6h^2 - \frac{1}{3}h^3)$ is the volume of water in the bowl when the height of the water is h cm.

Differentiating, $\dfrac{dV}{dh} = \pi(12h - h^2)$.

When h = 3, $\dfrac{dV}{dh} = 27\pi$.

We are told that the water is flowing into the bowl at a constant rate, i.e. $\dfrac{dV}{dt} = k$ (a constant).

So, $V = kt$ (since $V = 0$ when $t = 0$).
Volume of the bowl $V = \pi(6 \times 6^2 - \frac{1}{3}6^3)$ (since h = radius = 6), i.e. $V = 144\pi$ cm^3

So, $k = 144\pi$ cm^3/min i.e. $\dfrac{dV}{dt} = 144\pi$ cm^3/min.

We need to find the rate at which h is increasing when h = 3, i.e. $\dfrac{dh}{dt}$ when h = 3.

Use $\dfrac{dh}{dt} = \dfrac{dh}{dV} \times \dfrac{dV}{dt}$ (chain rule)

$= \dfrac{1}{27\pi} \times 144\pi$ cm/min

$= \dfrac{144}{27}$ cm/min

$= \dfrac{4}{45}$ cm/s.

GE *The period T of a simple pendulum is calculated using the formula $T = 2\pi\sqrt{\dfrac{l}{g}}$ where l is the length of the pendulum and g is a constant. Find the percentage change in the period if the pendulum is lengthened by 2%.*

Calculate $\dfrac{dT}{dl}$. Find δl as a percentage of l.

Use $\delta T \approx \dfrac{dT}{dl} \times \delta l$ to find δT.

Percentage change $= \dfrac{\delta T}{T} \times 100\%$.

---

 1 The volume of a sphere is given by $V = \frac{4}{3}\pi r^3$. An elastic spherical balloon is being blown up so that the radius is increasing at the rate of 1 cm per second. Calculate the rate at which the volume of the balloon is increasing when the radius is 5 cm.

*(W)

2 An elastic sphere filled with water has a small hole from which the water leaks. When the water leaks away at the rate of 16π cm^3/s, the radius is decreasing at the rate of $\frac{1}{4}$ cm/s. Assuming that the spherical shape is preserved as the water leaks away, calculate, at the instant given by the above data, (a) the radius, (b) the rate at which the surface area is decreasing.

*(O & C)

3 The radius of a circular oil slick is increasing at $1 \cdot 5$ m/s. Taking π to be $3 \cdot 14$, find, to 2 significant figures, the rate at which the area of the slick is increasing when its radius is 300 m.

*(L)

4 (a) The radius of a circular disc is increasing at a constant rate of $0 \cdot 003$ cm/s. Find the rate at which the area is increasing when the radius is 20 cm.
(b) The area of another circular disc increases from 100π to 101π cm^2. Use calculus to find the corresponding increase in the radius.

*(C)

5 A spherical balloon is inflated by gas being pumped in at the constant rate of 200 cm^3 per second. What is the rate of increase of the surface area of the balloon when its radius is 100 cm?

(Surface area of sphere $= 4\pi r^2$, volume of sphere $= \dfrac{4}{3}\pi r^3$.)

(S)

6 The area of the region enclosed between two concentric circles of radii x and y (x > y) is denoted by A. Given that x is increasing at the rate of 2 m s^{-1}, y is increasing at the rate of 3 m s^{-1} and, when $t = 0$, $x = 4$ metres and $y = 1$ metre, find: (i) the rate of increase of A when $t = 0$; (ii) the ratio of x to y when A begins to decrease, (iii) the time at which A is zero.

(J)

7 Two variables u and v are connected by the relation $\dfrac{1}{u} + \dfrac{1}{v} = \dfrac{1}{f}$, where f is a constant. Given that u and v both vary with time, t, find an equation connecting $\dfrac{du}{dt}, \dfrac{dv}{dt}$, u and v.

Given also that u is decreasing at a rate of 2 cm per second and that $f = 10$ cm, calculate the rate of increase of v when $u = 50$ cm.

(C)

8 A hemispherical bowl of radius a cm is initially full of water. The water runs out of a small hole at the bottom of the bowl at a constant rate which is such that it would empty the bowl in 24 s. Given that, when the depth of the water is x cm, the volume of water is $\frac{1}{3}\pi x^2(3a - x)$ cm^3, prove that the depth is decreasing at a rate of $a^3/\{36x(2a - x)\}$ cm/s. Find after what time the depth of water is $\frac{1}{2}a$ cm, and the rate at which the water level is then decreasing.

(O & C)

9 Apply the small increment formula $f(x + \delta x) - f(x) \approx \delta x f'(x)$, to tan x to find an approximate value of
$$\tan\left(\dfrac{100\pi + 4}{400}\right) - \tan\dfrac{\pi}{4}.$$

(L)

10 Fluid enters a leaky vessel through a valve. The valve admits fluid at a rate proportional to the volume of fluid already in the vessel, and the rate of leakage is proportional to the square of the volume already in the vessel. There is a balance between inflow and outflow when the volume in the vessel is V_0. Initially there is a volume $\frac{1}{4}V_0$ in the vessel, and the volume increases to $\frac{1}{2}V_0$ in time T. Find the time taken for the volume to increase from $\frac{1}{4}V_0$ to $\frac{3}{4}V_0$.

(O)

31 Special Points

Local maxima and minima, Points of inflexion, Tests for points, Applications.

Local maxima and minima

At a point of **local maximum** a function has a greater value than at points immediately on either side of it. At a point of **local minimum** a function has a smaller value than at points immediately on either side of it.

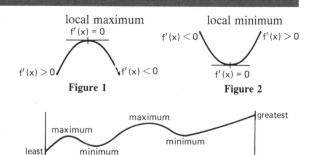

Figure 1   **Figure 2**

Local maxima and minima are also called **turning points**.
A function may have more than one turning point.
The local maxima and minima are not necessarily the greatest or least values of a function in a given range.

Figure 3

Points of inflexion

At a **point of inflexion**, the graph of a function changes the direction in which it is curving.

horizontal points of inflexion

Figure 4

general points of inflexion

Figure 5

Tests for points

A **stationary point** is a point at which $f'(x)=0$.
Local maxima, minima and horizontal points of inflexion are stationary points.
To test for stationary points
(a) Find $f'(x)$ and $f''(x)$.
(b) Put $f'(x)=0$ and solve the resulting equation to find the x-coordinate(s) of the point(s).
(c) Find $f''(x)$ at the stationary point(s).
(i)   If $f''(x)<0$, the point is a local maximum.
(ii)   If $f''(x)>0$, the point is a local minimum.
(iii)   If $f''(x)=0$, find the sign of $f'(x)$ for a value of x just to the left and just to the right of the point.

Sign to left	sign to right	type of point
+	−	maximum
−	+	minimum
+ −	+ −	} point of inflexion

To test for general points of inflexion
(a) Find $f''(x)$.
(b) Put $f''(x)=0$ and solve the resulting equation to find the possible x-coordinate(s).
(c) Find the sign of $f''(x)$ for a value of x just to the left and just to the right of the point. If $f''(x)$ changes sign, the point is a point of inflexion.

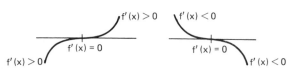 *Find the stationary points of $f(x)=\frac{1}{3}x^3-2x^2+3x$ and identify their nature.*

$$f(x)=\tfrac{1}{3}x^3-2x^2+3x$$
$$f'(x)=x^2-4x+3$$
$$f''(x)=2x-4$$

At stationary points $f'(x)=0$,
i.e. $x^2-4x+3=0$
$$(x-3)(x-1)=0$$
$$x=3 \text{ and } x=1.$$

When $x=3, f''(x)=2(3)-4>0$,
$$f(x)=\tfrac{1}{3}(3)^3-2(3)^2+3(3)=0.$$
Therefore $(3, 0)$ is a local minimum.

When $x=1, f''(x)=2(1)-4<0$,
$$f(x)=\tfrac{1}{3}(1)^3-2(1)^2+3(1)=\tfrac{4}{3}.$$
Therefore $(1, \tfrac{4}{3})$ is a local maximum.

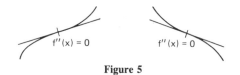 *Find any points of inflexion of $f(x)=\frac{1}{3}x^3-2x^2+3$.*

From above: $f''(x)=2x-4$
At a general point of inflexion $f''(x)=0$,
i.e. $2x-4=0 \Rightarrow x=2$.
For $x=2^+, f''(x)>0$ i.e. $f''(x)$ changes
For $x=2^-, f''(x)<0$ sign
So $x=2$ is a general point of inflexion.

Applications

The above methods can be applied to practical problems in which the **maximum** or **minimum** value of a quantity is required. The procedure is
(a) write an expression for the required quantity,
(b) use the given conditions to rewrite it in terms of a single variable,
(c) find the turning point(s) and their type(s). It is often obvious from the problem itself whether a maximum or minimum has been obtained.

A rectangle has perimeter 28m. What is its maximum area?

Let x and y metres be the sides of the rectangle.
Its perimeter $=2x+2y=28 \Leftrightarrow y=14-x$.
Its area $A=xy=x(14-x)$.
$$\frac{dA}{dx}=14-2x.$$

When A is a maximum, $\dfrac{dA}{dx}=0$, i.e. $14-2x=0 \Leftrightarrow x=7$
Since $y=14-x=7$
So the maximum area is $7^2 \text{ m}^2=49 \text{ m}^2$.

Special Points
Worked example, Guided example and Exam questions

 The lengths of the sides of a rectangular sheet of metal are 8 cm *and* 3 cm. *A square of side x cm is cut from each corner of the sheet and the remaining piece is folded to make an open box.*

(a) *Show that the volume V of the box is given by*
$V = 4x^3 - 22x^2 + 24x$ cm³.

(b) *Find the value of x for which the volume of the box is a maximum. Calculate the maximum volume.*

(a)

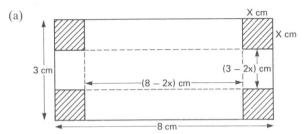

The volume of the box $V = (8-2x)(3-2x)x$ cm³
i.e. $V = 4x^3 - 22x^2 + 24x$ cm³. (1)

(b) Differentiating (1) with respect to x gives

$$\frac{dV}{dx} = 12x^2 - 44x + 24$$

For a maximum (or minimum) value of V, $\frac{dV}{dx} = 0$,

i.e. $12x^2 - 44x + 24 = 0$
$\Rightarrow (3x-2)(x-3) = 0$
i.e. $x = \frac{2}{3}$ or 3.

Clearly x cannot be 3 cm since the width of the sheet initially is only 3 cm. So $x = \frac{2}{3}$ cm.

Differentiating again gives,

$$\frac{d^2V}{dx^2} = 24x - 44.$$

When $x = \frac{2}{3}$, $\frac{d^2V}{dx^2} < 0$ i.e. V is a maximum.

So the maximum volume is given by substituting $x = \frac{2}{3}$ in (1) giving

$$V_{max} = \frac{200}{27} = 7\tfrac{11}{27} \text{ cm}^3.$$

 Find the co-ordinates of the stationary points on the curve $y = x^4 - 4x^3$. *Show that the curve has a point of inflexion at $x = 2$.*

Calculate $\frac{dy}{dx}$. Equate this to zero to find the two values of x where these stationary points occur.

Next calculate $\frac{d^2y}{dx^2}$ and use it to find the nature of the stationary points. (This check fails for one value of x so it is necessary to find the type of stationary point by considering the gradient $\frac{dy}{dx}$ on each side of this point.)

To show the curve has a point of inflexion at $x = 2$, show that $\frac{d^2y}{dx^2}$ is zero at $x = 2$ and the sign of the gradient of the curve on each side of the point is the same.

---

 1 The curve for which $\frac{dy}{dx} = a(x-p)(x-q)$, where a, p and q are constants, has turning-points at $(2, 0)$ and $(1, 1)$.
(i) State the value of p and q.
(ii) Using these values, determine the value of a.
*(C)

2 The curve $y = x^2 + ax + b$ has a turning point at $(1, 3)$. Find a and b. *(S)

3 An open rectangular box is made of very thin sheet metal. Its volume is 128 cm³, its width is x cm, and its length is $4x$ cm. Obtain an expression for its depth in terms of x. Show that the total surface area of its base, its ends and its sides is equal to $\left(4x^2 + \frac{320}{x}\right)$ cm². Calculate the dimensions of the box for which the surface area is a minimum; explain why your answer gives a minimum rather than a maximum.
*(W)

4 Differentiate the function $\frac{x+3}{\sqrt{(1+x^2)}}$ with respect to x. Find the value of x at which the function has a stationary value and determine the nature of the stationary value.

5 (a) The function f is defined by: $f(x) = \frac{4}{x-2} - \frac{1}{x+1}$.

Show that $f'(x) = -\frac{3x(x+4)}{(x-2)^2(x+1)^2}$.

Find the local maxima and minima of f, and sketch its graph.
(b) Show that the expression $\sin\theta + 2\cos\theta$, where $0 \leq \theta \leq \frac{1}{2}\pi$, has a maximum value equal to $\sqrt{5}$.
(W)

6 The parametric equations of a curve are $x = \log_e(1+t)$, $y = e^t$ for $t > -1$. Find $\frac{dy}{dx}$ and $\frac{d^2y}{dx^2}$ in terms of t. Prove that the curve has only one turning point and that it must be a minimum. (A)

7 Find the maximum point on the curve $y = \frac{1}{x^2 + 2x + 4}$ and show that there are points of inflexion at $(0, \frac{1}{4})$ and $(-2, \frac{1}{4})$.
(S)

8 Investigate the nature of the turning points of the curve $y = (x-1)(x-a)^2$ in the cases of (i) $a > 1$, (ii) $a < 1$. Sketch on separate diagrams the curves $y = (x-1)(x-a)^2$ when $a > 1$ and when $0 < a < 1$.
(L)

9 State the derivatives of $\sin x$ and $\cos x$, and use these results to show that the derivative of $\tan x$ is $\sec^2 x$.
Show further that
$$\frac{d}{dx}(\tan^{-1}x) = \frac{1}{1+x^2}.$$
A vertical rod AB of length 3 units is held with its lower end B at a distance 1 unit vertically above a point O. The angle subtended by AB at a variable point P on the horizontal plane through O is θ. Show that $\theta = \tan^{-1}x - \tan^{-1}\frac{x}{4}$, where $x = OP$. Prove that, as x varies, θ is a maximum when $x = 2$, and that the maximum value of θ can be expressed as $\tan^{-1}\frac{3}{4}$.
(J)

10 A right circular cone with semivertical angle θ is inscribed in a sphere of radius a, with its vertex and the rim of its base on the surface of the sphere. Prove that its volume is $\frac{2}{3}\pi a^3 \cos^4\theta \sin^2\theta$. If a is fixed and θ varies, find the limits within which this volume must lie.
(O & C)

63

32 Curve Sketching
Known curves, Unknown curves.

Known curves

A **sketch** of a curve shows its basic shape and main features. It is not an accurate drawing of the curve.

The student should be able to sketch the graphs of the basic functions considered elsewhere in this book, e.g. trigonometrical, exponential functions, etc.

The graphs of **simple transformations** of a known curve $y=f(x)$ can be easily sketched too.
The graph of:

$y=f(x)+c$ is $y=f(x)$ translated a distance c, parallel to the y-axis,

$y=f(x-c)$ is $y=f(x)$ translated to the right a distance c, parallel to the x-axis,

$y=cf(x)$ is $y=f(x)$ stretched by a scale factor c, parallel to the y-axis,

$y=f(cx)$ is $y=f(x)$ stretched by a scale factor $\dfrac{1}{c}$, parallel to the x-axis.

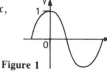 *From this sketch of $y=\cos x$, sketch:* (a) $y=\cos x+2$,
(b) $y=\cos(x-\pi)$,
(c) $y=2\cos x$,
(d) $y=\cos(2x)$.

Figure 1

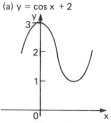
(a) y = cos x + 2

Figure 2

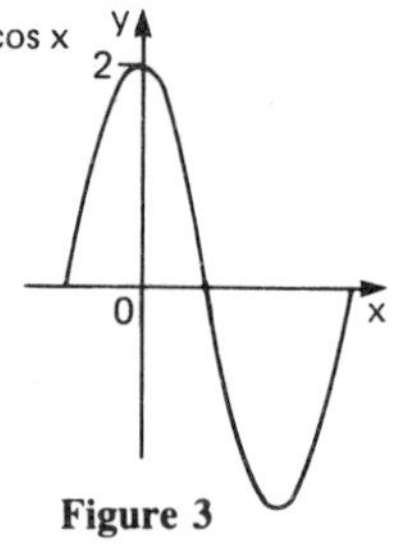

(c) y = 2 cos x

Figure 3

(b) y = cos (x−π)

Figure 4

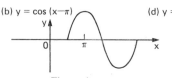
(d) y = cos (2x)

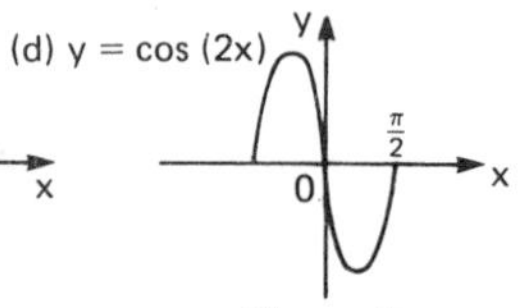

Figure 5

Unknown curves

Sketching a graph of an **unknown curve** $y=f(x)$ is like building up an identikit picture. In general, the important features can be built up by answering the following questions (as appropriate) and systematically adding the findings to a pair of axes.

(a) What is the **domain** of the function?
   If $f(x)$ is a polynomial, then the domain is the set of real numbers.
   If $f(x)$ is a rational function, say $P(x)/Q(x)$, then the values of x for which $Q(x)=0$ will give discontinuities, i.e. the curve will have asymptotes at these values of x.

(b) Does the curve cut the **y-axis**?
   If possible, let $x=0$ and find the corresponding value of y.

(c) Does the curve cut the **x-axis**?
   Let $y=0$ and, if possible, find the corresponding value(s) of x.

(d) Does the curve have **symmetry**?
   If $f(x)=f(-x)$, then $x=0$ is an axis of symmetry.
   If $f(-x)=-f(x)$, then the part of the graph for which $x<0$ is a half-turn about the origin of the part of the graph for which $x>0$.

(e) How does the curve behave as $x\to\pm\infty$?
   If $f(x)$ is a rational function, say $P(x)/Q(x)$, then divide $P(x)$ and $Q(x)$ by the highest power of x in $P(x)$ to find how $f(x)$ behaves as $x\to\pm\infty$.

(f) How does the curve behave near any **asymptotes**?
   If x_1 is a discontinuity, then look at the curve as:
   (i) $x\to x_1^-$, i.e. values of x just less than x_1,
   (ii) $x\to x_1^+$, i.e. values of x just greater than x_1.

(g) Has the curve any **special points** (maximum, minimum or inflexion)? (See Special Points, p. 62.)

(h) Are there any **regions** of the xy-plane where the curve does not exist?
   If $f(x)$ is a rational function, then rewrite $y=f(x)$ as a quadratic in x. For real x, the quadratic's discriminant ≥ 0 (see Quadratics, p. 6). This gives the regions of the plane where the curve does exist.

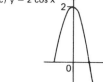 *Sketch the graph of $y=\dfrac{2x-1}{(x-2)^2}$.*

(a) Since $(x-2)^2$ is the denominator of the function, there is a discontinuity at $x=2$. So $x=2$ is an asymptote. The domain is the set of real numbers, excluding $x=2$.

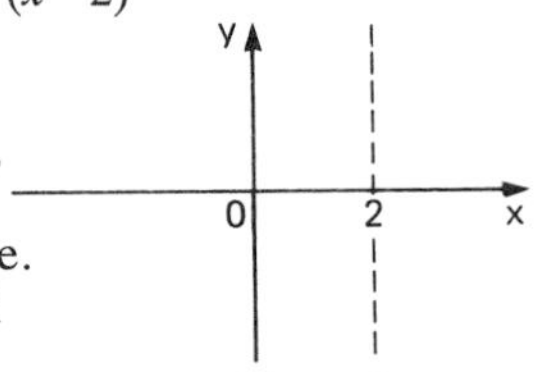

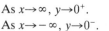

Figure 6

(b) When $x=0$, $y=-\frac{1}{4}$. Plot $(0, -\frac{1}{4})$.

(c) When $y=0$, $x=\frac{1}{2}$. Plot $(\frac{1}{2}, 0)$

(d) No symmetry since
$$f(x)\neq f(-x)$$
and $f(-x)\neq -f(x)$.

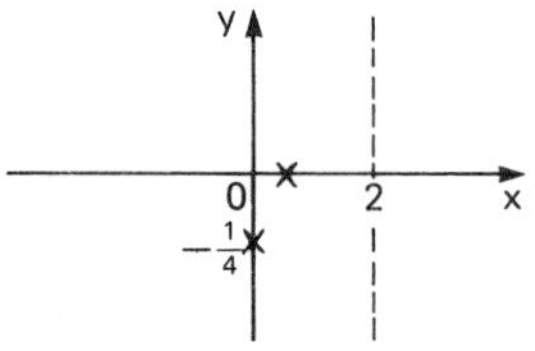

Figure 7

(e) $y=\dfrac{2x-1}{(x-2)^2}=\dfrac{2x-1}{x^2-4x+4}$

Dividing numerator and denominator by x (the highest power of x in the numerator) gives

$$y=\dfrac{2-\frac{1}{x}}{x-4+\frac{4}{x}}$$

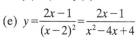

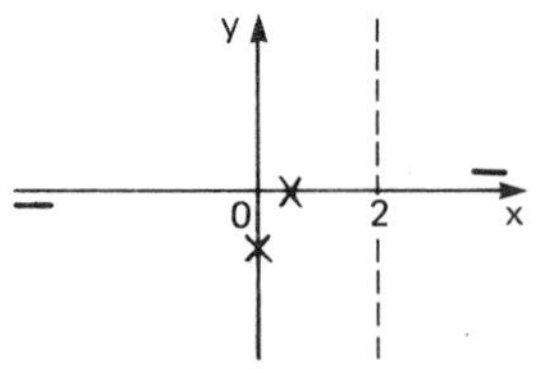

Figure 8

As $x\to\infty$, $y\to 0^+$.
As $x\to-\infty$, $y\to 0^-$.

(f) There is a discontinuity at $x=2$.
As $x\to 2^-$, $y\to +\infty$.
As $x\to 2^+$, $y\to +\infty$.

(g) $y'=-\dfrac{(x-2)(x+1)}{(x-2)^2}$

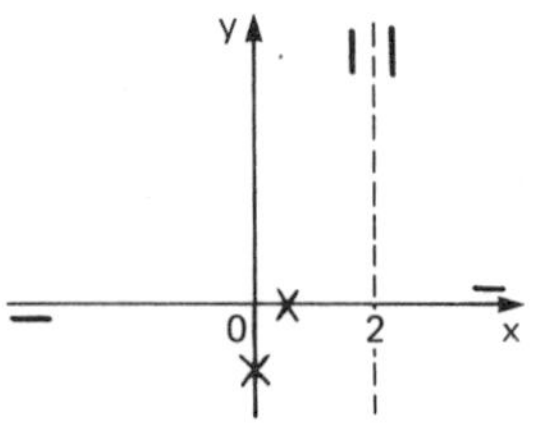

Figure 9

When $y'=0$, $x=-1$, i.e. $(-1, -\frac{1}{3})$ is a stationary point.

$$y''=\dfrac{(x-2)^2(2x+5)}{(x-2)^6}$$

When $x=-1$, $y''>0$, i.e. a local minimum.

(h) Obviously the curve does not exist below $(-1, -\frac{1}{3})$.

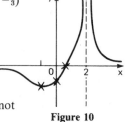

Figure 10

Curve Sketching
Worked example, Guided example and Exam questions

Sketch the curve $y = \dfrac{3(x-2)}{x(x+6)}$.

(a) Since $x(x+6)$ is the denominator of the function, there are discontinuities at $x=0$ and $x=-6$.
So, $x=0$ and $x=-6$ are asymptotes.
The domain of the function is the set of real numbers excluding $x=0$ and $x=-6$.
(b) Since $x=0$ is a discontinuity, the curve does not cut the y-axis.
(c) When $y=0$, $x=2$. Plot $(2, 0)$.
(d) No symmetry.
(e) $y = \dfrac{2x-6}{x^2+6x} = \dfrac{3-\frac{6}{x}}{x+6}$

As $x \to +\infty$, $y \to 0^+$
As $x \to -\infty$, $y \to 0^-$.
(f) Discontinuity at $x=0$.
As $x \to 0^-$, $y \to +\infty$.
As $x \to 0^+$, $y \to +\infty$.
Discontinuity at $x=-6$.
As $x \to -6^-$, $y \to -\infty$.
As $x \to -6^+$, $y \to -\infty$.
(g) $y' = \dfrac{-3(x^2-4x-12)}{x^2(x+6)^2} = \dfrac{-3(x+2)(x-6)}{x^2(x+6)^2}$,

So, $y'=0$ when $x=-2$ or 6.
As we pass through $x=-2$, y' goes $-$ve, 0, $+$ve, so $x=-2$ is a point of local minimum.
As we pass through $x=6$, y' goes $+$ve, 0, $-$ve, so $x=6$ is a point of local maximum.

(h) Rewrite $y = \dfrac{3(x-2)}{x(x+6)}$ as $yx^2 + 3(2y-1)x + 6 = 0$.

For real x, $9(2y-1)^2 \geq 4 \cdot y \cdot 6$.
i.e. $12y^2 - 20y + 3 \geq 0$.
$\Rightarrow y \leq \dfrac{1}{6}$ or $y \geq \dfrac{3}{2}$.

So, for $\dfrac{1}{6} < y < \dfrac{3}{2}$, the curve does not exist.

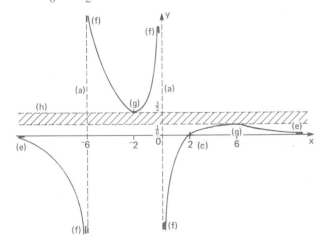

Sketch the curve $y^2 = x(x-3)^2$.

First sketch the graph of $f(x) = x(x-3)^2$.
Since $y = \pm\sqrt{(f(x))} = \pm(x-3)\sqrt{x}$, notice,
(a) $y^2 = x(x-3)^2$ is symmetrical about the x-axis,
(b) for $x<0$, y is non-real, i.e. graph does not exist,
(c) $(0, 0)$ and $(3, 0)$ lie on the graph,
(d) for $0 < f(x) < 1$, $|y| > f(x)$, since $\sqrt{(|f(x)|)} > f(x)$,
(e) for $f(x) > 1$, $|y| < f(x)$.
Hence sketch the curve $y^2 = x(x-3)^2$.

---

1 Sketch the curve $y = \dfrac{2}{x+1}$.
*(S)

2 Give that $f(x) = x - 1 + \dfrac{1}{x+1}$, x real, $x \neq -1$, find the values of x for which $f'(x) = 0$. Sketch the graph of f, showing the co-ordinates of the turning points and indicating clearly the form of the graph when $|x|$ becomes large. *(J)*

3 Sketch, on separate diagrams, the graphs of:
(i) $y = x^3(1-x)$; (ii) $y = (x+1)/(x+2)^2$.
(O & C)

4 Given that $y = \dfrac{x^2+1}{x^2+x+1}$,
(i) state the limiting value of y: (a) as x tends to ∞; (b) as x tends to $-\infty$,
(ii) show that y is finite for all real values of x,
(iii) determine the set of values which y must take for real values of x,
(iv) determine the values of x for which y has a stationary value. Sketch the curve which represents the given equation. *(A)*

5 Sketch (on separate diagrams) the graphs of:
(a) $y = x^2 - x^3$; (b) $y = 1 - e^x$; (c) $y = 1/(1-e^x)$.
You are only asked for rough sketches; details of maxima and minima are not required but you should indicate the behaviour of the curves for numerically large and small values of x. *(O & C)*

6 Find the turning points on the curve $y = e^x \sin x$ for $0 \leq x \leq 2\pi$. Sketch the curve over this interval. *(A)*

7 The function f is defined by $f(x) = \dfrac{4x+5}{x^2-1}$. Show that $f(x)$ cannot take values between -4 and -1. Sketch the graph of the function showing clearly the behaviour of f as $x \to \pm 1$ and as $x \to \pm\infty$.
(W)

8 If $y = 2x^4 - x^2 + 1$, find $\dfrac{dy}{dx}$ and deduce the three values of x for which $\dfrac{dy}{dx} = 0$. Find the corresponding values of y.
Distinguish between maximum and minimum values of y **either** by considering the signs of $\dfrac{dy}{dx}$ **or** by considering $\dfrac{d^2y}{dx^2}$. Hence sketch the curve for values of x from -1 to $+1$. *(O & C)*

9 (Throughout this question, $-\frac{1}{2}\pi \leq \sin^{-1} x \leq \frac{1}{2}\pi$.)
(i) Prove that, for $-1 < x < 1$, $\dfrac{d}{dx}(\sin^{-1} x) = (1-x^2)^{-\frac{1}{2}}$.
(ii) Given that $y = \dfrac{(1-x^2)^{\frac{1}{2}}}{x}$, $(-1 < x < 1, x \neq 0)$, prove that $\dfrac{dy}{dx}$ is always negative, and sketch the graph of y.
(iii) Given that $z = (1-x^2)^{\frac{1}{2}} \sin^{-1} x$, $(-1 < x < 1)$, find $\dfrac{dz}{dx}$.
Using your sketch in (ii), or otherwise, determine how many turning points there are on the graph of z.
(iv) Sketch the graph of z. (No co-ordinates of turning points are required.)
(C)

33 Integration
Definition, Standard integral, Definite integral, Recognising derivatives, Trigonometrical integrands.

Definition

Integration is the **inverse** of **differentiation**. It is the process of finding a function given its derivative.
If y is the derivative of $f(x)$ with respect to x, then the **indefinite integral** of y with respect to x is:
$\int y\,dx = f(x) + c$ where c is a **constant of integration**. The function to be integrated, y, is the **integrand**.

Standard integrals

	function	integral
algebraic	constant k	$kx + c$
	$x^n (n \neq -1)$	$\dfrac{x^{n+1}}{n+1} + c$
trigonometrical (x in radians)	$\sin x$	$-\cos x + c$
	$\cos x$	$\sin x + c$
	$\tan x$	$\ln\lvert\sec x\rvert + c$
	$\operatorname{cosec} x$	$\ln\lvert\tan \tfrac{1}{2}x\rvert + c$
	$\sec x$	$\ln\lvert\sec x + \tan x\rvert + c$
	$\cot x$	$\ln\lvert\sin x\rvert + c$

	function	integral
inverse trigonometrical	$\dfrac{1}{a^2 + x^2}$	$\dfrac{1}{a}\tan^{-1}\dfrac{x}{a} + c$
	$\dfrac{1}{\sqrt{(a^2 - x^2)}}$	$\sin^{-1}\dfrac{x}{a} + c$
logarithmic and exponential	$\dfrac{1}{x}$	$\ln\lvert x\rvert + c$
	e^x	$e^x + c$

Definite integral

The **definite integral** from a to b of $f(x)$ is:
$$\int_a^b f(x)\,dx = [F(x)]_a^b \text{ where } F(x) = \int f(x)\,dx$$
$$= F(b) - F(a)$$
a and b are called the **limits of integration**.
Note: The constant c is eliminated in the subtraction.

ℹ️
$$\int_2^6 \frac{1}{x}\,dx = [\ln x]_2^6$$
$$= \ln 6 - \ln 2$$
$$= \ln \frac{6}{2}$$
$$= \ln 3$$

Recognising derivatives

An integrand may be recognised as the result of differentiation by the chain rule.

Look for products of the form $f(x)[g(x)]^n$.
If $f(x)$ is a multiple of the derivative of $g(x)$, then the integrand is of this type, $f(x)$ and $g(x)$ are usually polynomials or trigonometrical functions.

A special case is an integrand of the form $\dfrac{f'(x)}{f(x)}$.

In this case: $\int \dfrac{f'(x)}{f(x)}\,dx = \ln\lvert f(x)\rvert + c$.

ℹ️
$\int \cos 2x\,dx = \tfrac{1}{2}\sin 2x + c$
since if $f(x) = \sin 2x$, then $f'(x) = 2\cos 2x$.

ℹ️
$\int x(x^2 + 1)^2\,dx = \tfrac{1}{6}(x^2 + 1)^3 + c$
Here $g(x) = (x^2 + 1)$ and $g'(x) = 2x = 2f(x)$.

ℹ️
$\int \dfrac{x}{3 + x^2}\,dx = \tfrac{1}{2}\ln\lvert 3 + x^2\rvert + c$
Here $f(x) = 3 + x^2$ and $f'(x) = 2x$.

Trigonometrical integrands

1. Even powers of $\sin x$ or $\cos x$
The integrand is rewritten using the double angle formulae (see Trigonometrical Identities p. 40).

2. Odd powers of $\sin x$ or $\cos x$
If the integrand is an odd power of:
(a) $\sin x$, rewrite it as (even power of $\sin x$) $\sin x$,
(b) $\cos x$, rewrite it as (even power of $\cos x$) $\cos x$.
Then use $\cos^2 x + \sin^2 x = 1$ to rewrite the:
(a) (even power of $\sin x$) in terms of $\cos x$,
(b) (even power of $\cos x$) in terms of $\sin x$.

3. Powers of $\tan x$
The identity $1 + \tan^2 x = \sec^2 x$ may be used to rewrite the power of $\tan x$.

4. Products of the form $\sin mx \cos nx$
The integrand may be rewritten as a sum or difference using the factor formulae (see Trigonometrical Identities p. 40).

ℹ️
$\int \cos^2 x\,dx = \int \tfrac{1}{2}(1 + \cos 2x)\,dx$
$= \tfrac{1}{2}x + \tfrac{1}{4}\sin 2x + c$

ℹ️
$\int \cos^3 x\,dx = \int \cos^2 x \cos x\,dx$
$= \int (1 - \sin^2 x)\cos x\,dx$
$= \int \cos x\,dx - \int \sin^2 x \cos x\,dx$
$= \sin x - \tfrac{1}{3}\sin^3 x + c$

ℹ️
$\int \tan^2 x = \int (\sec^2 x - 1)\,dx$
$= \tan x - x + c$

ℹ️
$\int \sin 3x \cos x\,dx = \int \tfrac{1}{2}(\sin 4x + \sin 2x)\,dx$
$= -\tfrac{1}{8}\cos 4x - \tfrac{1}{4}\cos 2x + c$

Integration
Worked example, Guided example and Exam questions

WE *Evaluate each of the following integrals:*

(a) $\int_1^4 \left(\dfrac{3}{x} - \sqrt{x}\right)^2 dx$

(c) $\int_0^1 (3+e^x)(2+e^{-x})\, dx$

(b) $\int_0^{\frac{\pi}{6}} \sin 3x\, dx$

(d) $\int_2^4 \dfrac{5}{x}\, dx$

(a) $\int_1^4 \left(\dfrac{3}{x} - \sqrt{x}\right)^2 dx = \int_1^4 \left(\dfrac{9}{x^2} - \dfrac{6}{\sqrt{x}} + x\right) dx$

$\qquad = \int_1^4 (9x^{-2} - 6x^{-\frac{1}{2}} + x)dx$

$\qquad = \left[\dfrac{9x^{-1}}{-1} - \dfrac{6x^{\frac{1}{2}}}{\frac{1}{2}} + \dfrac{x^2}{2}\right]_1^4$

$\qquad = \left(-\dfrac{9}{4} - 24 + 8\right) - \left(-9 - 12 + \dfrac{1}{2}\right)$

$\qquad = 2\frac{1}{4}.$

(b) $\int_0^{\frac{\pi}{6}} \sin 3x\, dx = \left[-\dfrac{1}{3}\cos 3x\right]_0^{\frac{\pi}{6}}$

$\qquad = \left(-\dfrac{1}{3}\cos\dfrac{\pi}{2}\right) - \left(-\dfrac{1}{3}\cos 0\right)$

$\qquad = \dfrac{1}{3}$

(c) $\int_0^1 (3+e^x)(2+e^{-x})\, dx = \int_0^1 (6 + 3e^{-x} + 2e^x + 1)\, dx$

$\qquad = [7x + (-1)3e^{-x} + 2e^x]_0^1$

$\qquad = (7 - 3e^{-1} + 2e) - (-3 + 2)$

$\qquad = 8 - \dfrac{3}{e} + 2e.$

(d) $\int_2^4 \dfrac{5}{x}\, dx = [5\ln x]_2^4$

$\qquad = (5\ln 4) - (5\ln 2)$

$\qquad = 5\ln\left(\dfrac{4}{2}\right)$

$\qquad = 5\ln 2.$

GE *Calculate the values of these integrals.*

(a) $\int_4^9 \left(\dfrac{3}{\sqrt{x}} - \dfrac{1}{x\sqrt{x}}\right) dx$

(c) $\int_0^{\frac{\pi}{4}} \sin^2 x\, dx$

(b) $\int_0^1 \dfrac{1}{1+x^2}\, dx$

(d) $\int_0^2 \dfrac{1}{2x+1}\, dx$

(a) Rewrite the expression remembering that $\dfrac{1}{\sqrt{x}} = x^{-\frac{1}{2}}$ and

$\dfrac{1}{x\sqrt{x}} = x^{-\frac{3}{2}}$. Integrate term by term.

(b) This is a standard integral.

(c) Use the identity $\sin^2 x = \dfrac{1}{2}(1 - \cos 2x)$.

(d) Use $\dfrac{d}{dx}(\ln x) = \dfrac{1}{x}$.

EX 1 Evaluate: $\int_1^2 \dfrac{(x^4-1)}{x^3}\, dx.$ 　　　　*(S)*

2 (a) Given that $f(x) = \left(2x - \dfrac{1}{x^2}\right)^2$, $x \neq 0$, find $f'(x)$.

　(b) Evaluate $\int_1^8 \left(\sqrt[3]{x} + \dfrac{1}{2\sqrt[3]{x}}\right) dx.$ 　　*(H)*

3 (a) Integrate with respect to x:
　　(i) $(2-x^2)^2$; (ii) $(x+1)x^{-\frac{1}{4}}$.

　(b) Evaluate $\int_0^2 x(x-2)(x+1)\, dx.$ 　　*(A)*

4 (i) Integrate with respect to x:
　　(a) $\dfrac{(x+3)^2}{x^4}$;　(b) $(1-3x)^{\frac{1}{4}}$;　(c) $\sin^2\dfrac{3x}{2}$.

　(ii) Evaluate:

　　(a) $\int_0^1 x^2(1-x)^2\, dx$;　(b) $\int_0^{\frac{\pi}{2}} \sin 2x\, dx$;

　　(c) $\int_0^{\frac{\pi}{6}} \sec^2 2x\, dx.$ 　　*(O & C)*

5 Evaluate each of the following:

　(i) $\int \left(e^{3x} + \dfrac{1}{x^2}\right) dx$;　(ii) $\int_1^e \left(x^2 + \dfrac{2}{x}\right) dx$;

　(iii) $\int_0^{\frac{\pi}{4}} (\cos x + \sin x)^2\, dx.$ 　　*(W)*

6 (a) Calculate the value of each of the following definite integrals:

　(i) $\int_{-1}^2 (x+1)(2x-3)\, dx$;　(ii) $\int_3^{27} \sqrt{(3x)}\, dx$;

　(iii) $\int_1^2 \sin(3\pi x)\, dx.$

　(b) Draw a sketch of the part of the curve $y = \sqrt{(x^2-9)}$ between $x=3$ and $x=6$. Without attempting to evaluate the integral, deduce that $4 < \int_5^6 \sqrt{(x^2-9)}\, dx < 3\sqrt{3}$, explaining your reasoning. 　　*(OLE)*

7 Integrate the following with respect to x:
　(i) $x^3(x^4-3)^5$;　(ii) $\tan x$;　(iii) xe^{x^2}. 　　*(S)*

8 Find $\int \sin x(1+\cos^2 x)\, dx.$ 　　*(C)*

9 Determine the values of p and q for which $x^2 - 4x + 5 \equiv (x-p)^2 + q$ and hence evaluate

$$\int_2^3 \dfrac{1}{x^2-4x+5}\, dx.$$

Calculate also $\int_2^3 \dfrac{2x-4}{x^2-4x+5}\, dx$ and deduce, or find

otherwise, the value of $\int_2^3 \dfrac{2x}{x^2-4x+5}\, dx.$

(Your answers need not be given as decimals.) 　　*(J)*

34 Methods of Integration
Use of partial fractions, Substitution, Parts.

Use of partial fractions

An integrand may be a rational function, not of the type $\dfrac{f'(x)}{f(x)}$

If it is a **proper fraction**, then express it in terms of partial fractions if possible (see Rational Functions p. 4). The solutions of the resulting integrals are often log and/or inverse trig functions.

If it is an **improper fraction**, then divide to obtain a quotient and proper fraction, which can then be expressed in partial fractions if necessary.

$\boxed{\mathbf{i}}$ $\displaystyle\int \frac{3x^2+2x+4}{(2x+1)(1+x^2)}\,dx = \int\left[\frac{3}{(2x+1)}+\frac{1}{(1+x^2)}\right]dx$

$\qquad\qquad = \tfrac{3}{2}\ln|2x+1|+\tan^{-1}x+c$

$\boxed{\mathbf{i}}$ $\displaystyle\int \frac{x^3+2x^2-10x-9}{(x-3)(x+3)}$

$\qquad = \displaystyle\int\left[x+2+\frac{1}{(x-3)}-\frac{2}{(x+3)}\right]dx$

$\qquad = \tfrac{1}{2}x^2+2x+\ln|x-3|-2\ln|x+3|+c$

Substitution

Some integrals may be found more easily by using a **substitution**, i.e. **changing the variable**, and evaluating the transformed integral.

If the evaluation of $\displaystyle\int f(x)\,dx$ requires a substitution, then there are two basic methods.

1. Let $x=g(\theta)$
(a) Find $\dfrac{dx}{d\theta}$.
(b) Replace x by $g(\theta)$ in $f(x)$ to give $F(\theta)$.
(c) Use $\displaystyle\int f(x)\,dx = \int F(\theta)\frac{dx}{d\theta}\,d\theta$.

2. Let $u=h(x)$
(a) Find $\dfrac{du}{dx}$ and hence $\dfrac{dx}{du}$.
(b) Change $f(x)$ to $F(u)$.
(c) Use $\displaystyle\int f(x)\,dx = \int F(u)\frac{dx}{du}\,du$.

When evaluating an **indefinite** integral by substitution, the answer must be transformed back to a function of x.

When evaluating a **definite** integral by substitution, change the limits of integration to the corresponding limits for the new variable and find the answer in this form.

$\boxed{\mathbf{i}}$ Using the substitution $x=\tan\theta$, find $\displaystyle\int_0^1 \frac{dx}{(1+x^2)^2}$.

Let $x=\tan\theta$ Limits:

x	θ
1	$\tfrac{\pi}{4}$
0	0

$\dfrac{dx}{d\theta}=\sec^2\theta$

$\displaystyle\int_0^1 \frac{dx}{(1+x^2)^2} = \int_0^{\frac{\pi}{4}} \frac{1}{(1+\tan^2\theta)^2}\sec^2\theta\,d\theta$

$\qquad = \displaystyle\int_0^{\frac{\pi}{4}} \frac{\sec^2\theta}{\sec^4\theta}\,d\theta = \int_0^{\frac{\pi}{4}}\frac{1}{\sec^2\theta}\,d\theta$

$\qquad = \displaystyle\int_0^{\frac{\pi}{4}} \cos^2\theta\,d\theta = \frac{1}{2}\int_0^{\frac{\pi}{4}}(1+\cos2\theta)\,d\theta$

$\qquad = \left[\dfrac{\theta}{2}+\dfrac{1}{4}\sin2\theta\right]_0^{\frac{\pi}{4}} = \dfrac{\pi}{8}+\dfrac{1}{4}$

$\boxed{\mathbf{i}}$ By a suitable substitution, find $\displaystyle\int \frac{x}{\sqrt{(x-3)}}\,dx$.

Let $u=\sqrt{(x-3)}$, so $\dfrac{du}{dx}=\dfrac{1}{2}(x-3)^{-\frac{1}{2}}=\dfrac{1}{2(x-3)^{\frac{1}{2}}}$

$\qquad\therefore \dfrac{dx}{du}=2(x-3)^{\frac{1}{2}}=2u$

and $x=u^2+3$.

$\displaystyle\int \frac{x}{\sqrt{(x-3)}}\,dx = \int \frac{(u^2+3)}{u}\cdot2u\,du = 2\int(u^2+3)\,du$

$\qquad = \tfrac{2}{3}u^3+6u+c$

$\qquad = \tfrac{2}{3}[\sqrt{(x-3)}]^3+6\sqrt{(x-3)}+c$

Parts

The integration of some products require **integration by parts** using

$$\int u\frac{dv}{dx}\cdot dx = uv - \int v\frac{du}{dx}\,dx$$

The product is usually of a 'power of x' and either a logarithmic, trigonometrical or exponential function. Care must be taken in the choice of the factor to be:
    u (the one to be differentiated)
and $\dfrac{dv}{dx}$ (the one to be integrated).

$v\dfrac{du}{dx}$ must be simpler to integrate than $u\dfrac{dv}{dx}$.

If it is not, make the other factor u.

$\boxed{\mathbf{i}}$ Evaluate $\displaystyle\int x\sin x\,dx$.

Let $u=x$ and $\dfrac{dv}{dx}=\sin x$

so $\dfrac{du}{dx}=1$ so $v=-\cos x$

$\displaystyle\int x\sin x\,dx = x\cdot(-\cos x)-\int(-\cos x)\cdot1\cdot dx$

$\qquad = -x\cos x+\displaystyle\int\cos x\,dx$

$\qquad = -x\cos x+\sin x+c$

Methods of Integration
Worked examples, Guided example and Exam questions

 Find the following integrals.

(a) $\int x^2 \ln x \, dx$ (b) $\int \dfrac{2x^2+2x+3}{(x+2)(x^2+3)} \, dx$

(a) This requires integration by parts.

Let $u = \ln x$, so $\dfrac{du}{dx} = \dfrac{1}{x}$,

and $\dfrac{dv}{dx} = x^2$, so $v = \dfrac{1}{3}x^3$.

Now $\int x^2 \ln x \, dx = (\ln x)\left(\dfrac{1}{3}x^3\right) - \int \dfrac{1}{3}x^3 \cdot \dfrac{1}{x} \, dx$

$= \dfrac{1}{3}x^3 \ln x - \dfrac{1}{3}\int x^2 \, dx$

$= \dfrac{1}{3}x^3 \ln x - \dfrac{1}{9}x^3 + c$

(b) Notice that the integrand can be written in partial fractions.

i.e. $\dfrac{2x^2+2x+3}{(x+2)(x^2+3)} = \dfrac{1}{x+2} + \dfrac{x}{x^2+3}$ (see unit 5)

So, $\int \dfrac{2x^2+2x+3}{(x+2)(x^2+3)} \, dx = \int \dfrac{1}{x+2} \, dx + \int \dfrac{x}{x^2+3} \, dx$

$= \ln (x+2) + \tfrac{1}{2} \ln (x^2+3) + c$

$= k \ln \{(x+2) \sqrt{(x^2+3)}\}$

 Using the substitution $u = \sqrt{(5x+1)}$, evaluate

$\int_0^3 2x\sqrt{(5x+1)} \, dx.$

If $u = \sqrt{(5x+1)}$ then $x = \dfrac{u^2-1}{5}$.

Hence $\dfrac{dx}{du} = \dfrac{2u}{5}$

and $\begin{cases} x = 3 \Rightarrow u = 4 \\ x = 0 \Rightarrow u = 1 \end{cases}$

Hence $\int_0^3 2x\sqrt{(5x+1)} \, dx = \int_1^4 \dfrac{2}{5}(u^2-1) \cdot u \cdot \dfrac{2u}{5} \, du$

$= \dfrac{4}{25}\int_1^4 (u^4-u^2) \, du$

$= \dfrac{4}{25}\left[\dfrac{u^5}{5} - \dfrac{u^3}{3}\right]_1^4$

$= \dfrac{4}{25}\left[\left(\dfrac{4^5}{5} - \dfrac{4^3}{3}\right) - \left(\dfrac{1}{5} - \dfrac{1}{3}\right)\right]$

$= \dfrac{4}{25}\left[204\dfrac{4}{5} - 21\dfrac{1}{3} + \dfrac{2}{15}\right]$

$= \dfrac{4}{25}\left[183\dfrac{9}{15}\right] = 29 \cdot 4$

 Find the following integrals

(a) $\int x^2 e^{4x} \, dx$ (b) $\int \dfrac{x^2-2}{x^2-1} \, dx$ (c) $\int \dfrac{1}{\sqrt{(1+x^2)}} \, dx$

(a) Since the integrand is a product use integration by parts.

Let $u = x^2$ and $\dfrac{dv}{dx} = e^{4x}$. After first application, you will get

---

an integral of the form $\int xe^{4x} \, dx$. This now requires integrating by parts. For this integral let $u = x$ and $\dfrac{dv}{dx} = e^{4x}$. Complete the integration.

(b) Notice the integrand is a rational function with numerator and denominator of equal degree. This cannot be expressed in partial fractions immediately. Divide out the integrand first, and express in partial fractions. Now integrate.

(c) Use the substitution $x = \tan u$ (since $1 + \tan^2 u = \sec^2 u$). Now integrate the function and substitute for $u = \tan^{-1}x$ at the end.

EX 1 Evaluate $\int_1^2 x(2-x)^7 \, dx$.   (L)

2 Evaluate, correct to two decimal places, $\int_0^1 (1-x) \sin x \, dx$.   (J)

3 Evaluate:

(a) $\int_1^4 \left(\sqrt{x} + \dfrac{1}{\sqrt{x}}\right)^3 \, dx$;

(b) $\int_0^{\frac{\pi}{2}} \cos 2x \sin 4x \, dx$;

(c) $\int_0^{\frac{3}{4}} \dfrac{1-x}{(x+1)(x^2+1)} \, dx$.   (L)

4 Given that $y = \dfrac{2x-1}{(x-2)(5-x)}$,

(i) express y in partial fractions;

(ii) evaluate $\int_3^4 y \, dx$.   (A)

5 (a) Evaluate (i) $\int_0^{\frac{\pi}{6}} \tan 2x \, dx$;   (ii) $\int_1^e x^2 \ln x \, dx$.

[Logarithms and powers of e need not be evaluated.]

(b) Find $\int \dfrac{e^{-\sqrt{x}}}{\sqrt{x}} \, dx$.   (C)

6 Evaluate the integrals

$\int_0^{\frac{1}{4}} \dfrac{dx}{\sqrt{(1-4x^2)}}$ and $\int_0^{\frac{1}{4}} \dfrac{x \, dx}{\sqrt{(1-4x^2)}}$.

7 Evaluate the indefinite integrals:

(i) $\int \dfrac{1}{\sqrt{(2x+1)}} \, dx$;   (ii) $\int \dfrac{x}{2x+1} \, dx$;   (iii) $\int \dfrac{e^x}{4-e^{2x}} \, dx$.

(You may find the substitution $e^x = t$ helpful in (iii).)   (O & C)

8 (a) Using the substitution $y = x+1$, or otherwise, evaluate:

$\int_{-1}^2 \dfrac{3}{x^2+2x+10} \, dx$.

(b) Find $\int 2xe^{2x} \, dx$.

(c) Find $\int \dfrac{x-4}{(x-1)^2(2x+1)} \, dx$.   (W)

35 Applications of Integration
Areas, Volumes of revolution, Mean value.

Areas

Before calculating areas, sketch the curve.

1. Area between a curve and the x-axis

The area bounded by
the curve $y=f(x)$,
the x-axis and the
lines $x=a$, $x=b$
is given by

$$\int_a^b y\,dx.$$

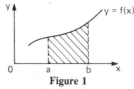

Figure 1

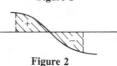

This is positive if the
area is above the x-axis
and negative if below.

Figure 2

If the curve cuts the x-axis between the given limits
(a) find areas above and below the x-axis separately,
(b) add their numerical values.

2. Area between a curve and the y-axis

The area bounded by
the curve $x=g(y)$,
the y-axis and the
lines $y=c$, $y=d$
is given by

$$\int_c^d x\,dy.$$

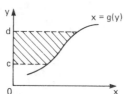

Figure 3

3. Area between two curves

To find the area between two curves, calculate
(a) the x-coordinates of the points of intersection of
the curves to give the limits of integration,
(b) the area under each curve separately,
(c) the difference between the areas.

 Find the area enclosed by $y=x^3-4x^2+3x$ and the x-axis between $x=0$ and $x=3$.

$$y=x^3-4x^2+3x$$
$$=x(x^2-4x+3)$$
$$=x(x-1)(x-3)$$

So the curve cuts the
x-axis at $x=0$,
$x=1$ and $x=3$.

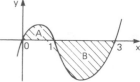

Figure 4

$$\text{Area A}=\int_0^1 (x^3-4x^2+3x)\,dx$$

$$=\left[\frac{x^4}{4}-\frac{4x^3}{3}+\frac{3x^2}{2}\right]_0^1$$

$$=\left(\frac{1}{4}-\frac{4}{3}+\frac{3}{2}\right)-0$$

$$=\frac{5}{12}\text{ square units}$$

$$\text{Area B}=\int_1^3 (x^3-4x^2+3x)\,dx$$

$$=\left[\frac{x^4}{4}-\frac{4x^3}{3}+\frac{3x^2}{2}\right]_1^3$$

$$=\left(\frac{81}{4}-\frac{108}{3}+\frac{27}{2}\right)-\left(\frac{5}{12}\right)$$

$$=-\frac{27}{12}-\frac{5}{12}$$

$$=-\frac{32}{12}\text{ square units}$$

The negative sign confirms that B is below the x-axis.

$$\text{Total area}=\frac{5}{12}+\frac{32}{12}=\frac{37}{12}\text{ square units}$$

Volumes of revolution

Before calculating volumes, sketch the curve.

1. Rotation about the x-axis

The area bounded by
the curve $y=f(x)$,
the x-axis and the
lines $x=a$, $x=b$ is
rotated once about
the x-axis.

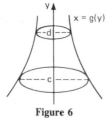

Figure 5

The volume of the solid formed is given by $\int_a^b \pi y^2\,dx$.

2. Rotation about the y-axis

The area bounded by
the curve $x=g(y)$,
the y-axis and the
lines $y=c$, $y=d$ is
rotated once about
the y-axis.

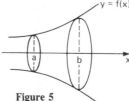

Figure 6

The volume of the solid formed is given by $\int_c^d \pi x^2\,dy$.

 Find the volume of the solid formed when the area between the x-axis, the lines $x=2$ and $x=4$ and the curve $y=x^2$ is rotated once about the x-axis. Leave your answer as a multiple of π.

Volume of revolution

$$=\int_2^4 \pi y^2\,dx$$

$$=\int_2^4 \pi (x^2)^2\,dx$$

$$=\int_2^4 \pi x^4\,dx$$

$$=\left[\frac{\pi x^5}{5}\right]_2^4=\frac{\pi}{5}(4^5-2^5)$$

$$=\frac{\pi}{5}(1024-32)$$

$$=\frac{992}{5}\pi\text{ cubic units}$$

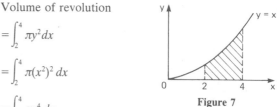

Figure 7

Mean value

The **mean value** of $y=f(x)$ over a closed interval
$a\le x\le b$ is defined to be

$$\frac{1}{b-a}\int_a^b y\,dx.$$

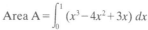 *Find the mean of $y=\sin x$ over the interval 0 to π.*

$$\text{Mean}=\frac{1}{\pi-0}\int_0^\pi \sin x\,dx=\frac{1}{\pi}[-\cos x]_0^\pi$$

$$=\frac{1}{\pi}\{[-(-1)]-[-(1)]\}=\frac{2}{\pi}$$

Applications of Integration
Worked examples, Guided example and Exam questions

WE

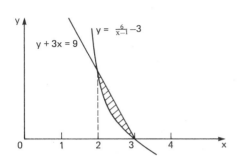

Calculate the shaded area shown between the curve $y = \dfrac{6}{x-1} - 3$

and the straight line $y + 3x = 9$.

The curve $y = \dfrac{6}{x-1} - 3$ and the line $y + 3x = 9$ intersect at $x = 2$
and $x = 3$.

We first calculate Area$_1$ between $y + 3x = 9$, $x = 2$, $x = 3$ and the
x-axis.

Since the shape formed is a trapezium,

$$\text{Area}_1 = \frac{1}{2}\left[(9 - 3 \times 2) + 0\right]\,1 = \frac{3}{2} \text{ square units.}$$

Area$_2$ between the curve, $x = 2$, $x = 3$ and the x-axis is

$$\begin{aligned}
\text{Area}_2 &= \int_2^3 \left(\frac{6}{x-1} - 3\right) dx \\
&= \left[6 \ln(x-1) - 3x\right]_2^3 \\
&= (6 \ln 2 - 9) - (6 \ln 1 - 6) \\
&= (6 \ln 2 - 3) \text{ square units.}
\end{aligned}$$

$$\begin{aligned}
\text{Required area} &= \text{Area}_1 - \text{Area}_2 \\
&= \frac{3}{2} - (6 \ln 2 - 3) \\
&= \frac{3}{2}(3 - 4 \ln 2) \text{ square units.}
\end{aligned}$$

WE *Sketch, on the same axes, that part of the curve $y = 16 - x^2$ and
the line $y = 6x$ which lies in the same quadrant. Shade the area
which satisfies $y \leqslant 16 - x^2$, $y \geqslant 6x$ and $x \geqslant 0$. Find the volume
generated when this area is rotated completely about the x-axis,
leaving your answer as a multiple of π.*

The required volume is
obtained by rotating the area
below the curve and between
the lines $x = 0$ and $x = 2$ and the
x-axis around the x-axis and
then subtracting the volume
obtained by rotating the
triangle with base 2 and height
12 around the x-axis.

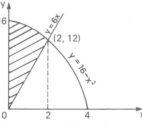

So, volume required $= \displaystyle\int_0^2 \pi(16 - x^2)^2\, dx - \int_0^2 \pi(6x)^2\, dx$

$$= \pi \int_0^2 (256 - 32x^2 + x^4 - 36x^2)\, dx$$

$$= \pi \int_0^2 (256 - 68x^2 + x^4)\, dx$$

$$= \pi\left[256x - \frac{68}{3}x^3 + \frac{x^5}{5}\right]_0^2$$

$$= \pi\left[(512 - 68.\tfrac{8}{3} + \tfrac{32}{5}) - (0)\right]$$

$= 337\frac{9}{15}\pi$ cubic units.

GE *A bowl is formed by rotating about the y-axis the area contained
between that part of the curve $2y = x^2$ from $x = 0$ to $x = 4$, the line
$x = 4$ and the x-axis. Calculate (a) the maximum volume of
water the bowl could hold (b) the volume of material forming
the bowl.*

First sketch the required area.
To find the volume of water the bowl could hold evaluate the

integral $\displaystyle\int_0^8 \pi\, x^2\, dy$.

To find the volume of material forming the bowl, calculate the
volume of a circular cylinder of radius 4 and height 8 units.
Subtract the first volume from this to give the required volume
of material.

EX

1 Calculate the area enclosed between the curve $y = 3x^2 + 2$,
the lines $x = 1$, $x = 3$ and the x-axis. *(A)*

2 (i) Sketch the curve $y = (x-1)(x-2)(x-3)$, indicating
clearly the co-ordinates of the points where it cuts
the axes.
(ii) Find the area, *above* the x-axis, bounded by the curve
and the axis of x. *(W)*

3 A function $f(x)$ is defined by the formula $f(x) = 2x^4 + 2x^3$,
$x \in R$, where R is the set of real numbers.
(a) Find the stationary points of $f(x)$ and determine their
nature, justifying your answers.
(b) Find where the graph of $f(x)$ meets the x and y axes and
make a rough sketch of the graph.
(c) Calculate the finite area bounded by the x-axis and the
graph of $f(x)$. *(H)*

4 Sketch the curve: $y = 1 - \dfrac{4}{x^2}$. The region R is bounded by

the curve $y = 1 - \dfrac{4}{x^2}$, the x-axis and the lines $x = \frac{1}{2}$ and

$x = 1$. Find the volume generated when R is rotated
completely about the x-axis, leaving your answer as a
multiple of π. *(S)*

5 Sketch the graphs of $y^2 = 16x$ and $y = x - 5$. Find
(i) the co-ordinates of their points of intersection;
(ii) the area of the finite region enclosed between the
graphs. *(A)*

6 Draw a rough sketch of the circle $x^2 + y^2 = 100$, and the
curve $9y = 2x^2$; find the co-ordinates of the points A and B
where they meet. Calculate the area bounded by the minor
arc AB of the circle and the other curve, and the volume
obtained by rotating this area about the axis Oy. *(OLE)*

7 The region R in the first quadrant is bounded by the y-axis,
the x-axis, the line $x = 3$ and the curve $y^2 = 4 - x$.
(i) Draw a sketch showing the region R and calculate
its area.
(ii) Calculate the volume formed when R is rotated
about the y-axis through one revolution. *(C)*

8 The function g is defined by $g(x) = \sqrt{(1 - x^2)}$. Find the area
of the region bounded by the graph of g, the lines $x = 0$,
$x = \frac{1}{2}$ and the x-axis, by using the substitution $x = \sin\theta$, or
otherwise, to evaluate the integral. *(W)*

9 The area bounded by the curve $y = \tan x$, the x-axis and the

ordinate $x = \dfrac{\pi}{3}$ is rotated about the x axis. Calculate the

volume of the solid of revolution so formed. (Give your
answer to 3 significant figures.) *(S)*

36 Differential Equations
Definitions, Solution, Formation.

Definitions

A **differential equation** is an equation which contains at least one differential coefficient.
Only **first order** differential equations will be considered here.

The **solution** of a differential equation is an equation relating the variables involved but containing no differential coefficients.
The **general solution** contains an arbitrary constant.
A **particular solution** may be obtained if an 'x value' and a corresponding 'y value' are given. These values are called **boundary conditions** or **initial conditions** and enable the constant to be calculated.

Graphically a differential equation describes a property of a **family of curves**. Its general solution is the equation of any member of that family. Its particular solution is the equation of one particular member of that family.

$\boxed{i}$ $\dfrac{dy}{dx}=3x$ is a first order differential equation

because the only differential coefficient is $\dfrac{dy}{dx}$.

Its general solution is $y=\frac{3}{2}x^2+c$.

If $x=0$ when $y=1$, $c=1$.
So the particular solution is $y=\frac{3}{2}x^2+1$.

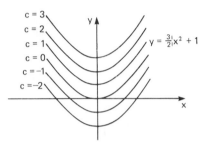

Figure 1

Solution

1. Equations of the form $\dfrac{dy}{dx}=f(x)$

Integrate directly so that $y=\displaystyle\int f(x)\,dx$.

$\boxed{i}$ $Solve\ \dfrac{dy}{dx}=\dfrac{1}{x}$

$$y=\int \frac{1}{x}\,dx$$
$$y=\ln|x|+c$$

2. Equations of the form $\dfrac{dy}{dx}=f(y)$

(a) Rewrite the equation in the form $\dfrac{1}{f(y)}\cdot\dfrac{dy}{dx}=1$.

(b) Integrate directly $\displaystyle\int\frac{1}{f(y)}\cdot dy=\int dx$.

$\boxed{i}$ $Solve\ \dfrac{dy}{dx}=\operatorname{cosec} y$

$$\frac{1}{\operatorname{cosec} y}\cdot\frac{dy}{dx}=1$$
$$\int \sin\cdot dy=\int dx$$
$$-\cos y=x+c$$

3. Equations in which the variables are separable

i.e. of the form $\dfrac{dy}{dx}=\dfrac{f(x)}{g(y)}$ or $f(x)\cdot g(y)$

(a) Separate the variables and rewrite in the form

$$g(y)\cdot\frac{dy}{dx}=f(x).$$

(b) Integrate each side separately

$$\int g(y)\,dy=\int f(x)\,dx.$$

$\boxed{i}$ $Solve\ y(1+x^2)\dfrac{dy}{dx}-2(1+y^2)=0.$

Separate the variables $\dfrac{y}{(1+y^2)}\cdot\dfrac{dy}{dx}=\dfrac{2}{(1+x^2)}$

Integrate $\displaystyle\int\frac{y}{(1+y^2)}dy=\int\frac{2}{(1+x^2)}dx$

$$\tfrac{1}{2}\ln|(1+y^2)|=2\tan^{-1}(1+x^2)+c$$

Formation

Many situations in science and nature are concerned with the rate at which a quantity changes. These can be expressed mathematically as differential equations which can then be solved.
Remember that

'rate of decrease' $=-$'rate of increase'

$\boxed{i}$ 1. The rate of decay of the nuclei of radioactive substances is proportional to the number x of nuclei remaining.
Rate of decay of x is $\dfrac{-dx}{dt}\propto x$.

So $\dfrac{-dx}{dt}=kx$ or $\dfrac{dx}{dt}=-kx$ (where k is a constant)

2. Suppose that the velocity v of a particle is proportional to the square of its displacement s.
This gives $v\propto s^2$
    i.e. $v=ks^2$ (where k is a constant).

But $v=\dfrac{ds}{dt}$, so $\dfrac{ds}{dt}=ks^2$.

Differential Equations
Worked examples, Guided example and Exam questions

 WE *Solve the differential equation* $\dfrac{dy}{dx} = \dfrac{\sin^2 x}{y^2}$ *given that y = 1 when x = 0.*

$\dfrac{dy}{dx} = \dfrac{\sin^2 x}{y^2}$ is a variable separable differential equation.

It can be rewritten as

$$y^2 \dfrac{dy}{dx} = \sin^2 x.$$

Integrating, $\displaystyle\int y^2 \, dy = \int \sin^2 x \, dx$

giving, $\dfrac{y^3}{3} = \dfrac{1}{2}\displaystyle\int (1 - \cos 2x) \, dx$

i.e. $\dfrac{y^3}{3} = \dfrac{1}{2}\left(x - \dfrac{\sin 2x}{2}\right) + c.$

$$\left[\text{Since } \sin^2 x = \dfrac{1}{2}(1 - \cos 2x)\right]$$

Since $y = 1$ when $x = 0$,

$$\dfrac{1^3}{3} = \dfrac{1}{2}\left(0 - \dfrac{\sin 0}{2}\right) + c$$

$$\Rightarrow \quad c = \dfrac{1}{3}.$$

So the solution is

$$\dfrac{y^3}{3} = \dfrac{1}{4}(2x - \sin 2x) + \dfrac{1}{3},$$

or $4y^3 = 3(2x - \sin 2x) + 4.$

WE *Newton's law of cooling states that the rate at which a body, at T°C above the temperature of its surroundings, cools is proportional to T.*
A body at 68°C is placed in a room at 16°C and after 5 minutes it has cooled to 55°C. What will be its temperature after a further five minutes?

Newton's law gives $\dfrac{dT}{dt} = -kT$, where k is a constant to be determined.

Integrating, $\displaystyle\int \dfrac{dT}{T} = -k\int 1 \, dt,$

i.e. $\ln T = c - kt$, c is the constant of integration.

So, $T = e^c \cdot e^{-kt}$.

When $t = 0$, $T = 68 - 16 = 52 \Rightarrow e^c = 52$

$\therefore \quad T = 52e^{-kt}$

When $t = 5$, $T = 55 - 16 = 39$

$$\therefore \quad 39 = 52e^{-5k} \Rightarrow e^{-5k} = \dfrac{39}{52} = \dfrac{3}{4}$$

After a further 5 minutes, i.e. $t = 10$,

$$T = 52e^{-10k}$$
$$= 52(e^{-5k})^2$$
$$= 52(\tfrac{3}{4})^2$$
$$= 29\tfrac{1}{4}$$

The temperature of the body is therefore $16 + 29\tfrac{1}{4} = 45\tfrac{1}{4}°C.$

GE *Find the solution of the differential equation* $(x+1)\dfrac{dy}{dx} = y$, *given that y = 4 when x = 1.*

Rewrite the differential equation in the form of $\dfrac{dy}{dx} = \dfrac{f(x)}{g(y)}$.

Separate the variables and integrate both sides. Include an arbitrary constant of integration. Use the conditions $y = 4$ when $x = 1$ to find the constant of integration.

EX

1 Find y in terms of x given that $x\dfrac{dy}{dx} = y(y+1)$ and $y = 4$ when $x = 2$.

(L)

2 Find the solution of the differential equation $\dfrac{dy}{dx} = xy \ln x$ which satisfies the initial conditions $x = 1$, $y = 1$, giving $\ln y$ in terms of x.

(O & C)

3 Find the solution of the differential equation

$2\dfrac{dy}{dx} = 2xe^{-2y} + e^{-2y}$ for which $y = 0$ when $x = 0$.

(S)

4 Solve the differential equation $(1 + e^y)\dfrac{dy}{dx} = e^{2y} \cos^2 x$, given that $y = 0$ when $x = 0$.

(A)

5 Solve the differential equation

$(1 + x)\dfrac{dy}{dx} = 1 - \sin^2 y$ for which $y = \dfrac{\pi}{4}$ when $x = 0$.

6 Find y in terms of x given that $\dfrac{dy}{dx} = y(1-y)$ and that $y = \tfrac{1}{2}$ when $x = 0$.

(C)

7 During a chemical reaction two substances A and B decompose. The number of grams, x, of substance A present at time t is given by $x = \dfrac{10}{(1+t)^3}$.

There are y grams of B present at time t and $\dfrac{dy}{dt}$ is directly proportional to the product of x and y. Given that $y = 20$ and $\dfrac{dy}{dt} = -40$ when $t = 0$, show that $\dfrac{dy}{dt} = \dfrac{-2y}{(1+t)^3}$.

Hence determine y as a function of t. Determine the amount of substance B remaining when the reaction is essentially complete.

(A)

8 A plant grows in a pot which contains a volume V of soil. At time t the mass of the plant is m and the volume of soil utilised by the roots is αm, where α is a constant. The rate of increase of the mass of the plant is proportional to the mass of the plant times the volume of soil not yet utilised by the roots. Obtain a differential equation for m, and verify that it can be written in the form

$$V\beta \dfrac{dt}{dm} = \dfrac{1}{m} + \dfrac{\alpha}{V - \alpha m}, \text{ where } \beta \text{ is a constant.}$$

The mass of the plant is initially $\dfrac{V}{4\alpha}$. Find, in terms of V and β, the time taken for the plant to double its mass. Find also the mass of the plant at time t.

(J)

37 Numerical Solution of Equations
Introduction, Initial values, Iterative methods.

Introduction Many equations cannot be solved exactly, but various methods of finding approximate numerical solutions exist. The most commonly used methods have two main parts:
(a) finding an initial approximate value   (b) improving this value by an iterative process.

Initial values The roots of $f(x)=0$ can be located approximately by either a graphical or an algebraic method.

Graphical method
Either (a) Plot (or sketch) the graph of $y=f(x)$. The real roots are at the points where the curve cuts the x-axis.

or   (b) Rewrite $f(x)=0$ in the form $F(x)=G(x)$. Plot (or sketch) $y=F(x)$ and $y=G(x)$. The real roots are at the points where these graphs intersect.

Algebraic method
Find two values a and b such that $f(a)$ and $f(b)$ have different signs. At least one root must lie between a and b if $f(x)$ is continuous.
If more than one root is suspected between a and b, sketch a graph of $y=f(x)$.

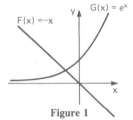 *Locate an approximate value for the root of*
$$f(x)=x+e^x=0.$$

Graphical method
Rewrite $f(x)=0$ as $e^x=-x$.
Let $F(x)=-x$ and $G(x)=e^x$.
Sketch $F(x)$ and $G(x)$

When $x=0$, $G(x)>F(x)$
    $x=-1$, $G(x)<F(x)$
$\therefore f(x)=0$ has a root between -1 and 0.

Figure 1

Algebraic method
$f(x)=x+e^x$
$f(1)=1+e^1=3.72$
$f(0)=0+e^0=1$
$f(-1)=-1+e^{-1}=-0.63$ $\left.\right\}$ different signs
$\therefore f(x)=0$ has a root between -1 and 0.

Iterative methods All **iterative methods** follow the same basic pattern.
A sequence of approximations $x_1, x_2, x_3, x_4, \ldots$ is found, each one closer to the root α of $f(x)=0$.
Each approximation is found from the one before it using a specified method.
The process is continued until the required accuracy is reached.

General iteration method
Rewrite the equation $f(x)=0$ in the form $x=g(x)$.
If the initial approximation is x_1
then calculate $x_2=g(x_1)$
        $x_3=g(x_2)$
        $x_4=g(x_3)$
        and so on . . .

This method fails if $|g'(x)|>1$ near the root.

Newton–Raphson method
If x_1 is an approximation to a root α of $f(x)=0$, then a better approximation x_2 is given by
$$x_2=x_1-\frac{f(x_1)}{f'(x_1)}$$
Repeat this process as required.

This method fails if (a) $f'(\alpha)$ is near to zero,
            or (b) $f''(\alpha)$ is very large.

⬛ Find the solution of $f(x)=x+e^x$ near $x=-1$ to three decimal places.

General iteration method
Write the given equation as $x=-e^x$.

$x_1 =-1$
$x_2 =-e^{-1}   =-0.368$
$x_3 =-e^{-0.368}=-0.692$
$x_4 =-e^{-0.692}=-0.500$
$x_5 =-e^{-0.500}=-0.607$
$x_6 =-e^{-0.607}=-0.545$
$x_7 =-e^{-0.545}=-0.580$
$x_8 =-e^{-0.580}=-0.560$
$x_9 =-e^{-0.560}=-0.571$
$x_{10}=-e^{-0.571}=-0.565$
$x_{11}=-e^{-0.565}=-0.568$
$x_{12}=-e^{-0.568}=-0.567$
$x_{13}=-e^{-0.567}=-0.567$ is the required solution.

Newton–Raphson method
$$f(x)=x+e^x$$
So $f'(x)=1+e^x$

$x_1=-1$

$$x_2=-1-\frac{(-1+e^{-1})}{(1+e^{-1})}=-1-\frac{(0.632)}{1.368}=-0.538$$

$$x_3=-0.538-\frac{(-0.538+e^{-0.538})}{(1+e^{-0.538})}=-0.567$$

$$x_4=-0.567-\frac{(-0.567+e^{-0.567})}{(1+e^{-0.567})}=-0.567$$

The required solution is $x=-0.567$.

Note: In this example although the Newton–Raphson method involves fewer steps, the calculation by the general iteration method is so simple that there is not much difference in the calculation time of each method.

Numerical Solution of Equations
Worked examples and Exam questions

 Show that the equation $x^3+x-6=0$ has a root between 1 and 2. Using Newton's approximation with starting point $1\cdot6$ determine, by means of two iterations, an approximation to this root, giving your answer to two decimal places.

Let $f(x)=x^3+x-6$

$f(1)=1+1-6=-4<0$

$f(2)=8+2-6=\ \ 4>0$

$\therefore\ f(x)=0$ between $x=1$ and $x=2$

i.e. $x^3+x-6=0$ between $x=1$ and $x=2$

$\therefore\ x^3+x-6=0$ has a root between $x=1$ and $x=2$.

Let $x_2=x_1-\dfrac{f(x_1)}{f'(x_1)}$ where $f'(x)=3x^2+1$

$x_2=1\cdot6-\dfrac{(1\cdot6^3+1\cdot6-6)}{3\times1\cdot6^2+1}$, using $x_1=1\cdot6$

$=1\cdot6-(-0\cdot0350)=1\cdot6350$

$x_3=1\cdot6350-\dfrac{(1\cdot6350^3+1\cdot6350-6)}{3\times1\cdot6350^2+1}$

$=1\cdot6350-(0\cdot0006)=1\cdot6344$

$\therefore$ to two decimal places the root is $1\cdot63$.

 Show that $x^3+3x-12=0$ has only one real root α, and that $1<\alpha<2$. Use linear interpolation to determine the number k, expressed to one decimal place, such that $k<\alpha<k+0\cdot1$.

Given that α is a root of $x^3+3x-12=0$

Then $(x-\alpha)$ is a factor of $x^3+3x-12$

Dividing $(x-\alpha)$ into $x^3+3x-12$ gives the quadratic factor $x^2+\alpha x+(3+\alpha^2)$

$\therefore\ x^3+3x-12=(x-\alpha)[x^2+\alpha x+(3+\alpha^2)]=0$

The factor $x^2+\alpha x+(3+\alpha^2)$ does not give rise to real roots because its discriminant is negative.

$\therefore\ x=\alpha$ is the only real root of $x^3+3x-12=0$

$f(1)=1+3-12=-8<0$

$f(2)=8+6-12=\ \ 2>0$

$\therefore\ 1<\alpha<2$

From the similar triangles shown in the diagram

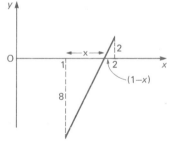

$\dfrac{8}{x}=\dfrac{2}{1-x}$, i.e. $8-8x=2x$

$\therefore\ x=0\cdot8$

Let $k=1\cdot8$, $f(1\cdot8)=-0\cdot768<0$

$\qquad\quad f(1\cdot9)=\ \ 0\cdot559>0$

$\therefore$ if $k=1\cdot8$ then $1\cdot8<\alpha<1\cdot8+0\cdot1=1\cdot9$

---

 1 Show graphically, or otherwise, that the equation $\ln x=4-x$ has only one real root and prove that this root lies between $2\cdot9$ and 3. By taking $2\cdot9$ as a first approximation to this root and applying the Newton–Raphson process once to the equation $\ln x-4+x=0$, or otherwise, find a second approximation, giving your answer to 3 significant figures.

(L)

2 Give, on the same diagram, a sketch of the graph of $y=3e^{\frac{1}{2}x}$ and of the graph $y=4x+6$. State the number of roots of the equation $3e^{\frac{1}{2}x}=4x+6$. Taking 4 as a first approximation to one root α of this equation, find a second approximation to α, giving three significant figures in your answer and showing clearly how this answer has been

obtained. Using a suitable integer as first approximation to another root β of the equation, find a second approximation to β, again giving three significant figures in your answer and showing clearly how this answer has been determined.

(C)

3 Use Newton's method once, starting with the approximation $x=2$, to obtain a second approximation x_1 for a root of the equation $x^5=x^3+25$.
What does it mean to claim that x_1 is correct to two decimal places? State briefly how you would prove this (no detailed working is expected).

$(O\ \&\ C)$

4 Show that the equation $x(x^2+2)-4=0$ has only one real root, and that this root lies between 1 and $1\cdot5$. Taking $1\cdot2$ as a first approximation to this root, use Newton's method to obtain a second approximation, giving your result to two places of decimals (or to six places if you use a machine).

(J)

5 Show, by means of a sketch graph, or otherwise, that the equation $e^{2x}+4x-5=0$ has only one real root, and that this root lies between 0 and 1. Starting with the value $0\cdot5$ as a first approximation to this root, use the Newton–Raphson method to evaluate successive approximations, showing the stages of your work and ending when two successive approximations give answers which, when rounded to two decimal places, agree.

(C)

6 Find, by *any* method, the solution of the equation $x^2+20\ln x=400$, giving your answer correct to 3 significant figures.

$(O\ \&\ C)$

7 Given that $y=e^x$, copy and complete the following table, giving the two required values of y to 2 decimal places.

x	0	$0\cdot5$	1	$1\cdot5$	2	$2\cdot5$
y	1	$1\cdot65$	$2\cdot72$			$12\cdot18$

On graph paper draw the graph of $y=e^x$ for $0\le x\le2\cdot5$. Using the same axes, draw a straight line graph and hence obtain an estimate to the root of the equation $e^x=5(1-x)$.

Evaluate $\displaystyle\int_1^2(e^x-2)\ dx$. (You may leave your answer in terms of e.)

On your graph show, by shading, a region whose area is given by the definite integral $\displaystyle\int_1^2(e^x-2)\ dx$.

(L)

75

38 Numerical Integration
Introduction, Trapezium rule, Simpson's rule.

Introduction

Sometimes it is impossible to evaluate the integral $\int_a^b f(x)\,dx$ exactly. Since this integral gives the area bounded by the curve $y=f(x)$, the x-axis and the lines $x=a$ and $x=b$, an approximate value for the integral can be found by estimating this area by another method.

Two common methods are the **trapezium rule** and **Simpson's rule.**

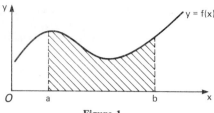

Figure 1

Trapezium rule

This method divides the area into n trapezia, each of width h.

Area under curve ≈ sum of areas of trapezia.

This gives

$$\int_a^b f(x)\,dx \approx \frac{h}{2}\{(y_0+y_n)+2(y_1+y_2+\ldots+y_{n-1})\}$$

where $h=\dfrac{b-a}{n}$.

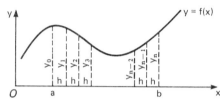

Figure 2

Simpson's rule

This method divides the area into an even number of parallel strips n, of width h, and approximates the area of pairs of strips using parabolas.

This gives

$$\int_a^b f(x)\,dx \approx \frac{h}{3}\{(y_0+y_n)+4(y_1+y_3+\ldots y_{n-1})+2(y_2+y_4+\ldots+y_{n-2})\}$$

Note: the number of strips must be even.

[i] *Evaluate $\int_1^9 \log_e x\,dx$ using 8 strips (a) by the trapezium rule (b) by Simpson's rule.*

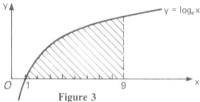

Figure 3

The integration interval $(b-a)=9-1=8$ units

So $h=\dfrac{b-a}{n}=\dfrac{8}{8}=1$.

The values of x at which y is calculated are:
1, 2, 3, 4, 5, 6, 7, 8, 9.

(a) Trapezium rule

Tabulating the results as follows helps the final calculation:

x	y	First and last ordinates	Remaining ordinates
1	y_0	0	
2	y_1		0.693
3	y_2		1.099
4	y_3		1.386
5	y_4		1.609
6	y_5		1.792
7	y_6		1.946
8	y_7		2.079
9	y_8	2.197	
Totals		2.197	10.604

$$\int_1^9 \log_e x\,dx \approx \frac{1}{2}\{(y_0+y_8)+2(y_1+\ldots+y_7)\}$$

$$=\frac{1}{2}\{2.197+2(10.604)\}$$

$$=11.703$$

(b) By Simpson's rule

The working can be arranged as follows to aid calculation:

x	y	First and last ordinates	'Odd' ordinates	Remaining ordinates
1	y_0	0		
2	y_1		0.693	
3	y_2			1.099
4	y_3		1.386	
5	y_4			1.609
6	y_5		1.792	
7	y_6			1.946
8	y_7		2.079	
9	y_8	2.197		
Totals		2.197	5.950	4.654

$$\int_1^9 \log_e x\,dx$$

$$\approx \frac{1}{3}\{(y_0+y_8)+4(y_1+\ldots+y_7)+2(y_2+\ldots+y_6)\}$$

$$=\frac{1}{3}\{2.197+4(5.950)+2(4.654)\}$$

$$=11.768$$

It can be shown by exact methods that the integral is 11.775 021 . . .

Tablulate, to two decimal places, the values of the expression $\sqrt{1+x^2}$ at unit intervals from $x=2$ to $x=7$ inclusive. Use these values to find an estimate by the trapezoidal rule of the mean value of the expression for $2 \leq x \leq 7$.

x	2	3	4	5	6	7
$f(x)$	$\sqrt{5}$	$\sqrt{10}$	$\sqrt{17}$	$\sqrt{26}$	$\sqrt{37}$	$\sqrt{50}$
	2·24	3·16	4·12	5·10	6·08	7·07
	y_0	y_1	y_2	y_3	y_4	y_5

Using the trapezium rule with six ordinates and h = 1 gives

$$\int_2^7 \sqrt{1+x^2}\, dx \approx \tfrac{1}{2}[(y_0+y_5)+2(y_1+y_2+y_3+y_4)]$$

$$\therefore \int_2^7 \sqrt{1+x^2}\, dx \approx \tfrac{1}{2}[(2\cdot24+7\cdot07)$$
$$+2(3\cdot16+4\cdot12+5\cdot10+6\cdot08)]$$
$$= \tfrac{1}{2}(9\cdot31+2\times18\cdot46)$$
$$= 23\cdot115$$

$$\therefore \text{ mean value} = \frac{1}{7-2}\int_2^7 \sqrt{1+x^2}\, dx$$

$$\approx \frac{1}{5}\times23\cdot115 = 4\cdot62 \text{ (2dp)}$$

 Use Simpson's rule with 7 ordinates (6 strips) to find an approximate value of
$$\int_0^6 xe^{-x}\, dx$$

Give the answer correct to two decimal places.

x	0	1	2	3	4	5	6
$f(x)$	0	e^{-1}	$2e^{-2}$	$3e^{-3}$	$4e^{-4}$	$5e^{-5}$	$6e^{-6}$
	0	0·368	0·271	0·149	0·073	0·034	0·015
	y_0	y_1	y_2	y_3	y_4	y_5	y_6

Simpson's rule with seven ordinates and h = 1 gives

$$\int_0^6 xe^{-x}\, dx \approx \tfrac{1}{3}[(y_0+y_6)+4(y_1+y_3+y_5)+2(y_2+y_4)]$$

$$\therefore \int_0^6 xe^{-x}\, dx \approx \tfrac{1}{3}[(0+0\cdot015)+4(0\cdot368+0\cdot149+0\cdot034)$$
$$+2(0\cdot271+0\cdot073)]$$
$$= \tfrac{1}{3}(0\cdot015+4\times0\cdot551+2\times0\cdot344)$$
$$= \tfrac{1}{3}(0\cdot015+2\cdot204+0\cdot688)$$
$$= 0\cdot969 = 0\cdot97 \text{ (2dp)}$$

[EX]

1 Use Simpson's rule with three ordinates to find an approximate value of
$$\int_{\frac{1}{2}}^{3} \frac{1}{x}\, dx.$$
(L)

2 The integral $\int_0^{\frac{1}{2}} \sqrt{(1-x^2)}\, dx$ is denoted by I. The value of I is to be estimated by using the trapezoidal rule, and T_1, T_2 denote the estimates obtained when one and two strips respectively are used. Calculate T_1 and T_2, giving your answers correct to three decimal places. Assuming that the error when the trapezoidal rule is used is approximately proportional to h^3, where h denotes the width of a strip,

show that an improved estimate of I is given by $(8T_2 - T_1)/7$ and evaluate this expression correct to three decimal places. Given that $y^2 = 1 - x^2$ is the equation of the circle whose centre is the origin and whose radius is 1 unit, show that $I = \tfrac{1}{2}\pi + \tfrac{1}{4}\sqrt{3}$. Hence calculate an estimate for the value of π. (C)

3 Use Simpson's rule with five ordinates to estimate
$$\int_0^{\frac{2\pi}{9}} \log_{10}(\cos x)\, dx$$
giving your answer to 3 decimal places. (A)

4 Use Simpson's Rule with five ordinates (i.e. 4 strips of equal width), working to four significant figures, to obtain an approximate value for
$$\int_0^{90} \sin \frac{\pi x}{180}\, dx.$$
Evaluate the integral directly. (J)

5 By considering suitable areas, or otherwise, show that, for any $n>0$,
$$\tfrac{1}{2} \leq \int_0^1 (1+x^n)^{-1}\, dx \leq 1.$$
When $n=4$, find a value (to three significant figures) for the integral, using Simpson's rule with five ordinates. $(O\ \&\ C)$

6 Values of a continuous function f were found experimentally as given below.

t	0	0·3	0·6	0·9	1·2	1·5	1·8
$f(t)$	2·72	3·00	3·32		4·06	4·48	4·95

Use linear interpolation to estimate $f(0\cdot9)$. Then use Simpson's rule with seven ordinates to estimate
$$\int_0^{1\cdot8} f(t)\, dt,$$ tabulating your working and giving your answer to two places of decimals. (J)

7 Tabulate, to three places of decimals, the values of $(1+x^4)^{\frac{1}{4}}$ for $x = 0, 0\cdot2, 0\cdot4, 0\cdot6, 0\cdot8$. Using Simpson's rule with five ordinates, estimate, to 3 significant figures, the value of
$$\int_0^{0\cdot8} (1+x^4)^{\frac{1}{4}}\, dx.$$
By expanding $(1+x^4)^{\frac{1}{4}}$ in powers of x as far as and including the term in x^8, obtain, to 3 significant figures, a second estimate for the value of this integral. (L)

8 The region defined by the inequalities $0 \leq x \leq \pi$, $0 \leq y \leq \log_{10}(1+\sin x)$ is rotated completely about the x-axis. Using any appropriate rule for approximate integration with five ordinates, find the volume of the solid of revolution formed, giving your answer to 3 significant figures. (A)

9 Given that $x \geq 4$, show that $e^{-\frac{1}{2}x^2} \leq e^{-2x}$ and hence show that
$$\int_4^8 e^{-\frac{1}{2}x^2}\, dx < 0\cdot0002.$$
[Take e^{-8} to be $0\cdot0003$.] Use Simpson's rule with 5 ordinates to estimate the value of $\int_0^4 e^{-\frac{1}{2}x^2}\, dx$ and hence obtain an estimate of $\int_{-8}^8 e^{-\frac{1}{2}x^2}\, dx$. (J)

Definitions

A **function** f from set A to set B, written $f: A \to B$, is a rule which associates with each element $x \in A$, one and only one element in B. This element of B is usually denoted by $f(x)$.

$f(x)$ is called the **image** of x, under f, or, more commonly, **the value of f at x**.

Set A is called the **domain** of the function and set B the **image set** of the function. It is not necessary for all the elements of B to be the image of some $x \in A$ under f.

The **range** of the function is that subset of the image set B which consists of all the possible images under f of all the elements of the domain A. It is denoted by $f(A)$.

The function f is called **one-one** if the images of distinct points of A, under f, are distinct points of B. Functions can be **many-one** or **one-one relations**.

ℹ️ $x^2 + 1$ is the function value at x of the function which 'squares x and adds 1'.
We sometimes write 'the function $f(x) = x^2 + 1$'.
Strictly $f(x)$ is not the function but the value of the function at x. However this $f(x)$ notation is the most common way of identifying a function.

ℹ️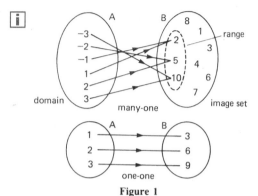

Figure 1

Graphs

The **graph of a function** is usually represented by using rectangular cartesian coordinates and plotting $f(x)$ against x.

To find the range for a given domain and function, it is safer to sketch the cartesian graph of the function over its domain.

Note: The end points of the domain do not necessarily give the end points of the range.

ℹ️ The graph of the function $f(x) = x^2 - 2x + 3$, x real, $0 \leq x \leq 3$, is shown.

From the graph it is clear that the range of $f(x)$ is $\{y : 2 \leq x \leq 6\}$

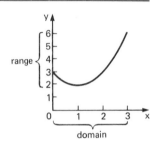

Figure 2

Inverse function

If a function f is one-one and maps an element x in the domain to an element y in the range, then the function that maps y back to x is the **inverse** of f, f^{-1}.

Figure 3

It can be seen from the graph of a function whether its inverse function exists, since one and only one value of x must correspond to any one value of y. The graph of $f^{-1}(x)$ is the reflection of the graph of $f(x)$ in the line $y = x$. So if the graph of $f(x)$ is known, then the graph of $f^{-1}(x)$ can be sketched easily.

ℹ️ Let $f(x) = x^3$ i.e. $y = x^3$ so $x = y^{\frac{1}{3}}$. Then the inverse function is $f^{-1}(y) = y^{\frac{1}{3}}$.

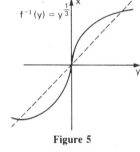

Figure 4   **Figure 5**

Composite function (or 'function of a function')

To find the **composite function**, $fg(x)$, of two functions f and g acting on suitably defined sets:
first find the image of x under g, i.e. $g(x)$, then find the image of $g(x)$ under f.
Note: The order of a composite function is important.
Alternative notation for $fg(x)$ is fg or $f \circ g$.

The inverse of fg is $(fg)^{-1} = g^{-1}f^{-1}$.

ℹ️ If $f(x) = x^3$ and $g(x) = x^{\frac{1}{3}}$, find $gf(x)$.

$gf(x) = g(x^3) = x$ This is expected since $g = f^{-1}$ and so maps y back to x.

ℹ️ $f(x) = 5x + 4$ and $g(x) = 3x - 2$, find $f \circ g$ and $g \circ f$.

$f \circ g = 5(3x - 2) + 4 = 15x - 6$ } Note:
$g \circ f = 3(5x + 4) - 2 = 15x + 10$ } $f \circ g \neq g \circ f$

Even and odd functions

An **even function** f is one for which $f(-a) = f(a)$, for all values of x.
The graph of an even function is symmetrical about the y-axis.
An **odd function** f is one for which $f(-a) = -f(a)$, for all values of x.
The graph of an odd function is symmetrical about the origin.

ℹ️
$y = x^2$
an even function

$y = x^3$
an odd function

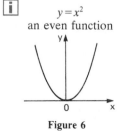   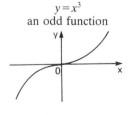

Figure 6   **Figure 7**

Functions
Worked examples and Exam questions

 The functions f and g are defined by
$$f:x \to 5x+4,$$
$$g:x \to 6x-k,$$
where $x \in \mathbb{R}$ and k is a constant.
(a) Find the value of k for which fg = gf
(b) Prove that fff:$x \to 125x+124$

(a) $fg:x \to 5(6x-k)+4 = 30x-5k+4$
   $gf:x \to 6(5x+4)-k = 30x+24-k$
   If $fg = gf$ then $30x-5k+4 = 30x+24-k$
   $$k = -5$$

(b) $f:x \to 5x+4$
   $ff:x \to 5(5x+4)+4 = 25x+24$
   $fff:x \to 5(25x+24)+4 = 125x+124,$
   as required.

 Let $f(x) = \dfrac{px+q}{x+r}$ where x, p, q, r are real and $x \neq \pm r$.

Find the condition for f to be an even function. Deduce that if f is an even function then f(x) must reduce to the form $f(x) = k$, where k is constant.

If $f(x)$ is an even function then $f(-x) = f(x)$

i.e. $\dfrac{p(-x)+q}{(-x)+r} = \dfrac{px+q}{x+r}$

$(x+r)(-px+q) = (r-x)(px+q)$
$-px^2+qx-rpx+rq = rpx+rq-px^2-qx$
$\qquad 2qx-2rpx = 0$
i.e. $\quad 2x(q-rp) = 0$
i.e. $q = rp$, since x is generally non-zero.

If $q = rp$,
$$f(x) = \frac{px+rp}{x+r} = \frac{p(x+r)}{(x+r)} = p,$$

which is of the form $f(x) = k$, k a constant.

 A function f is defined by $f(x) = 4x^2+5$ where $x \in \mathbb{R}^+$ and g is the inverse function of f.
Obtain a formula for g(x).
Show $[(g' \circ f)(1)] \times [f'(1)] = 1$

Let $y = 4x^2+5$, then $x = \sqrt{\dfrac{y-5}{4}} = \dfrac{\sqrt{y-5}}{2}$

i.e. $g:y \to \dfrac{\sqrt{y-5}}{2}$

$\therefore$ the inverse function of f is g where $(x) = \dfrac{\sqrt{x-5}}{2}$

$g'(x) = \frac{1}{4}(x-5)^{-\frac{1}{2}} = \dfrac{1}{4\sqrt{x-5}}$

$f'(x) = 8x$

$(g' \circ f)(x) = \dfrac{1}{4\sqrt{4x^2+5-5}} = \dfrac{1}{8x}$

$\therefore (g' \circ f)(1) = \dfrac{1}{8}$ and $f'(1) = 8$

$\therefore [(g' \circ f)(1)] \times [f'(1)] = \dfrac{1}{8} \times 8 = 1$, as required.

---

 1 The function f is defined by $f:x \to 4-2x-x^2$, where $x \in \mathbb{R}$.
   (a) Find the maximum value of $f(x)$.
   (b) State the range of f.   *(L)*

2 Express in terms of the functions $f:x \to \sqrt{x}$ and $g:x \to x+5$:
   (i) $x \to \sqrt{x+5}$;   (iv) $x \to \sqrt{x}+10$;
   (ii) $x \to x-5$;   (v) $x \to x^2+5$.
   (iii) $x \to x+10$;   *(C)*

3 The function f is defined by
$$f:x \to \frac{1}{2-x},$$
where $x \in \mathbb{R}$ and $x \neq 2$.
(a) Define in a similar way the inverse function f^{-1} and state its domain;
(b) Evaluate (i) $ff(3)$; (ii) $f^{-1}(3)$.   *(L)*

4 Sketch:
   (i) $y = |x|$ for the domain $-5 \leq x \leq 5$;
   (ii) $y^2 = x+1$ for the domain $-1 \leq x \leq 3$;
   (iii) $y = [x]$ for the domain $0 < x \leq 5$, where $[x]$ denotes the greatest integer less than x.   *(C)*

5 Let $f(n) = 9^{2n}-5^{2n}$, where n is a non-negative integer.
   (a) Evaluate $f(0)$ and $f(1)$.
   (b) Write down the value of $f(n+1)$.
   (c) Prove that $f(n+1)-25f(n) = 56(9^{2n})$.
   (d) Hence, using induction, prove that $f(n)$ is always divisible by 7.   *(O & C)*

6 Functions f and g are defined by
$$f:x \to \log_a x, \quad (x \in \mathbb{R}_+, a>1),$$
$$g:x \to \frac{1}{x}, \quad (x \in \mathbb{R}_+).$$

State the ranges of f and g, and show that if h denotes the composite function $f \circ g$, then $h(x)+f(x) = 0$.
Explain briefly why the composite function $g \circ f$ cannot be properly defined unless the domain is restricted to a subset of $\mathbb{R}_+$, and state a possible subset which would be suitable. Define fully the inverses of f and g, and determine whether or not $h^{-1}(x)+f^{-1}(x) = 0$.   *(C)*

7 The function φ is defined by $\varphi(x) = x^3+2x-1$, and the inverse function φ^{-1} is denoted by ψ. Find the values of $\psi(2)$ and $\psi'(2)$.   *(W)*

8 The functions f and g are defined by:
   $f:x \to \sin 2x; \quad x \in \mathbb{R}$;
   $g:x \to \cot x; \quad x \in \mathbb{R}, x \neq k\pi \ (k \in \mathbb{Z})$.
   State the periods of f and g. Find the period of the function of $f.f3g$. On separate axes, sketch the graphs of f, g and $f.g$ for the interval $\{x:-\pi < x < \pi, x \neq 0\}$. Find the range of the function $f.g$.   *(J)*

9 The functions f, g are defined for $x>0$ by;
   $f:x \to x^2; \quad g:x \to \log_e x$.
   Sketch and label the graphs of g, $f \circ g$ and g^{-1} on the same axes, using the same scale.   *(A)*

10 A function f is defined on the set S, where
$$S = \{x:x \in R, x \neq 3\}, \text{ by } f:x \to \frac{3x+b}{x-3}, \quad (b \neq -9).$$
   (i) Show that the inverse of f is f;
   (ii) Determine the range of values of b for which there are two invariant values of x and find these values of x when $b = 55$.
   A function g is defined on the set T, where $T = \{x:x \in R\}$, by
   $g:x \to x+2$. Determine whether the functions f and g are communitative under composition of functions on the set V, where $V = \{x:x \in S \cap T, x \neq 1\}$.   *(A)*

11 The functions f and g, each with domain D, where $D = \{x:x \in \mathbb{R} \text{ and } 0 \leq x \leq \pi\}$, are defined by $f:x \to \cos x$ and $g:x \to x-\frac{1}{2}\pi$. Write down and simplify an expression for $f[g(x)]$, giving its domain of definition. Sketch the graph of $y = f[g(x)]$.   *(L)*

40 Matrices

Definitions, Operations, Transformation of points, Transformation of lines.

Definitions

A **matrix** may be considered as a rectangular array of numbers. The entries in a matrix are called **elements.**
The **order** of a matrix is the number of rows × the number of columns.
A **row matrix** has only one row of elements. A **column matrix** has only one column of elements.
A **square matrix** has the same number of rows as columns, i.e. its order is of the form $(n \times n)$.
Matrices are equal if and only if they are of the same order and corresponding elements are equal.
A **zero** or **null matrix, 0,** is a matrix in which every element is zero.
The **identity** or **unit matrix, I,** is a square matrix in which each element in the leading diagonal is 1 and every other element is zero.

The **determinant** of a 2×2 matrix $A = \begin{pmatrix} a & b \\ c & d \end{pmatrix}$ is the number $\det A = \begin{vmatrix} a & b \\ c & d \end{vmatrix} = ad - bc$.

If $\det A = 0$, then A is called a **singular** matrix.
Every non-singular $n \times n$ matrix A has an **inverse** A^{-1} such that $AA^{-1} = A^{-1}A = I$.

If $A = \begin{pmatrix} a & b \\ c & d \end{pmatrix}$, then $A^{-1} = \dfrac{1}{\det A} \begin{pmatrix} d & -b \\ -c & a \end{pmatrix}$.

Operations

Matrices may be **added** (or **subtracted**) if and only if they are of the **same order**. Add (or subtract) corresponding elements.
Matrix addition is commutative and associative.

To **multiply** a matrix by a **scalar**, multiply each element of the matrix by the scalar.

Two matrices A and B may be **multiplied** together if and only if they are **compatible**, i.e. if the number of columns of A equals the number of rows of B. Each element of AB comes from a row in A and a column in B.
In general, matrix multiplication is not commutative. However, it is associative.

[i] If $A = \begin{pmatrix} 2 & 3 & 5 \\ 4 & 7 & 1 \end{pmatrix}$ and $B = \begin{pmatrix} 0 & -1 & 3 \\ 8 & 2 & -5 \end{pmatrix}$, then

$$A + B = \begin{pmatrix} 2 & 2 & 8 \\ 12 & 9 & -4 \end{pmatrix} \text{ and } A - B = \begin{pmatrix} 2 & 4 & 2 \\ -4 & 5 & 6 \end{pmatrix}$$

[i] $3 \begin{pmatrix} 5 & -3 & 6 \\ -1 & 0 & 7 \end{pmatrix} = \begin{pmatrix} 15 & -9 & 18 \\ -3 & 0 & 21 \end{pmatrix}$

[i] $\begin{pmatrix} a & b & c \\ d & e & f \end{pmatrix} \begin{pmatrix} p & q \\ r & s \\ t & u \end{pmatrix} = \begin{pmatrix} ap+br+ct & aq+bs+cu \\ dp+er+ft & dq+es+fu \end{pmatrix}$

$(2 \times 3 \text{ matrix}) \times (3 \times 2 \text{ matrix}) \rightarrow (2 \times 2 \text{ matrix})$

Transformations of points

Transformations in the plane, other than translations, can be produced and described using 2×2 matrices.

Any point (x, y) can be mapped to (x_1, y_1) using a 2×2 matrix M, where $\begin{pmatrix} x_1 \\ y_1 \end{pmatrix} = M \begin{pmatrix} x \\ y \end{pmatrix}$.

To find the matrix describing a given transformation,
(a) find the image of $P(1, 0)$, say $P_1(a, b)$,
(b) find the image of $Q(0, 1)$, say $Q_1(c, d)$,
(c) the required matrix is $\begin{pmatrix} a & c \\ b & d \end{pmatrix}$.

If M is the matrix which represents a transformation in the plane and $\det M \neq 0$, then M^{-1} is the matrix which represents the **inverse transformation.**

If M and N are two matrices representing two transformations for which the origin is an invariant point, then NM is the matrix which represents the result M **followed by** N.

[i] If $M = \begin{pmatrix} 0 & -1 \\ -1 & 0 \end{pmatrix}$, a reflection in $y = -x$, then

$(2, 3)$ is mapped to $(-3, -2)$ by M, since

$$\begin{pmatrix} 0 & -1 \\ -1 & 0 \end{pmatrix} \begin{pmatrix} 2 \\ 3 \end{pmatrix} = \begin{pmatrix} -3 \\ -2 \end{pmatrix}$$

[i] For rotation of $+90°$ about 0,
$P(1, 0) \rightarrow P_1(0, 1)$
$Q(0, 1) \rightarrow Q_1(-1, 0)$.

The matrix is $M = \begin{pmatrix} 0 & -1 \\ 1 & 0 \end{pmatrix}$.

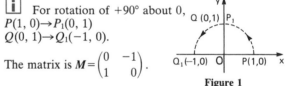

Figure 1

[i] $M = \begin{pmatrix} 0 & -1 \\ 1 & 0 \end{pmatrix}$ and $M^{-1} = \begin{pmatrix} 0 & 1 \\ -1 & 0 \end{pmatrix}$.

M^{-1} is a rotation of $-90°$ about 0.

[i] $NM = \begin{pmatrix} -1 & 0 \\ 0 & 1 \end{pmatrix} \begin{pmatrix} 0 & 1 \\ 1 & 0 \end{pmatrix} = \begin{pmatrix} 0 & -1 \\ 1 & 0 \end{pmatrix}$

M, a reflection in $y = x$, followed by N, a reflection in $x = 0$, is equivalent to a rotation of $+90°$ about 0.

Transformations of lines

The **linear transformation** T of the plane defined by

$$\begin{pmatrix} x \\ y \end{pmatrix} \rightarrow T \begin{pmatrix} x \\ y \end{pmatrix},$$

where $T = \begin{pmatrix} a & b \\ c & d \end{pmatrix}$ and $ad - bc \neq 0$,

maps any line in the plane to a line in the plane.

[i] *Find the image of $y = 3x$ under the mapping* $\begin{pmatrix} 2 & 3 \\ 1 & 2 \end{pmatrix}$.

Let any point on $y = 3x$ be $(\lambda, 3\lambda)$, where λ is a parameter. The image of $(\lambda, 3\lambda)$ is given by

$$\begin{pmatrix} 2 & 3 \\ 1 & 2 \end{pmatrix} \begin{pmatrix} \lambda \\ 3\lambda \end{pmatrix} = \begin{pmatrix} 11\lambda \\ 7\lambda \end{pmatrix}$$

This is the position vector of any point on $11y = 7x$.
So $11y = 7x$ is the required image of $y = 3x$.

Matrices
Worked example, Guided example and Exam questions

WE (a) *A transformation T is equivalent to a shear parallel to the x-axis (the invariant line) which takes (1, 2) to (7, 2), followed by a reflection in the line $y = x$. Find the matrix which defines T.*
(b) *A linear transformation P of the plane maps the points (1, 3), (−2, −3) to the points (2, 4), (−3, −11), respectively. Find the matrix of this transformation.*

(a) The matrix $\begin{pmatrix} 1 & k \\ 0 & 1 \end{pmatrix}$ represents a shear parallel to the x-axis (the invariant line).
Since $(1, 2) \rightarrow (7, 2)$, we have
$$\begin{pmatrix} 1 & k \\ 0 & 1 \end{pmatrix}\begin{pmatrix} 1 \\ 2 \end{pmatrix} = \begin{pmatrix} 1+2k \\ 2 \end{pmatrix} = \begin{pmatrix} 7 \\ 2 \end{pmatrix} \Rightarrow k = 3.$$

So $S = \begin{pmatrix} 1 & 3 \\ 0 & 1 \end{pmatrix}$ defines the shear.

The matrix R which defines reflection in $y = x$ is $\begin{pmatrix} 0 & 1 \\ 1 & 0 \end{pmatrix}$.

Hence, the matrix which represents the shear followed by the reflection is $RS = \begin{pmatrix} 0 & 1 \\ 1 & 0 \end{pmatrix}\begin{pmatrix} 1 & 3 \\ 0 & 1 \end{pmatrix} = \begin{pmatrix} 0 & 1 \\ 1 & 3 \end{pmatrix}$.

(b) Let $\begin{pmatrix} a & b \\ c & d \end{pmatrix}$ be the matrix which defines P.

Now $\begin{pmatrix} a & b \\ c & d \end{pmatrix}\begin{pmatrix} 1 \\ 3 \end{pmatrix} = \begin{pmatrix} a+3b \\ c+3d \end{pmatrix} = \begin{pmatrix} 2 \\ 4 \end{pmatrix}$,

and $\begin{pmatrix} a & b \\ c & d \end{pmatrix}\begin{pmatrix} -2 \\ -3 \end{pmatrix} = \begin{pmatrix} -2a-3b \\ -2c-3d \end{pmatrix} = \begin{pmatrix} -3 \\ -11 \end{pmatrix}$.

So, $a+3b = 2$,
and $2a+3b = 3$,
$\Rightarrow a = 1$, $b = \frac{1}{3}$.
Also, $c+3d = 4$,
and $2c+3d = 11$,
$\Rightarrow c = 7$, $d = -1$.

Hence $\begin{pmatrix} 1 & \frac{1}{3} \\ 7 & -1 \end{pmatrix}$ is the matrix which defines P.

GE *Prove that the map T of the plane defined by*
$$\begin{pmatrix} x \\ y \end{pmatrix} \rightarrow \begin{pmatrix} a & b \\ c & d \end{pmatrix}\begin{pmatrix} x \\ y \end{pmatrix}, \text{ where } ad-bc \neq 0, \text{ maps any line in the}$$
plane to a line in the plane.
If the line $l(m_1)$ of slope m_1 through the origin is mapped onto the line $l(m_2)$ of slope m_2 through the origin, prove that
$$bm_1m_2 - dm_1 + am_2 - c = 0.$$
Given that there is a pair of distinct lines $l(m_1)$, $l(m_2)$ such that T maps $l(m_1)$ onto $l(m_2)$ and maps $l(m_2)$ onto $l(m_1)$ where
$$m_1 \neq m_2,$$
(a) prove that $a+d = 0$,
(b) prove that $T^2 = kI$, giving k in terms of a, b, c.

Let $\begin{pmatrix} \lambda \\ m\lambda+n \end{pmatrix}$ be the position vector of any point on $y = mx+n$.

Find $\begin{pmatrix} a & b \\ c & d \end{pmatrix}\begin{pmatrix} \lambda \\ m\lambda+n \end{pmatrix}$ and equate the result to $\begin{pmatrix} X \\ Y \end{pmatrix}$, say.

Write down two equations in λ and eliminate λ. Hence get $Y = f(X)$, the equation of a straight line.
Let $\begin{pmatrix} \lambda \\ m_1\lambda \end{pmatrix}$ be the position vector of any point on $l(m_1)$.

Find $\begin{pmatrix} a & b \\ c & d \end{pmatrix}\begin{pmatrix} \lambda \\ m_1\lambda \end{pmatrix} = \begin{pmatrix} \alpha \\ \beta \end{pmatrix}$, say.

Since (α, β) lies on $l(m_2)$, $\beta = m_2\alpha$. Use this to obtain the required equation.

Since T maps $l(m_2)$ onto $l(m_1)$, write down a corresponding equation to that obtained. Use the two equations to establish $a+d = 0$.
Evaluate T^2, remembering $a+d = 0$. Hence obtain the required result and give k in terms of a, b, c.

EX **1** (i) If $M = \begin{pmatrix} 2 & -1 \\ 1 & 3 \end{pmatrix}$, find the values of M^2, M^3 and M^{-1}.

Find x and y, given that $M\begin{pmatrix} x \\ y \end{pmatrix} = \begin{pmatrix} 3 \\ 5 \end{pmatrix}$.

(ii) A transformation T is equivalent to an enlargement with centre at the origin, scale factor 2, followed by a reflection in the line $x+y = 0$. What matrix defines T? If T maps a point P onto $(6, 2)$, what are the co-ordinates of P?
(O & C)

2 A transformation M is represented by the matrix **M** where
$$\mathbf{M} = \begin{pmatrix} 4 & 1 \\ 2 & 3 \end{pmatrix}.$$
(i) Find the image of the point $(-2, 5)$ under **M**.
(ii) Find the inverse of **M**.
(iii) Given that the point $(11, 13)$ is the image of the point (a, b) under **M**, find the value of a and of b.
(iv) Find, in terms of α, the image of the point (α, α) under **M**.
(v) State the equation of the invariant line under **M**.
(C)

3 If $\mathbf{A} = \begin{pmatrix} 2 & 1 & 1 \\ 1 & 0 & 1 \\ 0 & -1 & 1 \end{pmatrix}$ and $\mathbf{B} = \begin{pmatrix} 1 & -1 & 1 \\ 0 & 0 & -1 \\ -1 & 2 & -1 \end{pmatrix}$, find
(a) **AB**;
(b) a matrix **X** such that $\mathbf{AX} + \mathbf{B} = \mathbf{A}$.
(O & C)

4 Let $\mathbf{A} = \begin{pmatrix} 1 & 1 \\ 0 & 1 \end{pmatrix}$.
(a) The plane is mapped onto itself by the map under which the point P of co-ordinates (x_1, y_1) is mapped to the point Q of co-ordinates (x_2, y_2), where $\begin{pmatrix} x_2 \\ y_2 \end{pmatrix} = \mathbf{A}\begin{pmatrix} x_1 \\ y_1 \end{pmatrix}$.

By considering $\mathbf{A}\begin{pmatrix} x \\ m_1x+c \end{pmatrix}$, prove that the line $y = m_1x+c$ is mapped onto a line of slope m_2, determining m_2 in terms of m_1. Hence or otherwise determine whether any line through the origin is mapped onto itself, and find any such line.
(b) Prove that there is no non-singular matrix **P** such that
$$\mathbf{P}^{-1}\mathbf{AP} = \begin{pmatrix} k_1 & 0 \\ 0 & k_2 \end{pmatrix} \text{ for real } k_1, k_2.$$
(O & C)

5 The transformation with matrix T, where $T = \begin{pmatrix} 2 & 1 \\ 2 & -2 \end{pmatrix}$, maps the point (x, y) into the point (x', y') so that
$$T\begin{pmatrix} x \\ y \end{pmatrix} = \begin{pmatrix} x' \\ y' \end{pmatrix}.$$

Find the equation of the image of the line $y = 3x$ under this transformation. Find also the equations of the lines through the origin which are turned through a right angle about the origin under this transformation.
(J)

41 Force Diagrams
Definitions, Types of forces, Drawing force diagrams.

Definitions

Mechanics is concerned with the action of forces on bodies.
In mechanics a **body** is any object to which a force can be applied.
A **rigid body** is a body whose shape is unaltered by any force applied to it.
A **particle** is a body whose dimensions, except mass, are negligible.
A **lamina** is a flat body having area but negligible thickness.
A **hollow body** is a three-dimensional shell having negligible thickness.

Types of forces

Forces occur in mechanics in various ways. Some of the most common are described below.

Weight W
The weight of a body is the force with which the earth attracts the body. It acts at the body's centre of gravity and is always vertically downwards.

Figure 1

A light body is considered to be weightless.

Push and pull P
Pushes and pulls are forces which act on a body at the point(s) where they are applied.

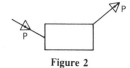

Figure 2

Normal reaction R
A normal reaction is a force which acts on a body in contact with a surface. It acts in a direction at right angles to the surfaces in contact.

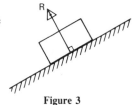

Figure 3

Friction F
Friction is a force which acts on a body in contact with a rough surface. It acts tangentially between the two surfaces and in a direction to resist the motion of the body.

Figure 4

Smooth surfaces are considered to be frictionless. Air resistance is ignored unless stated otherwise.

Tension T
The tension in a string is a force which acts on a body to which the string is attached. Tensions can also come from springs, rods, etc.

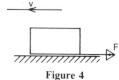

Figure 5

Thrust S
The thrust from a spring or rod is similar to a tension but acts in the opposite direction.

Figure 6

Drawing force diagrams

Drawing a clear **force diagram** is an essential first step in the solution of any problem in mechanics which is concerned with the action of forces on a body.
The following are important points to remember when drawing such a force diagram.

1. Make the diagram large enough to show clearly all the forces acting on the body and to enable any necessary geometry and trigonometry to be done.

2. Show only forces which are acting on the body being considered. A common fault is including forces which the body is applying to its surroundings (including other bodies).

3. Weight always acts on a body unless the body is described as light.

4. Contact with another object or surface gives rise to a normal reaction and sometimes friction.

5. Attachment to another object (by a string, spring, hinge, etc.) gives rise to a force on the body at the point of attachment.

6. Forces acting on a particle act at the same point. Forces acting on other bodies may act at different points.

7. Check that no forces have been omitted or included more than once.

ℹ️ Some simple force diagrams illustrate these points.

(a) Forces acting on a block on a smooth horizontal plane:

W – weight (vertically down)
R – normal reaction (at right angles to the surfaces in contact)

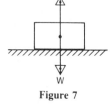

Figure 7

(b) Forces acting on a block at rest on a rough inclined plane:

W – weight
R – normal reaction
F – friction (acting to resist motion)

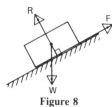

Figure 8

(c) Forces acting on a block being pulled (by a string) along a rough horizontal plane:

W – weight
R – normal reaction
F – friction
T – tension in string (acting away from body)

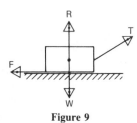

Figure 9

82

Force Diagrams
Exercises

The drawing of force diagrams is a skill it is essential to acquire during the study of mechanics. Although it does not appear by itself as an examination question it is often the first step in the solution of problems in this subject.

The following questions provide practice in this important skill.

EX In this exercise, identify clearly each type of force marked on your force diagrams.

1 Draw a diagram to show the force acting on a uniform ladder resting on horizontal rough ground and leaning against a smooth vertical wall.

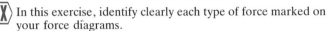

2 The sketch shows a uniform ladder resting on rough horizontal ground and leaning against a rough vertical wall with a man standing one quarter the way up the ladder. Draw diagrams to show the forces,
   (a) on the ladder,
   (b) on the man.

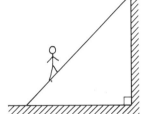

3 This diagram shows a bead resting on a rough inclined plane being acted on by the force P which is about to move it up the plane. Sketch the diagram and show all the forces acting on the bead.

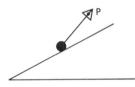

4 This diagram shows a bead resting on a rough inclined plane and just being prevented from moving down the plane by the force Q. Sketch the diagram and show all the forces acting on the bead.

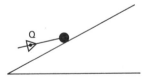

5 This sketch shows a large smooth sphere of weight W resting inside a smooth cylinder and being held in place by a small smooth sphere of weight w. Draw diagrams to show the forces acting on,
   (a) the large sphere,
   (b) the small sphere.

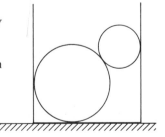

6 Draw two diagrams showing the forces acting on a block of wood which is
   (a) sliding down a rough inclined plane at steady speed,
   (b) accelerating down a rough inclined plane.

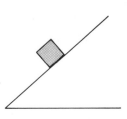

7 Draw a sketch showing the forces acting on a car which is being driven up an incline at steady speed.

8 This diagram shows a car travelling at steady speed on a level road and pulling a caravan. Draw diagrams to show the forces acting on,
   (a) the car,
   (b) the caravan.

9 A particle is suspended from a fixed point by a light inextensible string. Draw a force diagram showing the forces acting on the particle when it is moving with steady speed in a horizontal circle below the fixed point.

10 This sketch shows a rough rod resting against a rough cylinder with its lower end on rough ground. Draw diagrams showing the forces acting on:
   (a) the rod;
   (b) the cylinder.

11 A stone is thrown through the air. Draw a force diagram,
   (a) if air resistance is ignored,
   (b) if air resistance is present.

12 Two housebricks, one resting exactly on top of the other, stand on horizontal ground. Draw sketches to show the forces acting on,
   (a) the top brick, (b) the bottom brick.

13 A man is standing alone in a moving lift. Draw diagrams to show the forces acting on,
   (a) the man, (b) the lift, when the lift is:
   (i) accelerating upwards,
   (ii) travelling at steady speed,
   (iii) accelerating downwards.

14 A railway engine is pulling a train up an incline against frictional resistances. If the combined engine and train is experiencing a retardation, draw diagrams showing the forces acting on,
   (a) the engine, (b) the train.

42 1-D Kinematics

Definitions, Motion in one dimension, Equations of motion, Vertical motion under gravity.

Definitions

Kinematics is the study of displacement, velocity and acceleration.

Displacement is the position of a point relative to an origin O. It is a **vector**.
S.I. unit is the **metre** (m). Other metric units are centimetre (cm), kilometre (km).
Distance is the magnitude of the displacement. It is a **scalar**.

Velocity is the rate of change of displacement with respect to time. It is a **vector**.
S.I. unit: **metre per second** (m s^{-1} or m/s). Other metric units: cm s^{-1} or cm/s, km h^{-1} or km/h.
Speed is the magnitude of the velocity. It is a **scalar**.
Uniform velocity is constant speed in a fixed direction.

Average velocity is $\dfrac{\text{change in displacement}}{\text{time taken}}$. **Average speed** is $\dfrac{\text{total distance travelled}}{\text{time taken}}$.

Acceleration is the rate of change of velocity with respect to time. It is a **vector**.
S.I. unit: **metre per second squared** (m s^{-2} or m/s^2). Other metric units: cm s^{-2} or cm/s^2, km h^{-2} or km/h^2.
Negative acceleration is sometimes called **retardation**.
Uniform acceleration is constant acceleration in a fixed direction.

Motion in one dimension

When a particle moves in **one dimension**, i.e. along a straight line, it has only two possible directions in which to move. Positive and negative signs are used to identify the two directions.

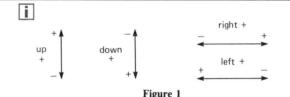

Figure 1

Equations of motion

The **equations for uniform acceleration** in a straight line are:

$$v = u + at$$
$$s = \tfrac{1}{2}(u+v)t$$
$$v^2 = u^2 + 2as$$
$$s = ut + \tfrac{1}{2}at^2$$

The notation used is:
s – displacement; t – time; u – initial velocity;
v – velocity at time t; a – acceleration.
Units must be consistent, e.g. if s is in metres and t in seconds, then u and v must be in m s^{-1} and a in m s^{-2}.

To solve problems using these equations of motion:
(a) Choose the positive direction.
(b) List the five quantities (s, t, u, v, a), fill in known values and mark which are to be found.
(c) Use the appropriate equation(s) to find the required unknown(s). If any three of the quantities are known, then the other two can always be found.

Note: These equations do not apply to acceleration which is not uniform.
Problems about non-uniform acceleration must be solved by graphical methods or by calculus.

ℹ️ *A particle moves in a straight line with constant acceleration. Its initial velocity is 6 m s^{-1} and its velocity after 8 s is 10 m s^{-1}. Find the acceleration and the displacement of the particle after 16 s.*

Assume the motion is horizontal.
Choose the direction of motion as positive.

1st stage
$s = ?$ m, $t = 8$ s, $u = 6$ m s^{-1}, $v = 10$ m s^{-1}, $a = ?$ m s^{-2}

Use $v = u + at$
  $10 = 6 + a(8)$
  $a = 0.5$ m s^{-2}
So the uniform acceleration is 0.5 m s^{-2}.

2nd stage
$s = ?$ m, $t = 16$ s, $u = 6$ m s^{-1}, $v = ?$ m s^{-1}, $a = 0.5$ m s^{-2}.

Use $s = ut + \tfrac{1}{2}at^2$
  $= (6)(16) + \tfrac{1}{2}(0.5)(16)^2$
  $= 160$ m
So the displacement after 16 s is 160 m.

Vertical motion under gravity

The motion of a body thrown **vertically upward** or falling **freely downward** (ignoring air resistance) is a special case of uniform acceleration in a straight line.

This uniform acceleration is due to **gravity** and acts vertically downwards towards the centre of the earth. It is denoted by g and common approximate values are **10 m s^{-2}** and **9.8 m s^{-2}**.

ℹ️ A body is thrown vertically upward.

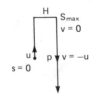

Figure 2

Taking up as positive,
$a = -g$.
At highest point H
s is maximum, $v = 0$.
At point of projection P
$s = 0$, $v = \pm u$.

1-D Kinematics
Worked example, Guided example and Exam questions

 A stone is projected vertically upwards from the top of a cliff 20 m high. After a time of 3 s it passes the edge of the cliff on its way down. Calculate

(a) *the speed of projection,*
(b) *the speed when it hits the ground,*
(c) *the times when it is 10 m above the top of the cliff,*
(d) *the time when it is 5 m above the ground.*

Take up as positive.
(a) $s = 0$, $t = 3$ s, $u = ?$, $v = ?$, $a = -10$ ms^{-2}
Use $s = ut + \frac{1}{2}at^2$
so, $0 = u.3 + \frac{1}{2}(-10) \, 3^2$
$\Rightarrow u = 15$ ms^{-1}.
(b) $s = -20$ m, $t = ?$, $u = 15$ ms^{-1}, $v = ?$, $a = -10$ ms^{-2}
Use $v^2 = u^2 + 2as$
so, $v^2 = 15^2 + 2 \, (-10)(-20)$
$\quad = 625$
$\Rightarrow v = -25$ ms^{-1} — particle is moving downwards with speed 25 ms^{-1}.
(c) $s = 10$ m, $t = ?$, $u = 15$ ms^{-1}, $v = ?$, $a = -10$ ms^{-2}
Use $s = ut + \frac{1}{2}at^2$
so, $10 = 15t + \frac{1}{2}(-10)t^2$
i.e. $t^2 - 3t + 2 = 0$
$\Rightarrow t = 1$ s or $t = 2$ s — the two times when it is 10 m above the top of the cliff.
(d) $s = -15$ m, $t = ?$, $u = 15$ ms^{-1}, $v = ?$, $a = -10$ ms^{-2}
Use $\quad s = ut + \frac{1}{2}at^2$
so, $-15 = 15t + \frac{1}{2}(-10)t^2$
i.e. $t^2 - 3t - 3 = 0$
i.e. $t = \dfrac{3 \pm \sqrt{[9 - 4.1.(-3)]}}{2} = \dfrac{3 \pm \sqrt{21}}{2}$
$\Rightarrow t \approx 3 \cdot 7$ s (discounting the negative root).

 A stone is thrown vertically upwards with a speed of 20 ms^{-1}. A second stone is thrown vertically upwards from the same point and with the same initial speed 20 ms^{-1} but 2 s later than the first one. Show that the two stones collide at a distance of 15 m above the point of projection.

Take up as positive. Let $s = h$ m be the displacement of each stone when they collide.
For first stone, $s = h$ m, $t = T$ s (say), $u = 20$ ms^{-1}, $v = ?$, $a = -10$ ms^{-2}.
For second stone, $s = h$ m, $t = (T-2)$ s, $u = 20$ ms^{-1}, $v = ?$, $a = -10$ ms^{-2}.
Use $s = ut + \frac{1}{2}at^2$ for each stone. Equate the two expressions for h and so find T. Use formula again with value for T found to calculate h.

---

 1 A train is uniformly retarded from 35 m/s to 21 m/s over a distance of 350 m. Calculate:
(a) the retardation;
(b) the total time taken under this retardation to come to rest from a speed of 35 m/s.
*(L)

2 Two particles, X and Y, are moving in the same direction on parallel horizontal tracks. At a certain point O, the particle X, travelling with a speed of 16 m/s and retarding uniformly at 6 m/s^2, overtakes Y, which is travelling at 8 m/s and accelerating uniformly at 2 m/s^2. Calculate:
(i) the distance of Y from O when the velocities of X and Y are equal;
(ii) the velocity of X when Y overtakes X.
*(C)

3 A car is moving along a straight horizontal road at constant speed 18 m/s. At the instant when the car passes a lay-by, a motor-cyclist leaves the lay-by, starting from rest, and moves with constant acceleration $2 \cdot 5$ m/s^2 in pursuit of the car. Given that the motor-cyclist overtakes the car T seconds after leaving the lay-by, calculate:
(a) the value of T;
(b) the speed of the motor-cyclist at the instant of passing the car.
*(L)

4 Two points A and B lie on a horizontal plane. A particle is projected vertically upwards from A with an initial speed of 20 m/s. One second later another particle is projected vertically upwards from B with an initial speed of $17 \cdot 5$ m/s. Calculate, at the instant at which the two particles are at the same vertical height above the plane:
(a) the time which has elapsed since the first particle was projected from A;
(b) the speeds of the two particles.
*(A)

5 A balloon is ascending at a constant speed of 3 m/s. The crew release some gas and as a result the balloon experiences a constant downward acceleration of $0 \cdot 25$ m/s^2. How much farther will the balloon ascend, and how long will it be before the balloon returns to its original height? If this height is 80 m above the ground and the balloon continues to descend with the same acceleration, how much longer will it be before the balloon strikes the ground and what will be its velocity at impact?
*(C)

6 A particle moves in a straight line with uniform acceleration α. Its initial velocity was u. Prove that the distance x travelled in time t is given by $x = ut + \frac{1}{2}\alpha t^2$. A motor-car is timed between three successive points X, Y and Z, where $XY = YZ = 2$ km. It takes 100 seconds to travel from X to Y and 150 seconds to travel from Y to Z. Given that the retardation of the car is uniform from the point X onwards, calculate the value of this retardation. Find also how far the car travels beyond Z before it stops.
*(W)

7 A particle X is projected vertically upwards from the ground with a velocity of 80 m/s. Calculate the maximum height reached by X. A particle Y is held at a height of 300 m above the ground. At the moment when X has dropped 80 m from its maximum height, Y is projected downwards with a velocity of v m/s. The particles reach the ground at the same time. Calculate the value of v.
*(C)

43 Graphs in Kinematics
Displacement-time graph, Velocity-time graph, Acceleration-time graph.

Displacement-time graph

A **displacement-time** graph (or *s-t graph*) for a body moving in a straight line shows its displacement s from a fixed point on the line plotted against time t. The **velocity** v of the body at time t is given by the **gradient** of the *s-t* graph at t, since $v = \dfrac{ds}{dt}$.

The *s-t* graph for a body moving with **constant velocity** is a **straight line**. The velocity v of the body is given by the gradient of the line.

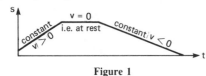

Figure 1

The *s-t* graph for a body moving with **variable velocity** is a **curve**.

The velocity at any time may be estimated from the gradient of the tangent to the curve at that time. The average velocity between two times may be estimated from the gradient of the chord joining them.

The *s-t* graph below shows the displacement s of a car from its starting point at given instants t.

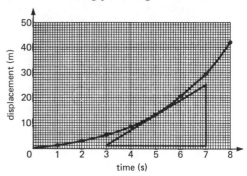

Figure 2

To estimate the velocity of the car after 5 s, draw the tangent to the graph at $t = 5$ s.

$$\therefore \text{velocity after } 5 \text{ s} \approx \frac{25-1}{7-3} = \frac{24}{4} = 6 \text{ m s}^{-1}.$$

Velocity-time graph

A **velocity-time graph** (or *v-t graph*) for a body moving in a straight line shows its velocity v plotted against time t.

The **acceleration** a of the body at time t is given by the **gradient** of the *v-t* graph at t, since $a = \dfrac{dv}{dt}$.

The **displacement** s in a time interval is given by the **area** under the *v-t* graph for that time interval, since $s = \displaystyle\int v \, dt$.

The *v-t* graph for a body moving with **uniform acceleration** is a **straight line**. The acceleration a of the body is given by the gradient of the line.

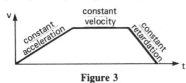

Figure 3

The total displacement s of a body can be found from a *v-t* graph of the above type by calculating the area of the trapezium.

The *v-t* graph for a body moving with **variable acceleration** is a **curve**.

The acceleration a of the body at any time may be estimated from the gradient of the tangent to the curve at that time.

The displacement s of the body in a given time interval may be estimated by finding the area under the *v-t* graph in that interval by a numerical method (see Numerical Integration p. 76).

The velocity-time graph for a moving vehicle is shown below.

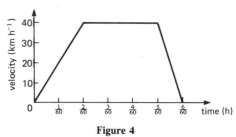

Figure 4

Use the graph to calculate:
(a) *the acceleration during the first two minutes*,
(b) *the retardation during the last minute*,
(c) *the total displacement of the vehicle.*

(a) The acceleration during the first two minutes
= change in velocity ÷ time taken
$$= 40 \div \frac{2}{60} = 1200 \text{ km h}^{-2}$$

(b) The retardation during the last minute
= change in velocity ÷ time taken
$$= 40 \div \frac{1}{60} = 2400 \text{ km h}^{-2}$$

(c) The total displacement of the vehicle is given by the area under the graph. This area is a trapezium.
$$\text{Total displacement} = \frac{1}{2}\left(\frac{3}{60} + \frac{6}{60}\right) \cdot 40 = 3 \text{ km.}$$

Acceleration-time graph

An **acceleration-time graph** for a body moving in a straight line shows its acceleration a plotted against time t.

The **velocity** v of the body in a time interval is given by the **area** under the acceleration-time graph for that time interval, since $v = \displaystyle\int a \, dt$.

Use the trapezium rule, with intervals of 1 s, to verify from this graph that the velocity after 4 s is 8.1 m s^{-1} approximately.

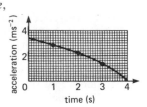

Figure 5

Graphs in Kinematics
Worked example, Guided example and Exam questions

 Two points P and Q are x metres apart in the same straight line. A particle starts from rest at P and moves directly towards Q with an acceleration a ms^{-2} until it acquires a speed of V ms^{-1}. It maintains this speed for a time T seconds and is then brought to rest at O under a retardation a ms^{-2}. Prove that

$$T = \frac{x}{V} - \frac{V}{a}.$$

Sketch the velocity–time graph.

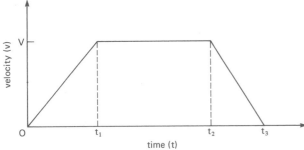

Using the definitions of acceleration and retardation,

$$a = \frac{V}{t_1} \Rightarrow t_1 = \frac{V}{a} \qquad [1]$$

and $$a = \frac{V}{t_3 - t_2} \Rightarrow t_3 - t_2 = \frac{V}{a} \qquad [2]$$

Also, we know that,
$$t_2 - t_1 = T \qquad [3]$$

$[1] + [2] + [3]$ gives $t_3 = T + \dfrac{2V}{a}$ $\qquad [4]$

We know that the total distance travelled between P and Q is x m and this is represented by the area under the graph.

i.e. $x = \frac{1}{2}Vt_1 + V(t_2 - t_1) + \frac{1}{2}V(t_3 - t_2)$
$x = \frac{1}{2}V(t_3 + t_2 - t_1)$.

Using the values of t_3 and $t_2 - t_1$ from [4] and [3] we have,

$$x = \frac{1}{2}V\left(T + \frac{2V}{a} + T\right)$$

$$= V\left(T + \frac{V}{a}\right).$$

i.e. $T = \dfrac{x}{V} - \dfrac{V}{a}$

 Two cars, C and D, travel equal distances of 9 km in the same time of 18 minutes, finishing at rest. Car C starts from rest, accelerates uniformly to a speed of 45 km/h, travels steadily at this speed and is then brought to rest with uniform retardation. Car D moves at a constant speed for the first 3·6 km and is then brought to rest with uniform retardation. Sketch speed–time graphs for the motion of each car and hence, or otherwise, calculate

(a) *the distance, in km C travels at steady speed,*
(b) *the initial speed in km/h of D,*
(c) *the retardation of D, stating the units.*

Sketch the two graphs. (Remember to change km/h to m/s, km to m, minutes to seconds, etc,)

(a) Let T (say) be the time for which C travels at constant speed. Work out the area under the graph (a trapezium) put it equal to 9 km ($= 9 \times 10^3$ m) the distance gone. Hence T.

(b) Let V m/s (say) be the initial speed of D and T_1 seconds be the time for which D travels at V m/s. Then $VT_1 = 3·6 \times 10^3$ — the distance travelled at V m/s. Work out the total area under the graph (a trapezium) and get a second equation connecting V and T_1. Solve the two equations for V.

(c) Now find T_1 from (b).
Retardation is $V \div (18 \times 60 - T_1)$ m/s^2.

---

1 A train travels between two stations, 3·9 km apart, in 6 minutes, starting and finishing at rest. During the first $\frac{3}{4}$ minute the acceleration is uniform, for the next $3\frac{3}{4}$ minutes the speed is constant and for the remainder of the journey the train is retarded uniformly. Sketch a speed–time graph of the journey and hence, or otherwise, calculate:
(i) the maximum speed, in km/h, attained by the train;
(ii) the acceleration of the train during the first $\frac{3}{4}$ minute, stating the units. *(A)

2 A motorist starting a car from rest accelerates uniformly to a speed of v m/s in 9 seconds. He maintains this speed for another 50 seconds and then applies the brakes and decelerates uniformly to rest. His deceleration is numerically equal to three times his previous acceleration.
(i) Sketch a velocity–time graph.
(ii) Calculate the time during which deceleration takes place.
(iii) Given that the total distance moved is 840 m calculate the value of v.
(iv) Calculate the initial acceleration. *(C)

3

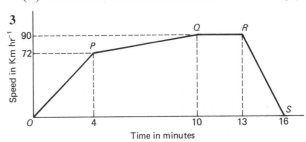

The figure, not drawn to scale, shows the speed/time diagram for a train journey taking 16 minutes. The train reaches a speed of 72 km hr^{-1} after 4 minutes and 90 km hr^{-1} after a further 6 minutes. The train maintains a steady speed for the next 3 minutes and then decelerates to rest in the last 3 minutes. Calculate:
(i) the acclerations in m s^{-2} during those parts of the journey corresponding to OP and PQ;
(ii) the total length of the journey;
(iii) the average speed for the whole journey in km hr^{-1}.
(O & C)

4 Trials are being undertaken on a horizontal road to test the performance of an electrically powered car. The car has a top speed V. In a test run the car moves from rest with uniform acceleration a and is brought to rest with uniform retardation r.
(i) If the car is to achieve top speed during a test run, by using a velocity–time sketch, or otherwise, show that the length of the test run must be at least
$$\frac{V^2(a+r)}{2ar}.$$
(ii) Find the least time taken for a test run of length
$$\text{(a) } \frac{2V^2(a+r)}{9ar}, \quad \text{(b) } \frac{2V^2(a+r)}{3ar}.$$
(iii) Find, in terms of V, the average speed of the car for the test run described in (ii) (b). (A)

5 Starting from rest at the point A, a particle moves, in a straight line, with constant acceleration until it reaches the point B. The particle then moves with constant retardation until it comes to rest at C, where $AB = 3BC$. The time taken to travel from A to B is T and the speed at B is V. Find, in terms of V and T;
(i) the time taken for the whole journey from A to C;
(ii) the distance AC. (C)

44 Relative Motion

Velocity triangle, Relative velocity, Problem solving.

Velocity triangle

When an aircraft flies through the air, its motion over the ground is affected by the way the air is moving.

To describe the motion of an aircraft we need:
ground-speed (GS) – speed of aircraft over ground,
track (T) – direction in which aircraft moves,
airspeed (AS) – speed of aircraft through air,
course (C) – direction in which aircraft points,
wind-speed (WS) – speed of wind,
wind direction (WD) – direction in which wind blows.

These give three vectors connected by the law of **vector addition:**

$(GS, T) = (AS, C) + (WS, WD)$
This gives the **velocity triangle.**

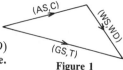

Figure 1

Similarly, when a body travels through water, its motion is affected by the way the water is moving. Instead of wind speed and direction, there is the speed and direction of the flow of water. The speed of the body in still water replaces the airspeed.

ℹ️ *A river, 0.4 km wide, flows from E to W at a steady speed of 1 km h⁻¹. A swimmer, whose speed in still water is 2 km h⁻¹, starts from the S bank and heads N across the river. Find his speed over the river bed and how far downstream he is when he reaches the N bank.*

For swimmer: speed in still water 2 km h⁻¹, course N.
For river: speed 1 km h⁻¹, direction E.

We need to find GS and T for the swimmer.
$\tan \alpha = \frac{1}{2}$
$\Rightarrow \alpha \approx 26.6°$.

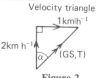

Figure 2

Speed over river bed $= \sqrt{2^2 + 1^2} = \sqrt{5} = 2.24$ km h⁻¹.

Displacement diagram   Distance, d, downstream
$= 0.4 \tan \alpha$
0.4 km $= 0.2$ km

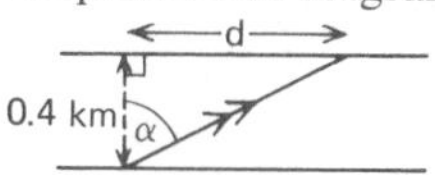

Figure 3

Relative velocity

When we say that A is moving with constant velocity v_A we mean that A is moving with constant velocity v_A relative to a fixed observer on earth.

If A and B are moving with constant velocities v_A and v_B respectively, then, to an observer on B, A will have a velocity $_A v_B$, where:

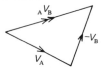

Figure 4

$_A v_B = v_A - v_B$.
$_A v_B$ is often called the **velocity of A relative to B.**

A and B may be two bodies, e.g. ships, aircraft, cars, cyclists, etc. or one of them could be one of the elements, e.g. rain, wind, etc.

ℹ️ *To a motor-cyclist travelling due N at 50 km h⁻¹ the wind appears to come from NW at 60 km h⁻¹. What is the true velocity of the wind?*

Imagine yourself to be the motor-cyclist.

$_w v_c = v_w - v_c$
and $_w v_c = 60$ km h⁻¹
$v_c = 50$ km h⁻¹

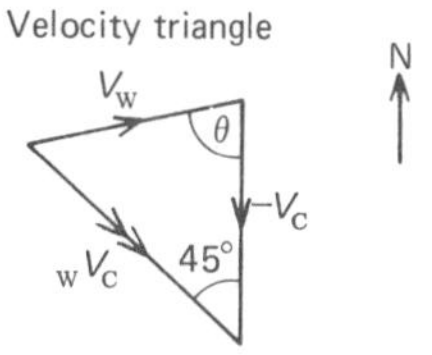

Figure 5

In the velocity triangle:
by the cosine rule, $v_w^2 = 60^2 + 50^2 - 2 . 60 . 50 \cos 45°$
$\Rightarrow v_w \approx 43.1$ km h⁻¹
by the sine rule, $\dfrac{\sin \theta}{60} = \dfrac{\sin 45°}{43.1} \Rightarrow \theta \approx 79.9°$

∴ true wind velocity is 43.1 km h⁻¹ from S 79.9° W.

Problem solving

Two bodies A and B, moving with constant velocities v_A and v_B respectively, will reach either a position of **interception** or of **closest approach.**

When solving such relative motion problems:
1. Draw an initial sketch using the given information.
2. Imagine yourself to be on one of the bodies, B say.
In an interception problem, B should be the body being intercepted.
3. State the relative velocity rule, i.e. $_A v_B = v_A - v_B$.
4. Draw the correct velocity triangle for $_A v_B$.
5. To find the magnitude and direction of $_A v_B$, use the trigonometry of the velocity triangle.
If it is an 'interception' problem:
the time, t, at which interception occurs is given by

$$t = \frac{\text{initial distance apart}}{_A v_B}.$$

If it is a 'closest approach' problem:
(a) draw a displacement diagram showing the initial positions A_0 and B_0 of A and B respectively and $_A v_B$.

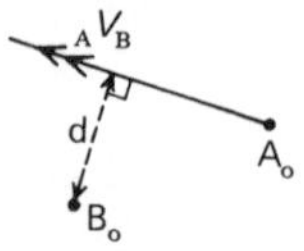

Figure 6

(b) find d, the shortest distance between A and B during motion, by trigonometry or scale drawing.

ℹ️ *A dinghy in distress is 6 km S 50° W of a lifeboat and drifting S 20° E at 5 km h⁻¹. In what direction should the lifeboat travel to reach the dinghy as quickly as possible if the maximum speed of the lifeboat is 35 km h⁻¹?*

Initial sketch

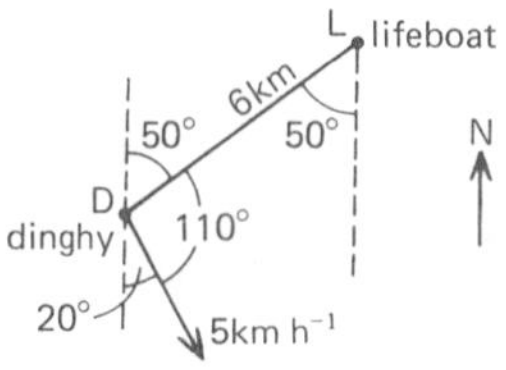

Imagine yourself on the dinghy D. The lifeboat will appear to travel directly towards you, i.e. $_L v_D$ will be in a direction S 50° W.

Figure 7

Velocity triangle

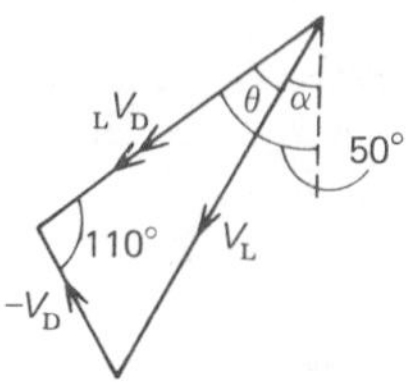

In the velocity triangle:
$_L v_D = v_L - v_D$
with $v_L = 35$ km h⁻¹
$v_D = 5$ km h⁻¹.

By the sine rule:
$$\frac{\sin \theta}{5} = \frac{\sin 110°}{35}$$
$\Rightarrow \theta \approx 7.7°$
So $\alpha \approx 50° - 7.7° = 42.3°$.

Figure 8

∴ the lifeboat must travel S 42.3° W.

Relative Motion
Worked example, Guided example and Exam questions

 Two straight-roads, one running North-South and the other running East-West, intersect at a crossroads O. Two men A and B are cycling at steady speeds towards O. At a certain instant, A is 20 m from O and travelling due North at 3 ms⁻¹ and B is 20 m from O and cycling due West at 4 ms⁻¹. Calculate the shortest distance apart of the two cyclists and the time which elapses before this is attained.

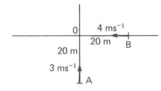

We view the situation from A, i.e. we calculate $_BV_A$.

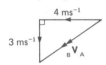

$_BV_A = 5$ ms⁻¹ (3:4:5 $\triangle$)

$\tan\theta = \dfrac{4}{3}$, so $\sin\theta = \dfrac{3}{5}$,

$\cos\theta = \dfrac{4}{5}$.

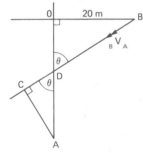

A sees B travelling along the direction BC.
The cyclists are closest together when A sees B to be at a position C, where AC is perpendicular to BC.
The shortest distance apart is, therefore, AC.

In $\triangle OBD$, $OD = 20\cot\theta = 15$ m.
Since $OA = 20$ m, $AD = 5$ m.
In $\triangle ACD$, $AC = AD\sin\theta = 4$ m.
Hence the shortest distance between the cyclists is 4 m.

To find the time taken to reach the 'closest position' we first calculate the distance BC.
$BC = BD + DC$
$\quad = 20\operatorname{cosec}\theta + 5\cos\theta$
$\quad = 25 + 3 = 28$ m.
A sees B travelling along BC at $_BV_A = 5$ ms⁻¹.
So the time taken to travel the 28 m is $28 \div 5 = 5\frac{3}{5}$ s.
Hence, the cyclists are at the 'closest position' after $5\frac{3}{5}$ s.

GE *At 12 noon a battleship, whose maximum speed is 30 knots, sights a submarine which is moving due North at 10 knots. When first sighted, the submarine is 15 nautical miles North-East of the battleship. Calculate,*
(a) the direction in which the battleship must be steered in order to intercept the submarine as quickly as possible,
(b) the time at which they meet,
(c) their distance apart at 1215 hours.

Make an initial sketch. The problem is best solved by viewing the interception from the submarine. The battleship appears to come directly towards you with a velocity $_BV_S = V_B - V_S$. You know the direction of $_BV_S$ is 045°. You know also the speed and direction of V_S and the speed V_B.
(a) Draw the velocity triangle.
(b) Use the sine rule to calculate the direction in which the battleship must be steered.
(c) Calculate $_BV_S$. The battleship covers the 15 nm distance to intercept the submarine at a speed of $_BV_S$. Hence find the time taken to meet.
(d) At 1215 hours, the battleship has travelled for 15 minutes at a speed of $_BV_S$. Find the distance covered in this time and subtract it from the initial separation of 15 nm. This gives their distance apart at 1215 hours.

EX 1 To a cyclist riding due North at 3 ms⁻¹ the wind appears to be blowing from the East. If the cyclist doubles his speed, but does not change his direction, the wind appears to be blowing from N60°E. Find, by drawing or calculation, the true wind speed and direction. The cyclist now turns around and cycles due South at 3 ms⁻¹. Calculate the apparent wind direction. *(S)*

2 A ship A, steaming in a direction 030° with a steady speed of 12 knots, sights a ship B. The relative velocity of B to A is 10 knots in a direction 270°. Find the magnitude and direction of the velocity of B. A changes direction but the magnitude of its velocity does not change so that the relative velocity of A to B is in the direction due North. Find the new direction of A. *(O & C)*

3

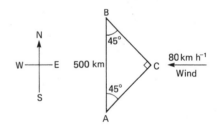

Three cities A, B and C are at the vertices of an isosceles triangle. B is 500 km due north of A, and C is due south-east from B and due north-east from A. A steady wind blows from east to west at a speed of 80 km h⁻¹ (see diagram). An aircraft, whose speed in still air is 200 km h⁻¹, flies direct from A to B then from B to C and then from C to A. Using a graphical method, or otherwise, find the time taken for each part of the journey, giving each answer in hours, to one place of decimals. *(C)*

4 A ship P is travelling due East at 30 kilometres per hour and a ship Q is travelling due South at 40 kilometres per hour. Both ships keep constant speed and course. At noon they are each 10 km from the point of intersection, O, of their courses, and moving towards O.
Find the co-ordinates, with respect to axes Ox eastwards and Oy northwards, of P and Q at time t hours past noon, and find the distance PQ at this time. Find the time at which P and Q are closest to one another. Find the magnitude and direction of the velocity of Q relative to P, indicating the direction on a diagram. Show that, at the position of closest approach, the bearing of Q from P is $\theta°$ South of East, where $\tan\theta° = \frac{3}{4}$.

(J)

45 1-D Particle Dynamics
Force, Mass and weight, Newton's laws of motion, Problem solving.

Force

A **force** is necessary to cause a body to **accelerate**. More than one force may act on a body. If the forces on a body are in **equilibrium**, i.e. balance out, then the body may be at rest or moving in a straight line at constant speed.

If there is a resultant force on the body, then the body will accelerate.

Force is a **vector**, i.e. it has magnitude and direction.

S.I. unit of force is the **newton** (N).

1 newton is the force needed to give a body of mass 1 kg an acceleration of $1\,\text{m s}^{-2}$.

Mass and weight

Mass and **weight** are different.

The **mass** of a body is a measure of the matter contained in the body. A massive body will need a large force to change its motion. The mass of a body may be considered to be constant, whatever the position of the body, provided that none of the body is destroyed or changed.

Mass is a **scalar**, i.e. it has magnitude only.

S.I. unit of mass is the **kilogram** (kg).

The **weight** of a body is the force with which the earth attracts it. It is dependent upon the body's distance from the earth, so a body 'weighs' less at the top of Everest than it does at sea level.

Weight is a **vector**, since it is a force.

S.I. unit of weight is the **newton** (N).

The weight W, in newtons, and mass m, in kilograms, of a body are connected by the relation $W = mg$, where g is the acceleration due to gravity, in m s^{-2}. Common approximate values for g are $10\,\text{m s}^{-2}$ and $9.8\,\text{m s}^{-2}$.

Newton's laws of motion

Newton's three laws of motion are the fundamental basis of the study of mechanics at this level. Although there is no direct proof of these laws, predictions made using them agree very closely with observations.

statement	consequence(s)
1st law Every body will remain at rest or continue to move in a straight line at constant speed unless an external force acts on it.	(a) If a body has an acceleration, then there must be a force acting on it. (b) If a body has no acceleration, then the forces acting on it must be in equilibrium.
2nd law The rate of change of momentum of a moving body is proportional to the external force acting on it and takes place in the direction of that force. So when an external force acts on a body of constant mass, the force produces an acceleration which is directly proportional to the force.	(a) The basic equation of motion for constant mass is: **Force = mass × acceleration** (in N) (in kg) (in m s^{-2}) (b) The force and acceleration of the body are both in the same direction. (c) A constant force on a constant mass gives a constant acceleration.
3rd law If a body A exerts a force on a body B, then B exerts an equal and opposite force on A.	These forces between bodies are often called reactions. In a rigid body the internal forces occur as equal and opposite pairs and the net effect is zero. So only external forces need to be considered.

Problem solving

The following are important points to remember when solving problems using Newton's laws of motion.

1. Draw a clear force diagram (see Force Diagrams p. 82).

2. If there is no acceleration, i.e. the body is either at rest or moving with uniform velocity, then the forces balance in any direction.

3. If there is an acceleration:
(a) mark it on the diagram using ──→ a **Figure 4**
(b) write down, if possible, an expression for the force,
(c) use Newton's 2nd law, i.e. write the equation of motion:
force = mass × acceleration

ℹ️ Body at rest on a rough inclined plane.

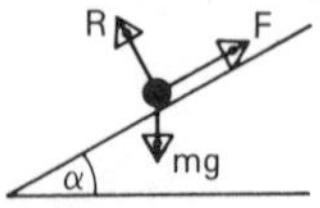

No acceleration so forces balance:
∥ to plane $\Rightarrow F = mg \sin \alpha$
⊥ to plane $\Rightarrow R = mg \cos \alpha$.

Figure 1

Body sliding down rough plane at constant speed.

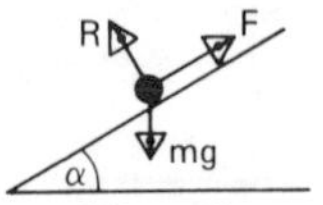

No acceleration so forces balance:
∥ to plane $\Rightarrow F = mg \sin \alpha$
⊥ to plane $\Rightarrow R = mg \cos \alpha$.

Figure 2

Body sliding down rough plane with acceleration.

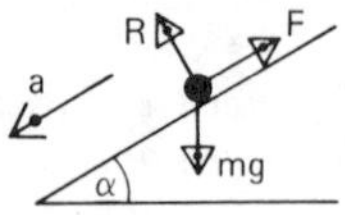

No 'a' ⊥ to plane
$\Rightarrow R = mg \cos \alpha$.
Net force down plane is
$mg \sin \alpha - F$.
2nd law $\Rightarrow mg \sin \alpha - F = ma$.

Figure 3

1-D Particle Dynamics
Worked example, Guided example and Exam questions

WE *A train of total mass 300 tonnes is travelling along a straight horizontal track at a constant speed of 54 km/h. The resistances to the motion are 50 Newtons per tonne. The rear coach of mass 50 tonnes becomes detached but the tractive force of the engine remains the same. Calculate*

(a) the acceleration of the rest of the train

(b) the distance the rear coach travels, after becoming detached, before it stops.

Force diagram for complete train.

50 × 300
(resistive force) ◄────── | 300 × 10³ kg | ─────► T (tractive force)

Since speed is constant, $T = 50 \times 300$ N
    i.e. $T = 15\,000$ N.

Force diagram for remainder of train.

50 × 250 ◄────── | 250 × 10³ kg | ─────► T = 15 000

Resultant accelerating force $= 15\,000 + 50 \times 250$ N
                              $= 2500$ N

Using Newtons 2nd law:
    Accelerating force $=$ mass $\times$ acceleration
    $2500 = (250 \times 10^3) \times a$
$\Rightarrow \qquad a = 0 \cdot 01$ ms^{-2}, the acceleration.

The only force acting on the rear coach when it becomes detached is the resistance force.

50 × 50 ◄────── | 50 × 10³ kg |

If f is the acceleration of the coach (to the right is positive) Newtons 2nd law gives:
    $(-50 \times 50) = (50 \times 10^3) \times f$
$\Rightarrow \qquad f = -0 \cdot 05$ ms^{-2} i.e. a retardation.

Now $s = ?$, $u = 54 \times \dfrac{5}{18}$ ms^{-1}, $t = ?$, $v = 0$, $a = -0 \cdot 05$ ms^{-2}

Using $v^2 = u^2 + 2as$
    $0 = 15^2 + 2(-0 \cdot 05)\, s$
$\Rightarrow \qquad s = 2250$ m — the distance the coach travels.

GE *A lift of mass 500 kg is descending with an acceleration of $1 \cdot 5$ ms^{-2}. A man of mass 80 kg is inside the lift. Calculate*
(a) the tension in the cable connected to the lift,
(b) the force between the man and the floor of the lift.

First consider the man and lift to be one mass subject to forces of T (say) upwards (the tension in the cable) and the weight force downwards. Write down the equation of motion. Next consider the force between the man's feet and the floor as (say), F. This force can be considered to be acting on either the man's feet upwards (if you consider forces on the man), or on the floor of the lift downwards (if you consider forces acting on the lift). Choose one of these only, draw a force diagram for either the man or the lift and then write down the appropriate equation of motion. Hence F.

EX 1 State Newton's Laws of Motion. A miniature engine of mass 110 kg is coupled to and pulls a miniature carriage of mass 30 kg along a horizontal track. The resistance to the motion of the engine is $\dfrac{1}{100}$ of its weight; the resistance to the motion of the carriage is $\dfrac{1}{150}$ of its weight. Given that the whole tractive force exerted by the engine is equal to the weight of 3 kg, find the tension in the coupling.

*(W)

2 A constant force of 35 N, always acting in the same horizontal direction, causes a particle of mass 2 kg to move over a rough horizontal plane. The particle passes two points X and Y, 4 m apart, with speeds of 5 m/s and 10 m/s respectively. The frictional resistance to motion is constant. Calculate:
   (i) the acceleration of the particle;
   (ii) the magnitude of the frictional resistance;
   (iii) the distance of the particle from X, 4 s after it has passed X.

*(A)

3 A water skier of mass 95 kg is towed by a horizontal rope behind a boat. His body is straight, and the thrust of the water acts along the line of his body. When moving with uniform velocity, he is leaning back at 10° from the vertical. Find the tension in the rope. The boat begins to accelerate, and the skier leans back at 15°. The tension in the rope now becomes 500 N. Find the acceleration of the boat.

*(OLE)

4 A breakdown truck of mass 2000 kg is towing a car of mass 1000 kg by means of a rope, up an incline of 1 in 20. The resistances due to friction on each vehicle are proportional to the masses of the vehicles. The engine of the truck exerts a tractive force of 3600 N when moving up the hill at a steady speed of 18 km/h. Show that the tension in the rope is 1200 N. The rope breaks and the two vehicles continue to move up the hill. Calculate:
   (i) how much time elapses before the car comes momentarily to rest;
   (ii) how far the car travels in this time.

*(OLE)

5 A heavy particle is suspended by a spring balance from the ceiling of a lift. When the lift moves up with constant acceleration f m/s^2 the balance shows a reading $1 \cdot 8$ kg. When the lift descends with constant acceleration $\frac{1}{3}f$ m/s^2 the balance shows a reading 1 kg. Find the mass of the particle and the value of f.

(L)

6 When pulled by a horizontal force of 16 N, a particle of mass 2 kg moving on a rough horizontal plane has an acceleration of 4 ms^{-2}. Find the coefficient of friction. When the same plane is inclined at 23° to the horizontal and the force of 16 N no longer acts, the particle slides down with an acceleration of a ms^{-2}. Find a.

(O & C)

46 Connected Particles
Problem solving, Common situations.

Two particles connected by a light inextensible string which passes over a fixed light smooth frictionless pulley are called **connected particles.**
The tension in the string is the same throughout its length, so each particle is acted upon by the same tension.

Problems concerned with connected particles usually involve finding the acceleration of the system and the tension in the string.

To solve problems of this type:
1. Draw a clear diagram showing the forces on each particle and the common acceleration.
2. Write down the equation of motion,
i.e. force = mass × acceleration
for each particle separately.
3. Solve the two equations to find the common acceleration, a, and/or the tension, T, in the string.

i *Two particles mass m_1 and m_2, with $m_1 > m_2$, are connected by a light inextensible string which passes over a fixed light smooth frictionless pulley. Find the common acceleration, a, of each mass and the tension, T, in the string when the system is moving freely.*

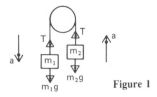

Figure 1

For m_1, with acceleration $a \downarrow$, the equation of motion, i.e. force = mass × acceleration

is $\qquad m_1g - T = m_1a \qquad (1)$

For m_2, with acceleration $a \uparrow$, the equation of motion

is $\qquad T - m_2g = m_2a \qquad (2)$

Adding (1) and (2) to eliminate T:
$$m_1g - m_2g = m_1a + m_2a$$
i.e. $\qquad a = \dfrac{(m_1 - m_2)}{(m_1 + m_2)}g$

Substituting for a in (1):
$$m_1g - T = m_1\frac{(m_1 - m_2)}{(m_1 + m_2)}g$$

Rearranging gives:
$$T = \frac{2m_1m_2}{(m_1 + m_2)}g$$

Common situations

The simplest situation in which connected particles occur is illustrated above. There are several other situations in which the motion of connected particles is considered. The most common are shown below.

One particle on a **smooth horizontal table** as shown.

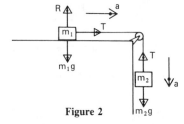

Figure 2

For m_1:
Resolving ⊥ to table $\qquad R = m_1g$
Equation of motion: $\qquad T = m_1a$
For m_2:
Equation of motion: $m_2g - T = m_2a$

One particle on a **rough horizontal table** as shown

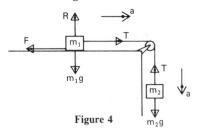

Figure 4

For m_1:
Resolving ⊥ to table: $\qquad R = m_1g$
Equation of motion: $\quad T - F = m_1a$
For m_2:
Equation of motion: $m_2g - T = m_2a$

One particle on a **smooth inclined plane** as shown.

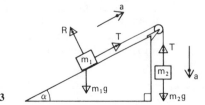

Figure 3

For m_1:
Resolving ⊥ to plane: $\qquad R = m_1g \cos \alpha$
Equation of motion: $T - m_1g \sin \alpha = m_1a$
For m_2:
Equation of motion: $\quad m_2g - T = m_2a$

One particle on a **rough inclined plane** as shown.

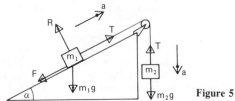

Figure 5

For m_1:
Resolving ⊥ to plane: $\qquad R = m_1g \cos \alpha$
Equation of motion: $T - F - m_1g \sin \alpha = m_1a$
For m_2:
Equation of motion: $\quad m_2g - T = m_2a$

Connected Particles
Worked example and Exam questions

 A smooth plane and a rough plane, both inclined at 45° to the horizontal, intersect in a fixed horizontal ridge. A particle P of mass 1 kg is held on the smooth plane by a light inextensible string which passes over a small frictionless pulley A on the ridge and is attached to a particle Q of mass 3 kg which rests on the rough plane. The plane containing P, Q and A is perpendicular to the ridge. The system is released from rest with the string taut. Given that the acceleration of each particle is of magnitude $\sqrt{2}$ ms^{-2}, find

(a) *the tension in the string,*

(b) *the coefficient of friction between Q and the rough plane,*

(c) *the magnitude and direction of the force exerted by the string on the pulley.*

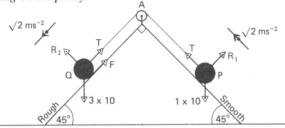

Let T be the tension in the string,
and F be the frictional force on Q.
Let R_1 and R_2 be the normal reactions on P and Q respectively.

For P

Equation of motion:
$T - 10 \cos 45° = 1 \times \sqrt{2}$
$\Rightarrow \quad T = 5\sqrt{2} + \sqrt{2}$
   i.e. $T = 6\sqrt{2}$ N, the tension in the string.

For Q

Since friction is limiting (particle is in motion)
   $F = \mu R_2$, μ coefficient of friction.
Resolving perpendicular to slope:
   $R_2 = 3 \times 10 \cos 45°$,
$\Rightarrow R_2 = 15\sqrt{2}$ N.
Equation of motion:
   $3 \times 10 \cos 45° - T - F = 3 \times \sqrt{2}$
$\Rightarrow \quad F = 15\sqrt{2} - 3\sqrt{2} - 6\sqrt{2}$
   i.e. $F = 6\sqrt{2}$ N.
Hence $6\sqrt{2} = \mu 15\sqrt{2}$
$\Rightarrow \mu = \dfrac{6\sqrt{2}}{15\sqrt{2}} = \dfrac{2}{5}$, the coefficient of friction.

Forces on the pulley are:

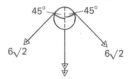

The resultant force is
$6\sqrt{2} \cos 45° + 6\sqrt{2} \cos 45°$,
   i.e. 12 N vertically downwards.

 1 Masses of 1 kg and 2 kg are attached to the ends of a long light string which passes over a light pulley supported by a frictionless horizontal axis. If the tension in the string is T newtons, write down the equation of motion of each mass, and hence find:
   (i) the tension in the string;
   (ii) the time taken for the heavier mass to fall from rest a distance of 1·5 m.
(O & C)

2 Two particles X and Y, of mass 150 g and 100 g respectively, are attached to the ends of a light inextensible string. The particle X is held on a rough horizontal table, with the string passing along the table and over a small smooth pulley which is fixed on the edge of the table. The string is in a plane perpendicular to the edge of the table and Y hangs freely. The coefficient of friction between X and the table is $\frac{1}{3}$. The particle X is released and after $1\frac{1}{2}$ s, before X reaches the edge of the table, the string breaks. Calculate:
   (a) the acceleration with which X moves before the string breaks;
   (b) the speed attained by X at the instant the string breaks;
   (c) the total distance X moves before coming to rest, assuming it still has not reached the edge of the table.
(A)

3 Two particles A and B, of masses 0·4 kg and 0·3 kg respectively, are connected by a light inextensible string. The particle A is placed near the bottom of a smooth plane inclined at 30° to the horizontal. The string passes over a small smooth light pulley which is fixed at the top of the inclined plane and B hangs freely. The system is released from rest, with each portion of the string taut and in the same vertical plane as a line of greatest slope of the inclined plane. Calculate:
   (a) the common acceleration, in m/s^2, of the two particles;
   (b) the tension, in N, in the string.
Given that A has not reached the pulley, find:
   (c) the time taken for B to fall 6·3 m from rest;
   (d) the speed that B has then acquired.
[Take the value of g to be 9·8 ms^{-2}]
(L)

4 (In this question, you may assume that the strings are of such a length, and the pulley so positioned, that at no time during the motion is the pulley hit by either mass.) Masses of $3m$ and m are connected by a light inextensible string passing over a light smooth pulley. The system is released from rest with the string taut and with both masses at a height h above the ground. Find:
   (i) the acceleration with which they both move;
   (ii) the tension in the string before the $3m$ mass hits the ground;
   (iii) the loss of kinetic energy at the impact of the $3m$ mass with the ground, (the $3m$ mass does not rebound from the ground);
   (iv) the greatest height reached by the smaller mass, and the total time taken to get there;
   (v) the speed with which the $3m$ mass leaves the ground again.
(S)

5 A light inextensible string passes over a smooth light fixed pulley and masses of 3 kg and 7 kg are attached to its ends. The system is held at rest with the string taut, those parts not in contact with the pulley being vertical, and then released. Find the acceleration of each mass and the tension in the string, stating units. After the 7 kg mass has descended a distance of one metre it strikes an inelastic horizontal table. Show that the time taken for this to happen is 5/7 s from the start of the motion, assuming that the acceleration due to gravity is 9·8 ms^{-2}. Find the time during which the 7 kg mass is at rest on the table. (Assume throughout that the 3 kg mass does not reach the pulley.)
(O & C)

47 Work and Energy

Definitions, Hooke's law, Kinetic and potential energy, Mechanical energy, Conservation of mechanical energy.

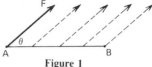

Definitions	**Work** may be done by or against a force (often gravity). It is a **scalar**. When a constant force F moves its point of application from A to B, the **work done** by F is $F\cos\theta.AB$. S.I. unit of work is the **joule** (J).

Figure 1

1 joule is the work done by a force of 1 N in moving its point of application 1 m in the direction of the force.

Energy is the capacity to do work. It is a scalar.
S.I. unit of energy is the **joule** (the same as work).
A body possessing energy can do work and lose energy. Work can be done on a body and increase its energy.
i.e. Work done = change in energy.

Hooke's law

Hooke's law for an elastic string or spring is $T=\dfrac{\lambda x}{l}$,

where T is the tension in string or spring,
λ is its **modulus of elasticity**,
l is its natural (unstretched) length,
x is the extension. (Note: a negative extension of a spring is a compression.)

Work done in stretching the string or spring is $\dfrac{\lambda x^2}{2l}$.

ℹ️ *An elastic string AB of natural length* 3 m *and modulus* 12 N *has its end A attached to a fixed point. A force of* 4 N *is applied to the end B. Calculate the work done by the force in producing the extension.*

Hooke's law, $T=\dfrac{\lambda x}{l}$, gives $4=\dfrac{12x}{3}$

$\Rightarrow x=1\,m$, the extension.

Work done by the force $=\dfrac{12\times1^2}{2\times3}=2\,\text{J}$.

Kinetic and potential energy

Kinetic energy and **potential energy** are types of **mechanical energy**.
(a) Kinetic energy (K.E.) is due to a body's motion. The K.E. of a body of mass m, moving with velocity v, is $\frac{1}{2}mv^2$.
(b) Potential energy (P.E.) is due to a body's position. Gravitational P.E. is a property of height.
The P.E. of a body of mass m at a distance h,
(i) above an initial level is mgh,
(ii) below an initial level is $-mgh$.
The initial level can be any level you choose and the P.E. at the initial level is zero.
Elastic P.E., a property of stretched elastic strings and springs or compressed springs, is $\dfrac{\lambda x^2}{2l}$, where λ is the modulus of the elasticity of the 'string', l is its natural length and x the extension (see Hooke's law).

ℹ️ A stone of mass 2 kg is thrown horizontally with speed 3 m s^{-1}.
K.E. of the moving stone is $\frac{1}{2}mv^2=\frac{1}{2}\times2\times3^2\,\text{J}=9\,\text{J}$.

ℹ️ A body of mass 2 kg is suspended 3 m above a floor. The P.E. of the body relative to:
(a) the floor is $mgh=2\times10\times3\,\text{J}=60\,\text{J}$.
(b) a table 1 m high is $mgh=2\times10\times2\,\text{J}=40\,\text{J}$.
(in both cases taking $g=10\text{ m s}^{-2}$.)

ℹ️ An elastic string of natural length 0.5 m and modulus 1 N is stretched to a length 0.75 m.

Elastic P.E. stored in the string $=\dfrac{\lambda x^2}{2l}=\dfrac{1\times0.25^2}{2\times0.5}\text{J}$

$=0.0625\,\text{J}$

Mechanical energy

The **mechanical energy** (M.E.) of a particle (or body) = P.E. + K.E. of the particle (or body).

If a system includes one or more elastic strings, then:
total M.E. of the system = P.E. + K.E. + elastic P.E.

M.E. is lost (as heat energy or sound energy) when we have:
resistances (friction) or
impulses (collisions or strings jerking taut).

ℹ️ A particle of mass 3 kg is moving with a speed of 5 m/s, 0.5 m above ground level (P.E. = 0).
Total mechanical energy
$=\text{P.E.}+\text{K.E.}$
$=mgh+\frac{1}{2}mv^2$
$=3\times10\times0.5+\frac{1}{2}\times3\times5^2=52.5\,\text{J}$

ℹ️ A 5 kg mass, moving horizontally on a smooth table at 9 m/s, hits a vertical plane barrier and rebounds at 4 m/s.
Loss in K.E. $=\frac{1}{2}mv_1^2-\frac{1}{2}mv_2^2$
$=\frac{1}{2}\times5\times9^2-\frac{1}{2}\times5\times4^2=162.5\,\text{J}$.

Conservation of mechanical energy

The total mechanical energy of a body (or system) will be **conserved** if
(a) no external force (other than gravity) causes work to be done, and
(b) none of the M.E. is converted to other forms.
Given these conditions: P.E. + K.E. = constant
or loss in P.E. = gain in K.E.
or loss in K.E. = gain in P.E.

ℹ️ *A particle falls freely from rest until its speed is* 9 m s^{-1}. *How far has it fallen? (Use $g=10\text{ m s}^{-2}$)*

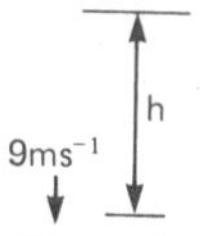

9ms^{-1}

Figure 2

Initially, P.E. $=mgh$, K.E. $=0$.
So total M.E. $=mgh$.
Finally, P.E. $=0$, K.E. $=\frac{1}{2}m.9^2$.
So total M.E. $=\frac{1}{2}m.9^2$.

Total M.E. is conserved, so $mgh=\frac{1}{2}m9^2\Rightarrow h=4.05\,\text{m}$.

Work and Energy
Worked example, Guided example and Exam questions

WE (*a*) *A particle of mass m is projected directly up a rough plane of inclination α with velocity V. If μ is the coefficient of friction between the particle and the plane, calculate how far up the plane the particle travels before coming to rest.*

(*b*) *A light elastic string OA of natural length l and modulus 2mg has its end O fixed to a point on a ceiling. A particle of mass m is attached to the end A of the string and is held as close as possible to O. If the particle is released from rest, find the maximum length of OA in the subsequent motion.*

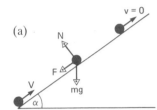

(a)

F = frictional force
N = normal reaction
Let l be the distance along the slope the particle travels.

Since the particle is moving along the plane, friction is limiting, so $F = \mu N$.
Resolving perpendicular to the plane:

$$N = mg \cos \alpha,$$

so frictional force $F = \mu mg \cos \alpha$.
At bottom of slope, total energy $= \frac{1}{2} mV^2$.
At top of slope, total energy $= mgl \sin \alpha$.
Work done against friction $= mgl \cos \alpha$.

Now,
initial total energy
= final total energy + work done against friction,
i.e. $\frac{1}{2}mV^2 = mgl \sin \alpha + mgl \cos \alpha$

$$\Rightarrow l = \frac{V^2}{2g(\sin \alpha + \mu \cos \alpha)}$$

and is the distance travelled by the particle along the slope before it comes to rest.

(b) Let L be the maximum length of OA.
OA will be a maximum when the particle comes to rest at its lowest point.
When the particle is at O,

$$P.E. = mgL$$
$$E.P.E. = 0$$
$$K.E. = 0.$$

So, total energy $= mgL$
When particle is at lowest point,

$$P.E. = 0$$
$$E.P.E. = \frac{2mg(L-l)^2}{2l}$$
$$K.E. = 0.$$

So, total energy $= \frac{2mg(L-l)^2}{2l}$

By conservation of mechanical energy,

$$mgl = \frac{2mg(L-l)^2}{2l}$$
$$\therefore L^2 - 3lL + l^2 = 0$$
$$\Rightarrow L = \frac{l}{2}(3 + \sqrt{5}), \text{ the maximum length of } OA.$$

GE *The top of a chute whose length is 12 m is 3 m vertically above its lowest point. A parcel of mass 1·6 kg slides from rest from the top of the chute and reaches the lowest point with a speed of 5 m/s. Calculate, for the parcel,*

(*i*) *the gain in kinetic energy,*
(*ii*) *the loss in potential energy,*

(*iii*) *the work done in overcoming the frictional resistance,*
(*iv*) *the average value of this resistance.*

After reaching the lowest point of the chute, the parcel slides along a horizontal floor the resistance to motion being 4 N. Calculate how far the parcel travels before coming to rest.

(i) Find the kinetic energy at the top and bottom of the slope. Hence find gain.
(ii) Find the potential energy at the top and bottom of the slope. Hence find loss.
(iii) Find the loss of mechanical energy. (This is the difference between the total energy at the top and bottom of the slope.) Hence work done.
(iv) Work done by the average resistance is $R \times 12$ m, where R is the average resistance. Equate this to work done calculated in (iii). Hence find R.

For final part, use
loss in K.E. = work done against 4 N resistive force
to find the distance travelled.

EX 1 A block of mass 6·5 kg is projected with a velocity of 4 m/s up a line of greatest slope of a rough plane. Calculate the initial kinetic energy of the block. The coefficient of friction between the block and the plane is $\frac{2}{3}$ and the plane makes an angle θ with the horizontal where $\sin \theta = \frac{5}{13}$. The block travels a distance of d m up the plane before coming instantaneously to rest. Express in terms of d:
   (i) the potential energy gained by the block in coming to rest;
   (ii) the work done against friction by the block in coming to rest.
   Hence calculate the value of d.

*(C)

2 A boy on a sledge slides down a hill of variable gradient. In so doing he travels a distance of 168 m, measured along the surface of the track, and descends a vertical distance of 30 m. The combined mass of the boy and the sledge is 80 kg. If the initial speed is 2 ms^{-1} and the final speed is 16 ms^{-1}, find in the same units:
   (i) the increase in the kinetic energy of the combined mass of boy and sledge;
   (ii) the work done by gravity.
   Hence find the average resistance to motion (defined as the work done against the resisting forces divided by the distance travelled).

*(O & C)

3 A fixed plane is inclined at an angle α to the horizontal, where $\tan \alpha = \frac{4}{3}$. A particle of mass m is projected, from the point A on the plane, up a line of greatest slope. The coefficient of friction between the particle and the plane is $\frac{1}{3}$. The particle has moved a distance d up the plane when it comes instantaneously to rest at the point B.
   (i) Find the total work done against the external applied forces during the motion from A to B.
   (ii) Find the speed of projection from A.
   (iii) Find the total work done against the external applied forces during the motion from A to B and back to A again.
   (iv) Find the kinetic energy of the particle when it has passed through A and moved a further distance $4d$ down the plane from A.

(C)

4 An elastic string, of natural length l and modulus of elasticity λ, is stretched to a length $l+x$. As a result, the tension in the string is mg and the energy stored in it is E. Find x and λ in terms of E, g, l and m.

(L)

---

48 Power
Definitions, Moving vehicles, Common situations.

Definitions

Power is the rate at which a force does work.
It is a **scalar**.

S.I. unit of power is the **watt** (W).

1 watt (W) = 1 joule per second ($J\,s^{-1}$).

The **kilowatt** (1 kW = 1000 W) is often used.

When a body is moving in a straight line with velocity $v\,m\,s^{-1}$ under a tractive force F newtons, the power of the force is Fv watts.

ℹ️ *A pump raises water at a rate of 500 kg per minute through a vertical distance of 3 m. If the water is delivered at $2.5\,m\,s^{-1}$, find the power developed.*

The pump does work to create both P.E. and K.E.
P.E. created is $500 \times 10 \times 3 = 15\,000$ J per min.
K.E. created is $\frac{1}{2} \times 500 \times 2.5^2 = 1562.5$ J per min.
So, power developed is $(15\,000 + 1562.5) \div 60 \approx 276$ W

Moving vehicles

The power of a moving vehicle is supplied by its engine. The **tractive force** of an engine is the pushing force it exerts.

To solve problems involving moving vehicles:

1. Draw a clear force diagram.
Note: 'non-gravitational resistance' means 'frictional force'.
2. Resolve forces perpendicular to the direction of motion.
3. If the velocity is:
(a) constant (steady speed), then resolve forces parallel to the direction of motion (Newton's first law),
(b) not constant (accelerating), then find the net force acting and write down the equation of motion in the direction of motion.
4. Use Power = tractive force × speed.

ℹ️ *A train of total mass 200 tonnes is moving at a steady speed of $72\,km\,h^{-1}$ on a straight level track. If the non-gravitational resistance is 10^4 newtons, at what rate is the engine working? Take $g = 10\,m\,s^{-2}$.*

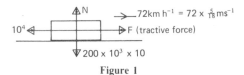

Figure 1

Resolve ($\uparrow$): $N = 200 \times 10^3 \times 10$
$= 2 \times 10^6$ newtons
Resolve ($\rightarrow$): $F = 10^4$ newtons
Power $= Fv$
$= 10^4 \times 72 \times \dfrac{5}{18} = 2 \times 10^5$ W = 200 kW

Common situations

The following illustrate some common situations which arise in problems.

ℹ️ **1. Vehicles on the level**
(a) moving with **steady speed** v

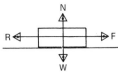

Figure 2

Resolve ($\uparrow$): $N = W$
Resolve ($\rightarrow$): $F = R$
Power: $P = Fv$

(b) moving with **acceleration** a and instantaneous speed v

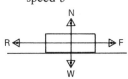

Figure 5

Resolve ($\uparrow$): $N = W$
Equation of motion:
$F - R = ma$
Power: $P = Fv$

Note: if the vehicle is retarding, i.e. $R > F$, then a will be in the opposite direction.

2. Vehicles on a slope of angle α
(a) moving with **steady speed** v
(i) moving **up**

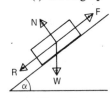

Figure 3

Resolve $\perp$ to plane:
$N = W \cos \alpha$
Resolve $\parallel$ to plane:
$F = R + W \sin \alpha$
Power: $P = Fv$

(b) moving with **acceleration** a and instantaneous speed v
(ii) moving **up**

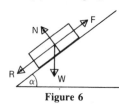

Figure 6

Resolving $\perp$ to plane:
$N = W \cos \alpha$
Equation of motion $\parallel$ to plane:
$F - R - W \sin \alpha = ma$
Power: $P = Fv$

Note: if the vehicle is moving up the plane but retarding, then $F < R + W \sin \alpha$.
(ii) moving **down**

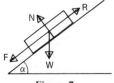

Figure 7

Resolve $\perp$ to plane:
$N = W \cos \alpha$
Equation of motion $\parallel$ to plane:
$F + W \sin \alpha - R = ma$
Power: $P = Fv$

(ii) moving **down**

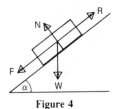

Figure 4

Resolve $\perp$ to plane:
$N = W \cos \alpha$
Resolve $\parallel$ to plane:
$F + W \sin \alpha = R$
Power: $P = Fv$

Note: if the vehicle is retarding, using the engine for braking, then F acts up the plane and $F + R > W \sin \alpha$.

Power
Worked example and Exam questions

 A bus of mass 5 tonnes freewheels down a slope of inclination $\sin^{-1}\left(\dfrac{1}{40}\right)$ *to the horizontal at constant speed. Assuming that the non-gravitational resistances remain the same, find the rate at which the engine must work in order to drive the bus up the same incline at a steady speed of 12 km/h. If the power is suddenly increased to 10 kW find, in m/s², the immediate acceleration of the bus.*

Stage 1, *bus freewheeling down slope.*
F is the non-gravitational resistance. Since speed is steady down the slope,
forces down slope = forces up slope
i.e. $5 \times 10^3 \times 10 \times \dfrac{1}{40} = F$

$\Rightarrow F = 1250$ N, the non-gravitational resistance.

Stage 2, *bus being driven up the slope at steady speed.*
T is the tractive force of the engine.
Since speed is steady,
forces up slope = forces down slope
i.e. $T = 1250 + 5 \times 10^3 \times 10 \times \dfrac{1}{40}$

$= 1250 + 1250$
So, $T = 2500$ N.
Now, power = TV watts, V being the speed up the slope.

$V = 12 \times \dfrac{5}{18}$ m/s.

$= \dfrac{10}{3}$ m/s

So, power $= 2500 \times \dfrac{10}{3}$ watts

$= \dfrac{25}{3}$ kW

$= 8\frac{1}{3}$ kW.

When the power is suddenly increased to 10 kW, a new tractive force T_1 acts, but the speed at this instant is still $\dfrac{10}{3}$ m/s.

Using Power = TV watts,

$10 \times 10^3 = T_1 \times \dfrac{10}{3}$

$\Rightarrow \qquad T_1 = 3 \times 10^3$ N.

Stage 3, *bus accelerating up the slope.*
We now have a resultant accelerating force acting up the slope equal to,

$T_1 - F - 5 \times 10^3 \times 10 \times \dfrac{1}{40}$

$= 3 \times 10^3 - 1250 - 1250$
$= 500$ N.

Using, resultant accelerating force = mass × acceleration, up the slope, we have,
$500 = 5 \times 10^3 \times a$, where a m/s² is the acceleration

$\Rightarrow \qquad a = \dfrac{500}{5 \times 10^3}$ m/s²

i.e. $a = \dfrac{1}{10}$ m/s².

1 The engine of a car is working at a constant rate of 6 kW in driving the car along a straight horizontal road at a constant speed of 54 km/h. Find, in N, the resistance to the motion of the car.

(L)

2 A motor car of mass 800 kg is towing a trailer of mass 300 kg along a straight horizontal road, Resistances, which are constant, are 600 N for the car and 240 N for the trailer. Calculate the tractive force exerted by the motor and the tension in the coupling between the car and the trailer in each of the following cases:
(i) when both are travelling at constant velocity;
(ii) when both are accelerating at $2 \cdot 5$ m/s².
Calculate the power developed by the motor when the car and trailer are travelling at a constant velocity of 15 m/s.

(C)

3 (Take g as 10 ms⁻².) The three parts of this questions are all concerned with a car of mass 3000 kg.
(i) The car will *just* run down a slope of inclination α ($\sin \alpha = \frac{1}{20}$) under its own weight (i.e. the engine is *not* switched on). Find the resistance to motion in newtons.
(ii) The car is driven along a level road, its engine exerting a constant tractive force of T newtons. Given that the force resisting motion is the same (in magnitude) as when the car just runs down the plane (Part (i)) and that, starting from rest, the car reaches a speed of 15 ms⁻¹ in a time of 5 minutes, calculate:
(a) the acceleration of the car;   (b) the value of T;
(c) the power developed by the engine.
(iii) When the car is driven *up* the slope of inclination α ($\sin \alpha = \frac{1}{20}$) its engine exerts such a tractive force that the car travels at a constant speed of 40 km/hour. Given that the resistance to motion is the same, in magnitude, as in Parts (i) and (ii), calculate the rate of working in kW, giving your answer correct to two decimal places.

(W)

4 A car of mass 1000 kg moves with its engine shut off down a slope of inclination α, where $\sin \alpha = \frac{1}{20}$, at a steady speed of 15 ms⁻¹. Find the resistance, in newtons, to the motion of the car. Calculate the power delivered by the engine when the car ascends the same inclination at the same steady speed, assuming that resistance to motion is unchanged. [Take g as 10 ms⁻².]

(L)

5 The frictional resistance to the motion of a car of mass 1000 kg is kv newtons, where v ms⁻¹ is its speed and k is constant. The car ascends a hill of inclination $\arcsin(\frac{1}{10})$ at a steady speed of 8 ms⁻¹, the power exerted by the engine being $9 \cdot 76$ kW. Prove that the numerical value of k is 30. Find the steady speed at which the car ascends the hill if the power exerted by the engine is $12 \cdot 8$ kW.
When the car is travelling at this speed, the power exerted by the engine is increased by 2 kW. Find the immediate acceleration of the car. (Take $9 \cdot 8$ ms⁻² as the acceleration due to gravity.)

(O & C)

6 A car of mass $1 \cdot 2$ tonnes is travelling up a slope of 1 in 150 at a constant speed of 10 ms⁻¹. If the frictional and air resistances are 100 newtons, calculate the power exerted by the engine. The car descends the same slope working at a rate of 2 kW. What will be its acceleration when its speed is 20 ms⁻¹ if the resistances are the same? If the engine is shut off when the speed of the car is 25 ms⁻¹ as it descends the slope how long will it be before the car comes to rest?

(S)

49 Impulse and Momentum

Definitions, Relation between impulse and momentum, Conservation of momentum, Problem solving, Impulses in strings.

Definitions

Impulse is the time effect of a force. It is a **vector**. For a **constant force** F, acting for time t,

$$\text{impulse} = Ft.$$

For a **variable force** F, acting for time T,

$$\text{impulse} = \int_0^T F\,dt.$$

S.I. unit of impulse is the **newton second** (Ns).

The **momentum** of a moving body is the product of its mass m and velocity v, i.e. momentum $= mv$. It is a **vector** whose direction is that of the velocity. S.I. unit of momentum is the **newton second** (Ns).

ℹ️ *A constant force acts on a particle of mass 0.5 kg changing its speed from 3 m s^{-1} to 7 m s^{-1}, the force acting in the direction of motion. What is its impulse?*

Impulse $I = Ft = mat$ since $F = ma$
$v = u + at \Rightarrow at = 7 - 3 = 4$
So, impulse $I = 0.5 \times 4 = 2\,Ns$

ℹ️ *Find the momentum of a particle of mass 1.5 kg moving in a straight line at 5 m s^{-1}.*

Momentum $= 1.5 \times 5 = 7.5\,Ns$

Relation between impulse and momentum

The **impulse** of a force, constant or variable, is equal to the **change of momentum** is produces. If a force F acts for a time t on a body of mass m, changing its velocity from u to v, then
$$\text{impulse} = mv - mu.$$

ℹ️ *A golf ball of mass 0.06 kg resting on a tee is given a horizontal impulse of 1.8 Ns. Calculate the velocity v with which it moves off.*

Using impulse $=$ change of momentum
$$1.8 = 0.06\,v - 0.06 \times 0 \Rightarrow v = 30\text{ m s}^{-1}.$$

Conservation of momentum

The **principle of conservation of momentum** states that the total momentum of a system is constant in any direction provided no external force acts in that direction,

i.e. initial momentum $=$ final momentum.

In this context a system is usually two bodies.

ℹ️ *A pile driver of mass 2 tonnes, moving with velocity 7 m s^{-1} before impact, hits a stationary pile of mass 0.5 tonne. Find their common velocity after impact.*

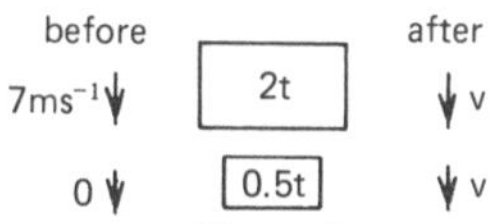

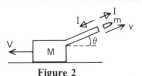

before   after
7ms^{-1} ↓   2t   ↓ v
0 ↓   0.5t   ↓ v

Figure 1

By conservation of momentum:
$$2000 \times 7 = 2000\,v + 500\,v$$
$$\Rightarrow \quad v = 5.6\text{ m s}^{-1}.$$

Problem solving

Problems concerning impulse and momentum usually involve finding the impulse acting or the velocity or mass of a body in a system.

To find an impulse for such a system: write down the impulse equation for each body.

To find a velocity or mass for such a system: write down the equation of conservation of momentum.

ℹ️ *A gun of mass M, whose barrel is at an angle of elevation θ, fires a shell of mass m and recoils horizontally with speed V. The shell travels at speed v relative to the barrel of the gun.*

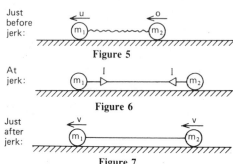

Figure 2 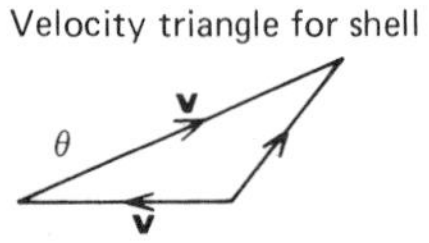   **Figure 3**

Velocity triangle for shell

Velocity of shell, relative to the ground, as it leaves the barrel is the resultant of **v** and **V**.

For the gun ($\rightarrow$): $-I \cos\theta = -MV$
For the shell ($\rightarrow$): $I\cos\theta = m(v\cos\theta - V)$
For the shell ($\uparrow$): $I\sin\theta = mv\sin\theta$
By conservation of momentum ($\rightarrow$):
$$0 = m(v\cos\theta - V) - MV$$

Impulses in strings

When a string jerks taut, impulses, which are equal in magnitude but opposite in direction, act at the two ends. If two particles are attached by a string which jerks taut, then the two particles will experience the **equal and opposite impulses**.

Impulse problems for other connected particles may be solved in the same way if the string is considered to be straight and the particles move in a straight line

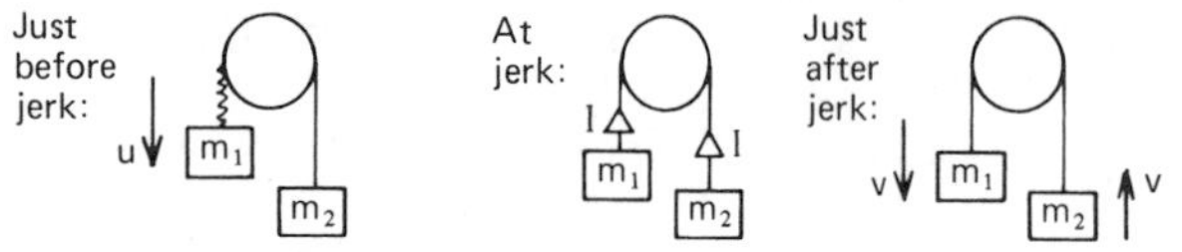

Just before jerk:

At jerk:

Just after jerk:

Figure 4

So this system is equivalent to that shown in ℹ️

ℹ️ Consider this system involving two masses m_1 and m_2.

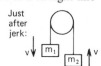

Just before jerk:

Figure 5

At jerk:

Figure 6

Just after jerk:

Figure 7

For mass m_1: $-I = m_1 v - m_1 u$
For mass m_2: $I = m_2 v - m_2 \times 0$
By conservation of momentum: $m_1 u = m_1 v + m_2 v.$

Impulse and Momentum
Worked example, Guided example and Exam questions

A shell of mass 20 kg is travelling horizontally at 100 ms⁻¹ when it suddenly explodes into three pieces A, B, C of masses 12 kg, 6 kg and 2 kg respectively. The diagram shows the direction of travel of the shell before the explosion and the directions of the three pieces A, B, C after the explosion.

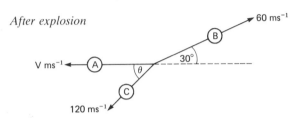

Calculate
(a) the angle θ made by the direction of C with the backward horizontal direction,
(b) the speed of A.

Since no external impulse acts on the system, momentum is conserved in any chosen direction.
Using conservation of momentum at right angles to the original direction of travel, we have:

$$6 \times 60 \cos 60° - 2 \times 120 \sin \theta = 0$$
$$\Rightarrow \quad \sin \theta = \tfrac{3}{4}$$
$$\therefore \quad \theta = \sin^{-1}\left(\tfrac{3}{4}\right)$$

Using conservation of momentum parallel to the original direction of travel we have:

$$6 \times 60 \cos 30° - 12 \times V - 2 \times 120 \cos \theta = 0$$
$$\therefore \quad V = 15\sqrt{3} - 20 \cos \theta.$$

Since $\sin \theta = \tfrac{3}{4}$, $\cos \theta = \dfrac{\sqrt{7}}{4}$,

hence, $V = 15\sqrt{3} - 5\sqrt{7}$
  i.e. $V \approx 12 \cdot 75$ ms⁻¹.

 A pile driver of mass 3 tonnes falls through a distance of 5 m onto a pile of mass 1 tonne without rebounding. If the pile is driven 15 cm into the ground find the average resistance of the ground and the time of penetration.

Use conservation of momentum to find the common velocity of the pile and driver combination just after impact. Use constant acceleration formulae to find the value of the retardation over 15 cm. Write down the equation of motion for the pile and driver combination. (Remember that the accelerating force downwards will be $(4 \times 10^4 - R)$ Newtons, where R Newtons is the average resistance of the ground). Finally use constant acceleration formulae to find the time of penetration.

---

 1 A bullet of mass 50 g, moving horizontally, strikes a stationary target at 486 m/s and becomes embedded in it. The target is of mass 4 kg and is free to move. Calculate, ignoring the time taken by the bullet to become embedded in the target:
  (i) the speed at which the target and the embedded bullet move initially;
  (ii) the impulse imparted to the target by the bullet,
  (iii) the kinetic energy lost in the impact.   *(A)*

2 A railway engine of mass 5300 kg, moving on horizontal rails at $0 \cdot 4$ m/s, strikes the buffers in a siding and is brought to rest from this speed in $0 \cdot 2$ s. Calculate:
  (a) the impulse, in N s, of the force exerted by the buffers on the engine in bringing the engine to rest;
  (b) the magnitude, in N, of this force, assuming it to be constant.   *(L)*

3 Two particles A and B, of masses $3m$ and m respectively, are connected by a light inextensible string and are free to move on a smooth horizontal table. Initially, A is at rest and B has speed u in the direction AB. The first diagram shows the situation just before the string tightens. Find the impulse in the string at the instant when it tightens and the common speed of A and B afterwards.

Later, B is brought instantaneously to rest while A continues towards B. Immediately after the impact between the two particles, the speeds of A and B are x and y respectively, as indicated in the second diagram. In this impact, two-ninths of the kinetic energy of the system just before the impact is lost.

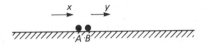

Prove that $3x + y = \tfrac{3}{4}u$ and that $3x^2 + y^2 = \tfrac{7}{48}u^2$. Hence determine the values of x and y in terms of u.   *(C)*

4 When a body of mass $3m$ is moving in a straight line with speed u, it explodes. As a result it splits into two bodies A and B, of respective masses m and $2m$, which move in the same straight line as before, but in opposite directions. Given that the extra energy created by the explosion is $3mu^2$, find the speeds of A and B. The body B then immediately strikes a body of mass M which is at rest and also free to move in the same straight line. If the impact is perfectly elastic, prove that B will subsequently strike A if $M > 6m$.   *(OLE)*

5 A gun of mass 600 kg is free to move along a horizontal track and is connected by a light inelastic rope to an open truck containing sand whose total mass is 1490 kg. The truck is free to move along the same track as the gun. A shell of mass 10 kg is fired from the gun towards the truck and when it leaves the barrel has a horizontal velocity of 915 ms⁻¹ relative to the gun and parallel to the track. The shell lodges in the sand where it comes to relative rest before the rope tightens. Find:
  (i) the speeds of the gun and shell just after the shell leaves the barrel;
  (ii) the speed of the truck before the rope tightens when the shell is at relative rest inside the truck;
  (iii) the speed of the gun and truck just after the rope tightens;
  (iv) the loss in kinetic energy due to the rope tightening;
  (v) the magnitude of the impulsive tension in the rope.
    (S)

6 A particle of mass m, initially at rest, is subjected to an impulse I. In the ensuing motion the only force on the particle is a force directly opposing the motion of the particle and of magnitude k times the square of the velocity of the particle. Show that at time t after the impulse the particle has velocity $v = \dfrac{mI}{Ikt + m^2}$.   *(J)*

50 Impact

Definitions, Direct impact of spheres, Direct impact with a wall, Oblique impact with a wall.

Definitions

If two bodies **rebound** on collision, then the impact is **elastic.**

If two bodies **coalesce** on collsion, then the impact is **inelastic.**

Newton's experimental law for an elastic impact of two bodies can be written as:
   speed of separation after impact $=e\times$ speed of approach before impact
where e is the **coefficient of restitution.**

The value of e depends on the materials of the colliding bodies. For different materials $0\leqslant e\leqslant 1$.

If $e=0$, the impact is inelastic. If $e=1$, the impact is perfectly elastic (not realisable in practice).

Direct impact of spheres

Direct impact takes place when **two similar spheres** moving along the same straight line collide.

To solve problems involving the direct impact of two smooth spheres of masses m_1 and m_2, moving with initial velocities u_1 and u_2 ($u_1>u_2$) and final velocities v_1 and v_2 and coefficient of restitution e:

1. Draw a clear diagram.

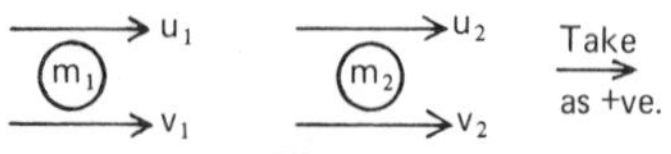

before impact
after impact

Figure 1

2. Use conservation of momentum in the chosen direction,
i.e. $m_1u_1+m_2u_2=m_1v_1+m_2v_2$

3. Apply Newton's experimental law,
i.e. $v_2-v_1=e(u_1-u_2)$

ℹ *A sphere of mass 3 kg moving at 5 m s⁻¹ strikes a similar sphere of mass 2 kg travelling in the opposite direction at 2 m s⁻¹. The coefficient of restitution is ⅖. Find the velocities after impact.*

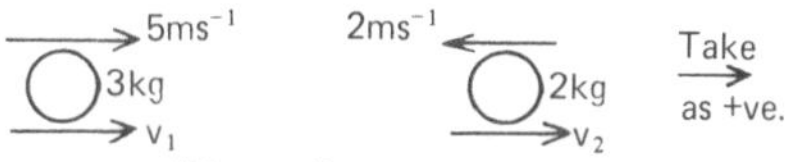

before impact   → 5ms⁻¹ 3kg   2ms⁻¹ ← 2kg   Take as +ve.
after impact   → v_1 → v_2

Figure 2

By the conservation of momentum:
$$(3\times5)+(2\times-2)=3v_1+2v_2$$
i.e. $11=3v_1+2v_2$ (1)

By Newton's experimental law:
$$v_2-v_1=\tfrac{2}{7}(5-(-2))$$
i.e. $v_2-v_1=2$ (2)

Solving (1) and (2): $v_1=1.4\text{ m s}^{-1}$, $v_2=3.4\text{ m s}^{-1}$.

Direct impact with a wall

When a smooth sphere collides **directly with a smooth wall,** the sphere's direction of motion is perpendicular to the wall. The sphere receives an impulse perpendicular to the wall.

To solve problems involving the direct impact with a vertical wall of a smooth sphere of mass m moving with velocity u before impact and v after impact:

1. Draw a clear diagram.

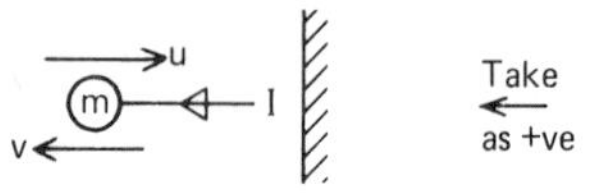

before impact
after impact

Figure 3

2. Apply Newton's experimental law,
i.e. $v=eu$

3. Use impulse = change in momentum if needed,
i.e. $I=mv-(-mu)$

ℹ *A ball of mass 2 kg travelling along a horizontal floor at 5 m s⁻¹ collides directly with a vertical wall. The coefficient of restitution is 0.3. Calculate the speed of the rebound and the impulse given to the ball in the collision.*

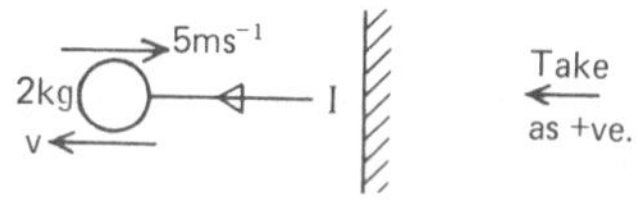

before impact   2kg → 5ms⁻¹ ← I   Take as +ve.
after impact   v ←

Figure 4

By Newton's experimental law:
$$v=0.3\times5$$
$$=1.5\text{ m s}^{-1}\quad\text{the rebound speed.}$$

Using Impulse = change in momentum:
$$I=(2\times1.5)-(2\times-5)$$
$$=13\text{ N s}$$

Oblique impact with a wall

When a smooth sphere collides **obliquely with a smooth wall,** the sphere's direction of motion is at an angle ($\neq90°$) to the wall. The sphere receives an impulse perpendicular to the wall. The component of velocity parallel to the wall is unchanged since both surfaces are smooth.

To solve problems in which a smooth sphere mass m strikes a wall at an angle α with velocity u and rebounds at an angle β with velocity v:

1. Draw a clear diagram.

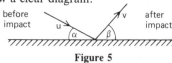

Figure 5

2. Equate components of velocity parallel to the wall,
i.e. $u\cos\alpha=v\cos\beta$

3. Use Newton's experimental law, perpendicular to the wall,
i.e. $v\sin\beta=eu\sin\alpha$

4. Use Impulse = change of momentum, perpendicular to the wall if needed,
i.e. $I=mv\sin\beta-(-mu\sin\alpha)$

ℹ *A smooth sphere travelling along the ground at 3 m s⁻¹ strikes a smooth wall at 60° and rebounds at 45°. Calculate the velocity after impact and the value of the coefficient of restitution.*

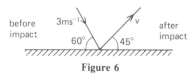

Figure 6

Equate components of velocity parallel to the wall:
$$3\cos60°=v\cos45°$$
$$\therefore v=\frac{3\sqrt{2}}{2}\text{ m s}^{-1}$$

By Newton's experimental law, perpendicular to the wall
$$v\sin45°=e\times3\sin60°$$
$$\therefore e=\frac{1}{\sqrt{3}}$$

Impact
Worked example, Guided example and Exam questions

WE *Three smooth spheres A, B and C of masses 2m, 7m and 14m respectively and of equal size, are at rest on a smooth horizontal floor. The centres of the spheres lie in a straight line. The co-efficient of restitution between each pair of spheres is $\frac{1}{2}$. Sphere A is projected towards B with speed u. Show that after two impacts B is at rest and A and C are each moving with equal speeds in opposite directions. Calculate the total kinetic energy of the spheres after the two impacts.*

(a) Diagram showing the first impact between A and B:

(b) Conservation of momentum (take to the right as positive):
$$(2m \times u) + (7m \times 0) = 2m \times v_1 + 7m \times v_2$$
i.e. $\qquad 2u = 2v_1 + 7v_2$ $\qquad\qquad$ [1]

(c) Newton's experimental law:
$$v_2 - v_1 = \tfrac{1}{2}u \qquad\qquad [2]$$
[1] and [2] give $v_1 = -\frac{1}{6}u$ and $v_2 = \frac{1}{3}u$.

Sphere B now goes on to hit sphere C.

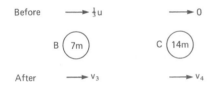

Conservation of momentum:
$$(7m \times \tfrac{1}{3}u) + (14m \times 0) = (7m \times v_3) + (14m \times v_4)$$
i.e. $\qquad\qquad u = 3v_3 + 6v_4$ $\qquad\qquad$ [3]

Newton's experimental law:
$$v_4 - v_3 = \tfrac{1}{2}(\tfrac{1}{3}u)$$
i.e. $6v_4 - 6v_3 = u$ $\qquad\qquad$ [4]

[3] and [4] give: $v_3 = 0$ and $v_4 = \frac{1}{6}u$.
So A and C are now each moving with speed $\frac{1}{6}u$ in opposite directions and B is at rest.
Total KE after two impacts is
$$\tfrac{1}{2}.2m(\tfrac{1}{6}u)^2 + \tfrac{1}{2}.(14m).(\tfrac{1}{6}u)^2 = \tfrac{2}{9}mu^2.$$

GE *A ball is thrown vertically upwards from the floor with velocity V. It rebounds from the ceiling, which is a height h above the floor, and then rebounds from the floor. After this second rebound, it just reaches the ceiling again.*
If the coefficient of restitution between the ball and ceiling is e, and between the ball and floor is f, prove that
$$V^2 = \frac{2gh(1 - f^2 + e^2 f^2)}{e^2 f^2}.$$

First treat the ball as a particle moving freely under gravity and calculate its velocity just before it hits the ceiling. Consider impact with ceiling knowing the initial velocity (just found) and find its velocity just as it comes off the ceiling. Now find the velocity of the ball just before it hits the floor. Use impact again to find the new upward velocity just as it begins to rise. The final velocity, since the ball just reaches the ceiling for a second time, is zero. Using the results you have obtained, derive the required equation.

EX

1 A railway truck A of mass 4000 kg travelling at 2 m/s collides with another truck B of mass 6000 kg travelling at 1 m/s in the same direction. The speed of truck A after the collision is 1·25 m/s in the same direction. Calculate the speed of truck B after the collision. A and B are now brought to rest by frictional forces which are in each case 50 N per 1000 kg mass. Calculate:
 (i) for how long A and B are each in motion after the collision;
 (ii) the final distance between them.

$\qquad\qquad\qquad\qquad\qquad$ *(C)

2 Two identical smooth spheres, S and T, moving in opposite directions with speeds u, 3u respectively, collide directly. The sphere T is reduced to rest. Find the coefficient of restitution between the spheres.

$\qquad\qquad\qquad\qquad\qquad$ (L)

3 A particle moving along a smooth horizontal floor hits a smooth vertical wall and rebounds in a direction at right angles to its initial direction of motion. The coefficient of restitution is e. Find, in terms of e, the tangent of the angle between the initial direction of motion and the wall. Prove that the kinetic energy after the rebound is e times the initial kinetic energy.

$\qquad\qquad\qquad\qquad\qquad$ (J)

4 (i) A sphere of mass m moving along a smooth horizontal table with speed V collides directly with a stationary sphere of the same radius and of mass 2m. Obtain expressions, in terms of V and the coefficient of restitution e, for the speeds of the two spheres after impact. Half of the kinetic energy is lost in the impact. Find the value of e.
 (ii) A particle of mass m moving in a straight line with speed u receives an impulse of magnitude I in the direction of its motion. Show that the increase in kinetic energy is given by $I(I + 2mu)/(2m)$.

$\qquad\qquad\qquad\qquad\qquad$ (L)

5 Three beads A, B and C, of masses 3m, 2m and m respectively, are threaded in that order on a smooth horizontal straight wire. Initially, A, B and C are separated and at rest, and A is then projected towards B with speed 10V. The coefficient of restitution between A and B is $\frac{1}{2}$. Show that the speed of B after the collision is 9V. Find the velocity of A after the collision and find also the kinetic energy lost in the collision. After A and B have collided, C is projected towards B with speed 6V. Given that the collision between B and C is perfectly elastic (e = 1), find the velocities of B and C after this collision.

$\qquad\qquad\qquad\qquad\qquad$ (C)

6 Two particles, A of mass 2m and B of mass m, moving on a smooth horizontal table in opposite directions with speeds 5u and 3u respectively, collide directly. Find their velocities after the collision in terms of u and the coefficient of restitution e. Show that the magnitude of the impulse exerted by B on A is $\frac{16}{3}mu(1 + e)$. Find the value of e for which the speed of B after the collision is 3u. Moving at this speed B subsequently collides with a stationary particle C of mass km, and thereafter remains attached to C. Find the velocity of the combined particle and find the range of values of k for which a third collision will occur.

$\qquad\qquad\qquad\qquad\qquad$ (J)

51 Projectiles
Definition, Analysis of motion, Standard results.

Definition

A **projectile** is a particle which is given an initial velocity and then moves freely under gravity. It is assumed that gravity is the only force acting on the particle, i.e. air resistance is negligible.
If its **initial velocity** is **vertical**, then the particle will move in a **straight line** under gravity.
If its **initial velocity** is **not vertical**, the particle will move in a **curve** (a parabola).

Analysis of motion

Consider a particle projected with initial velocity u at an angle α to the horizontal and has velocity v at time t.

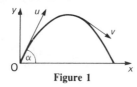

Figure 1

Its flight can be analysed by considering **horizontal** and **vertical motion separately** and using the **equations for uniform acceleration** in a straight line,

i.e. $v = u + at$
$s = ut + \frac{1}{2}at^2$
$v^2 = u^2 + 2as$

	horizontal motion	**vertical** motion
u	$u_x = u \cos \alpha$	$u_y = u \sin \alpha$
a	$\ddot{x} = 0$	$\ddot{y} = -g$
v	$v_x = u \cos \alpha$	$v_y = u \sin \alpha - gt$
s	$x = (u \cos \alpha)t$	$y = (u \sin \alpha)t - \frac{1}{2}gt^2$

$\boxed{i}$ *A particle is projected from ground level with speed $30\ m\,s^{-1}$ at an angle of $30°$ to the horizontal. Calculate: (a) the time of flight, (b) the range.*

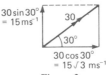

Figure 2

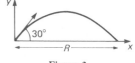

Figure 3

(a) Consider vertical motion:
When the particle reaches the ground $s = 0$.
$u = 15$, $v = ?$, $a = -10$, $s = 0$, $t = ?$

Using $s = ut + \frac{1}{2}at^2$
$0 = 15t - \frac{1}{2}(10)t^2$

i.e. $5t(3 - t) = 0$
$\Rightarrow t = 0$ or 3 s
$t = 0$ is the starting time, $t = 3$ s is time of flight.

(b) The horizontal velocity $15\sqrt{3}$ m s^{-1} is constant. Since the particle travels for 3 s at this velocity, range $R = 15\sqrt{3} \times 3 = 45\sqrt{3}$ m.

Standard results

Time of flight
i.e. the time taken for the projectile to travel along its path from O to A.

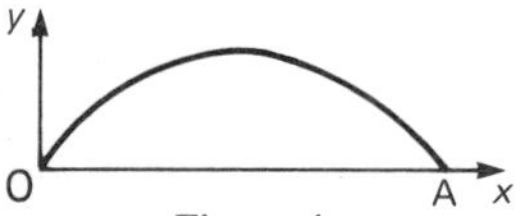

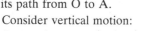

Figure 4

Consider vertical motion:
At any time t, $y = (u \sin \alpha)t = \frac{1}{2}gt^2$
At A, $y = 0 \therefore 0 = (u \sin \alpha)t - \frac{1}{2}gt^2$
$\Rightarrow t = 0$ (initial position)
or $t = \dfrac{2u \sin \alpha}{g}$ (time of flight)

Range
i.e. the horizontal distance OA travelled by the projectile.

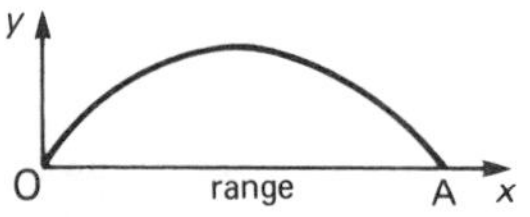

Figure 5

Consider horizontal motion:
At any time t, $x = (u \cos \alpha)t$

When $t = \dfrac{2u \sin \alpha}{g}$,

Then $x = (u \cos \alpha)\dfrac{2u \sin \alpha}{g}$

i.e. $x = \dfrac{u^2 \sin 2\alpha}{g}$ (the range)

This is a maximum when $\sin 2\alpha = 1$, i.e. $\alpha = \dfrac{\pi}{4}$ or $45°$.

So the maximum range is $\dfrac{u^2}{g}$.

Angle of projection for a given range

Let the range be $\dfrac{ku^2}{g}$.

i.e. $\dfrac{u^2 \sin 2\alpha}{g} = \dfrac{ku^2}{g}$

Figure 6

$\Rightarrow \sin 2\alpha = k$
If $k = \sin \theta$, then $2\alpha = \theta$ or $\pi - \theta$,
i.e. $\alpha = \frac{1}{2}\theta$ or $\frac{1}{2}(\pi - \theta)$
So there are two possible angles of projection.

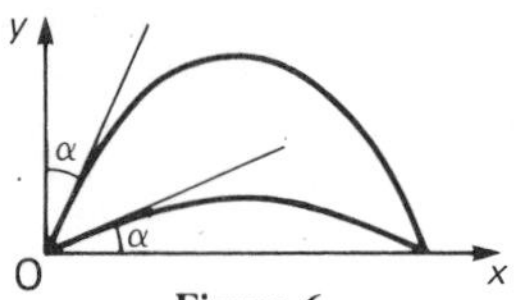

Greatest height
The projectile reaches its greatest height when the vertical velocity is zero.

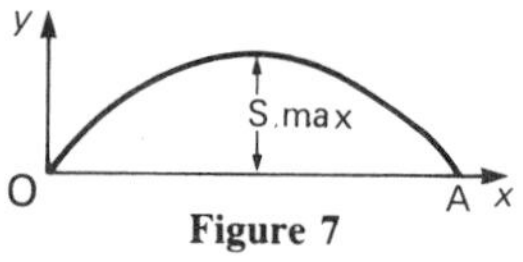

Figure 7

Consider vertical motion:
using $v^2 = u^2 + 2as$
$0 = u^2 \sin^2 \alpha - 2gs$
$\Rightarrow s_{max} = \dfrac{u^2 \sin^2 \alpha}{2g}$ (greatest height)

Direction and velocity at any time
At any time the projectile is always moving along a tangent to its path.
Horizontal velocity: $u \cos \alpha$

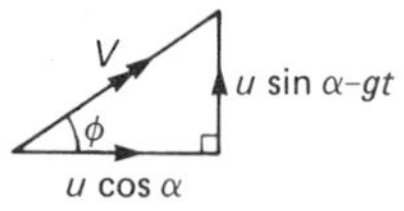

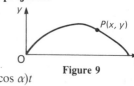

Figure 8

Vertical velocity: $u \sin \alpha - gt$
From the diagram: $V^2 = (u \cos \alpha)^2 + (u \sin \alpha - gt)^2$
$\tan \phi = \dfrac{(u \sin \alpha - gt)}{(u \cos \alpha)}$

Equation of the path of the projectile
Take x and y axes through the point of projection O. If t is the time for the projectile to travel from O to $P(x, y)$, then:

Figure 9

Horizontal distance: $x = (u \cos \alpha)t$
Vertical distance: $y = (u \sin \alpha)t - \frac{1}{2}gt^2$
Eliminating t between these equations gives:

$$y = x \tan \alpha - \dfrac{gx^2}{2u^2} \sec^2 \alpha$$

the equation of the path or trajectory of the projectile.

Projectiles
Worked example and Exam questions

 A stone is thrown from a point O which is at the top of a cliff 50 m above a horizontal beach. The speed with which the stone is thrown is 40 ms^{-1} and it hits the beach at a point R which is at a horizontal distance of 200 m from O. If the stone was thrown at an angle of elevation α, show that one of the possible values of tan α is $\frac{3}{5}$ and find the other possible value.

Given that tan $\alpha = \frac{3}{5}$, calculate

(i) *the time the stone was in the air,*
(ii) *the angle at which the stone hits the beach at R.*

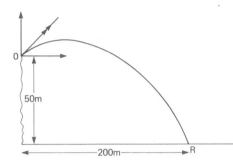

Consider the horizontal motion of the stone.
The horizontal velocity 40 cos α ms^{-1} is constant,
so, 40 cos $\alpha \times T = 200$ (T is the time of flight)
$\Rightarrow$ $T = 5$ sec α.   [1]
Consider the vertical motion (take up as positive).
$s = -50$, $u = 40$ sin α, $a = -10$, $t = T$
Using $s = ut + \frac{1}{2}at^2$ gives
    $-50 = 40$ (sin $\alpha)T + \frac{1}{2}(-10)T^2$,
and substituting for T from [1] gives
    $-50 = 200$ tan $\alpha - 125$ sec^2 α.
So, $-2 = 8$ tan $\alpha - 5(1 + \tan^2 \alpha)$
    i.e. $5 \tan^2 \alpha - 8$ tan $\alpha + 3 = 0$
    i.e. $(5$ tan $\alpha - 3)($tan $\alpha - 1) = 0$
$\Rightarrow$ tan $\alpha = \frac{3}{5}$ or 1.

(i) If tan $\alpha = \frac{3}{5}$, sin $\alpha = \frac{3}{\sqrt{34}}$ and cos $\alpha = \frac{5}{\sqrt{34}}$.

  Using [1],
        $T = 5$ sec α
    i.e. $T = 5 \cdot \frac{\sqrt{34}}{5}$,

So $T = \sqrt{34}$ s — the time of flight.
(ii) To find the angle the stone hits the beach at R, we need to find the vertical component of velocity at R.

  Consider the vertical motion (take up as positive).

$u = 40$ sin $\alpha = \frac{60\sqrt{34}}{17}$, $v = ?$, $a = -10$, $t = T = \sqrt{34}$

  Using $v = u + at$ gives,
        $v = \frac{60\sqrt{34}}{17} + (-10)\sqrt{34}$
          $= \frac{-110}{17}\sqrt{34}$ ms^{-1}.
Let θ be the angle to the horizontal at which the stone hits the beach then the vector triangle is

$\tan \theta = \dfrac{\frac{110}{17}\sqrt{34}}{\frac{100}{17}\sqrt{34}} = \frac{110}{100} = \frac{11}{10}$

    i.e. the stone hits the beach at an angle $\tan^{-1}\left(\frac{11}{10}\right)$ to the horizontal.

---

1 A body is projected upwards from a point on a horizontal plane, with a velocity of 40 m/s, at an angle of 60° to the horizontal. The point of projection is at a horizontal distance of 40 m from the foot of a vertical wall which is 10 m high, and the motion takes place in a plane perpendicular to the wall. Calculate:
   (i) the vertical height by which the body clears the wall;
   (ii) the greatest height above the horizontal plane reached by the body;
   (iii) the time of flight of the body;
   (iv) the horizontal distance beyond the wall at which the body strikes the plane.

(A)

2 A ball was thrown from a balcony above a horizontal lawn. The velocity of projection was 10 m/s at an angle of elevation α, where tan $\alpha = \frac{3}{4}$. The ball moved freely under gravity and took 3 s to reach the lawn from the instant when it was thrown. Calculate:
(a) the vertical height above the lawn from which the ball was thrown;
(b) the horizontal distance between the point of projection and the point A at which the ball hit the lawn;
(c) the angle, to the nearest degree, between the direction of the velocity of the ball and the horizontal at the instant when the ball reached A.

(L)

3 A batsman strikes a ball at a height of $1 \cdot 5$ m above the ground, giving it an initial speed of 29 ms^{-1} at an angle of 30° to the horizontal. What is the minimum distance of the boundary from the batsman if he scores a 'six' (i.e. the ball passes over the boundary line without first bouncing)? A fielder, who is capable of catching a ball at a height of $2 \cdot 75$ m or below, goes to the boundary line. What speed must the batsman give to the ball if he is to hit it at the same height and elevation as before and ensure that he will score a six and not be caught by the fielder?

(S)

4 The muzzle speed of a gun is V and it is desired to hit a small target at a horizontal distance a away and at a height b above the gun. Show that this is impossible if $V^2(V^2 - 2gb) < g^2a^2$, but that, if $V^2(V^2 - 2gb) > g^2a^2$, there are two possible elevations for the gun. Show that, if $V^2 = 2ga$ and $b = \frac{3}{4}a$, there is only one possible elevation, and find the time taken to hit the target.

(OLE)

5 A particle is projected from a point O with speed 5 ms^{-1} at an angle of elevation θ, where $\theta \neq \pi/2$, and moves freely under gravity. Taking the acceleration due to gravity to be 10 ms^{-2}, show that the equation of the path of the projectile referred to horizontal and upward vertical axes

Ox, Oy is $y = x \tan \theta - \dfrac{x^2}{5}(1 + \tan^2 \theta)$.

By considering this equation as a quadratic equation in tan θ, show that there are two distinct values of θ for which the projectile passes through a given point (X, Y), where $X > 0$, provided that $20Y < 25 - 4X^2$.
Given that the two values of θ are α and β and that (X, Y) is a point whose co-ordinates satisfy this inequality, write down expressions for tan $\alpha + \tan \beta$ and tan α tan β in terms of X and Y, and deduce an expression for tan $(\alpha + \beta)$. If $Y = X$, show that $\alpha + \beta = 3\pi/4$.

(J)

Definitions

Consider a particle P of mass m moving in a **horizontal circle,** centre O, radius r, with **constant speed** v.

Figure 1

The **linear velocity** v of P is directed along the tangent to the circle at P.

The constant **angular velocity** ω of P is $\omega = \dfrac{v}{r}$.

ω is measured in **radians per second** (rad. s^{-1}). There is no acceleration along the tangent since the particle moves with constant speed around the circle. The **acceleration** a of P is in the direction $\overrightarrow{PO}$, i.e.

towards the centre of the circle, and $a = \dfrac{v^2}{r}$ or $r\omega^2$.

By Newton's 2nd law, this acceleration must be produced by a **force** which is also directed **towards the centre of the circle.**
So the **equation of motion** for the particle is:

$$\text{force} = \frac{mv^2}{r} \text{ or } mr\omega^2.$$

This force may be the tension in a string, a fractional force, a gravitational force, etc.

ℹ️ *Find the velocity and acceleration of a particle moving in a horizontal circle, radius 20 cm, at a constant angular velocity of 30 revolutions per minute.*

Angular velocity $= 30$ revolutions per minute

$$= 30 \times \frac{2\pi}{60} \text{ rad. s}^{-1} = \pi \text{ rad. s}^{-1}$$

Velocity $v = r\omega = 0.2 \times \pi \text{ m s}^{-1} \approx 0.628 \text{ m s}^{-1}$

Acceleration $a = r\omega^2 = 0.2 \times \pi^2 \text{ m s}^{-2} \approx 1.97 \text{ m s}^{-2}$

ℹ️ *A particle of mass 0.5 kg is attached by a light inextensible string, length 2 m, to a fixed point O on the top of a smooth horizontal table. The particle is made to rotate in a horizontal circle with the string taut at a constant speed of $8\,ms^{-1}$. Calculate the tension in the string.*

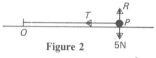

Figure 2 5N

Equation of motion along PO: $T = 0.5 \times \dfrac{8^2}{2} = 16$ N

Problem solving

When solving problems in which a particle P describes a horizontal circle, centre O, with constant speed or constant angular velocity:
1. Draw a clear force diagram.
2. Resolve vertically, (since the particle does not move up or down, forces must balance in this direction).
3. Write down the equation of motion along the radius $\overrightarrow{PO}$.

ℹ️ *A conical pendulum consists of a particle of mass 2 kg attached to one end B of a light inextensible string AB of length 1 m. A is a fixed point. The particle describes a horizontal circle whose centre O is 0.5 m vertically below A. Calculate the tension in the string and the angular velocity of the particle.*

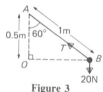

Figure 3

Since $OA = 0.5$ m and $AB = 1$ m,

$O\hat{A}B = 60°$ and $OB = \dfrac{\sqrt{3}}{2}$ m.

Let ω be angular velocity.

Resolve (↑): $T \cos 60° = 20 \Rightarrow T = 40$ N (1)
Equation of motion along BO:

$$T \sin 60° = 2 \times \frac{\sqrt{3}}{2} \times \omega^2 \qquad (2)$$

From (1) and (2): $\omega^2 = 20 \Rightarrow \omega = 2\sqrt{5}$ rad. s^{-1}.

Common situations

The following illustrate some other common situations which arise in problems.
1. Particle P moving inside a **hollow cone,** with friction
(a) P about to move **up** cone (ω—a maximum)

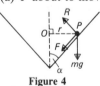

Figure 4

Resolve (↑):
$$R \sin \alpha = mg + F \cos \alpha$$
Equation of motion along $\overrightarrow{PO}$:

$$R \cos \alpha + F \sin \alpha = \frac{mv^2}{r}$$

(b) P about to move **down** cone (ω—a minimum)

Figure 6

Resolve (↑):
$$R \sin \alpha + F \cos \alpha = mg$$
Equation of motion along $\overrightarrow{PO}$:

$$R \cos \alpha - F \sin \alpha = \frac{mv^2}{r}$$

2. Car rounding a bend on a **banked track**
(a) car cornering at **maximum speed**

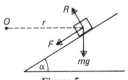

Figure 5

resolve (↑):
$$R \cos \alpha = mg + F \sin \alpha$$
Equation of motion along $\overrightarrow{PO}$:

$$R \cos \alpha + F \cos \alpha = \frac{mv^2}{r}$$

(b) car cornering at **minimum speed**

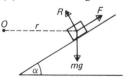

Figure 7

Resolve (↑):
$$R \cos \alpha + F \sin \alpha = mg$$
Equation of motion along $\overrightarrow{PO}$:

$$R \sin \alpha - F \cos \alpha = \frac{mv^2}{r}$$

Motion in a Horizontal Circle
Worked example, Guided example and Exam questions

WE *Two rigid light rods AB, BC each of length $\frac{1}{2}$ m are smoothly jointed at B and the rod AB is smoothly jointed at A to a fixed smooth vertical rod. The joint at B has a particle of mass 2 kg attached. A small ring, of mass 1 kg is smoothly jointed to BC at C and can slide on the vertical rod below A. The ring rests on a smooth horizontal ledge fixed to the vertical rod at a distance*

$\dfrac{\sqrt{3}}{2}$ *m below A. The system rotates about the vertical rod with constant angular velocity 6 radians per second.*

Calculate
(a) the forces in the rods AB and BC,
(b) the force exerted by the ledge on the ring.

Notice that the angles at A and C are 30°.
Force diagram for the system.

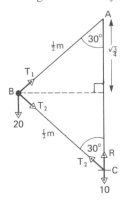

Let T_1 and T_2 be the forces in the rods AB and BC respectively.
Let R be the force exerted on the ring by the ledge.
Resolving vertically for C:
$T_2 \cos 30° + R = 10$ [1]
Resolving vertically for B:
$T_1 \cos 30° = T_2 \cos 30° + 20$ [2]
Equation of motion for B:
$T_1 \cos 60° + T_2 \cos 60°$
$\qquad\qquad = 2 \times \frac{1}{4} \times 6^2$ [3]

Solving [2] and [3] gives

$T_1 = 18 + \dfrac{20\sqrt{3}}{3}$ N and $T_2 = 18 - \dfrac{20\sqrt{3}}{3}$ N.

Substituting for T_2 into [1] gives
$R = (20 - 9\sqrt{3})$ N.

GE *A body of mass M moves in a circle of radius R under a force kM which is directed towards the centre, k being constant. Show that the velocity of the body is of constant magnitude and that T, the time for a revolution, is given by*

$$T = 2\pi \sqrt{\left(\frac{R}{k}\right)}.$$

The gravitational attraction of the earth on a satellite of mass M in a circular orbit of radius R about the centre of the earth is gMr^2/R^2, where r is the radius of the earth. If its time of revolution is T, show that

$$R^3 = g\left(\frac{rT}{2\pi}\right)^2.$$

A satellite's orbit keeps it always vertically above a fixed point on the equator. Given that the radius of the earth is $6\cdot378 \times 10^6$ m and that the earth takes $8\cdot616 \times 10^4$ s to turn on its axis, show that the height of the satellite above the surface of the earth is approximately $3\cdot58 \times 10^7$ m.
(Take the acceleration due to gravity to be $9\cdot8$ m/s².)

Write down the equation of motion using $\dfrac{v^2}{R}$ for the

acceleration. Hence show v is constant. Use $v = R\omega$ to find ω.

$T = \dfrac{2\pi}{\omega}$ gives required result for T. For the second part, use

$k = \dfrac{gr^2}{R^2}$ in the expression obtained for T. Hence get the

required expression for R^3. The final part of the answer is obtained by using the given values for r and T to find R. Height of the satellite above the earth's surface is $R - r$.

EX 1 In a conical pendulum the inelastic string is of length $1\frac{1}{2}$m and a particle of mass M kg is attached to the end of the string. When the string makes an angle of 60° with the vertical the tension in the string is 10 N. Calculate the value of M. The string will break if the tension in it exceeds 48 N. Calculate the greatest number of revolutions per minute the particle can attain.   *(A)

2 (a) A small smooth ring P of mass m is threaded on to a light inextensible string of length $2a$, whose ends are tied to fixed points A and B where A is distant a vertically above B. Show that it is not possible for the particle to move in a horizontal circle at constant speed with AP and BP equally inclined to the vertical, but that it can move in a horizontal circle with BP horizontal and find its speed in this case.

(b) A smooth hollow right circular cone with vertex downwards is rotating with angular velocity ω about its axis which is vertical. A particle is at relative rest on the inside of the cone at a vertical height h above the vertex. Prove that $\omega = \sqrt{(g/h)} \cot \alpha$ where α is the semi-vertical angle of the cone.   *(S)*

3 Prove that the acceleration of a point moving in a fixed circle of radius r with constant angular speed ω is directed towards the centre of the circle and has magnitude $\omega^2 r$.

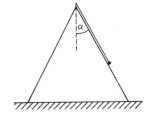

One end of a light inextensible string of length l is attached to the vertex of a smooth cone of semi-vertical angle α. The cone is fixed to the ground and its axis is vertical. The other end of the string is attached to a particle of mass m which can rotate in a horizontal circle in contact with the outer surface of the cone, as shown in the diagram. Given that the angular speed of the particle is ω, find an expression for the tension in the string. Find an expression for the greatest value of ω for which the motion as described can occur.   *(C)*

4 A light inextensible string of length $5a$ has one end fixed at a point A and the other end fixed at a point B which is vertically below A and at a distance $4a$ from it. A particle P of mass m is fastened to the midpoint of the string and moves with speed u, and with the parts AP and BP of the string both taut, in a horizontal circular path whose centre is the midpoint of AB. Find, in terms of m, u, a and g, the tensions in the two parts of the string, and show that the motion described can take place only if $8u^2 \geqslant 9ga$.   *(J)*

5 An artificial satellite of mass m moves under the action of a gravitational force which is directed towards the centre, O, of the earth and is of magnitude F. The orbit of the satellite is a circle of radius a and centre O. Obtain an expression for T, the period of the satellite, in terms of m, a and F. Show that, if the gravitational force acting on a body of mass m at a distance r from O is $m\mu/r^2$, where μ is a constant, then $T^2\mu = 4\pi^2 a^3$.
Assuming that the radius of the earth is 6400 km and that the acceleration due to gravity at the surface of the earth is 10 ms^{-2}, show that $\mu = (6\cdot4)^2 10^{13}$ m³ s^{-2}.
Hence, or otherwise, find the period of revolution, in hours to 2 decimal places, of the satellite when it travels in a circular orbit 600 km above the surface of the earth.   *(L)*

53 Motion in a Vertical Circle
Definitions, Types of motion, Problem solving, Simple pendulum.

Definitions

When a particle P, of mass m, is moving in a **vertical circle**, centre O, radius r, its **speed** v is **variable**.

The particle P has an **acceleration** a in the direction $\overrightarrow{PO}$, i.e. **towards the centre of the circle**, given by $a = \dfrac{v^2}{r}$ or $r\omega^2$. So this **acceleration** is **variable** too.

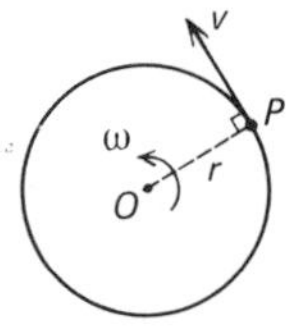

Figure 1

(Note: The tangential acceleration need not be considered at this level.) By Newton's 2nd law, the acceleration towards O must be produced by a **force** which is also directed **towards O**.

So the **equation of motion** for the particle along the radius $\overrightarrow{PO}$ is: force $= \dfrac{mv^2}{r}$ or $mr\omega^2$.

This variable force will be the resultant of the weight force mg and at least one other force.

Types of motion

The **two main types** of motion in a vertical circle for a particle with initial speed u are described below.

1. The particle **cannot leave the circular path**, e.g. a bead threaded on a vertical wire.
The particle can do one of these three things.

(i) **Complete the circle** if $v > 0$ at top

(ii) Come to **rest** at top if $v = 0$ at top

(iii) **Oscillate** if $v = 0$ for $0 < \theta < \pi$

2. The particle **can leave the circular path** and become a projectile, e.g. a particle attached to a string.
The particle can do one of these three things.
(T is the tension in the string or a normal reaction)

(i) **Complete the circle** if $T \geq 0$ for all values of θ

(ii) Become a **projectile** if $T = 0$ for $\pi/2 < \theta \leq \pi$

(iii) **Oscillate** if $v = 0$ for $\theta \leq \pi/2$

 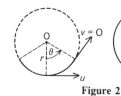   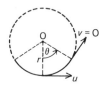

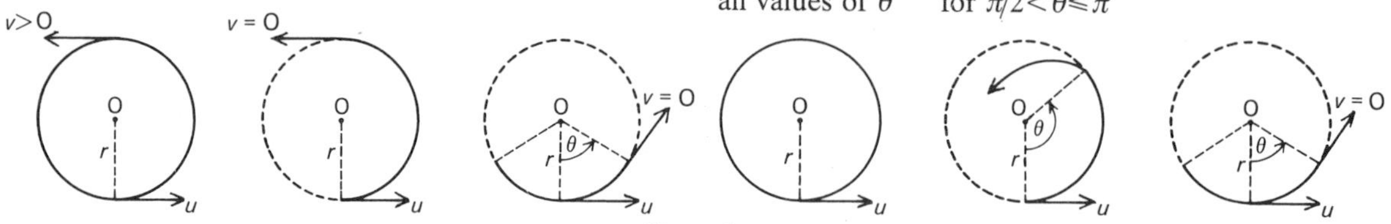

Figure 2

and $u^2 > 4gr$. \qquad and $u^2 = 4gr$. \qquad and $u^2 < 4gr$. \qquad and $u^2 \geq 5gr$. \qquad and $2gr < u^2 < 5gr$. and $u^2 \leq gr$.

Problem solving

When solving problems in which a particle P describes a vertical circle, centre O:

1. Draw a clear force diagram.
2. Use conservation of mechanical energy, i.e. Initial (P.E. + K.E.) = (P.E. + K.E.) at any point.
3. Write down the equation of motion for the particle along the radius $\overrightarrow{PO}$.

Note: The given conditions of the problem will indicate whether the particle completes a circle, oscillates, becomes a projectile, etc. (see 'Types of motion' above).

ℹ *A particle of mass m is suspended from a fixed point O by a light inextensible string of length l. When the particle is hanging freely in equilibrium it is given a horizontal speed of $\sqrt{3gl}$. Find the tension T in the string when the angle between the string and the downward vertical is 60°.*

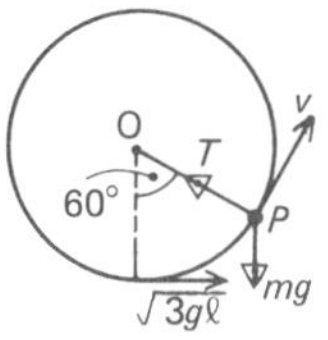

Figure 3

By conservation of M.E.:
$$\tfrac{1}{2}m(\sqrt{3gl})^2 = \tfrac{1}{2}mv^2 + mgl(1 - \cos 60°)$$
Equation of motion along:
$$\overrightarrow{PO}$$
$$T - mg\cos 60° = \frac{mv^2}{l}$$

Eliminating v gives $T = \tfrac{5}{2}mg$.

Simple pendulum

A **simple pendulum** is a special case of a particle making **small oscillations** in a vertical circle.
When the pendulum has swung through an angle θ from the downward vertical, $v = l\dot{\theta}$.

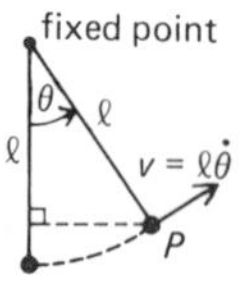

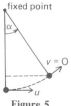

Figure 4

By conservation of M.E. at P:
$$mg(l - l\cos\theta) + \tfrac{1}{2}m(l\dot{\theta})^2 = \text{constant}$$
Differentiating with respect to time gives:
$$(mgl\sin\theta)\dot{\theta} + ml^2\dot{\theta}\ddot{\theta} = 0$$
i.e. $\ddot{\theta} = -\dfrac{g}{l}\sin\theta$

Since θ is small, $\sin\theta \approx \theta$, so approximately
$$\ddot{\theta} = -\frac{g}{l}\theta$$

Compare this with the basic S.H.M. equation $(\ddot{x} = -\omega^2 x)$.

So the motion is **angular S.H.M.** with period $2\pi\sqrt{\dfrac{l}{g}}$.

ℹ *A simple pendulum consists of a particle of mass m making small oscillations on the end of a light inextensible string of length l attached to a fixed point. When hanging in equilibrium, the particle is given a small horizontal speed u. Calculate α, the maximum angular displacement.*

Figure 5

By conservation of M.E.:
$$\tfrac{1}{2}mu^2 = mg(l - l\cos\alpha)$$
$$\Rightarrow \cos\alpha = 1 + \frac{u^2}{2gl}$$
For small u, α will be small
and $\cos\alpha \approx 1 - \tfrac{1}{2}\alpha^2$
So $1 + \tfrac{1}{2}\alpha^2 \approx 1 + \dfrac{u^2}{2gl}$
$$\Rightarrow \alpha \approx \frac{u}{\sqrt{gl}}$$

Motion in a Vertical Circle
Worked example, Guided example and Exam questions

 WE *A particle of mass m is suspended from a fixed point O by a light inextensible string of length l. The particle is hanging freely in equilibrium when it is given a horizontal speed of $\sqrt{(3gl)}$. Find the height of the particle above its equilibrium position when the string becomes slack.*

First draw a diagram showing the forces on the particle P when OP makes an angle θ with the downward vertical.
Let T be the tension in the string and v be the speed of the particle at P.

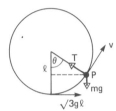

Conservation of energy gives:
$$\tfrac{1}{2}m(\sqrt{3gl})^2 = \tfrac{1}{2}mv^2 + mg(l - l\cos\theta)$$
i.e. $3gl = v^2 + 2gl(1 - \cos\theta)$ [1]
Equation of motion along PO:
$$T - mg\cos\theta = m\frac{v^2}{l} \quad [2]$$

When the string goes slack, $T = 0$. Use this in [2] to get:
$$v^2 = -gl\cos\theta.$$
Use this value of v^2 in [1]:
$$3gl = -gl\cos\theta + 2gl - 2gl\cos\theta$$
$$\Rightarrow \quad \cos\theta = -\frac{1}{3}.$$

The height of the particle when the string becomes slack is
$$l - l\cos\theta = l + \frac{1}{3}l = \frac{4}{3}l.$$

 GE *A smooth sphere with centre O and radius a is fixed with a point B of its surface in contact with a vertical wall. A particle P of mass m rests at the highest point A of the sphere. It is slightly disturbed so that it moves from rest towards the wall in the plane OAB. If at any instant in the subsequent motion the line OP makes an angle θ with the line OA and the particle is still in contact with the sphere, find expressions for the velocity of P at this instant and for the reaction of the sphere upon it in terms of m, g, a and θ.*

Prove that the particle leaves the sphere when $\cos\theta = \dfrac{2}{3}$ and that its speed is then $\sqrt{(2ga/3)}$.
Show that P hits the wall at a height $\tfrac{1}{8}a(5\sqrt{5} - 9)$ above B.

Draw a clear force diagram. Use conservation of energy (take OB as the initial level for P.E. and consider the initial velocity of P to be zero).
Write down the equation of motion along PO. Particle leaves sphere when the reaction is zero. When particle leaves the sphere, consider it to be a projectile moving freely under gravity (*see* unit 51).

 EX

1 A heavy particle connected to a fixed point O by a light inelastic string of length a is moving in a vertical circle about O. Its speed when at the lowest point of the circle is $(\tfrac{7}{2}ga)^{\frac{1}{2}}$. Find the inclination of the string to the vertical when it becomes slack, and show that the speed of the particle is then $(\tfrac{1}{2}ga)^{\frac{1}{2}}$. Find also the maximum height above O reached by the particle.

(OLE)

2 A particle P is projected horizontally with speed u from the lowest point A of the smooth inside surface of a fixed hollow sphere of internal radius a.
(i) In the case when $u^2 = ga$ show that P does not leave the surface of the sphere. Show also that, when P has

moved halfway along its path from A towards the point at which it first comes to rest, its speed is
$$\sqrt{\{ga(\sqrt{3} - 1)\}}.$$
(ii) Find u^2 in terms of *ga 1in the case when P* leaves the surface at a height $\dfrac{3a}{2}$ above A, and find, in terms of a and g, the speed of P as it leaves the surface.

(J)

3 A particle of mass m slides down the smooth outside surface of a fixed sphere of radius a. At the top of the sphere its velocity is horizontal and of magnitude u. If θ is the angle that the radius to the particle makes with the upward vertical, show that the reaction between the particle and the sphere is $mg(3\cos\theta - 2) - mu^2/a$. Show that the particle, when just displaced from rest at the top, leaves the surface when $\theta = \cos^{-1}\tfrac{2}{3}$, and find its speed at that instant.

(W)

4 A small bead is free to move on a smooth circular wire of radius a which is fixed in a vertical plane. The bead is projected with speed $(4ag)^{\frac{1}{2}}$ from the lowest point of the wire. Prove that the radius from the centre of the wire to the bead makes an angle θ with the vertical at time t after the instant of projection, where
$t = (a/g)^{\frac{1}{2}} \ln(\sec\tfrac{1}{2}\theta + \tan\tfrac{1}{2}\theta)$, and the reaction between the wire and the bead vanishes when $\theta = \pi - \cos^{-1}(\tfrac{2}{3})$.

(OLE)

5 One end of a light inextensible string of length a is attached to a particle of mass m. The other end is attached to a fixed point O which is at a height $\tfrac{5}{2}a$ above the horizontal ground. Initially the string is taut and horizontal. The particle is then projected vertically downward with velocity $(2ag)^{\frac{1}{2}}$. When the string has turned through an angle θ ($<\pi$) find the velocity v of the particle and show that the tension in the string is $mg(2 + 3\sin\theta)$.
(a) If the string can withstand a tension of at least $5mg$, prove that the string will not break.
(b) If the string can withstand a tension of at most $\tfrac{7}{2}mg$, find the values of θ and v when the string breaks. In this case find the time to reach the vertical through O. Hence show that the particle strikes the ground at the point vertically below O.

(O & C)

6 Show that small oscillations of a simple pendulum of length l are simple harmonic with period $2\pi\sqrt{l/g}$. A pendulum clock beats seconds (i.e. one half-period = 1 second) at a point where $g = 9 \cdot 812$ ms^{-2}. Find the length of the pendulum correct to 3 significant figures. If the clock is moved to a place where $g = 9 \cdot 921$ ms^{-2}, will the clock gain or lose? Find how much it would gain or lose during one day. To what length should the pendulum be altered if it is to register correctly?

(S)

54 Variable Forces

Introduction, Force as a function of time, Force as a function of velocity, Force as a function of displacement.

Introduction	If the **force** acting on a body of constant mass is **variable**, then the **acceleration** of the body will also be **variable**. This is a consequence of Newton's second law of motion which states that: force = mass × acceleration. Under these conditions the **acceleration** of the body must be expressed as a **function of time or velocity or displacement** and calculus used to solve the problem.

Force as a function of time

The **force** or **acceleration** may be given as a **function of time** t.

To find v in terms of t:

use $a = \dfrac{dv}{dt}$.

To find s in terms of t:
(a) find v in terms of t,

(b) use $v = \dfrac{ds}{dt}$.

> ℹ️ A particle of mass 8 kg is acted upon by a force $4(1 - e^{-\frac{t}{6}})N$. If the body is initially at rest, find the velocity of the particle after 3 s.

Use $a = \dfrac{dv}{dt}$ for acceleration.

The equation of motion ($F = ma$) gives:

$$4(1 - e^{-\frac{t}{6}}) = 8\frac{dv}{dt}$$

i.e. $\dfrac{dv}{dt} = \frac{1}{2}(1 - e^{-\frac{t}{6}})$

Integrating gives: $v = \frac{1}{2}(t + 6e^{-\frac{t}{6}}) + c$
When $t = 0$, $v = 0$, so $0 = \frac{1}{2}(6) + c \Rightarrow c = -3$
Hence, $v = \frac{1}{2}(t + 6e^{-\frac{t}{6}}) - 3$
After $3s$, $v = \frac{1}{2}(3 + 6e^{-\frac{1}{2}}) - 3 \simeq 3.45 \text{ m s}^{-1}$

Force as a function of velocity

The **force** or **acceleration** may be given as a **function of the velocity** v.

To find v in terms of t:

use $a = \dfrac{dv}{dt}$.

To find v in terms of s:

use $a = v\dfrac{dv}{ds}$.

> ℹ️ A particle of unit mass moves from rest along a straight line under the action of a force $(2 - 0.1\,v)N$, where v is the velocity in $m\,s^{-1}$. Find the displacement when the velocity is $10\,m\,s^{-1}$.

Use $a = v\dfrac{dv}{ds}$ for acceleration

The equation of motion ($F = ma$) gives:

$$(2 - 0.1v) = 1 \cdot v\frac{dv}{ds}$$

So $\displaystyle\int ds = \int \frac{v\,dv}{(2 - 0.1v)}$

i.e. $\displaystyle\int ds = \int \left(-10 + \frac{20}{(2 - 0.1v)}\right) dv$

$$s = -10v - 200 \ln|2 - 0.1v| + c$$

When $s = 0$, $v = 0$, so $c = 200 \ln 2$.

Hence, $s = 200 \ln \left|\dfrac{2}{2 - 0.1v}\right| - 10v$

When $v = 10$, $s = 200 \ln 2 - 100 \approx 38.6 \text{ m}$.

Force as a function of displacement

The **force** or **acceleration** may be given as a **function of the displacement** s.

To find v in terms of s:

use $a = v\dfrac{dv}{ds}$.

To find s in terms of t:
(a) find v in terms of s,

(b) use $v = \dfrac{ds}{dt}$.

> ℹ️ A particle of unit mass, moving in a straight line, is acted upon by a force equal to $(-4x)N$, where $x\,m$, is the displacement of the particle from a fixed point O in the line. If initially the particle is at rest when $x = 3\,m$, find the velocity when $x = 1\,m$.

Use $a = v\dfrac{dv}{dx}$ for acceleration.

The equation of motion ($F = ma$) gives:

$$1 \cdot v\frac{dv}{dx} = -4x$$

So $\displaystyle\int v\,dv = -4\int x\,dx$
$$\frac{1}{2}v^2 = -2x^2 + c$$

When $x = 3$, $v = 0$, so $c = 18$.
Hence, $v^2 = 4(9 - x^2)$
When $x = 1$, $v^2 = 4.8 \Rightarrow v = \pm 2\sqrt{2} \text{ m s}^{-1}$.

A particle of unit mass falls from rest under gravity through the air. The resistance of the air is kv^2 where k is a constant and v is the speed of the particle after it has fallen for time t. Calculate the time taken for the particle to acquire a speed V.

Equation of motion for the particle is

$$\frac{dv}{dt} = g - kv^2.$$

Rearranging this differential equation we have,

$$\int \frac{dv}{g - kv^2} = \int dt$$

i.e. $\dfrac{1}{g} \displaystyle\int \dfrac{dv}{1 - \dfrac{k}{g}v^2} = \int dt$

Let $\dfrac{k}{g} = w^2$, then $1 - \dfrac{k}{g}v^2 = 1 - w^2v^2 = (1 + wv)(1 - wv).$

Using partial fractions we have

$$\frac{1}{2g} \int \left[\frac{1}{1 + wv} + \frac{1}{1 - wv} \right] dv = \int dt$$

i.e. $\dfrac{1}{2gw} \ln \left| \dfrac{1 + wv}{1 - wv} \right| = t + c$, c is a constant of integration.

When $t = 0$, $v = 0 \Rightarrow c = 0$.

So, $t = \dfrac{1}{2gw} \ln \left| \dfrac{1 + wv}{1 - wv} \right|$.

But $w = \sqrt{\dfrac{k}{g}}$, so the time taken for the particle to acquire a speed of V is, say, T where,

$$T = \frac{1}{2\sqrt{kg}} \ln \left| \frac{\sqrt{g} + V\sqrt{k}}{\sqrt{g} - V\sqrt{k}} \right|.$$

 A body of mass m falls from rest under gravity through a resisting medium. The resistance is kv^2 per unit mass where k is constant and v is the velocity when the body has fallen a distance x.

(i) *Given that u is the limiting value of the velocity, i.e. the velocity for which there would be zero acceleration, establish the differential equation*

$$v\frac{dv}{dx} = k(u^2 - v^2).$$

(ii) *By solving this differential equation obtain an expression for v in terms of x.*

(iii) *Show that the work done by the resistance when the body has fallen a distance x is*

$$\frac{mu^2}{2}(2kx + e^{-2kx} - 1).$$

(i) Set up the equation of motion for the particle using $v\dfrac{dv}{dx}$ as the expression for acceleration. As acceleration approaches zero i.e. $v\dfrac{dv}{dx} \rightarrow 0$, velocity approaches u

i.e. $g - ku^2 \rightarrow 0$.

Hence get the required differential equation.

(ii) Solve the equation using the method of variables separable.

(iii) The resistance is a variable force so use

$$\text{Work done} = \int_0^x F dx \text{ where } F = kv^2$$

i.e. Work done $= \displaystyle\int_0^x m\, kv^2\, dx.$

Use the expression found for v in terms of x (from (ii)) in the integral to obtain the work done.

 1 A particle of mass m, subject to a resistance mk times the square of its speed, is projected vertically downwards with speed w, where $kw^2 < g$. Find the speed of the particle when it has descended a distance x.

(J)

2 Show that the acceleration of an object moving along a straight line may be written as $v\dfrac{dv}{ds}$. A vehicle of mass 2500 kg moving on a straight course is subject to a single resisting force in the line of motion of magnitude kv newtons, where v metres per second is the velocity and k is constant. At 100 km/h this force is 2000 N.
The vehicle is slowed down from 100 km/h to 50 km/h. Find:
(i) the distance travelled;
(ii) the time taken.

(A)

3 A particle, of mass m, moves in a horizontal straight line under the action of a resisting force of magnitude mkv^2, where v is the velocity and k is a positive constant. When $t = 0$, $v = U$ and $x = a$, where x is the displacement from the origin at time t. Find expressions for:
(i) v in terms of x;
(ii) v in terms of t;
(iii) x in terms of t.

(C)

4 At time t a particle is moving vertically downwards with speed v in a medium which exerts a resistance to the motion proportional to the square of the speed of the particle. If its terminal velocity is V, prove that the equation of motion of the particle is $\dfrac{dv}{dt} = g\left(1 - \dfrac{v^2}{V^2}\right)$, and write down the equation of motion when the particle is moving vertically upwards. A particle is projected vertically upwards in the medium with speed equal to the terminal velocity V. Prove that when it returns to the point of projection its speed is $V/\sqrt{2}$.

(OLE)

5 The motion of a particle is such that its speed v at time t is given by $\dfrac{dv}{dt} = \dfrac{1}{2}(v - v^2)$ and $v = 0 \cdot 2$ when $t = 0$. By solving the differential equation:
(i) find the value of t when $v = 0 \cdot 5$, giving your answer correct to two decimal places;
(ii) express v in terms of t.
By considering the differential equation in the form $v\dfrac{dv}{dx} = \dfrac{1}{2}(v - v^2)$, or otherwise, where x is the distance travelled when the speed is v, and $x = 0$ when $v = 0 \cdot 2$, show that the value of x when $v = 0 \cdot 8$ is double its value when $v = 0 \cdot 6$.

(J)

6 A particle of mass m is projected vertically upwards under gravity with speed u in a medium in which the resistance is mk times the speed. If the particle reaches its greatest height H in a time T, show that $u = gT + kH$. If the particle returns to its original position with speed w after a further time T', show that $w = gT' - kH$. Find the particle's speed as a function of time during the upward motion and show that $kT = \log_e(1 + ku/g)$.

(W)

109

Definitions

Simple Harmonic Motion (S.H.M.) is a special type of oscillation.

In linear S.H.M. a particle oscillates in a straight line with a **linear acceleration** which is **proportional to the linear displacement from a fixed point** and always **directed towards that fixed point.**

Consider a particle P, oscillating with S.H.M. along the line AA′, as shown in the diagram.

O is a fixed point, the **centre** or **mean position.**

a is the **amplitude** of the S.H.M., i.e. the maximum distance of the particle from the centre O.

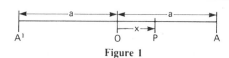

Figure 1

Note:
At O, where $x=0$,
   (a) the speed is a maximum ($=na$),
   (b) the acceleration is zero.
At A and A', where $x=\pm a$,
   (a) the speed is zero,
   (b) the acceleration is maximum ($=\omega^2 a$).

Equations

The **basic equation of S.H.M.** is:

$$\frac{d^2x}{dt^2}=-\omega^2x \text{ where } \omega \text{ is a constant.}$$

From it can be derived:

$$v^2=\left(\frac{dx}{dt}\right)^2=\omega^2(a^2-x^2)$$

$$x = a \sin \omega t, \text{ if } x=0 \text{ when } t=0$$

$$x = a \cos \omega t, \text{ if } x=a \text{ when } t=0$$

The **period**, T, of the S.H.M. is the time for a complete oscillation, i.e. to travel a total distance $4a$.

The **frequency**, f, of the S.H.M. is the number of oscillations made per unit time.

$$T=\frac{2\pi}{\omega}=\frac{1}{f}$$

Note: T and f are independent of the amplitude a.

ℹ️ *A particle is executing S.H.M. with period $\pi/4$ s and amplitude 0.5 m. Calculate:*

(a) the speed when the displacement is 0.25 m,

(b) the magnitude of the acceleration when the displacement is 0.1 m.

(a) $T=\dfrac{2\pi}{\omega}\Rightarrow\omega=\dfrac{2\pi}{T}=\dfrac{2\pi}{\pi/4}=8$

Using $v^2=\omega^2(a^2-x^2)$
$$=8^2[(0.5)^2-(0.25)^2]=12$$
$$\therefore v = \sqrt{12}\approx 3.46 \text{ m s}^{-1}$$

(b) Using acceleration $=-\omega^2x$
$$=-8^2(0.1)=-6.4 \text{ m s}^{-2}$$
$\therefore$ the magnitude of the acceleration is 6.4 m s^{-2}.

Forces producing S.H.M.

A **force directed towards a fixed point** and **proportional to the displacements from that point** produces S.H.M.

A simple example of a force producing S.H.M. is the tension in a stretched elastic string or spring.

To show that the motion produced is S.H.M.:

1. Draw a clear force diagram showing the particle in equilibrium.

2. Use Hooke's law to find the static extension.

3. Draw a diagram showing the particle at a point between the equilibrium position and an extreme.

4. Write down the equation of motion measuring displacement from the equilibrium position.

5. Compare this equation with the basic equation of S.H.M., i.e. $\dfrac{d^2x}{dt^2}=-\omega^2x$.

Once a motion has been shown to be S.H.M. then the equations of S.H.M. can be used to find ω, v, a, etc.

If only part of the motion of a particle is S.H.M., then each part of the motion must be dealt with separately. The motion of a particle on an elastic string, for example, is S.H.M. only while the string is in tension. When the string becomes slack the only force acting on the particle is its weight. So the particle moves in a vertical line under gravity, its critical speed being obtained from $v^2=\omega^2(a^2-x^2)$.

ℹ️ *A particle of mass m hanging on the end of an elastic string of natural length l and modulus λ is pulled down a distance a ($<mgl/\lambda$) below its equilibrium position and then released. Prove that the subsequent oscillations are simple harmonic, find the period of oscillations and state the amplitude.*

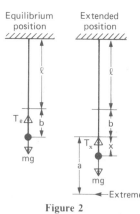

Figure 2

In equilibrium position:
b is the static extension.

By Hooke's law: $T_e=\dfrac{\lambda b}{l}$

Resolve ($\uparrow$): $T_e=mg$

So $mg=\dfrac{\lambda b}{l}$ (1)

In extended position:
By Hooke's law:
$$T_x=\frac{\lambda(b+x)}{l} \quad (2)$$

Equation of motion:
$$mg-T_x=m\frac{d^2x}{dt^2} \quad (3)$$

Equations (1), (2) and (3) give $\dfrac{d^2x}{dt^2}=\dfrac{-\lambda}{ml}x.$

This is of the form $\dfrac{d^2x}{dt^2}=-\omega^2x$ with $\omega=\sqrt{\dfrac{\lambda}{ml}}$,

i.e. S.H.M. about the static equilibrium position.

Period of oscillations $T=\dfrac{2\pi}{\omega}=2\pi\sqrt{\dfrac{ml}{\lambda}}.$

Amplitude is a.

Simple Harmonic Motion
Worked example, Guided example and Exam questions

 A particle is executing simple harmonic motion with amplitude 2 metres and period 12 seconds. Calculate the maximum speed of the particle.
Initially, the particle is moving at maximum speed. Show that the distance moved by the particle until its speed is half the maximum value is $\sqrt{3}$ metres and find the time taken by the particle to travel this distance.
(Standard formulae relating to simple harmonic motion may be quoted without proof.)

Using $T = \dfrac{2\pi}{\omega}$, we have

$$12 = \frac{2\pi}{\omega} \Rightarrow \omega = \frac{\pi}{6}.$$

Now, $v^2 = \omega^2(a^2 - x^2)$

so, $v^2 = \dfrac{\pi^2}{36}(4 - x^2)$, since $a = 2$ m.

v is a maximum when $x = 0$,

i.e. $v_{\max} = \dfrac{\pi}{3}$ m s^{-1} is the maximum speed.

When $v = \dfrac{1}{2}v_{\max}$ i.e. $v = \dfrac{\pi}{6}$ m s^{-1},

we have, $\left(\dfrac{\pi}{6}\right)^2 = \dfrac{\pi^2}{36}(4 - x^2)$,

i.e. $x^2 = 3$

$\Rightarrow$ $x = \sqrt{3}$ m — the distance travelled.

To find the time taken to travel this distance,
use, $x = a\sin \omega t$,

i.e. $\sqrt{3} = 2\sin\left(\dfrac{\pi}{6}t\right)$

$\Rightarrow$ $\dfrac{\pi}{6}t = \dfrac{\pi}{3}$

$\Rightarrow$ $t = 2$ seconds.

 One end of an elastic string of modulus mg and natural length a is attached to a fixed point O. To the other end A are attached two particles P and Q, P having mass 2m and Q having mass m. The particles hang down in equilibrium under gravity. If Q falls off, show that P subsequently performs simple harmonic motion and state the period and amplitude of this motion. If on the other hand P falls off, find the distance from O of the highest point reached by Q.

First use Hooke's Law for the elastic string attached to a particle of mass $3m$ to find the total length OA. Now use Hooke's Law for a particle of mass $2m$ to find a new length OA. The difference between the two lengths will be the amplitude of the oscillations.
Consider the $2m$ particle to be moving and displaced x metres for its equilibrium position. Write down the equation of motion and show this is an equation of simple harmonic motion. Hence ω and T. For the last part of the question, use conservation of energy (*see* unit 45).

---

 1 A particle P of mass 8 kg describes simple harmonic motion with O as centre and has a speed of 6 m/s at a distance of 1 m from O and a speed of 2 m/s at a distance of 3 m from O.
  (i) Find:
    (a) the amplitude of the motion;
    (b) the period of the motion;
    (c) the maximum speed of P;
    (d) the time taken to travel from O directly to one extreme point B of the motion.
  (ii) Determine the magnitude of:

(a) the acceleration of P when at a distance of 2 m from O;
(b) the force acting on P when at a distance of 2 m from O;
(iii) Write down an expression for the displacement of P from O at any time t, given that P is at O at $t = 0$. Hence, or otherwise, find the time taken to travel directly from O to a point C between O and B and at a distance of 1 m from O. Find also the time taken to go directly from C to the point D between O and B and at a distance of 2 m from O. [Answers may be left in a form involving inverse trigonometric functions.]

(*A*)

2 A particle moves on the line Ox so that after time t its displacment from O is x, and $\dfrac{d^2x}{dt^2} = -9x$.

When $t = 0$, $x = 4$ and $dx/dt = 9$. Find:
  (i) the position and velocity of the particle when $t = \pi/6$;
  (ii) the maximum displacement of the particle from O.

(*J*)

3 A particle of mass m is attached to one end of a light elastic string of length a and modulus $\frac{3}{2}mg$. The other end of the string is attached to a fixed point O and the particle hangs in equilibrium under gravity at the point E. Find the distance OE. If the particle is a *further* distance x below E show that the resultant force acting on the particle is proportional to x. The particle is pulled down to the point at a distance a below E and released from rest. Show that, in the subsequent motion and while the string is taut, the particle executes Simple Harmonic Motion and that its distance below E at time t after being released is

$$a\cos\left\{\left(\frac{3g}{2a}\right)^{\frac{1}{2}}t\right\}$$

(*O & C*)

4 (Take g as 10 m s^{-2} in this question.) A light elastic spring has natural length a and modulus of elasticity λ. Prove that the energy stored in the spring when it is stretched is $\lambda x^2/(2a)$, where x is the extension. A light elastic spring of natural length 0.2 m and modulus of elasticity 50 N hangs vertically with one end attached to a fixed point and with a particle of mass 2 kg attached to the lower end.
  (i) Calculate the extension of the spring when the particle is in equilibrium.
  (ii) The particle is pulled down below its equilibrium position until the *total* extension of the spring is 0.2 m and it is then released from rest in this position. Calculate the speed of the particle when it passes the equilibrium position, and find the maximum compression of the spring in the resulting motion.

(*C*)

5 A mass rests on a horizontal platform which is moving horizontally to and fro with simple harmonic motion of amplitude 0.5 m, making twenty complete oscillations per minute. If the mass remains at rest relative to the platform throughout the motion, show that the coefficient of friction must not be less than 0.224 approximately. If $\mu = \frac{1}{2}$, the mass is 10 kg, and the platform is stopped abruptly when 0.3 m from its mean position, write down the equation describing the subsequent motion of the mass, and derive the relation that holds between the kinetic energy of the mass and the work done against friction during that motion. How far will the mass slide before coming to rest?

(*W*)

Displacement, Velocity, Acceleration, Force, Momentum, Impulse, Kinetic energy, Work, Power.

Displacement, velocity, acceleration

The **position vector** r of a point $P(x, y, z)$ referred to the origin O can be written as:

$$\mathbf{r}=x\mathbf{i}+y\mathbf{j}+z\mathbf{k} \text{ or } \mathbf{r}=\begin{pmatrix} x \\ y \\ z \end{pmatrix} \text{ in three dimensions;}$$

$$\mathbf{r}=x\mathbf{i}+y\mathbf{j} \text{ or } \mathbf{r}=\begin{pmatrix} x \\ y \end{pmatrix} \text{ in two dimensions.}$$

The **velocity** v of P is $\dfrac{d\mathbf{r}}{dt}$, sometimes written as $\dot{r}$.

The acceleration a of P is $\dfrac{d\mathbf{v}}{dt}$ or $\dfrac{d^2\mathbf{r}}{dt^2}$,

sometimes written as $\ddot{r}$.
S.I. unit for $\mathbf{r}$ is the **metre**(m); for $\mathbf{v}$ the **metre per second** ($\mathrm{m\,s^{-1}}$); for $\mathbf{a}$ the **metre per second squared** ($\mathrm{m\,s^{-2}}$).

[i] *A particle moves so that its position vector r in metres at time t in seconds is $r=5i+t^3j+t^2k$. Find the velocity and acceleration of the particle at time t and its speed after 3 seconds.*

At time t: $\mathbf{v}=\dfrac{dr}{dt}=3t^2\mathbf{j}+2t\mathbf{k}$

$$\mathbf{a}=\dfrac{d\mathbf{v}}{dt}=6t\mathbf{j}+2\mathbf{k}$$

Speed is the magnitude of the velocity, i.e. $|\mathbf{v}|$.
$$|\mathbf{v}|=\sqrt{[(3t^2)^2+(2t)^2]}$$
$$=t\sqrt{(9t^2+4)}$$
When $t=3$, $\quad |\mathbf{v}|=3\sqrt{85}\approx 27.7\ \mathrm{m\,s^{-1}}$

Force

Newton's second law of motion:
i.e. **force = mass × acceleration**
can be given in vector form as
$$\mathbf{F}=m\mathbf{a}$$
or $\quad \mathbf{F}=m\ddot{\mathbf{r}}$
S.I. unit for force is the **newton**(N)

[i] *Find the force acting on a mass of 2 kg with position vector $r=t^3i+2t^2j+3k$ at time t.*
$$\mathbf{r}=t^3\mathbf{i}+2t^2\mathbf{j}+3\mathbf{k}$$
$$\dot{\mathbf{r}}=3t^2\mathbf{i}+4t\mathbf{j}$$
$$\ddot{\mathbf{r}}=6t\mathbf{i}+4\mathbf{j}$$

So $\mathbf{F}=m\ddot{\mathbf{r}}=2(6t\mathbf{i}+4\mathbf{j})=12t\mathbf{i}+8\mathbf{j}$

Momentum

The momentum of a particle mass m, velocity $\mathbf{v}$ is given by:

$$\text{momentum}=m\mathbf{v}$$
S.I. unit for momentum is the **newton second** (Ns).

[i] *Find the momentum of a particle mass 5 kg and velocity $v=2i-j$.*

momentum $=5(2\mathbf{i}-\mathbf{j})=10\mathbf{i}-5\mathbf{j}$ Ns.

Impulse

The **impulse I** produced by a force $\mathbf{F}$ acting for time T is defined as:

$$\mathbf{I}=\int_0^T \mathbf{F}dt.$$

S.I. unit for impulse is the **newton second** (Ns).
Since $\mathbf{F}=m\mathbf{a}$,

$$\mathbf{I}=\int_0^T m\mathbf{a}\,dt=m\int_0^T \dfrac{d\mathbf{v}}{dt}\,dt.$$

$$=m(\mathbf{v}-\mathbf{u}) \text{ i.e. change in momentum.}$$

[i] *Find the impulse produced by a force given by $F=\sin t\,i+\cos t\,j$ acting for the time interval $0\leqslant t\leqslant \pi/2$.*

Impulse $=\mathbf{I}=\int \mathbf{F}\,dt=\int_0^{\frac{\pi}{2}}(\sin t\mathbf{i}+\cos t\mathbf{j})\,dt$

$$=[-\cos t\mathbf{i}+\sin t\mathbf{j}]_0^{\frac{\pi}{2}}$$
$$=(\mathbf{j})-(-\mathbf{i})=\mathbf{i}+\mathbf{j}$$

Kinetic energy

The **kinetic energy, K.E.**, of a particle mass m, velocity $\mathbf{v}$ is

$$\tfrac{1}{2}m\mathbf{v}\cdot\mathbf{v}=\tfrac{1}{2}mv^2$$
S.I. unit for kinetic energy is the **joule** (J).

[i] *Find the K.E. of a particle of mass 4 kg and velocity $v=3i+2j$.*

K.E. $=\tfrac{1}{2}\cdot 4(3\mathbf{i}+2\mathbf{j})\cdot(3\mathbf{i}+2\mathbf{j})=26$ J

Work

The **work** done by a force $\mathbf{F}$ is defined as $\int \mathbf{F}\cdot\mathbf{v}\,dt$.
S.I. unit for work is the **joule** (J).

Since $\mathbf{F}=m\dfrac{d\mathbf{v}}{dt}$,

the work done is $m\displaystyle\int \mathbf{v}\cdot\dfrac{d\mathbf{v}}{dt}\cdot dt.$

$$=\tfrac{1}{2}mv^2-\tfrac{1}{2}mu^2 \text{ i.e. increase in K.E.}$$

[i] *Find the work done by a force $F=2ti+4j$ on a particle with velocity $v=5i-tj$ in the time interval $0\leqslant t\leqslant 2$.*

Work done $=\displaystyle\int_0^2 \mathbf{F}\cdot\mathbf{v}\,dt=\int_0^2 (2t\mathbf{i}+4\mathbf{j})\cdot(5\mathbf{i}-t\mathbf{j})\,dt$

$$=\int_0^2 6t\,dt=[3t^2]_0^2=12 \text{ J}$$

Power

The **power** exerted by a force $\mathbf{F}$ is the rate at which $\mathbf{F}$ does work.

So power is $\dfrac{d}{dT}\left(\displaystyle\int_0^T \mathbf{F}\cdot\mathbf{v}\,dt\right)=\mathbf{F}\cdot\mathbf{v}$

S.I. unit for power is the **watt** (W).

[i] *Find the power exerted by a force $F=5t^2i+2tj$ on a particle with velocity $v=ti-2t^2j$ at time t.*

Power $=\mathbf{F}\cdot\mathbf{v}=(5t^2\mathbf{i}+2t\mathbf{j})\cdot(t\mathbf{i}-2t^2\mathbf{j})$
$$=5t^3-4t^3=t^3$$

Vectors in Dynamics
Worked example and Exam questions

In this question the units of mass, length and time are the kilogram, metre and second respectively.
*A particle of unit mass moves so that its position vector **r** at time t is*
$$\mathbf{r} = \cos t\,\mathbf{i} + \sin t\,\mathbf{j} + \tfrac{1}{2}t^2\,\mathbf{k}.$$
Find (a) *the momentum at time t,*
    (b) *the kinetic energy at time t,*
    (c) *the work done on the particle in the time interval t = 0 to t = 4,*
    (d) *the force acting on the particle at time t,*
    (e) *the power exerted by this force at time t.*

Since $\mathbf{r} = \cos t\,\mathbf{i} + \sin t\,\mathbf{j} + \tfrac{1}{2}t^2\,\mathbf{k}$,
$$\dot{\mathbf{r}} = -\sin t\,\mathbf{i} + \cos t\,\mathbf{j} + t\mathbf{k},$$
and $\ddot{\mathbf{r}} = -\cos t\,\mathbf{i} - \sin t\,\mathbf{j} + \mathbf{k}.$

(a) The momentum at time t is $m\dot{\mathbf{r}}$ so,
momentum $= -\sin t\,\mathbf{i} + \cos t\,\mathbf{j} + t\mathbf{k}$ Ns.

(b) The kinetic energy at time t is $\tfrac{1}{2}mv^2 = \tfrac{1}{2}m\dot{\mathbf{r}}.\dot{\mathbf{r}}$ so,
kinetic energy
$$= \tfrac{1}{2}(-\sin t\,\mathbf{i} + \cos t\,\mathbf{j} + t\mathbf{k}).(-\sin t\,\mathbf{i} + \cos t\,\mathbf{j} + t\mathbf{k})$$
$$= \tfrac{1}{2}(\sin^2 t + \cos^2 t + t^2)$$
$$= \tfrac{1}{2}(1 + t^2) \text{ joules.}$$

(c) The work done on the particle in the time interval $t = 0$ to $t = 4$ is $\int_0^4 \mathbf{F}.\mathbf{v}\,dt$, so

Work done
$$= \int_0^4 (-\cos t\,\mathbf{j} - \sin t\,\mathbf{j} + \mathbf{k}).(-\sin t\,\mathbf{i} + \cos t\,\mathbf{j} + t\mathbf{k})dt$$
$$= \int_0^4 (\sin t \cos t - \sin t \cos t + t)dt$$
$$= \left[\frac{t^2}{2}\right]_0^4$$
$$= 8 \text{ joules.}$$

(d) The force acting on the particle at time t is $\mathbf{F} = m\ddot{\mathbf{r}}$, so
Force $= -\cos t\,\mathbf{i} - \sin t\,\mathbf{j} + \mathbf{k}$ N

(e) The power exerted by the force at time t is $\mathbf{F}.\mathbf{v}$, so
Power $= (-\cos t\,\mathbf{j} - \sin t\,\mathbf{j} + \mathbf{k}).(-\sin t\,\mathbf{i} + \cos t\,\mathbf{j} + t\mathbf{k})$
$$= (\sin t \cos t - \sin t \cos t + t)$$
$$= t \text{ watts.}$$

---

1 A particle initially at rest at the point $(2, 2)$ has acceleration $\begin{pmatrix} t \\ 3 \end{pmatrix}$ in m s^{-2} after t seconds. Find vector expressions for its velocity and position after t seconds.
Find its change in position in the first $^1\!/_{100}$ second. After how many seconds is it moving in a direction inclined at 45° to the x-axis?   *(O & C)*

2 A force $\mathbf{F}_1 = (4\mathbf{i} + 2\mathbf{j})$ N. State the magnitude of $\mathbf{F}_1$. A second force $\mathbf{F}_2$ has magnitude $8\sqrt{5}$ N and acts in a direction given by the vector $\mathbf{i} + 2\mathbf{j}$. State the force vector $\mathbf{F}_2$. Hence calculate the resultant force $\mathbf{F}_R$ of these two forces and the unit vector in the direction of this resultant. What is the acceleration that this resultant would produce on a mass of 3 kg? The mass was initially at rest at the point with position vector $2\mathbf{i} - 4\mathbf{j}$. If the two forces continue to act on the mass, show that the point with position vector $18\mathbf{i} + 20\mathbf{j}$ lies on the path traced out by the mass.   *(S)*

3 In this question distances are measured in metres and time in seconds. At time $t = 0$ two particles P and Q are set in motion in the x-y plane. Initially P is at $A(1, 0)$ and Q is

at $B(0, 8)$. The particle P moves with a constant speed of 5 m/s parallel to the line $3y = 4x$ and Q moves with a constant speed of 4 m/s parallel to the line $y = -\lambda x$, the sense of motion of both P and Q being that in which x is increasing. Given that $\mathbf{i}$ and $\mathbf{j}$ are the unit vectors in the directions of x increasing and y increasing, respectively, show that the unit vectors in the directions of motion of P and Q are $\dfrac{3}{5}\mathbf{i} + \dfrac{4}{5}\mathbf{j}$ and $\dfrac{1}{\sqrt{1+\lambda^2}}\mathbf{i} - \dfrac{\lambda}{\sqrt{1+\lambda^2}}\mathbf{j}$ respectively.

Determine, in the form $a\mathbf{i} + b\mathbf{j}$:
(i) the velocities of P and Q;
(ii) the vectors $\overrightarrow{AP}$, $\overrightarrow{BQ}$ and $\overrightarrow{PQ}$ at time t.
Show that, if P and Q meet, λ must satisfy the equation
$$7(1 + \lambda^2)^{\frac{1}{2}} = 8 - \lambda.$$
Verify that $\lambda = -\tfrac{3}{4}$ is a solution of this equation and for this value of λ find the time when P and Q meet.
  (A)

4 In this question $\mathbf{i}$ and $\mathbf{j}$ are vectors of magnitude 1 km in directions E and N respectively. Units of time and speed are hours and kilometres per hour. A and B move in a horizontal plane, A with constant velocity $4\mathbf{i} + 4\mathbf{j}$ and B with constant acceleration $2\mathbf{i} + 2\mathbf{j}$. At time $t = 0$, A is at the point with position vector $\mathbf{i} + 4\mathbf{j}$ and B is at $4\mathbf{i} + \mathbf{j}$ moving with velocity $2\mathbf{j}$.
(i) Find the position vectors of A and B at time t and hence show that $\mathbf{AB} = (t^2 - 4t + 3)\mathbf{i} + (t^2 - 2t - 3)\mathbf{j}$.
(ii) Find the time when B will be due S of A and the distance AB at that moment.
(iii) Show that A and B subsequently collide and give the time at which this happens.
(iv) Find the magnitude and direction of the velocity of B just before the collision occurs.
  (S)

5 At time t the position vector $\mathbf{r}$ of the point P with respect to the origin O is given by $\mathbf{r} = (a \sin pt)\mathbf{i} + a\mathbf{j}$, where a and p are constants. Show that the vector $\dfrac{d^2\mathbf{r}}{dt^2} + p^2\mathbf{r}$ is constant during the motion.   *(L)*

6 A particle of mass m moves in a horizontal plane Oxy with speed v along the x-axis in the positive direction. It is subjected to a horizontal impulse $\mathbf{I}$ which turns its direction of motion through 30° in an anticlockwise sense and reduces its speed to $v/\sqrt{3}$. Find the vector $\mathbf{I}$. At the same instant an impulse $-\mathbf{I}$ is applied to a particle of mass $3m$ which is at rest. Find the magnitude and direction of the resultant velocity of this particle.
  (J)

7 At time t a particle is in motion with velocity $\mathbf{v}$ and is being acted upon by a variable force $\mathbf{F}$. Write down expressions for (i) the power at time t, (ii) the work done by $\mathbf{F}$ during the time interval $0 \leqslant t \leqslant T$.
The particle, of mass m, moves in a plane where $\mathbf{i}$ and $\mathbf{j}$ are perpendicular unit vectors so that its position vector at time t is given by $\mathbf{r} = 2a \cos 2t\,\mathbf{i} + a \sin 2t\,\mathbf{j}$, where a is a positive constant. Derive expressions for the velocity $\mathbf{v}$ and the force $\mathbf{F}$ at time t. Obtain an expression in terms of t for the power at time t and show that the work done by $\mathbf{F}$ during the interval $0 \leqslant t \leqslant T$ is $3ma^2(1 - \cos 4T)$. If T varies, find the maximum value of the work done by $\mathbf{F}$ and determine also the smallest value of T for which this maximum value is reached.
  (J)

57 Coplanar Concurrent Forces

Definitions, Resultant of two forces, Resolving a force, Resultant of a system of forces.

Definitions

Coplanar forces are forces whose lines of action all lie in the same plane.
Concurrent forces act at the same point. If forces act on a particle, then they must be concurrent.

Resultant of two forces

If two forces **p** and **q** are **concurrent** then they act at a point, O say. Their **resultant r** may be found by **vector addition** using the **triangle law**, i.e. **p**+**q**=**r**.

Figure 1

The **magnitude** of **r** is given by the cosine rule:

i.e. $r=\sqrt{p^2+q^2+2pq\cos\theta}$
since $\cos(180°-\theta)=-\cos\theta$.

The resultant r also acts at the point O as shown.
The **direction** of r is given by the sine rule:

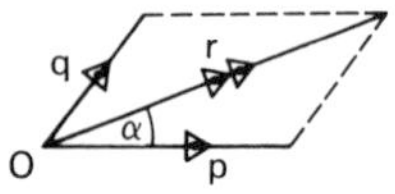

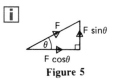

Figure 2

i.e. $\sin\alpha=\dfrac{q}{r}\sin\theta$

If **p** and **q** are at right angles, then $\theta=90°$ and

$r=\sqrt{p^2+q^2}$ and $\tan\alpha=\dfrac{q}{p}$.

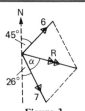

ℹ️ *A particle is acted upon by forces in a horizontal plane: 6 newtons in a direction NE, 7 newtons in a direction S26°E. Find the resultant force.*

Figure 3

The magnitude R of the resultant force is given by:
$$R=\sqrt{[6^2+7^2+(2\times6\times7\cos109°)]}$$
$$=\sqrt{[36+49-27.35]}$$
$$=\sqrt{57.65}$$
$$\Rightarrow R\approx7.59 \text{ newtons}$$

If α is the angle between the direction of the resultant and the 7 newton force, then

$$\sin\alpha\approx\frac{6}{7.59}\sin109°$$
$$\approx0.7474$$
$$\Rightarrow\alpha\approx48.4°$$

Resolving a force

A single force can be split into two **components** or **resolutes** by the converse of the triangle law for vector addition. This process is called **resolving the force**.

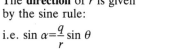

Figure 4

a=**b**+**c**
a is resolved into components **b** and **c**.

Although a force may be resolved in an infinite number of ways, the most useful way is when the two components are perpendicular to each other.

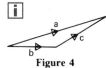

Figure 5

F is resolved into perpendicular components $F\cos\theta$ and $F\sin\theta$.

In problem solving it is often necessary to resolve a force in one or more directions, e.g. horizontally, vertically, parallel to a plane, perpendicular to a plane, etc.

In diagrams it is conventional to mark the magnitudes of a force and its resolutes.

ℹ️ A particle is acted upon by the coplanar concurrent forces **W**, **R** and **F** as shown. The table gives the resolutes of the forces in four directions.

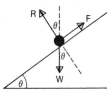

Figure 6

Direction / Force	Vertical	Horizontal	∥ to slope	⊥ to slope
W	W	0	$W\sin\theta$	$W\cos\theta$
R	$R\cos\theta$	$R\sin\theta$	0	R
F	$F\sin\theta$	$F\cos\theta$	F	0

Resultant of a system of forces

To find the **resultant of a system** of coplanar concurrent forces:

1. Resolve each force in a stated direction and find the sum of the resolutes, p say, in that direction.

2. Resolve each force in a direction perpendicular to the first stated direction and find the sum of these resolutes, q say.

3. Find the resultant r of these two concurrent forces p and q by vector addition using the triangle law.

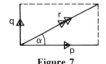

Figure 7

$$r=\sqrt{p^2+q^2} \text{ and } \tan\alpha=\frac{q}{p}$$

ℹ️ *Find the single force which is equivalent to the given system of forces.*

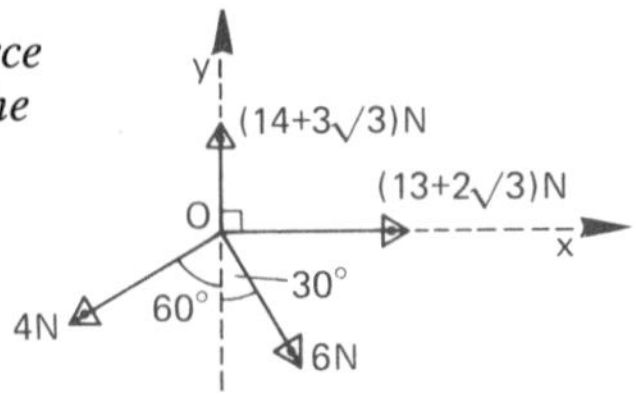

Figure 8

Total resolute in direction Ox
$=(13+2\sqrt3)-4\cos30°+6\cos60°=16$ N
Total resolute in direction Oy
$=(14+3\sqrt3)-4\cos60°-6\cos30°=12$ N

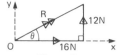

Figure 9

$R=20$ N$(3:4:5\triangle)$
$\theta=$ arc tan 0.75

Coplanar Concurrent Forces
Worked example, Guided example and Exam questions

PQRS is a square. Calculate the resultant of the following forces:
5 N acting along PQ,
3√2 N acting along PR,
3 N acting along PS.
Find also the angle the resultant makes with PQ.

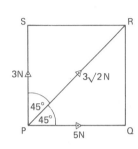

Let X and Y be the components of the resultant force in the directions of PQ and PS.

$$X = (5 + 3\sqrt{2} \cos 45°) \text{ N}$$
$$= \left(5 + 3\sqrt{2} \cdot \frac{1}{\sqrt{2}}\right) \text{ N}$$
$$= 8 \text{ N},$$
and
$$Y = (3 + 3\sqrt{2} \cos 45°) \text{ N}$$
$$= \left(3 + 3\sqrt{2} \cdot \frac{1}{\sqrt{2}}\right) \text{ N}$$
$$= 6 \text{ N}.$$

Hence, the resultant force **R** has magnitude R where,
$$R = \sqrt{(X^2 + Y^2)}$$
$$= \sqrt{(8^2 + 6^2)}$$
$$= 10 \text{ N}.$$

R makes an angle $\tan^{-1}\left(\dfrac{Y}{X}\right)$ with PQ,

i.e. R makes an angle 36° 52' with PQ.

The medians of an equilateral triangle PQR intersect at G. Forces of magnitude 2, 2 and 4 N act along GQ, GR and GP respectively. Calculate the magnitude and direction of the resultant force.

Choose two directions at right angles along which to resolve, say, RQ and GP.
Find the resolutes, say, X and Y of the resultant force in the two chosen directions.
Calculate $R = \sqrt{(X^2 + Y^2)}$, the magnitude of the resultant force.
State the direction of the resultant force.

---

1 A horizontal force **R** is of magnitude 12 N and acts due east from a point O. The horizontal forces **P** and **Q** act from O in the directions 030° and due south respectively. Given that **P** + **Q** = **R**, calculate the magnitudes of **P** and **Q**.

(L)

2 The following horizontal forces pass through a point O: 5 N in a direction 000°, 1 N in a direction 090°, 4 N in a direction 225° and 6 N in a direction 315°. Find the magnitude and direction of their resultant. Two further horizontal forces are introduced to act at O: P N in a direction 135° and Q N in a direction 225°. If the complete set of forces is now in equilibrium calculate the value of P and of Q.

(C)

3 The resultant of a force $2P$ N in a direction 060° and a force 10 N in a direction 180° is a force of $\sqrt{3}P$ N. Calculate the value of P and the direction of the resultant. A third force of 25 N, concurrent with the other two and in the same plane, is added so that the resultant of the system is in the direction 180°. Find the direction in which the third force is applied and find the magnitude of the resultant.

(C)

4 (i) Find the resultant of the system of coplanar forces shown in the figure, giving its magnitude and the angle it makes with the 400 N force.

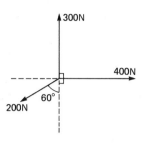

(ii) Two forces, P and Q, which are inclined at an angle of 120°, have a resultant of magnitude $P\sqrt{7}$. Calculate the magnitude of Q in terms of P.

(O & C)

5 Three forces act through the point with position vector $3\mathbf{i} + 2\mathbf{j}$.
$\mathbf{F}_1$ has magnitude 15 newtons and acts in the direction given by $3\mathbf{i} + 4\mathbf{j}$.
$\mathbf{F}_2$ has magnitude $3\sqrt{2}$ newtons and acts in the direction given by $\mathbf{i} - \mathbf{j}$.
$\mathbf{F}_3$ has magnitude $4\sqrt{5}$ newtons and acts in the direction given by $2\mathbf{i} + \mathbf{j}$.
Express the three forces in terms of the unit vectors $\mathbf{i}$ and $\mathbf{j}$. Hence find the resultant $\mathbf{F}_R$ of these three forces in terms of the unit vectors. If $\mathbf{i}$ and $\mathbf{j}$ are the unit vectors in the directions of the x and y axes, the unit of length being the metre, illustrate the resultant $\mathbf{F}_R$ graphically. Using the graph, calculate the magnitude of the moment of $\mathbf{F}_R$ about the origin.

(S)

6 (a) In the regular hexagon $ABCDEF$, $\overrightarrow{AB} = \mathbf{a}$ and $\overrightarrow{BC} = \mathbf{b}$. Express, in terms of $\mathbf{a}$ and $\mathbf{b}$, the vectors:
(i) $\overrightarrow{AC}$; (ii) $\overrightarrow{AD}$; (iii) $\overrightarrow{AE}$; (iv) $\overrightarrow{AF}$.

(b) The origin O, the point A with position vector $4\mathbf{i} + 3\mathbf{j}$ and the point C with position vector $3\mathbf{i} - 4\mathbf{j}$ are three vertices of a square $OABC$. Calculate the position vector of B. Forces of magnitudes 5 N, $10\sqrt{2}$ N and 10 N act along $\overrightarrow{OA}$, $\overrightarrow{OB}$ and $\overrightarrow{CO}$ respectively. Express each of these forces as a vector in terms of $\mathbf{i}$ and $\mathbf{j}$. Hence show that the resultant of these forces acts along $\overrightarrow{OA}$ and calculate the magnitude of this resultant.

(A)

58 Moments and Couples
Moment of a force, Resultant moment, Principle of moments, Parallel forces, Couple

---

Moment of a force

When a force acts on a rigid body it may cause the body to turn about an axis. This **turning effect** is measured by the moment of the force about this axis.

The **moment of a force about an axis** is defined as the product of the magnitude of the force and the perpendicular distance of the line of action of the force from the axis. It is usual to refer simply to the moment **'about a point'** instead of using the more correct description 'about an axis through a point . . .'.

The moment of force F about O is Fd.

the moment of force F about X is zero.

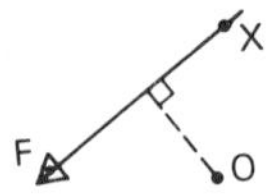

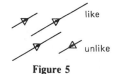

Figure 1

S.I. unit of moment is the newton-metre (Nm). Anticlockwise $\circlearrowleft$ moments are usually taken as positive, clockwise $\circlearrowright$ moments as negative.

ⓘ A heavy rod AB is acted upon by the coplanar forces X, Y, W and P as shown. The table gives the moments of the forces about A, B and G.

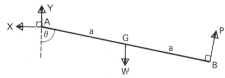

Figure 2

Force \ Point	X	Y	W	P
A	0	0	$-W.a\sin\theta$	$P.2a$
B	$X.2a\cos\theta$	$-Y.2a\sin\theta$	$W.a\sin\theta$	0
G	$X.a\cos\theta$	$-Y.a\sin\theta$	0	$P.a$

Resultant moment

When a set of coplanar forces acts on a body, the **resultant moment** about a point in the plane is the algebraic sum of the moments of the individual forces about that point.

ⓘ *Forces act along the sides of a square $ABCD$ of side 1 m as shown. Find the resultant moment about A.*

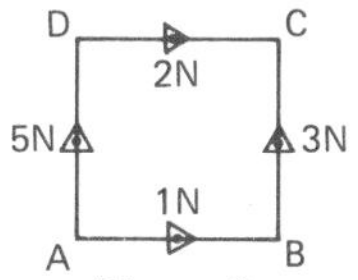

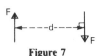

Figure 3

Resultant moment about A is:
$$(1\times0)+(3\times1)$$
$$+(-2\times1)+(-5\times0)$$
$$=1\,\text{Nm}$$

Principle of moments

The **principle of moments** states that:
the resultant moment of a set of coplanar forces about a point is equal to the moment of their resultant about the same point.

ⓘ *Forces of magnitude 4 N and 3 N act along the sides AB and AD respectively of a square $ABCD$ of side 2 m. Find the perpendicular distance d of the line of action of their resultant R from 0, the mid-point of DC.*

Figure 4

$$R=\sqrt{3^2+4^2}=5\,\text{N}$$

Moment of R about O is $-5d$
Resultant moment of forces about O is:
$$(3\times1)-(4\times2)=-5.$$

By the principle of moments:
$$-5=-5d\Rightarrow d=1\,\text{m}$$

Parallel forces

Like parallel forces act in the same direction.

Unlike parallel forces act in opposite directions.

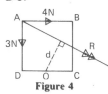

Figure 5

The **resultant** of a set of parallel forces is parallel to the original forces. Its magnitude is the algebraic sum of the magnitudes of the individual forces.
The location of the resultant may be found using the principle of moments.

ⓘ *In the given diagram, find the distance x of the resultant R from the point A.*

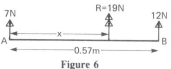

Figure 6

Moment of R about A is $19x$.
Resultant moment of 7 N and 12 N forces about A is:
$$(7\times0)+(12\times0.57)=6.84$$
By the principle of moments:
$$6.84=19x\Rightarrow x=0.36\,\text{m}$$

Couple

A **couple** is formed by two equal unlike parallel forces which are non-collinear.

Figure 7

It has zero resultant but is does have a moment.
The **moment of the couple** shown is Fd. This is constant about any point in the plane of the couple.

ⓘ *The diagram shows a uniform rod AB, pivoted at A and held horizontally by a couple of moment G. Find G.*

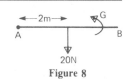

Figure 8

Moments about A give: $G=20\times2=40\,\text{Nm}$.

Moments and Couples
Worked example, Guided example and Exam questions

WE *Calculate the turning effect about O of the force $F_1 = 3i+j$ acting at the point $r_1 = i+j$ and the force $F_2 = 2i-5j$ acting at the point $r_2 = 2i-j$.*

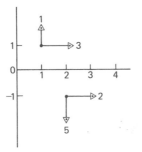

Total clockwise moment
about O
$= (3 \times 1) + (5 \times 2)$
$= 13$ Nm.

Total anticlockwise moment
about O
$= (1 \times 1) + (2 \times 1)$
$= 3$ Nm.

Resultant moment about $O = -13 + 3$
$\qquad = -10$ Nm,
i.e. a clockwise moment of 10 Nm.

GE *A uniform plank AB, 3 m long, of mass 10 kg, is supported in a horizontal position by two vertical strings attached at distances 1 m and 2 m from end A. Calculate the mass which should be placed at the end A of the plank so that*
(i) the tension in the string nearer end B just vanishes,
(ii) the tension in the string nearer A is three times that in the string nearer B.

Draw a diagram showing the forces acting on the plank. These will be the two tensions in the strings, the weight force of the plank and the unknown weight at A.
(i) Let the tension in the string nearer B be zero. Find the unknown weight by taking moments about the point where the other string is attached to the plank.
(ii) Use tensions $3T$ and T say. Resolve vertically and take moments once to find the unknown weight this time.

EX 1 Two forces F_1 and F_2 of magnitudes $3\sqrt{5}$ N and $\sqrt{5}$ N act through the point with position vector $2i+j$ in directions $i+2j$ and $i-2j$ respectively. Calculate F_1 and F_2 and hence state F_R the resultant of these forces. Draw a clear diagram to illustrate F_R in component form and hence calculate the magnitude of the moment of F_R about the point with position vector i. What do you deduce from this result? (Take the unit of length to be 1 metre.)
(S)

2 A uniform rod AB, of length 150 mm and mass 50 g, has a load of X grams attached at a point C of the rod such that $AC = 30$ mm. The rod rests horizontally on a smooth support placed at a point D where $AD = 60$ mm. Calculate the numerical value of X. A second smooth support is now placed at B and the load of X grams is removed. A new load of 60 g is attached to the rod at a point E such that $EB = 35$ mm. If the rod still rests in a horizontal position, calculate the thrust on each support.
(A)

3 A uniform bar AB of length 6 m and mass 15 kg lies horizontally on supports C and D where $AC = 1$ m and $DB = 2$ m. Find the magnitudes of the vertical forces at C and D.
(i) Find the value of the downward vertical force which, when applied at A, would just cause the bar to tilt.
(ii) Find the distance from A at which an upward vertical force of 60 N should be applied to cause the bar just to tilt about D.
(C)

4 A thin non-uniform beam AB, of length 6 m and mass 50 kg, is in equilibrium resting horizontally on two smooth supports which are respectively 2 m and $3 \cdot 5$ m from A. The thrusts on the two supports are equal. Find the position of the centre of gravity of the beam. The original supports are removed and a load of 10 kg is attached to the beam at B. The loaded beam rests horizontally on two new smooth supports at A and C, where C is a point on the beam 1 m from B. Calculate the thrusts on each of the new supports.
(A)

5 A straight uniform rigid rod AB is of length 8 m and mass 10 kg. The rod is supported at the point X, where $AX = 5$ m, and, when downward vertical forces of magnitudes P and $4P$ newtons are applied at A and B respectively, the rod rests in equilibrium with AB horizontal. Calculate:
(a) the value of P;
(b) the force, in N, exerted on the support at X.
(L)

6 A uniform rod AGB of weight w N rests horizontally on two supports C and D. $AG = GB = 6$ cm. The support C can be placed in any position from A to G, and the support D can be placed in any position from G to B. The reactions at the supports C and D are P N and Q N respectively. Denoting AC by x cm and DB by y cm, express P and Q in terms of x, y and w. Given that $P = 2Q$:
(i) show that $2x - y = 6$;
(ii) find the value of x when the support D is placed at B.
(C)

7 A uniform straight plank AB, of mass 12 kg and length 2 m, rests horizontally on two supports, one at C and the other at D, where $AC = CD = 0 \cdot 6$ m. A particle P of mass X kg is hung from B and the plank is on the point of tilting:
(a) Find the value of X.
The particle P is removed from B and hung from A.
(b) Find, in N, the magnitude of the force exerted on the plank at each support.
(L)

8 The figure shows a light horizontal beam AB, of length 9 m, supported at its ends by a force S acting vertically and a force R acting at an angle of α to the line of the beam. A force of 30 N is applied to the beam, at an angle of 30°, 3 m from B. If the beam is in equilibrium, calculate:
(a) S; (b) α; (c) R.
Calculate the magnitude and sense of the necessary moment that would have to be applied at A to reduce the reaction at B to zero.
(S)

9 A non-uniform rod AB of length $6a$ rests in a horizontal position on two pegs distant a from each end. (Draw your diagram with B to the right of A.) The rod will just tilt if a weight W is attached to A or a weight $2W$ is attached to B. Find the weight of the rod and the distance of its centre of gravity from A. When the rod is resting on the pegs (with neither of the weights attached) a clockwise couple of moment Wa is applied to the rod in the vertical plane containing the rod. Find the magnitudes of the reactions of the pegs on the rod.
(S)

59 Equilibrium

Particle in equilibrium, Rigid body in equilibrium, Conditions of equilibrium.

Particle in equilibrium

When a particle is in equilibrium under a system of coplanar concurrent forces the following condition is satisfied:
the total resolute of all the forces in any direction must be zero.

When solving problems about particles in equilibrium:

1. Draw a clear force diagram.

2. Choose a direction for resolving, remembering that the resolute of a force in a direction perpendicular to itself is zero.

3. Resolve the forces acting in this chosen direction and equate the total resolute to zero.

4. If necessary, resolve the forces acting in another suitable direction and equate the total resolute to zero.

ℹ️ *A particle on a slope is subject to the forces shown in the diagram and is in equilibrium. Find the forces R and P.*

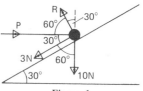

Figure 1

Resolving vertically (to eliminate force P):
$$R \cos 30° - 10 = 3 \cos 60° = 0$$
i.e.
$$R = \frac{23\sqrt{3}}{3} \text{ N}$$

Resolving parallel to slope (to eliminate force R):
$$P \cos 30° - 3 - 10 \cos 60° = 0$$
i.e.
$$P = \frac{16\sqrt{3}}{3} \text{ N}$$

Rigid body in equilibrium

When a rigid body is in equilibrium under a system of coplanar forces, the following two conditions are satisfied:

1. the total resolute of all the forces in any direction must be zero,

2. the total moment of all the forces about any point in the plane must be zero.

When solving problems about rigid bodies in equilibrium:

1. Draw a clear force diagram.

2. Choose two directions at right angles for resolving, remembering that the resolute of a force in a direction perpendicular to itself is zero.

3. Resolve the forces acting in the two chosen directions and equate each total resolute to zero.

4. Take moments about a suitable point in the plane, remembering that the moment of a force about a point on the line of action of the force is zero.

ℹ️ *A uniform ladder 5 m long, weight 200 N, rests on rough horizontal ground and against a smooth vertical wall. It is inclined at an angle of 30° to the vertical. Find the normal reactions at each end of the ladder.*

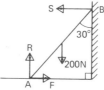

Figure 2

Ladder AB.
R – normal reaction at A.
S – normal reaction at B.
F – frictional force.

Moments about A give:
$$S \times 5 \cos 30° = 200 \times 2.5 \cos 60°$$
$$S = \frac{100\sqrt{3}}{3} \text{ N}$$

Resolving vertically gives:
$$R - 200 = 0 \Rightarrow R = 200 \text{ N}$$

Conditions for equilibrium

If a system of coplanar forces is in equilibrium, then:
1. no resultant force must act (or the system would have an acceleration),
and 2. no resultant turning effect must exist, i.e. it must not reduce to a couple.

Three sets of necessary and sufficient conditions for equilibrium are given below. Any of these may be used to test the equilibrium of a system of forces.

Set I
A system of coplanar forces is in equilibrium if:
1. the total resolutes of the forces in two perpendicular directions are each zero,
and 2. the resultant moment about any point in the plane is zero.

Set II
A system of coplanar forces is in equilibrium if:
1. the resultant moments about any two points in the plane, P and Q say, are each zero,
and 2. the total resolute of the forces in one direction, not perpendicular to PQ, is zero.

Set III
A system of coplanar forces is in equilibrium if: the resultant moments about three non-collinear points are each zero.

ℹ️ *ABCD is a square of side 2 m. Forces of magnitudes 7, 3, 3, 7 and $4\sqrt{2}$ N act along $\overrightarrow{BA}$, $\overrightarrow{BC}$, $\overrightarrow{DC}$, $\overrightarrow{DA}$, and $\overrightarrow{AC}$ respectively. Show that this system of forces is in equilibrium.*

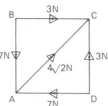

Figure 3

To show that this system of coplanar forces is in equilibrium it is necessary and sufficient to show that the resultant moments of the forces about three non-collinear points are each zero.

Moments about A give:
$$(3 \times 3) - (3 \times 2) = 0$$
Moments about B give:
$$(3 \times 2) - (7 \times 2) + (4\sqrt{2} \times \sqrt{2}) = 0$$
Moments about C give:
$$(7 \times 2) - (7 \times 2) = 0$$

Hence, the system of forces is in equilibrium.

Equilibrium
Worked example, Guided example and Exam questions

 The diagram shows a uniform rod AB of weight W and length 2a which is smoothly hinged at its midpoint to a fixed pivot M. A particle of weight 2W is attached to the rod at A. The other end B has a light string attached which is fastened to a fixed point C. The rod is in equilibrium with AB making an angle θ with the

horizontal, where $\cos \theta = \dfrac{3}{4}$. *The angle ABC is 90°. The points A, B and C are all in the same vertical plane.*

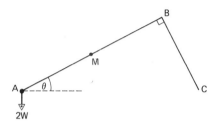

Calculate, in terms of W,
 (i) the tension in the string,
 (ii) the magnitude of the resultant force exerted by the pivot on the rod.

Draw a diagram showing the forces acting on the rod *AB*.

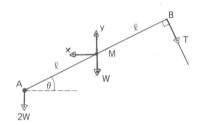

$$\cos \theta = \frac{3}{4}, \text{ so } \sin \theta = \frac{\sqrt{7}}{4}.$$

Let *T* be the tension in the string.
Let *X* and *Y* be the horizontal and vertical components of the reaction at the pivot.

(i) Moments about *M*:
$$Tl = 2Wl \cos \theta$$
$$\Rightarrow \quad T = \frac{3}{2}W \text{ — the tension in the string.}$$

(ii) Resolving vertically:
$$Y = 2W + W + T \cos \theta$$
i.e. $\quad Y = 3W + \dfrac{3}{2}W \cdot \dfrac{3}{4}$
$$\Rightarrow \quad Y = 4\tfrac{1}{8}W.$$
Resolving horizontally:
$$X = T \sin \theta$$
$$\Rightarrow \quad X = \frac{3\sqrt{7}}{8}W.$$

Magnitude of the force at the pivot is
$$\left[\left(4\tfrac{1}{8}\right)^2 + \left(\frac{3\sqrt{7}}{8}\right)^2\right]^{\frac{1}{2}} W \approx 4.24W.$$

 A smooth circular cylinder of weight 150 N rests on level ground DL, touching a vertical wall LM, as shown in the diagram. The axis of the cylinder is parallel to the ground and the wall.

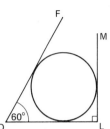

A uniform rod DF of weight 100 N and length 1·5 m, smoothly hinged at D so as to move in a vertical plane perpendicular to the axis of the cylinder, rests at an angle of 60° to the horizontal, touching the cylinder at a point 0·5 m from the end F. Calculate the magnitude of the forces on the cylinder at the points where it touches the rod, the ground and the wall.

First draw *two* separate force diagrams:
 (i) Showing the forces acting on the cylinder,
 (ii) Showing the forces acting on the rod.
Resolve horizontally and vertically for the cylinder. Take moments about *D* for the rod.
Solve the three equations to find the three required forces.

 1 A light inextensible string *ABCDE* has its ends *A* and *E* fixed at two points in the same horizontal line and has loads attached at the points *B, C* and *D*. If *AB, BC, CD* and *DE* make angles of 45°, 30°, 30° and 60° respectively with the horizontal and the load at the lowest point *C* is 3 kg, calculate, assuming the string and the loads are in equilibrium:
 (i) the tension in each of the four portions of the string,
 (ii) the loads at *B* and at *D*.   *(A)*

2 A uniform rod *AB*, of length 4*a* and weight *W*, is smoothly hinged to a fixed point at *A*. The rod is held at 60° to the horizontal with *B* above *A* by a horizontal force **F** acting at *B*. Calculate, in terms of *W*:
 (a) the magnitude of **F**;
 (b) the magnitude of the force exerted by the hinge on the rod and find, to the nearest degree, the direction of this force. The horizontal force acting at *B* is removed and the rod is held in the same position by resting against a fixed smooth peg at *C*, where *AC* = 3*a*.
 (c) Calculate, in terms of *W*, the magnitude of the force exerted by the peg on the rod.
 (d) Find, to the nearest degree, the direction of the force exerted by the hinge on the rod.   *(L)*

3 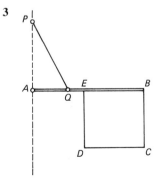 The figure shows a shop-sign consisting of a uniform metal bar *AB* weighing 10 kg and a uniform rectangular plate *BCDE* weighing 20 kg, the whole being supported by a light stay *PQ*. The joints at *P, Q* and *A* are smooth hinges. The dimensions are: *AB* = 1·5m, *AQ* = 0·5 m, *AE* = 0·7 m, *PQ* = 1 m.

Write down the equation of moments about *A*, and hence calculate the tension in the stay *PQ* in newtons. Find also the horizontal and vertical thrusts on the hinge at *A*.
        (O & C)

60 Three Force Problems

Three forces in equilibrium, Triangle of forces, Lami's theorem, Useful formulae.

Three forces in equilibrium

When a body is in equilibrium under a system of three coplanar forces only, several special results apply which make these 'three force problems' easier to solve.

The first special result is:
If a body is in equilibrium under the action of three coplanar forces, then these forces are either:

    (a) parallel
or (b) concurrent.

Spotting this in questions often makes their solution much easier.

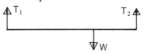 This horizontal bar held by two vertical strings is in equilibrium.

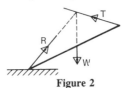

Figure 1

Three parallel coplanar forces.

This bar in contact with rough ground and held by a string is in equilibrium

Figure 2

Three concurrent coplanar forces.

Triangle of forces

The second special result is:
If a body is in equilibrium under the action of three concurrent coplanar forces, then these forces can be represented by the sides of a triangle taken in order, since their vector sum must be zero.

Force diagram   Triangle of forces

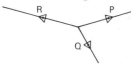

Figure 3

From the triangle of forces: $T_1 = W \cos \alpha$

$$T_2 = W \sin \alpha$$

Notice how the arrows 'chase each other round' in the triangle of forces, showing zero resultant, i.e. particle in equilibrium.

 A particle of weight W supported by two light strings as shown. Find the tensions T_1 and T_2.

Force diagram   Triangle of forces

Figure 4

Lami's theorem

The third special result is Lami's theorem which is a version of the sine rule.

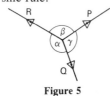

Figure 5

If three concurrent coplanar forces P, Q and R are in equilibrium and the angles between Q and R, R and P, P and Q are α, β and γ respectively as shown, then Lami's theorem states that:

$$\frac{P}{\sin \alpha} = \frac{Q}{\sin \beta} = \frac{R}{\sin \gamma}$$

This is particularly useful when one of the forces and the angles between pairs of forces are known.
It may also be used to solve the triangle of forces when this has been sketched.

The diagram shows a uniform ladder AB of weight 200 N resting in equilibrium with its foot on horizontal rough ground and its upper end against a rough vertical wall. Find R and S, the total reactions at A and B.

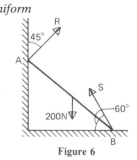

Figure 6

The ladder is in equilibrium under three non-parallel forces R, S and 200 N, so the forces are concurrent.

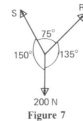

Figure 7

Lami's theorem gives:

$$\frac{200}{\sin 75°} = \frac{R}{\sin 150°} = \frac{S}{\sin 135°}$$

$$R = \frac{\sin 30°}{\sin 75°} \times 200 \approx 104 \text{ N}$$

$$S = \frac{\sin 45°}{\sin 75°} \times 200 \approx 146 \text{ N}$$

Useful formulae

The following trigonometrical formulae are often useful.
When D divides AB in the ratio $m:n$ then:

$$(m+n) \cot \theta = n \cot A - m \cot B$$
$$(m+n) \cot \theta = m \cot \alpha - n \cot \beta$$

When D is the mid-point of AB, then

$$2 \cot \theta = \cot A - \cot B = \cot \alpha - \cot \beta$$

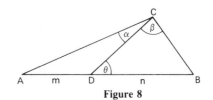

Figure 8

Three Force Problems
Worked example, Guided example and Exam questions

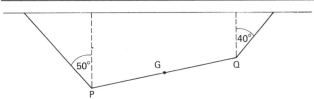

The diagram shows a uniform rod PQ of mass 10 kg which is held in equilibrium by two strings attached at the ends P and Q of the rod. The strings are inclined to the vertical as shown. Calculate:
(a) the tensions in the strings,
(b) the angle made by PQ with the vertical.

Draw the force diagram showing the forces acting on the rod PQ. θ is the angle of inclination of PQ to the vertical.

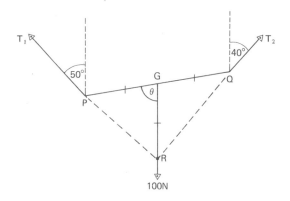

The rod is in equilibrium under the action of the three forces which pass through a common point, say, R.
Draw the triangle of forces.

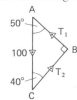

From the triangle,
$T_1 = 100 \cos 50° \text{ N},$
$T_2 = 100 \cos 40° \text{ N}.$

Since G is the midpoint of PQ and $\angle ABC = 90°$, P, Q and R must all lie on a circle (angle at the centre is 90°).
Hence $GP = GQ = GR$ so $\angle GPR = 50°$. Hence $\theta = 80°$ is the angle of inclination of PQ to the vertical.

 A smooth uniform sphere of radius 18 cm and weight 24 N rests against a smooth vertical wall and is supported by an inextensible wire 12 cm long tied to the wall and the surface of the sphere. Calculate the tension in the wire and the reaction of the wall on the sphere.

Sketch a diagram showing the three forces acting on the sphere. (They must be concurrent.)
Sketch the triangle of forces. One force, 24 N, is known.
Use the triangle to calculate the other two forces.

---

 1 The figure shows a uniform heavy bar AB of length 1 m and mass 8 kg which is freely hinged at A to a vertical wall. The bar rests in equilibrium with AB in contact with a smooth

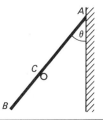

peg at C and makes an angle θ with the vertical where $\sin \theta = 0 \cdot 6$. The distance AC is 60 cm. Draw a figure showing the forces acting on the bar. Calculate:
(i) the magnitude of the reaction at C;
(ii) the angle which the reaction at A makes with the horizontal. *(C)*

2

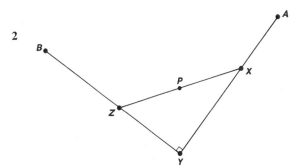

The diagram represents a uniform plane triangular lamina XYZ of weight 50 N and the angles YXZ and XYZ are 34° and 90° respectively. The lamina is suspended in equilibrium by two light strings AX and BZ. AXY and BZY are straight lines. If P is the mid-point of XZ, show that $PX = PY = PZ$. Why must PY be vertical? Hence calculate (i) the angle which XZ makes with the horizontal; (ii) the tensions in the strings.

(O & C)

3 A uniform rod XY of mass 10 kg rests in a vertical plane with the end X in contact with a smooth vertical wall. The end Y is below X. The rod is inclined at 60° to the vertical and is held in equilibrium by a light string attached to Y and to a point Z in the wall vertically above X. Show, in a diagram, the forces acting on the rod and hence that the distance ZX is half the length of the rod.
Find (i) the inclination of the string to the vertical; and (ii) the tension in the string.

(W)

4 Two light rings can slide on a rough horizontal rod. The rings are connected by a light inextensible string of length a to the mid-point of which is attached a weight W. Show that the greatest distance between the rings consistent with equilibrium is $\mu a/(1+\mu^2)^{\frac{1}{2}}$ where μ is the coefficient of friction between either ring and the rod.

5 A uniform body is in the form of a thick hemispherical shell having internal and external radii $2a$ and $3a$. Show that the distance of the centre of mass from the centre of the plane face is $\frac{195}{152} a$.

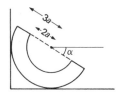

The body rests, as shown (in cross-section), on a fixed horizontal surface and against a fixed vertical surface, with its axis in a vertical plane perpendicular to both surfaces. The vertical surface is smooth and the coefficient of friction between the body and the horizontal surface is $\frac{1}{4}$. Given that the body is in limiting equilibrium, find the value of $\sin \alpha$, where α is the inclination of the plane face to the horizontal. *(C)*

61 Friction
Force of friction, Angle of friction, Problem solving.

Force of friction

When one body slides or attempts to slide over another, **forces of friction** usually exist between the two surfaces in contact.

Forces of friction act between **rough** surfaces in contact. **Smooth** surfaces in contact are frictionless. The following **experimental laws** describe the behaviour of frictional forces.

1. A frictional force only exists when one body slides or tries to slide over another.

2. A frictional force always **opposes** the tendency of one body to slide over another.

3. The magnitude of a frictional force may vary, always being just sufficient to prevent motion, until it reaches a **maximum value** called the **limiting value**.

4. The **limiting value** of the frictional force is μR, where μ is called the **coefficient of friction** and R is the normal reaction for the surfaces in contact. μ is a measure of the degree of roughness of the two surfaces in contact and is different for different pairs of surfaces.

5. When one body slides over another, the frictional force between them equals the limiting value μR. A consequence of laws (3) and (4) is that the frictional force F obeys the relation $F \leq \mu R$.

i A body, mass m, is resting on a rough horizontal plane and is being acted upon by a horizontal force P.

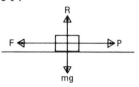

Figure 1

As the magnitude of P is gradually increased from zero, the magnitude of the frictional force F will also increase from zero in an attempt to prevent motion. When motion begins F has reached its maximum, called the limiting value μR, and cannot increase any more to prevent motion. So the frictional force remains constant, i.e. $F = \mu R$, whatever the increase in P.

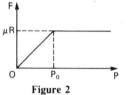

For:
$0 \leq P < P_0$ body stationary and $F < \mu R$.
$P = P_0$ limiting equilibrium and $F = \mu R$.
$P > P_0$ body accelerates and $F = \mu R$.

Figure 2

Angle of friction

The **resultant** S of the frictional force F and the normal reaction R is called the **total reaction**.

It makes an angle θ with the normal, where $\tan \theta = \dfrac{F}{R}$.

The normal reaction R is constant, but the frictional force F may vary.

As the frictional force F increases from zero to its maximum value F_L, the limiting value μR, the angle θ increases from zero to a maximum value λ, called the **angle of friction**.

$$\tan \lambda = \frac{\mu R}{R} = \mu$$

i.e. $\lambda = \tan^{-1} \mu$

When the frictional force has reached its limiting value F_L:

the **direction** of the total reaction S_L is at an angle λ to the normal reaction R,

the **magnitude** of the total reaction is $\sqrt{(R^2 + \mu^2 R^2)}$ $= R\sqrt{(1 + \mu^2)} = R \sec \lambda$.

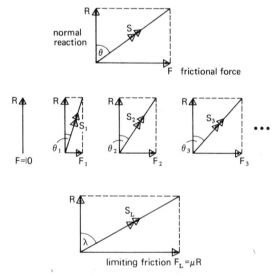

limiting friction $F_L = \mu R$

Figure 3

Problem solving

The following points are important when solving problems involving a frictional force F.

1. Draw a clear force diagram. Show the frictional force as F, do not use μR. Remember: F tends to oppose motion.

2. In general $F \leq \mu R$. If F has reached its limiting value, then $F = \mu R$ may be used in the solution.

3. 'Limiting equilibrium' indicates that the body is at rest but on the point of moving and that $F = \mu R$.

4. If the body is in equilibrium, then the equations of equilibrium (see p. 118) and $F \leq \mu R$ are used.

5. If λ is given, not μ, then it is often easier to solve the problem by considering the total reaction, rather than F and R separately. This is often the case in three force problems (see p. 120).

i *Find the least force P required to just prevent this particle from sliding down this inclined plane.*

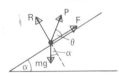

Figure 4

Resolve ∥ to slope: $P \cos \theta + F - mg \sin \alpha = 0$
$\Rightarrow F = mg \sin \alpha - P \cos \theta$
Resolve ⊥ to slope: $R + P \sin \theta - mg \cos \alpha = 0$
$\Rightarrow R = mg \cos \alpha - P \sin \theta$
Limiting friction, so $F = \mu R$
$\quad mg \sin \alpha - P \cos \theta = \mu(mg \cos \alpha - P \sin \theta)$

Using $\mu = \tan \lambda$, $P = \dfrac{mg \sin(\alpha - \lambda)}{\cos(\theta + \lambda)}$

P is a minimum when $\cos(\theta + \lambda)$ is a maximum, i.e. 1. Hence, minimum $P = mg \sin(\alpha - \lambda)$. This occurs when $(\theta + \lambda) = 0$, i.e. $\theta = -\lambda$.

Friction
Worked example, Guided example and Exam questions

WE *A uniform ladder of length 7 m leans against a vertical wall at an angle of 45° to the horizontal ground. The coefficients of friction between the ladder and the wall and the ladder and the ground are $\frac{1}{3}$ and $\frac{1}{2}$ respectively. How far up the ladder can a man, whose weight is half that of the ladder, ascend before the ladder slips?*

Assume that the man has climbed to such a position that the ladder is about to slip.
This diagram shows the forces acting on the ladder AB. M is the man.

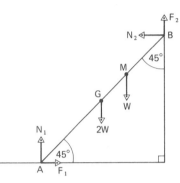

Let $AM = x$ metres.
$AG = GB = 3\frac{1}{2}$ metres.
F_1 and F_2 are the frictional forces.
N_1 and N_2 are the normal reactions.

Since ladder is about to slip
$F_1 = \frac{1}{2}N_1$ and $F_2 = \frac{1}{3}N_2$.
Resolving vertically:
$N_1 + F_2 = 3W$.
Resolving horizontally:
$\qquad F_1 = N_2$.

These four equations give,
$$N_1 = \frac{18}{7}W, \quad F_1 = \frac{9}{7}W, \quad N_2 = \frac{9}{7}W, \quad F_2 = \frac{3}{7}W.$$

Moments about A:
$2W \times 3\frac{1}{2} \cos 45° + W \times x \cos 45° = N_2 \times 7 \sin 45° + F_2 \times 7 \cos 45°$,
so, $7W + xW = 9W + 3W$

i.e. $x = 5$ metres — the distance up the ladder the man can safely climb.

GE *A particle P of weight W rests in limiting equilibrium on a rough plane which is inclined at an angle α to the horizontal. Prove that the coefficient of friction between the particle and the plane is $\tan \alpha$.*

(a) Figure (a) shows a horizontal force H, which is applied to P and acts in the vertical plane containing the line of greatest slope of the inclined plane which passes through P. If equilibrium is limiting with P on the point of moving up the plane, find H in terms of W and α.

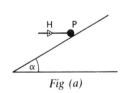 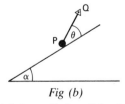

Fig (a)   *Fig (b)*

(b) Figure (b) shows a force Q which is applied to P in the vertical plane containing the line of greatest slope through P. The force is inclined at an angle θ to this line. If equilibrium is limiting with P on the point of moving up the plane show that

$$Q = \frac{W \sin 2\alpha}{\cos (\theta - \alpha)}.$$

Hence find, in terms of α, the value of θ for which Q is least.

Draw a diagram showing the three forces acting on P.
Since equilibrium is limiting, $F = \mu N$. Hence μ.
(a) Copy Figure (a) and add the missing three forces to it. (Remember friction will act *down* the plane).
Resolve parallel and perpendicular to the plane. Use $\mu = \tan \alpha$. Hence H.

(b) Copy Figure (b) and add the missing three forces to it. Use resolution in two directions at right angles and $\mu = \tan \alpha$. Hence Q. Q will be a minimum when $\cos (\theta - \alpha)$ is a maximum, i.e. 1. Hence state the value of θ for $\cos (\theta - \alpha) = 1$.

---

1 A particle P, of mass $7m$, is placed on a rough horizontal table, the coefficient of friction between P and the table being μ. A force of magnitude $2mg$, acting upwards at an acute angle α to the horizontal, is applied to P and equilibrium is on the point of being broken by the particle sliding on the table. Given that $\tan \alpha = \dfrac{5}{12}$, find the value of μ.

*(L)

2 A block of mass $1·5$ kg lying on a rough inclined plane is prevented from slipping down the plane by a string attached at one end to a point of the block and, at the other end, to a fixed point. The string is parallel to a line of greatest slope of the plane. The angle of inclination of the plane to the horizontal is α, where $\tan \alpha = \frac{4}{3}$, and the coefficient of friction between the block and the plane is $0·5$. Calculate the tension of the string in newtons. If the string is cut, find the acceleration with which the block slides down the plane.

(O & C)

3 The foot of a uniform ladder, of length l and weight W, rests on rough horizontal ground, and the top of the ladder rests against a smooth vertical wall. The ladder is inclined at 30° to the vertical. Find the magnitude of the force exerted by the ladder on the wall. Given that the coefficient of friction between the ladder and the ground is $\frac{1}{4}\sqrt{3}$, show that a man of weight $4W$ cannot climb to the top of the ladder without the ladder slipping, and find the least weight which when placed on the foot of the ladder would enable the man to climb to the top of the ladder.

(L)

4 A fixed hollow hemisphere has centre O and is fixed so that the plane of the rim is horizontal. A particle A of mass m can move on the inside surface of the hemisphere. The particle is acted on by a horizontal force of magnitude P, whose line of action is in the vertical plane through O and A. The diagram shows the situation when A is in equilibrium, the line OA making an acute angle θ with the vertical.

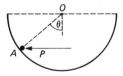

(i) Given that the inside surface of the hemisphere is smooth, find $\tan \theta$ in terms of P, m and g.
(ii) Given instead that the inside surface of the hemisphere is rough, with coefficient of friction μ between the surface and A, and that the particle is about to slip downwards, show that $\tan \theta = \dfrac{P + \mu mg}{mg - \mu P}$.

(C)

Force diagrams

Bodies in contact may be either simply **touching** each other or connected together with a **hinge**.

In both cases, if the bodies in contact are in **equilibrium** under a set of coplanar forces, then:

(a) the **complete system** is in **equilibrium**,

(b) each **separate body** is in **equilibrium**.

The complete system may be treated as if it were a single 'body'. Its force diagram must show the forces which act on this 'body' and originate from outside it, e.g. weight force, reactions and frictional forces between the 'body' and its surroundings.

Force diagrams for each separate body must show the forces which act on that body, e.g. weight force, reactions and any frictional forces at all points of contact.

Note: The force acting at a smooth hinge is usually shown resolved into its horizontal and vertical components in the force diagram.

ℹ A rough cylinder of weight W rests in equilibrium on rough horizontal ground with its axis horizontal. A uniform rod of weight w is smoothly hinged at its lower end to a point on the ground and rests in contact with the cylinder such that its upper end is above the point of contact, the plane containing the rod being perpendicular to the axis of the cylinder.

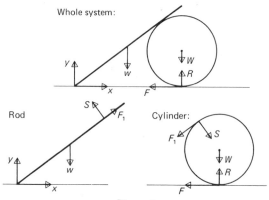

Figure 1

Problem solving

When solving problems involving bodies in contact:

1. Draw a clear force diagram for the complete system, i.e. treat the complete system as if it were a single 'body'.

2. Calculate any external forces required which act on the complete system by resolving and taking moments.

3. Draw individual force diagrams for the separate bodies which make up the system.

4. Resolve and take moments as necessary to calculate any required reactions.

Note: The **maximum number of independent equations** which can be obtained for a system consisting of n separate bodies is $3n$. In practice this maximum number of equations is usually not needed to solve a problem.

ℹ *Two uniform rods AB and BC, each of length 2a and of mass 2 kg and 3 kg respectively are smoothly hinged at B. The ends A and C are each smoothly hinged to two points in the same horizontal straight line and distance 2a apart. Find the horizontal and vertical components of the reactions at each hinge.*

For the whole system:

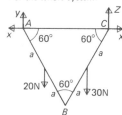

Figure 2

By symmetry the horizontal reactions at A and C must be equal.

Let X and Y be the horizontal and vertical components of the reaction at A.

Let X and Z be the horizontal and vertical components of the reaction at C.

Resolve ($\uparrow$): $Y+Z=50$ (1)

Moments about A: $Z.2a=20.\frac{1}{2}a+30.\frac{3}{2}a$

$$\Rightarrow Z=27.5\,\text{N}$$

From (1) $Y=22.5\,\text{N}$

For separate bodies:

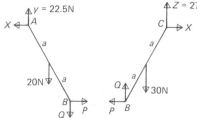

Let P and Q be the horizontal and vertical components of reaction at B.

Figure 3

Moments about B for BC:

$$30.\tfrac{1}{2}a+X.2a.\frac{\sqrt{3}}{2}=Z.2a.\tfrac{1}{2}\Rightarrow X\approx 7.22\,\text{N}$$

Resolve ($\rightarrow$) for BC: $P=X\approx 7.22\,\text{N}$

Resolve ($\uparrow$) for AB: $Q=30-Z=2.5\,\text{N}$.

Bodies in Contact
Worked example, Guided example and Exam questions

A rough circular cylinder is fixed with its axis horizontal. A uniform rod AB of length 2l and weight W is placed in contact with the cylinder at a point C, where A, B and C all lie in the same vertical plane. When a horizontal force P is applied at B which just prevents the rod from slipping downwards, the inclination of AB to the horizontal is θ and the mid point of AB is at a distance x from C. If μ(= tan λ) is the coefficient of friction between the rod and the cylinder, prove that,

(a) $P = W \tan(\theta - \lambda)$,

(b) $\mu < \tan \theta$,

(c) $x = l \sin \theta \sin(\theta - \lambda) \sec \lambda$.

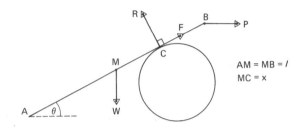

$AM = MB = l$
$MC = x$

The force diagram shows the forces acting on AB.
R is the normal reaction at C, F is the frictional force acting at C.

(a) Since AB is about to slip,
$$F = \mu R = (\tan \lambda) R \qquad [1]$$
Resolving parallel to AB:
$$F + P \cos \theta = W \sin \theta \qquad [2]$$
Resolving perpendicular to AB:
$$R = W \cos \theta + P \sin \theta. \qquad [3]$$
[1] and [2] give,
$$(\tan \lambda)R = W \sin \theta - P \cos \theta,$$
[3] × tan λ gives,
$$(\tan \lambda)R = W \tan \lambda \cos \theta + P \tan \lambda \sin \theta.$$
Eliminating R gives,
$$P(\cos \theta + \tan \lambda \sin \theta) = W(\sin \theta - \tan \lambda \cos \theta)$$
i.e. $P \cos(\theta - \lambda) \sec \lambda = W \sin(\theta - \lambda) \sec \lambda$
$\Rightarrow \qquad\qquad P = W \tan(\theta - \lambda).$

(b) If $\theta < \lambda$, $\tan(\theta - \lambda) < 0 \Rightarrow P < 0$; equilibrium impossible.
If $\theta = \lambda$, $\tan(\theta - \lambda) = 0 \Rightarrow P = 0$; not true.
Hence $\theta > \lambda \Rightarrow \tan \theta > \tan \lambda$
$\quad$ i.e. $\mu < \tan \theta$.

(c) Moments about C give:
$$Wx \cos \theta = P(l - x) \sin \theta.$$
Using the value of P obtained we get,
$x \cos \theta = (l - x) \sin \theta \tan(\theta - \lambda)$
$\quad$ i.e. $l \sin \theta \tan(\theta - \lambda) = x(\sin \theta \tan(\theta - \lambda) + \cos \theta)$
$\Rightarrow \quad l \sin \theta \sin(\theta - \lambda) = x(\sin \theta \sin(\theta - \lambda) + \cos \theta \cos(\theta - \lambda))$
$\qquad\qquad\qquad\qquad = x \cos(\theta - (\theta - \lambda))$
$\qquad\qquad\qquad\qquad = x \cos \lambda$
$\Rightarrow \qquad\qquad x = l \sin \theta \sin(\theta - \lambda) \sec \lambda.$

 Two equal rough cylinders are lying in contact with each other, with their axes parallel and horizontal, on a rough horizontal plane. A third equal cylinder is placed symmetrically on top of the other two. If equilibrium is about to be broken by the upper cylinder slipping between the other two, prove that the coefficient of friction between any two cylinders is $2 - \sqrt{3}$.

Draw two force diagrams, one showing the forces acting on the top cylinder and the other showing the forces acting on one of

---

the lower cylinders. (By symmetry, the forces acting on the other lower cylinder will be the same.)
Take moments about the point of contact of the lower cylinder with the horizontal plane. If the top cylinder is about to slip, the normal reaction between the two lower cylinders will be zero. Put this force equal to zero in the moments equation. Use '$F = \mu R$' for the forces acting at the point of contact of the upper and lower cylinders to obtain the required value of 'μ'.

---

1 Two uniform rods AB and BC of the same thickness and material, and of length 4 metres and 3 metres respectively, are freely hinged together and rest in a vertical plane with the ends A and C on a rough horizontal plane. The system is in limiting equilibrium when the angle ABC is 90°. Determine how equilibrium will be broken when the angle ABC is slightly increased beyond 90° and show that the coefficient of friction between the rods and the ground is $\frac{84}{163}$.

$\qquad\qquad\qquad\qquad\qquad\qquad (W)$

2 Two uniform rods, AB and AC, each of length $2a$ and weighing $3W$ and W respectively, are freely jointed at A and rest in a vertical plane with B and C on a rough horizontal floor. The coefficient of friction between either rod and the floor is μ. In limiting equilibrium,
(i) will slipping first occur at B or at C?
(ii) what will be the reaction at A (giving both magnitude and direction)?
(iii) at what angle to the horizontal will the rods be inclined?

$\qquad\qquad\qquad\qquad\qquad\qquad (S)$

3 A uniform sphere of radius a and weight $W/\sqrt{3}$ rests on a rough horizontal table. A uniform rod AB of weight $2W$ and length $2a$ is freely hinged at A to a fixed point on the table and leans against the sphere so that the centre of the sphere and the rod lie in a vertical plane. The rod makes an angle of 60° with the horizontal. Show that the frictional force between the rod and the sphere is $\frac{1}{3}W$. The coefficient of friction at each point of contact is μ. What is the smallest value of μ which makes equilibrium possible?

$\qquad\qquad\qquad\qquad\qquad\qquad (O \& C)$

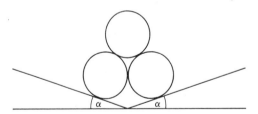

4 Two smooth uniform right circular cylinders, each of mass m and radius a, are placed symmetrically in contact with each other and with two planes, each inclined at angle α to the horizontal. The axes of the cylinders lie in the same horizontal plane, and are parallel to the line of intersection of the two inclined planes. Another smooth uniform circular cylinder, of mass $2m$ and of radius a, is placed symmetrically on top of the other two. If the two lower cylinders are forced apart, show that $\tan \alpha < 1/(2\sqrt{3})$.

$\qquad\qquad\qquad\qquad\qquad\qquad (OLE)$

63 Equivalent Systems of Forces
Equivalence, Reduction to a force or a couple, Reduction to a force and a couple.

Equivalence

Systems of coplanar forces which produce exactly the same linear and turning effects on a rigid body are **equivalent.**

To establish that two coplanar force systems are equivalent, show that:

    1. the total resolutes in two perpendicular directions are the same in both systems,

and 2. the resultant moment about a point in the plane containing the forces is the same in both systems.

A known system of coplanar forces may be replaced by an equivalent, often simpler, system using (1) and (2).

i These two systems of forces are equivalent:

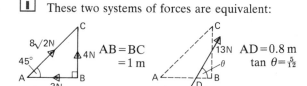

$AB=BC$
$=1\,\mathrm{m}$

$AD=0.8\,\mathrm{m}$
$\tan\theta=\frac{5}{12}$

Figure 1

	System I	System II
Resolve $\parallel AB$	$8\sqrt{2}\sin 45°-3=5$	$13\sin\theta=5$
Resolve $\perp AB$	$8\sqrt{2}\cos 45°+4=12$	$13\cos\theta=12$
Moments (A)	$4\times 1=4$	$13\times 0.8\sin\theta=4$

Reduction to a force or a couple

The **resultant** of a system of coplanar forces, which is not in equilibrium, is either a **single force or a couple.** This resultant is equivalent to the original system of coplanar forces.

A system **reduces to a single force** if: one or both of the total resolutes in two perpendicular directions is non-zero.

To locate the position of the line of action of the resultant force, take moments about any point in the plane and use the principle of moments.

A system **reduces to a couple** if one of the following sets of conditions is satisfied.

Set I

    1. The total resolutes in two perpendicular directions are each zero.

and 2. The resultant moment about any point in the plane is non-zero.

Set II

The resultant moments about three non-collinear points are each non-zero.

i *ABC is an isosceles triangle with $AB=AC=5$ cm and $BC=6$ cm. M is the mid-point of BC. Forces of 15, 12, 5 and 8 N act along $\overrightarrow{AB},\overrightarrow{BC},\overrightarrow{CA},\overrightarrow{MA}$ respectively.*
Show that this system of forces is equivalent to a couple and find its moment.

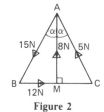

Figure 2

$AB=AC=5\,\mathrm{cm}=0.05\,\mathrm{m}$
$BM=MC=3\,\mathrm{cm}=0.03\,\mathrm{m}$
so $AM=4\,\mathrm{cm}=0.04\,\mathrm{m}$.

Let $\angle MAB=\angle MA\overset{\frown}{C}=\alpha$,
so $\sin\alpha=\frac{3}{5}$
$\cos\alpha=\frac{4}{5}$

Resolve $\parallel$ to BC:
$\qquad 12-15\sin\alpha-5\sin\alpha=12-15\times\frac{3}{5}-5\times\frac{3}{5}=0.$

Resolve $\perp$ to BC:
$\qquad 8-15\cos\alpha-5\cos\alpha=8-15\times\frac{4}{5}-5\times\frac{4}{5}=0$

Moments about A:
$\qquad 12\times 0.05\cos\alpha=0.48\,\mathrm{N\,m}$
Hence, the system of forces ≡ couple of moment $0.48\,\mathrm{N\,m}$.

Reduction to a force and a couple

Any system of coplanar forces acting on a rigid body may be replaced by an equivalent system which consists of a **single force**, acting at a particular point in the plane of the forces, **together with a couple.**

i *Forces 3 N and 5 N act along the sides $\overrightarrow{AC}$ and $\overrightarrow{BC}$ of an equilateral triangle, side a metres. A force of 2 N acts along the altitude $\overrightarrow{AD}$. Find the force at B and the couple which together are equivalent to this system.*

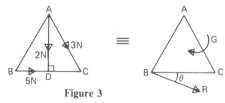

Figure 3

Let R be the required single force at angle θ to BC and G be the moment of the required couple

Resolve along BC for original system: $5+3\cos 60°$
                       for new system: $R\cos\theta$
So $R\cos\theta=5+3\cos 60°$ (1)
Resolve $\perp$ to BC for original system: $2+3\cos 30°$
                       for new system: $R\sin\theta$
So $R\sin\theta=2+3\cos 60°$ (2)
Solving (1) and (2) gives $R\approx 7.96\,\mathrm{N}$ and $\theta\approx 35.3°$

Moments about B: $(2\times\frac{1}{2}a)+(3\times a\sin 60°)=G$

i.e. $G=\left(\frac{1+3\sqrt{3}}{2}\right)a\,\mathrm{Nm}$

Equivalent Systems of Forces
Worked example, Guided example and Exam questions

WE *ABCD is a square of side 2 m. Forces of magnitudes 3 N, 5N, 7 N and 2 N act along the sides DA, AB, BC and CD respectively. Calculate:*

(a) the magnitude of the resultant of the forces and the angle made by the resultant with AD,

(b) the sum of the moments of the forces about A,

(c) the distance from A of the point where the line of action of the resultant of the forces cuts DA produced.

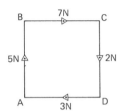

(a) Let X and Y be the resolutes of the resultant force in the directions AD and AB respectively.

$X = (7 - 3)$ N $= 4$ N
$Y = (5 - 2)$ N $= 3$ N
Hence, $R = \sqrt{(4^2 + 3^2)}$ N $= 5$ N
at an angle $\theta = \tan^{-1}\left(\frac{3}{4}\right)$ to AD.

(b) Moments about A for the system of forces gives
$G = (7 \times 2) + (2 \times 2)$ Nm
$\quad = 18$ Nm, in a clockwise sense.

(c) Let **R**, the resultant of the forces, cut DA produced at P.
Let $AP = x$ metres.

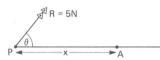

Moments about A for the resultant force gives $G = 5 \sin \theta \times x$.

Since the moments G must be equal,
$5 \sin \theta \times x = 18$

$\Rightarrow \qquad x = \dfrac{18}{5 \cdot \dfrac{3}{5}} = 6$ m $\left(\text{since } \sin \theta = \dfrac{3}{5} \right).$

Hence, the resultant of the forces cuts DA produced at 6 m from A.

GE *A regular hexagon ABCDEF has sides of length 2 m. Forces of magnitude 4, R, 2, 3, 6 and 5 newtons act along the sides AB, BC, DC, DE, EF and AF respectively. A seventh force of magnitude S newtons acts along EB. The directions of the forces are indicated by the order of the letters. Calculate R and S when the given system of forces is equivalent to*

(a) a single force of magnitude 6 N in a direction parallel to EB,

(b) a couple,

In (b) calculate also the moment of the couple.

(a) Find the sum of the resolutes of the forces
 (i) parallel to EB and equate to 6 N,
(ii) perpendicular to EB and equate to zero.
Solve the two resulting equations to find R and S.

(b) Repeat (a) (ii) and (a) (i) but equate the forces to zero. Solve the resulting equations to find the new values for R and S.
To find the moment of the couple, take moments about the centre of the hexagon for the total system of forces.

EX

1 Forces of 5 N and 3 N act along the sides $\overrightarrow{AB}$, $\overrightarrow{AC}$ respectively of an equilateral triangle ABC of side 12 m. Find the magnitude and direction of their resultant. The line of action of the resultant intersects BC at D. By taking moments about D, or otherwise, find the length of BD.
(C)

2 A rectangle is defined by the four points $A(0, 0)$, $B(5, 0)$, $C(5, 3)$ and $D(0, 3)$, distances being measured in metres. Forces of magnitude 6N, 8N, 4N, and 2N act along AB, BC, CD and DA respectively in directions indicated by the letters. Calculate:
(a) the magnitude of the resultant of this system of forces;
(b) the angle between the line of action of the resultant and the x-axis. The line of action of this resultant cuts the x-axis at $(a, 0)$.
(c) Find a value for a by consideration of moments about A. Hence determine the equation of the line of action of this resultant.
(S)

3 The square $ABCD$ has each side of length 6 m. Forces of magnitude 1, 2, 8, 5, $5\sqrt{2}$ and $2\sqrt{2}$ N act along AB, BC, CD, DA, AC and DB respectively, in the directions indicated by the order of the letters. Prove that these forces are equivalent to a couple. Calculate the magnitude and sense of this couple.
(A)

4 $OABC$ is a square of side 1 m. Forces of magnitude 2, 3, 4 and $5\sqrt{2}$ newtons act along OA, AB, CB and AC respectively in directions indicated by the order of the letters and a couple of moment 7 N m acts in the plane of the square in the sense $OCBA$. Find the magnitude and direction of the resultant of the system and the equation of its line of action referred to OA and OC as axes. What is the magnitude and direction of the least force introduced at A if the resultant of the original system and this new force is to pass through O?
(S)

5 A rectangle $ABCD$ has $AB = 3$ cm and $BC = 4$ cm. Forces, all measured in newtons and of magnitudes 2, 4, 6, 8 and k, act along AB, BC, CD, DA and AC respectively, the direction of each force being shown by the order of the letters. The resultant of the five forces is parallel to BD. Find k and show that the resultant has magnitude $\frac{8}{5}$ newtons. Find the distance from A of the line of action of the resultant.
(O & C)

6 A rigid rectangular lamina $ABCD$, with $AB = 4a$ and $BC = 3a$, is subject to forces of magnitudes $10P$, P, $2P$, $3P$ acting along CA, AD, DC, CB respectively in the directions indicated by the order of the letters.
 (i) Find the magnitude of the resultant of the four forces.
 (ii) Find the tangent of the acute angle between the line of action of the resultant and the edge AB of the lamina.
 (iii) Find the distance from A of the point where the line of action of the resultant meets AB.
 (iv) Indicate clearly on a diagram the line of action and the direction of the resultant.
 (v) Find the magnitude and sense of the couple G which, if added to the system, would cause the resultant force to act through E, the midpoint of CD.
 (vi) In the case when G is *not* applied, find forces S along AB, T along AD and U along BC which, when added to the system, would produce equilibrium.
(J)

64 Centre of Mass
Definitions, System of particles, Symmetry, Standard results, Composite bodies.

Definitions

The **centre of mass** of a body is the point at which the mass of the body may be considered to be acting.
The **centre of gravity** of a body is the point through which the line of action of its weight acts.
The **centroid** of a body is at its geometric centre.
The centre of mass and centre of gravity of a body coincide in a uniform gravitational field.
The centre of mass and the centroid of a body coincide in a uniform body.
A **uniform body** has uniform density.

System of particles

For a **set of particles** of masses $m_1, m_2, m_3, \ldots$ at the points $(x_1, y_1), (x_2, y_2), (x_3, y_3), \ldots$ in the x-y plane, the centre of mass $(\bar{x}, \bar{y})$ is given by

$$\bar{x} = \frac{m_1 x_1 + m_2 x_2 + m_3 x_3 + \ldots}{m_1 + m_2 + m_3 + \ldots} = \frac{\Sigma m_i x_i}{\Sigma m_i}$$

and $\bar{y} = \dfrac{m_1 y_1 + m_2 y_2 + m_3 y_3 + \ldots}{m_1 + m_2 + m_3 + \ldots} = \dfrac{\Sigma m_i y_i}{\Sigma m_i}$

[i] *Particles of masses 5, 3 and 8 kg are at (0, 0), (3, 4) and (6, 0) respectively. Find $(\bar{x}, \bar{y})$, their centre of mass.*

$\bar{x} = \dfrac{5(0) + 3(3) + 8(6)}{5 + 3 + 8} = \dfrac{57}{16}$ So $(\bar{x}, \bar{y})$ is $\left(\dfrac{57}{16}, \dfrac{3}{4}\right)$.

$\bar{y} = \dfrac{5(0) + 3(4) + 8(0)}{5 + 3 + 8} = \dfrac{3}{4}$

Symmetry

The centre of mass of a **uniform body** lies on every line or plane of **symmetry** of the body. It is at the point where any two lines, or three planes of symmetry intersect.
By symmetry the centres of mass of the following lie at their geometric centres:
uniform rod, circular lamina, rectangular lamina, sphere, cuboid.

Standard results

uniform body	centre of mass
triangular lamina	intersection of medians
circular arc, radius a, angle at centre 2α	$\dfrac{a \sin \alpha}{\alpha}$ from centre
circular sector, radius a, angle at centre 2α	$\dfrac{2a \sin \alpha}{3\alpha}$ from centre
solid hemisphere, radius a	$\dfrac{3a}{8}$ from plane face
hollow hemisphere, radius a	$\dfrac{a}{2}$ from plane face
solid cone, height h (tetrahedron, pyramid)	$\dfrac{h}{4}$ from base
hollow cone, no base, height h	$\dfrac{h}{3}$ from 'base'

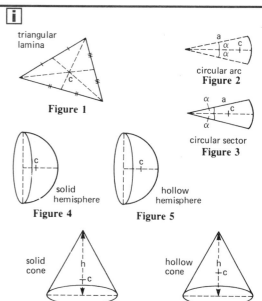

[i]

triangular lamina — Figure 1

circular arc — Figure 2

circular sector — Figure 3

solid hemisphere — Figure 4

hollow hemisphere — Figure 5

solid cone — Figure 6

hollow cone — Figure 7

Composite bodies

A composite body is one made from two or more parts (usually standard).

To find the centre of mass of a composite body
(a) Draw a clear diagram.
(b) Mark any lines of symmetry.
(c) Choose two axes at right angles to each other. If a line of symmetry exists choose this as an axis.
(d) Divide the body into known (standard) bodies.
(e) Tabulate the 'masses' and 'distances of centres of masses from the chosen axes'.
Note: In a uniform lamina, mass ∝ area.
        In a uniform solid, mass ∝ volume.
(f) Take moments about the chosen axes.
(g) Use the principle of moments:
Moment of total mass about an axis = sum of moments of separate masses about same axis.

This method can also be used to deal with a body from which a part has been removed.

[i]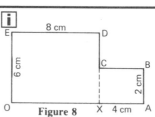

Figure 8

Find the centre of mass of this uniform lamina referred to as OA and OE.

Divide the lamina into two rectangles OXDE and XABC.

shape	mass (M is mass/unit area)	distance of centre of mass	
		from OE	from OA
OXDE	$(6 \times 8)M = 48M$	4	3
XABC	$(4 \times 2)M = 8M$	$(8+2) = 10$	1
OABCDE	$56M$	$\bar{x}$	$\bar{y}$

Moments about OE give: $48M \times 4 + 8M \times 10 = 56M \times \bar{x}$
$$\Rightarrow \bar{x} = 4\tfrac{6}{7} \text{ cm}$$
Moments about OA give: $48M \times 3 + 8M \times 1 = 56M \times \bar{y}$
$$\Rightarrow \bar{y} = 2\tfrac{5}{7} \text{ cm}$$

∴ Centre of mass is at a point which is $4\tfrac{6}{7}$ cm from OE and $2\tfrac{5}{7}$ cm from OA.

Centre of Mass
Worked example, Guided example and Exam questions

 A uniform right circular solid cylinder has a radius r and length 4r. A solid hemisphere of radius r is cut from one end of the cylinder, the plane face of which is one of the plane faces of the cylinder. The hemisphere so removed, is now attached by its plane face to the uncut plane face of the cylinder thus forming a new solid. Find the position of the centre of mass of the new solid.

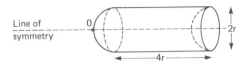

Let O be the point on the line of symmetry at the extreme end of the solid.
Let $\bar{x}$ be the distance of the centre of mass of the 'new solid' from O.

Shape	Mass	Distance of c of m from O
(1)	$4\pi r^3$	$r + 2r = 3r$
(2)	$\frac{2}{3}\pi r^3$	$\frac{5}{8}r$
(3)	$\frac{2}{3}\pi r^3$	$r = 4r - \frac{3}{8}r = \frac{37}{8}r$
(4)	$4\pi r^3$	$\bar{x}$

Notice that if we put together shapes (1) and (2) we get the same shape as we do by putting together shapes (3) and (4). We use this to write down our moments equation.
Moments about O give:

$$(4\pi r^3 \times 3r) + \left(\frac{2}{3}\pi r^3 \times \frac{5}{8}r\right) = \left(\frac{2}{3}\pi r^3 \times \frac{37}{8}r\right) + (4\pi r^3 \times \bar{x})$$

$$\Rightarrow \bar{x} = \frac{7}{3}r.$$

 A uniform lamina is formed by removing a circular disc of radius r from a circular disc of radius 2r as shown. Find the position of the centre of gravity of the lamina with respect to the two axes Ox and Oy as shown.

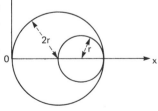

Notice that Ox is an axis of symmetry.
Tabulate:

(a) The masses of
    (i) the original circular disc,
    (ii) the small circular disc which was removed,
    (iii) the lamina.
(b) The distances of the centres of mass from O along Ox.

· Use the principle of moments, putting together shapes (ii) and (iii) to make (i). Hence find the position of the centre of mass of the lamina.

---

1 A uniform rectangular lamina $ABCD$ is of mass $3M$; $AB = DC = 4$ cm and $BC = AD = 6$ cm. Particles, each of mass M, are attached to the lamina at B, C and D. Calculate the distance of the centre of mass of the loaded lamina: (a) from AB, (b) from BC.   *(L)*

2 (a) The figure below shows a plate of uniform thickness, circular in shape, which has two circular holes drilled through it. The radius of the plate is 80 mm and AOB, COD are perpendicular axes of the plate.

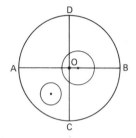

One hole, of radius 20 mm, has its centre on OB at a distance of 5 mm from O; the other, of radius 10 mm, has its centre at a distance of 40 mm from OA and 20 mm from OC. Show that the centre of mass of the plate is located on the axis COD and find its distance from O.
(b) Masses p, q and r are placed at the points whose cartesian co-ordinates are respectively $(0, 0)$, $(20, 0)$ and $(0, 16)$. If $p:q:r = 1:4:3$, find the co-ordinates of the centre of mass of the three masses.   *(W)*

3 A can in the form of a circular cylinder, without a lid, is made of thin metal sheeting of uniform thickness and with a mass per unit area of 1 g/cm². The radius of the can is 10 cm and its height is 20 cm. The can is placed with its base on a horizontal plane and is half-filled with a liquid of density $1 \cdot 5$ g/cm³. Calculate the height of the centre of gravity of the can together with the liquid, above the base of the can.   *(A)*

4 The diagram shows a square $OABC$ of side a. The mid-point of BC is D. Show that, with respect to OA and OC as axes, the co-ordinates of the centroid F of the triangular region ABD are $(\frac{5}{6}a, \frac{2}{3}a)$. Find the co-ordinates of the centre of mass of a uniform lamina in the form of the figure $OADC$.   *(J)*

5 A uniform solid right circular cone has its top removed by cutting the cone by a plane parallel to its base, leaving a truncated cone of height h, the radii of its ends being r and $4r$. Show that the distance of the centre of gravity of the truncated cone from its broader end is $\frac{9}{28}h$.   *(O & C)*

6 Prove that the centre of gravity of a uniform circular arc subtending an angle 2θ at the centre of a circle of radius a is $a\sin\theta/\theta$ from the centre. Deduce that the centre of gravity of a uniform sector bounded by that arc and the radii to its extremities is $\frac{2}{3}a\sin\theta/\theta$ from the centre. Show also that the centre of gravity of a segment of a circular lamina cut off by a chord subtending a right angle at the centre of the circle is $\frac{2}{3}\sqrt{2}a/(\pi-2)$ from the centre.   *(W)*

Suspended bodies

When a body is **freely suspended** from a frictionless pivot, P, it will rest in equilibrium with its centre of mass, M, directly below the point of suspension on the vertical through P.

For equilibrium:
the reaction R and the weight W act in opposite directions along the same vertical line, and $R = W$.

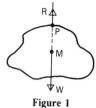

Figure 1

Problems concerned with freely suspended bodies in equilibrium often require you to find the angle made by a line in the body with the vertical. This angle is called the angle of inclination to the vertical.

To solve problems of this type
(a) Draw a clear diagram.
(b) Find the position of the centre of mass, M, of the body.
(c) Draw in the line through the point of suspension, P, and the centre of mass, M. This line will be vertical when the body is freely suspended from P. Your original diagram need not be redrawn.
(d) Obtain an expression for the angle of inclination to the vertical using the position of the centre of mass and geometry or trigonometry from the diagram.

ℹ️ *A uniform rectangular plate $ABCD$ of weight $2W$ has dimensions $AB = DC = 2a$ and $AD = BC = 2b$. A particle of weight W is attached to the plate at C. When the lamina, with the particle attached, is freely suspended from D, DC makes an angle θ with the downward vertical. Show that $\tan\theta = \dfrac{b}{2a}$.*

First find the centre of gravity, $G(\bar{x}, \bar{y})$, for the plate and particle.

Choose DC and DA as axes for 'moments'.

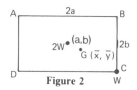

Figure 2

Consider the weight $2W$ of the plate to be at (a, b). The particle of weight W is at $(2a, 0)$.

Moments about DA give:
$$2W \times a + W \times 2a = 3W\bar{x} \Rightarrow \bar{x} = \frac{4a}{3}$$

Moments about DC give:
$$2W \times b + W \times 0 = 3W\bar{y} \Rightarrow \bar{y} = \frac{2b}{3}$$

Now show the vertical line through G and D, the point of suspension.

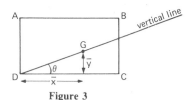

Figure 3

Since DC makes an angle θ with the vertical,
$$\tan\theta = \frac{\bar{y}}{\bar{x}} = \frac{2b}{3} \bigg/ \frac{4a}{3} = \frac{b}{2a}.$$

Toppling bodies

When a body rests in equilibrium on a plane, it will be **stable** provided the line of action of the weight force lies within the extreme points of contact between the body and the plane.

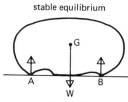

stable equilibrium

Figure 4

The body will **topple** if the line of action of the weight force lies outside one of these extreme points of contact. It will topple about the point of contact nearest to the line of action of the weight.

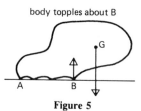

body topples about B

Figure 5

If the body is on a rough inclined plane, it will **topple** if the line of action of the weight force lies outside the lower point of contact of the body with the plane. On an inclined plane the equilibrium of a body may be broken by sliding rather than toppling.

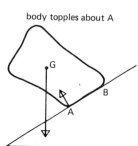

body topples about A

Figure 6

ℹ️ *$ABCD$ is a uniform square lamina, of side a, from which EBC is removed. Find the least length of AE which will allow the lamina $AECD$ to stand in a vertical plane with AE on a horizontal surface.*

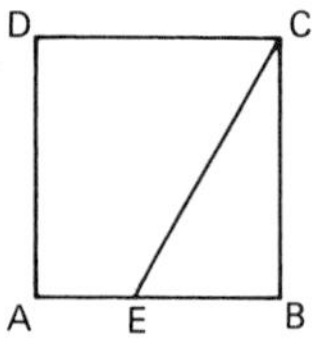

Figure 7

First find the distance, $\bar{x}$, of the centre of gravity of $AECD$ from AD.
Let $AE = x$ and let w be the weight per unit area

shape	weight	distance of c. of g. from AD
$ABCD$	$a^2 w$	$\frac{1}{2}a$
EBC	$\frac{1}{2}a(a-x)w$	$x + \frac{2}{3}(a-x) = \frac{1}{3}(2a+x)$
$AECD$	$\frac{1}{2}a(a+x)w$	$\bar{x}$

Moments about AD give:
$$a^2 w \times \tfrac{1}{2}a = \tfrac{1}{2}a(a-x)w \times \tfrac{1}{3}(2a+x) + \tfrac{1}{2}a(a+x)w \times \bar{x}$$
$$\Rightarrow \bar{x} = \frac{a^2 + ax + x^2}{3(a+x)}$$

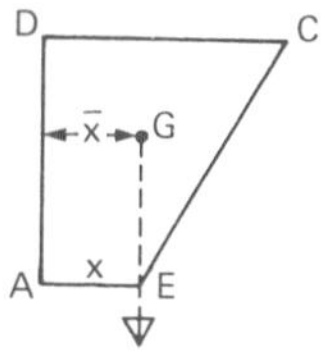

For no toppling about E:
$\bar{x} \leq AE$,

i.e. $\dfrac{a^2 + ax + x^2}{3(a+x)} \leq x.$

Figure 8

$$\Rightarrow x \geq \tfrac{1}{2}a(\sqrt{3} - 1).$$

So the least value of x for equilibrium is $\tfrac{1}{2}a(\sqrt{3} - 1)$.

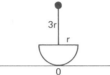

WE *A toy is constructed as follows.*
A particle of weight W is attached to one end of a light rod of length 3r. The other end of the rod is attached to the centre of the plane face of a uniform solid hemisphere of weight w and radius r, the angle between the rod and the plane face of the hemisphere being 90°. Find, in terms of W, the least value of w such that when the toy is knocked over on a horizontal surface it will always return to an upright position.

We first find the position of the centre of gravity of the toy.

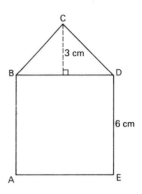

Shape	Weight	Height of c of g above O
	$W+w$	$\bar{h}$
	W	$4r$
	w	$\frac{5}{8}r$

Moments about O give:
$$(W+w)\bar{h}=4rW+\frac{5}{8}rw$$
$$\Rightarrow \quad \bar{h}=\frac{r}{8}\left[\frac{32W+5w}{W+w}\right]$$

Now consider the toy to be knocked over.

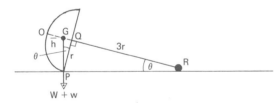

Let θ be the angle between the rod and the horizontal. If the toy is to return to the vertical position, the line of action of the weight force must lie to the left of P, the point of contact between the horizontal plane and the hemisphere.

In $\triangle PQR$, $\tan\theta=\dfrac{r}{3r}=\dfrac{1}{3}$.

In $\triangle PQG$, $\tan\theta=\dfrac{r-\bar{h}}{r}$.

So, $\dfrac{r-\bar{h}}{r}=\dfrac{1}{3}$, $\Rightarrow\bar{h}=\dfrac{2r}{3}$.

Hence, $\dfrac{2r}{3}=\dfrac{r}{8}\left[\dfrac{32W+5w}{W+w}\right]$

i.e. $w=48W$.

So, the least value of w such that the toy always returns to an upright position is $48W$.

GE *A thin uniform plate is formed by a square of side 6 cm being surmounted by an isosceles triangle of vertical height 3 cm, as shown below.*

(a) State the distance of the centre of gravity of the triangle BCD from its side BD.

(b) Calculate the distance of the centre of gravity of the whole plate from sides AB and AE.

(c) If the plate is freely suspended from A, find the inclination of AE to the horizontal.

(a) Do as the question says — state.

(b) Notice the shape has a line of symmetry.
Use this line of symmetry as one axis. Use AE as the other axis.
Draw the lamina *ABCDE*.
Tabulate the weights of the triangle, square and plate and the distances of their centres of gravity from AE.
Use the principle of moments to find the centre of gravity of the lamina.

(c) Mark the centre of gravity G of the lamina on your drawing.
Put in the 'vertical' through A and G. Mark as θ the angle between AE and the 'vertical'. Use trigonometry to find θ.
The inclination of AE to the horizontal is $90°-\theta$.

---

1 A uniform lamina of mass M is in the shape of a circular disc, centre O. Two points A and B on the circumference of the disc are such that the angle AOB is a right angle. A particle of mass kM is attached to the disc at B. When the loaded disc is in equilibrium suspended freely from A, AO is inclined to the downward vertical at an angle α, where $\tan\alpha=0\cdot4$. Calculate the value of k.

(A)

2 A uniform lamina $ABCD$ has the shape of a trapezium with DC parallel to AB and $\widehat{BAD}=90°$, $AB=3a$, $CD=2a$ and $AD=2b$. Find the distances of its centre of gravity from AB and AD. The lamina is freely pivoted at B to a fixed point and hangs down in equilibrium. If $a=7$ cm, $b=13$ cm, find the inclination of AB to the vertical.

(O & C)

3 A toy consists of a solid hemisphere of radius a to which is glued a solid circular cylinder of radius a and height $2a$ so that the plane end of the hemisphere is in complete contact with a plane end of the cylinder. The cylinder is made of uniform material of density ϱ, and the hemisphere is made of uniform material of density $k\varrho$. The toy is designed so that if placed on a horizontal table with the hemisphere downwards and then tilted to one side, it will return to the vertical position. Show that $k>8$. The toy is placed on a desk of slope α where $\sin\alpha=\frac{1}{8}$, sufficiently rough to prevent slipping. It rests in equilibrium with the hemisphere in contact with the desk. Find an expression giving the (acute) angle β made by its axis of symmetry with the vertical. Hence deduce that $k\geq13\frac{1}{2}$.

(O & C)

66 Pictorial Representation

Categorial data, Bar chart, Pie chart, Isotype diagram.

Categorical data

Information which can be put into categories is called **categorical data.**

Categorical data may be represented pictorially by **bar charts, pie charts** and **isotype diagrams** (or **pictograms**).

Table showing a man's annual expenditure.

Item	Expenditure (£)
House	3000
Food	2000
Fuel	800
Travel	800
Others	1100

Here the categories are the items of expenditure, e.g. House, etc.

Bar chart

In a **bar chart** data are represented by a series of parallel bars of equal width. The length of each bar is proportional to the frequency of the category it represents. The distance between adjacent bars should be equal.

Bars may be drawn horizontally or vertically but all must start from the same base line. If drawn vertically, the diagram is often called a column graph.

For the data above a suitable scale is 1 cm≏£500.

The length of the bar representing 'House' is $\frac{3000}{500}$ cm $=6$ cm

The lengths of the other bars, working down the table, are 4 cm, 1.6 cm, 1.6 cm and 2.2 cm.

Bar chart showing a man's annual expenditure

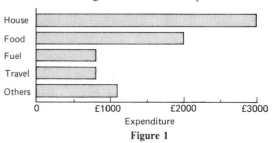

Figure 1

Pie chart

In a **pie chart,** the total data is represented by the area of a circle which is divided into sectors, one sector for each category.

Each sector angle is a fraction of 360° and may be found from:

$$\text{sector angle} = \frac{\text{category value}}{\text{sum of category values}} \times 360°.$$

Since the 'sum of the category values' is represented by the complete circle, the sum of the sector angles must be 360°. This is a useful calculation check.
A pie chart should not contain too many sectors (less than eight is usual). Each sector should be clearly labelled.

For the data above the 'sum of category values' is £7700.

The sector angle for 'House' is $\frac{3000}{7700} \times 360° = 140.3°$

The sector angles for the other sectors, in order, are 93.5°, 37.4°, 37.4° and 51.4°.

Pie chart showing a man's annual expenditure

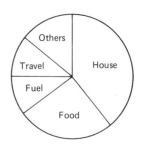

Figure 2

Isotype diagram (or pictogram)

In an **isotype diagram** (or **pictogram**) data are represented by symbols (usually pictures). A symbol is used to represent a stated number of units of the data. Appropriate fractions of the symbol are used to represent fractions of the basic number of units.
The symbol used is often associated in some way with the data it represents.

For the data above a suitable symbol would be

 to represent £500.

Figure 3

Isotype diagram showing a man's annual expenditure

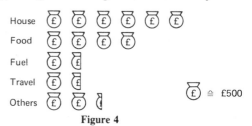

Figure 4

Pictorial Representation
Worms example and Exam questions

The approximate expenditure in the United Kingdom during an academic year on various sectors of education is given below:

Sector	Expenditure (millions of £)
Primary	523
Secondary	608
Special	46
Adult	260
Teacher training	57
Universities	258

Illustrate these data by a pie chart, showing clearly any necessary calculations.

Total expenditue is £1752 millions

Thus £1752 millions $\longleftrightarrow$ 360°

$$£1 \text{ million} \longleftrightarrow \left(\frac{360}{1752}\right)^\circ$$

and £523 millions $\longleftrightarrow \left(\frac{360 \times 523}{1752}\right)^\circ = 107 \cdot 5^\circ$, to the nearest half-degree.

Replacing 523 by the other entries in the table in turn and tabulating the appropriate sector angles gives the following table.

Sector	Sector angle in pie chart
Primary	107·5°
Secondary	125·0°
Special	9·5°
Adult	53·5°
Teacher training	11·5°
Universities	53·0°

The total of the sector angles is 360° as required.

The appropriate pie chart is shown below.

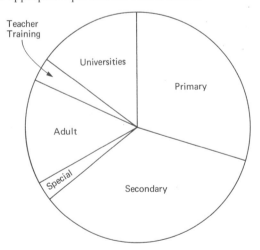

1 The table below, adapted from a magazine article, shows the approximate total number of deaths in England and Wales in 1975 from preventable external causes.

Cause	Number of deaths
Smoking	100 000
Traffic Accidents	6 000
Alcohol	5 000
Falls	4 000
Others	2 000

Represent this information on a pie chart, stating the size of the angles. Suggest briefly how the pie chart might be displayed on a poster to achieve immediate visual impact.

*(A)

2 In 1977, the group sales of ICI by geographical region were:

United Kingdom	£1 868 000 000
The rest of Western Europe	£865 000 000
The Americas	£628 000 000
Australasia	£505 000 000
Others	£807 000 000

Display this information on a pie chart, showing clearly any necessary calculations.

*(A)

3 The cash flow of BP during 1977 was reported as follows:

'Where the funds came from.'

Income before UK tax	£769 million
Depreciation	£304 million
Other	£19 million
Total	£1092 million

'How they were used.'

Capital investment	£721 million
UK tax	£35 million
Additional working capital	£65 million
Dividends	£78 million
Repayment of borrowings	£193 million
Total	£1092 million

Represent these data diagrammatically.

*(A)

4 Five companies form a group. The sales of each company during the year ending 5th April, 1978, are shown in the table below

Company	A	B	C	D	E
Sales (in £1 000s)	55	130	20	35	60

Draw a pie chart of radius 5 cm to illustrate this information. For the year ending 5th April, 1979, the total sales of the group increased by 20% and this growth was maintained for the year ending 5th April, 1980. If pie charts were drawn to compare the total sales for each of these years with the total sales for the year ending 5th April, 1978, what would be the radius of each of these pie charts? If the sales of company *E* for the year ending 5th April, 1980, were again £60 000, what would be the angle of the sector representing them? *(C)

5 Answer any TWO of the following:
(a) Explain the meaning of 'correlation'. Referring to your projects, discuss briefly two correlation coefficients which you have studied, and compare their uses.
(b) Discuss three methods of representing data diagrammatically, explaining the advantages and disadvantages of each method.
(c) You wish to study the prevalence of five different grass types on a very large grassy hillside. Explain the principles which would guide you in collecting random samples and, assuming that a table of random numbers is available, how this table could be used for the collection. *(L)*

Definition

When repeated observations are made on a variable the result is a **frequency distribution**. The variable may be discrete or continuous.

A frequency distribution is recorded in a **frequency table**. The values of the variable are often grouped into classes. The frequency table gives the possible values or **class intervals** of the variable and the corresponding frequencies.

ℹ The marks of 40 students in a test (discrete data)

Mark	–9	10–19	20–29	30–39	40–49
Frequency	4	8	17	9	2

Estimated U.K. population in 1930 (continuous data)

Age (years)	0–4	5–14	15–29	30–59	60–
Number (millions)	3.5	7.6	11.9	17.7	5.3

Class limits and class boundaries

For a **grouped frequency distribution** (continuous or discrete) the **lower** and **upper class limits** are the extreme values of each class in the frequency table.

Open classes have no well-defined class limits. They may be closed at arbitrary but convenient points.

For a continuous frequency distribution, the **lower class boundary** (l.c.b.) and **upper class boundary** (u.c.b.) are the smallest and largest values respectively that an item in the class can (theoretically) have.

For a grouped discrete frequency distribution the class boundaries are adjusted because of the continuous nature of the variable scale. They may be found using: $\frac{1}{2}$ (u.c. limit of one class + l.c. limit of next class).

The **class width** or **class-interval** is the difference between the two class boundaries, i.e. (u.c.b.–l.c.b.).

The **class mid-point** or **mid-value** is half-way between the two class boundaries, i.e. $\frac{1}{2}$(l.c.b + u.c.b). It is often used to represent the class-interval.

ℹ In the discrete data above, the open class '–9' may be closed at 0 (the lowest possible mark). It becomes '0–9' with lower and upper class limits of 0 and 9.

In the continuous data above, the open class '60–' may be closed at 109 (a possible maximum age!)

ℹ For the class '15–29' in the continuous data above: If the ages were measured 'to the nearest year', then the l.c.b. is 14.5 and the u.c.b. is 29.5, the class width is $(29.5 - 14.5) = 15$, the class mid-value is $\frac{1}{2}(14.5 + 29.5) = 22$.

Note: If the ages were measured as 'age in years last birthday', then the real class interval would be '15– less than 30', with class width 15.

For the class '10–19' in the discrete data above: l.c.b. is $\frac{1}{2}(9 + 10) = 9.5$ and u.c.b. is $\frac{1}{2}(19 + 20) = 19.5$, the class width is $(19.5 - 9.5) = 10$, the class mid-value is $\frac{1}{2}(9.5 + 19.5) = 14.5$.

Histogram

A **histogram** illustrates a frequency distribution. It consists of rectangles drawn on a continuous base. The area of each rectangle is proportional to the frequency of the class it represents. The rectangles do not have to be of equal width but their bases must be proportional to the class widths, i.e. the extremes of the base of each rectangle are at the l.c.b. and u.c.b. of the class it represents.

The ratio of the heights of each rectangle may be found

using $\frac{\text{class frequency(cf)}}{\text{class width(cw)}}$

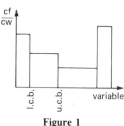

Figure 1

for each class. A suitable scale is selected for the vertical axis which should be labelled as shown.

Note: Since the base is continuous, there must be no space between rectangles. The u.c.b. of one class must coincide with the l.c.b. of the next.

ℹ *Draw a histogram for the continuous data above given that the ages are 'to the nearest year'.*

class widths:	4.5	10	15	30	50
$\frac{\text{class frequency}}{\text{class width}}$ (to 2 d.p.)	0.78	0.76	0.79	0.59	0.11

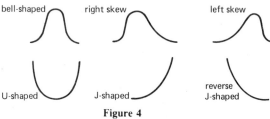

Figure 2

Frequency polygon and frequency curve

A **frequency polygon** is a line graph which may be drawn by joining the mid-points of the tops of the rectangles which would form a histogram.

It is extended to the next lower and higher classes, these having zero frequency.

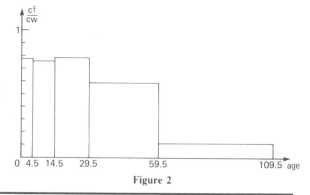

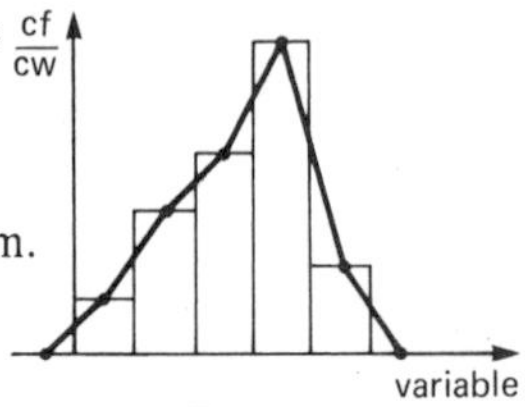

Figure 3

A smooth **frequency curve** may be drawn through the mid-points if the number of classes is large.

ℹ Some commonly occurring frequency curves

bell-shaped   right skew   left skew

U-shaped   J-shaped   reverse J-shaped

Figure 4

Frequency Distributions
Worked example, Guided example and Exam questions

WE *In 1959 the age distribution of the population of the United Kingdom was as follows:*

Age (years)	Number (in millions)
0–	15·6
20–	6·6
30–	7·4
40–	6·9
50–	6·8
60–	7·7
80–100	1·0

Draw a histogram for these data.

Let the linear scale for the age axis be 1 cm to 10 years. The calculation of the histogram column heights is set out in the table below.

Class	Class width	Freq.	Histogram col.width (cm)	$\frac{\text{Class freq.}}{\text{Class width}}$	Histogram col. height (cm)
0–	20	15·6	2	0·78	7·8
20–	10	6·6	1	0·66	6·6
30–	10	7·4	1	0·74	7·4
40–	10	6·9	1	0·69	6·9
50–	10	6·8	1	0·68	6·8
60–	20	7·7	2	0·385	3·85
80–100	20	1·0	2	0·05	0·5

The appropriate histogram is shown below.

Note: the column class frequency/class width gives the ratio of the column heights in the histogram, by taking the scale as 10 cm to one unit on the corresponding axis gives a set of column heights of reasonable dimensions.

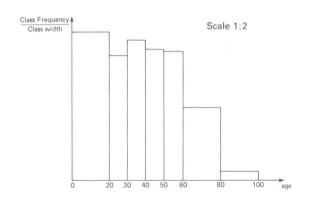

GE *Draw a frequency polygon for the following frequency distribution which gives the distribution of life of 300 electric light bulbs.*

Life (hours)	Frequency
950–	20
1200–	28
1300–	44
1400–	27
1450–	29
1500–	29
1550–	28
1600–	46
1700–	28
1800–2100	21

Since a frequency polygon is obtained by erecting ordinates at the centre of each class interval, the calculation procedure is almost identical with that for a histogram. Appropriate table headings together with the first row of the table completed are shown below.

Class	Class width	Class mid-point	Freq.	Freq. poly interval width (cm)	$\frac{\text{Class freq.}}{\text{Class width}}$	Freq. poly ordinate height (cm)
950–	250	1075	20	2·5	0·08	0·8

The scales used are 1 cm to 100 hours for the horizontal (life) axis and 10 cm to 1 unit for the vertical (class freq./class width) axis, giving rise to the frequency polygon shown below.

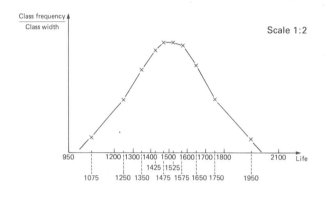

EX

1 The number of passengers on a certain regular weekday train service on each of 50 occasions was

165	141	163	153	130	158	119	187	185	209
177	147	166	154	159	178	187	139	180	143
160	185	153	168	189	173	127	179	163	182
171	146	174	149	126	156	155	174	154	150
210	162	138	117	198	164	125	142	182	218

Choose suitable class intervals and reduce these data to a grouped frequency table. Plot the corresponding frequency polygon on squared paper using suitable scales.

(A)

2 Eighty numbers are recorded below. Construct a grouped frequency table using class intervals: 0–, 0·2–, 0·4–, 0·6–, 0·8–, 0·9–1·000. Draw an accurate histogram, using suitable scales, which should be clearly indicated, to represent this table.

0·841	1·000	0·837	0·404	0·159	0·670	0·966	0·951
0·909	0·537	0·009	0·551	0·917	0·987	0·739	0·254
0·141	0·420	0·846	1·000	0·832	0·396	0·167	0·677
0·757	0·991	0·906	0·529	0·018	0·559	0·920	0·985
0·959	0·650	0·132	0·428	0·851	1·000	0·827	0·388
0·279	0·288	0·763	0·992	0·902	0·522	0·027	0·566
0·657	0·961	0·956	0·644	0·124	0·436	0·856	1·000
0·989	0·751	0·271	0·296	0·768	0·993	0·898	0·514
0·412	0·150	0·664	0·964	0·954	0·637	0·115	0·444
0·544	0·913	0·988	0·745	0·262	0·306	0·774	0·994

(A)

68 Mode and Means

Mode, Arithmetic mean, Weighted mean, Geometric mean, Harmonic mean.

Mode

The **mode** of a set of values is that value which occurs most frequently.

The **modal class** of a grouped frequency distribution is the class with the greatest frequency.

The value of the mode within the modal class may be estimated from the histogram of the distribution as shown in the diagram.

To calculate the value of this mode use:

$$\text{mode} = L + \left(\frac{\Delta_1}{\Delta_1 + \Delta_2}\right) c$$

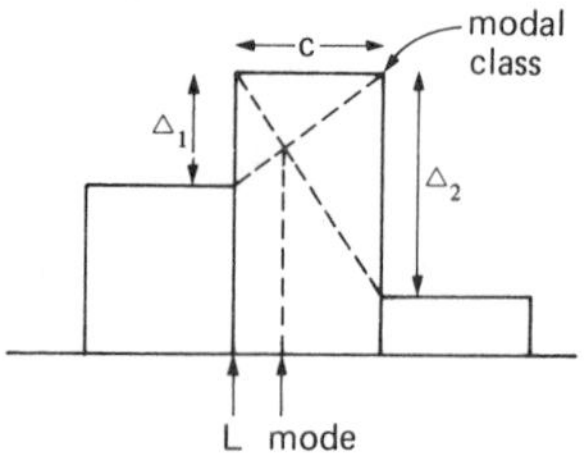

Figure 1

where L = lower boundary of modal class
Δ_1 = modal class $f-f$ of next lower class
Δ_2 = modal class $f-f$ of next higher class
c = width of modal class

ℹ The mode of 2, 3, 3, 1, 3, 2, 4, 5, 6, 3, 2, 4, 4, 3 is 3 since the value 3 has the largest frequency.

ℹ *Calculate the mode of this distribution*:

Class	10–	15–	20–	25–	30–	35–40
f	2	12	27	41	30	7

For this distribution use: $\text{mode} = L + \left(\dfrac{\Delta_1}{\Delta_1 + \Delta_2}\right) c$

The modal class is '25–'.
So $L = 25$
$\Delta_1 = 41 - 27 = 14$
$\Delta_2 = 41 - 30 = 11$
$c = 5$

$\therefore \text{mode} = 25 + \left(\dfrac{14}{14 + 11}\right) 5 = 27.8$

Arithmetic mean

The **arithmetic mean** of the n values $x_1, x_2, \ldots, x_n$ is

$$\bar{x} = \frac{(x_1 + x_2 + \ldots + x_n)}{n} = \frac{\Sigma x}{n}$$

For n values with respective frequencies $f_1, f_2, \ldots f_n$,

$$\bar{x} = \frac{(f_1 x_1 + f_2 x_2 + \ldots + f_n x_n)}{(f_1 + f_2 + \ldots + f_n)} = \frac{\Sigma fx}{\Sigma f}$$

If a is a **working mean** and $d_r = a - x_r$, then

$$\bar{x} = a + \frac{\Sigma fd}{\Sigma f}$$

For a grouped frequency distribution (continuous or discrete), the mean is given by:

$$\bar{x} = \frac{\Sigma fx}{\Sigma f}$$

where the class mid-value represents each class.

If a is a working mean and either the class intervals are of equal width c or the deviations d_r have a common factor c such that $d_r = cu_r$, then

$$\bar{x} = a + c\frac{\Sigma fu}{\Sigma f}$$

ℹ *Calculate the arithmetic mean of the discrete data given in this frequency distribution.*

x	1	2	3	4	5	6
f	1	3	5	3	1	1

Tabulating with a working mean 3 gives

x	f	d	fd
1	1	−2	−2
2	3	−1	−3
3 = a	5	0	0
4	3	1 ·	3
5	1	2	2
6	1	3	3
	$\Sigma f = 14$		$\Sigma fd = 3$

$\therefore$ arithmetic mean $= 3 + \dfrac{3}{14} = 3.21$ (2d.p.)

Weighted mean

If the numbers $x_1, x_2, \ldots, x_n$ are given the weights $w_1, w_2, \ldots, w_n$, then **the weighted mean** is

$$\bar{x}_w = \frac{(w_1 x_1 + w_2 x_2 + \ldots + w_n x_n)}{(w_1 + w_2 + \ldots + w_n)} = \frac{\Sigma wx}{\Sigma w}$$

ℹ A pupil's marks in four tests are 38%, 67%, 43%, 72%.
The weights for the tests are 1, 2, 2, 3 respectively.

$$\bar{x}_w = \frac{38 \times 1 + 67 \times 2 + 43 \times 2 + 72 \times 3}{1 + 2 + 2 + 3} = 59.25\%$$

Geometric mean

The **geometric mean** of the n values $x_1, x_2, \ldots, x_n$ is

$$\text{G.M.} = \sqrt[n]{(x_1 \times x_2 \times \ldots \times x_n)}$$

For n values with respective frequencies $f_1, f_2, \ldots, f_n$,

$$\text{G.M.} = \sqrt[\Sigma f]{(x_1^{f_1} \times x_2^{f_2} \times \ldots \times x_n^{f_n})}$$

ℹ The geometric mean of this distribution is:

x	1	2	3	4	5	6
f	2	3	5	3	4	2

$$\text{G.M.} = \sqrt[19]{(1^2 \times 2^2 \times 3^5 \times 4^3 \times 5^4 \times 6^2)}$$
$$= \sqrt[19]{(2.80 \times 10^9)} = 3.14$$

Harmonic mean

The **harmonic mean** of the n values $x_1, x_2, \ldots, x_n$ is $n/\Sigma(1/x)$

For n values with respective frequencies $f_1, f_2, \ldots, f_n$, harmonic mean is $n/\Sigma(f/x)$

ℹ The harmonic mean of the 5 values 2, 4, 5, 6, 9 is:

$$\text{H.M.} = \frac{5}{\frac{1}{2} + \frac{1}{4} + \frac{1}{5} + \frac{1}{6} + \frac{1}{9}} = 4.07$$

Mode and Means
Worked examples, Exercises and Exam questions

 (a) *(i)* *Obtain the arithmetic mean of the numbers*
1, 2, 3, 4, 5, 6, 7, 8.
(ii) *Hence show that the geometric mean of the numbers*
$9, 9^2, 9^3, 9^4, 9^5, 9^6, 9^7, 9^8$
is 3^9.

(b) *Given that $p > q > 0$, use the expansion of $(\sqrt{p} - \sqrt{q})^2$ to show that the arithmetic mean of p and q is greater than their geometric mean.*

(a) (i) Let the arithmetic mean be $\bar{x}$,

$$\bar{x} = \frac{1+2+3+4+5+6+7+8}{8} = \frac{36}{8} = \frac{9}{2}$$

(ii) Let the geometric mean be g

$$g = \sqrt[8]{9^1 \times 9^2 \times 9^3 \times 9^4 \times 9^5 \times 9^6 \times 9^7 \times 9^8}$$

$$= 9^{\frac{1+2+3+4+5+6+7+8}{8}}$$

$$= 9^{\bar{x}} = 9^{\frac{9}{2}} = (3^2)^{\frac{9}{2}} = 3^9$$

(b) Since $p > q$, $\sqrt{p} > \sqrt{q}$

$\therefore \sqrt{p} - \sqrt{q} > 0$

and $(\sqrt{p} - \sqrt{q})^2 > 0$

$\therefore (\sqrt{p})^2 + (\sqrt{q})^2 - 2\sqrt{p}\sqrt{q} > 0$

$$\frac{p+q}{2} > \sqrt{pq}$$

But $\frac{p+q}{2} = AM_{pq}$ and $\sqrt{pq} = GM_{pq}$

$\therefore AM_{pq} > GM_{pq}$

 The following distribution gives the estimated total population of the UK for the year 1977, in millions. Find the mean age.

Age	0–	10–	20–	30–	40–	50–	60–	70–	80–
Frequency	10·1	9·3	7·9	7·6	6·7	6·7	6·0	3·8	1·2

For the following tabular calculation note the points below:
(1) the last class has been closed at 90 years.
(2) the working mean a is 45 years.
(3) d is the deviation of the mid-class value from a.
(4) $c \, (= 10)$ is the scale factor for the mid-class deviations.

Class	Class mid-value	Frequency f	Deviation from a, d	$u = \dfrac{d}{c}$	fu
0–	5	10·1	−40	−4	−40·4
10–	15	9·3	−30	−3	−27·9
20–	25	7·9	−20	−2	−15·8
30–	35	7·6	−10	−1	−7·6
40–	45 = a	6·7	0	0	0
50–	55	6·7	10	1	6·7
60–	65	6·0	20	2	12·0
70–	75	3·8	30	3	11·4
80–	85	1·2	40	4	4·8
90					

$$\Sigma f = 59 \cdot 3 \qquad\qquad \Sigma fu = -91 \cdot 7 + 34 \cdot 9$$
$$= -56 \cdot 8$$

$$\bar{x} = a + \frac{c\Sigma fu}{\Sigma f} = 45 - \frac{10 \times 56 \cdot 8}{59 \cdot 3} = 35 \cdot 42 \text{ years}$$

---

 1 The wing spans of 1000 specimens of a type of insect were measured accurately and the results are summarised in the table below.

Wing span (cm)	Frequency
0–	0
1–	5
2–	20
3–	90
4–	215
5–	265
6–	240
7–	120
8–	35
9–	10
10–	0

Calculate the mean wing span. *

2 The percentage marks of 100 candidates in an examination are given in the table below.

Mark per cent	Frequency
0-19	3
20-29	4
30-39	9
40-49	27
50-59	22
60-69	18
70-79	8
80-99	6

Calculate the mean percentage mark. *

3 The atmospheric pressure at a weather station was measured once each day for a year. The results are recorded below.

Pressure (millibars)	Number of days
920-	3
980-	14
990-	44
1000-	77
1010-	150
1020-	63
1030-1040	14

Find the mean pressure. *

4 (a) Calculate: (i) the arithmetic mean; (ii) the geometric mean of the numbers 1, 2, 3, 4, 5, 6.
(b) Show, by means of clearly labelled sketches, the shape of the frequency curve of a distribution when it is: (i) skew with the mode less than the mean; (ii) bi-modal. *(A)

5 (a) For a set of positive real numbers $x_1, x_2, \ldots, x_n$, give the formula for: (i) their arithmetic mean; (ii) their geometric mean. Show that the arithmetic mean of two different positive real numbers x_1 and x_2 is always greater than their geometric mean.
(b) Define the mode of frequency distribution and give a real-life example of a multimodal distribution. *(A)

6 The arithmetic mean of the positive numbers a_1, a_2, a_3, is $\bar{a}$ and their geometric mean is A. Similarly the arithmetic mean of the positive numbers b_1, b_2, b_3, is $\bar{b}$ and their geometric mean is B. Find in terms of $\bar{a}$ and $\bar{b}$ only the arithmetic mean of:
(i) $(a_1 + b_1), (a_2 + b_2), (a_3 + b_3)$;
(ii) $a_1, b_1, a_2, b_2, a_3, b_3$
(iii) $(100 + 10a_1 + b_1), (100 + 10a_2 + b_2), (100 + 10a_3 + b_3)$.
Find in terms of A and B only the geometric mean of:
(iv) $a_1 b_1, a_2 b_2, a_3 b_3$;
(v) $a_1^2 b_1^3, a_2^2 b_2^3, a_3^2 b_3^3$. *(A)

7 (a) (i) Obtain the arithmetic mean of the numbers:
1, 2, 3, 4, 5, 6, 7, 8, 9.
(ii) Hence, or otherwise, show that the geometric mean of the numbers: $4, 4^2, 4^3, 4^4, 4^5, 4^6, 4^7, 4^8, 4^9$, is 2^{10}.
(b) Given that $a > b > 0$, use the expansion of $(\sqrt{a} - \sqrt{b})^2$ to show that the arithmetic mean of a and b is greater than their geometric mean. *(A)

69 Median and Quantiles

Definitions, Median for a discrete variable, Median and quantiles for grouped data.

Definitions

The **median** is that value of the variable which divides the distribution into two equal parts with equal frequencies.

Associated with the median are the quartiles, deciles and percentiles. These are called the **quantiles**.

The **quartiles** divide the distribution into four equal parts. The three corresponding values of the variable being denoted by Q_1, Q_2 and Q_3.

The **deciles**, $D_1, D_2, \ldots, D_9$, divide the distribution into ten equal parts.

The **percentiles**, $P_1, P_2, \ldots, P_{99}$, divide the distribution into one hundred equal parts.

If M is the median, $M = Q_2 = D_5 = P_{50}$.

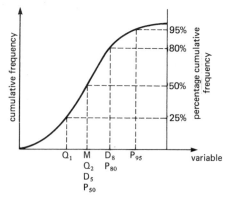

Figure 1

Median for a discrete variable

To find the median for a set of discrete variables
(a) Arrange the distribution in order of magnitude:
$$x_1, x_2, x_3, \ldots, x_n$$
(b) If n is odd, the median is the middle item, i.e. the $\frac{1}{2}(n+1)$th item.
(c) If n is even, the median is the arithmetic mean of the middle two items, i.e. the $\frac{1}{2}n$th and $\frac{1}{2}(n+2)$th items.

ⓘ *Find the median of (a) 3, 7, 4, 1, 2, 3, 6, 9, 8, (b) 9, 3, 1, 5, 8, 5, 7, 4, 8, 4, 7, 6*

(a) In order: 1, 2, 3, 3, 4, 6, 7, 8, 9
The median is 4.

(b) In order: 1, 3, 4, 4, 5, 5, 6, 7, 7, 8, 8, 9.
The median is $\frac{1}{2}(5+6) = 5.5$.

Median and quantiles for grouped data

For grouped data of a discrete or continuous variable the following **graphical method** can be used to **estimate the median and quantiles**.
(a) Form the **cumulative frequency distribution.** The cumulative frequency for any class is the sum of frequencies of that class and lower classes.
(b) Plot each cumulative frequency against the upper limit of the corresponding class interval.
(c) Join these points with either straight line segments to form the **cumulative frequency polygon** or a smooth curve to form the **cumulative frequency curve** (or ogive).
(d) The middle number of the distribution is located on the cumulative frequency axis and the corresponding value of the variable is the median. The quantiles can be found in a similar way.

A more accurate value of the median may be found using the formula:

$$\text{median} = L + \left(\frac{\frac{1}{2}N - (\Sigma f)_L}{f_{\text{median}}}\right)c$$

where
L = lower boundary of median class
N = total frequency
$(\Sigma f)_L$ = sum of frequencies below median class
f_{median} = frequency of median class
c = width of median class

ⓘ *The table below is the frequency distribution of marks obtained in a test by 200 students.*

mark	10–	20–	30–	40–	50–	60–	70–	80–90
f	18	34	58	42	24	10	6	8

Draw a cumulative frequency polygon to illustrate the data. Estimate the median from the graph and by calculation.

How many students would fail if the pass mark is 40? If the top 10% of students are to be given a grade I, what is the lowest mark which will achieve this?

The cumulative frequency distribution is:

mark	10–	20–	30–	40–	50–	60–	70–	80–90
cf	18	52	110	152	176	186	192	200

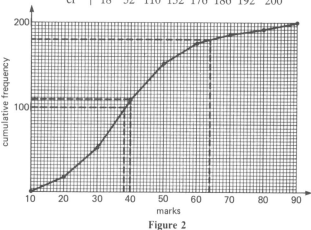

Figure 2

From the graph: $M \approx 38$ marks

By calculation: $M = \dfrac{30 + (\frac{1}{2}(200) - 52)10}{58} = 38.3$

From the graph : 110 students fail,
64 is the lowest mark for grade I.

Median and Quantiles
Worked example and Exam questions

WE *The lives of 80 circuit components were recorded in hours to the nearest hour and grouped as follows:*

Life (hours)	Frequency
660-669	3
670-679	7
680-689	10
690-699	15
700-709	24
710-719	12
720-729	5
730-739	4

(i) *State the limits between which the actual life of each component in the first group must lie.*

(ii) *Construct the cumulative frequency and draw the cumulative frequency curve.*

(iii) *Use the curve to estimate (a) the median, (b) the 90th percentile.*

(i) The true limits for the first class are $659 \cdot 5 - 669 \cdot 5$, since the life is recorded to the nearest hour.

(ii)

Life	660– 669	670– 679	680– 689	690– 699	700– 709	710– 719	720– 729	730– 739
f	3	7	10	15	24	12	5	4
cf	3	10	20	35	59	71	76	80

The appropriate cumulative frequency curve is shown below.

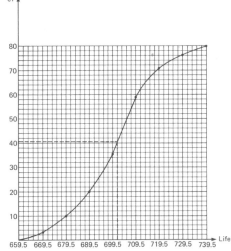

(iii) Since there are 80 observations the median corresponds to the $40 \cdot 5$th observation; from the graph the median value is estimated to be $701 \cdot 5$ hours.
The 90th percentile corresponds to the $\frac{90}{100} \times 80\text{th} = 72\text{nd}$ observation; from the graph this value is estimated to be $720 \cdot 5$ hours.

EX 1 The table shows the marks, collected into groups, of 400 candidates in an examination. The maximum mark was 99.

Marks	No. of candidates	
0–9	10	Compile the cumulative frequency table and draw the cumulative frequency curve. Use your curve to estimate: (i) the median; (ii) the 20th percentile. If the minimum mark for Grade A was fixed at 74, estimate from your curve the percentage of candidates obtaining Grade A. *(C)*
10–19	26	
20–29	42	
30–39	66	
40–49	83	
50–59	71	
60–69	52	
70–79	30	
80–89	14	
90–99	6	

2 The table below shows the average productivity of coal miners in Great Britain over a 12-month period, measured in tonnes per manshift.

Productivity	Miners (thousands)
0–1	14
1–1·5	31
1·5–2·0	51
2·0–2·5	62
2·5–3·0	44
3·0–3·5	11
3·5–5·0	20

(a) Illustrate these data by drawing a histogram;
(b) Find, in thousands, the total number of miners;
(c) Draw a cumulative frequency polygon on a separate sheet;
(d) Estimate, from your graph, the median productivity;
(e) Estimate the total coal produced per shift
(O & C)

3 On a particular date (30 June 1968), the age distribution of the population of the United Kingdom is given to have been as follows:

Age range	0–9	10–14	15–19	20–59	60–64	65–104
Millions of people	7·7	4·6	4·4	28·4	2·9	8·0

The categories are such that, for example, 10–14 consists of all people who have had their 10th birthday, but not their 15th birthday, and thus spans a range of 5 calendar years.

(i) Display the data on a histogram.
(ii) Estimate the median age, giving your answer to the nearest year.
(iii) Calculate an estimate of the mean age of the population of the United Kingdom on that date, showing your working clearly. *(C)*

4 An inspection of 34 aircraft assemblies revealed a number of missing rivets as shown in the following table.

Number of rivets missing	Frequency
0–2	4
3–5	9
6–8	11
9–11	6
12–14	2
15–17	1
18–20	0
21–23	1

Draw a cumulative frequency curve. Use this curve to estimate the median and the quartiles of the distribution.
(O & C)

5 30 specimens of sheet steel are tested for tensile strength, measured in $kN\ m^{-2}$. The table gives the distribution of the measurements.

Tensile strength	Number of specimens
405–415	4
415–425	3
425–435	6
435–445	10
445–455	5
455–465	2

Draw a cumulative diagram of this distribution. Estimate the median and the 10th and 90th percentiles. *(O & C)*

70 Measures of Dispersion
Range, Semi-interquartile range, Mean deviation, Variance and standard deviation.

Range

The **range** is the difference between the largest and smallest items of the distribution,

i.e.   largest item − smallest item.

ℹ️ *Calculate the range for this distribution.*

x	1	2	3	4	5	6
f	1	3	9	2	1	1

The range is $(6-1)=5$.

Semi-interquartile range

The **semi-interquartile range** $= \frac{1}{2}(Q_3-Q_1)$ where Q_1 and Q_3 are the first and third quartiles (see Median and Quantiles p. 138).

ℹ️ *Calculate the semi-interquartile range for the distribution with $Q_1=29.41$ and $Q_3=49.52$.*

Semi-interquartile range $= \frac{1}{2}(49.52-29.41)=10.05$.

Mean deviation

The **mean deviation** can be measured from the arithmetic mean, median, mode or any other specified value.
It is usually measured from the arithmetic mean.

The mean deviation from the mean of the n values $x_1, x_2, \ldots, x_n$ is:

$$\text{M.D.} = \frac{\Sigma|x-\bar{x}|}{n}$$

The mean deviation from the mean of a frequency distribution is:

$$\text{M.D.} = \frac{\Sigma f|x-\bar{x}|}{\Sigma f}$$

This formula is also used for continuous and grouped discrete frequency distributions but the class mid-value is used to represent each class.

ℹ️ *Calculate the mean deviation from the mean of this distribution.*

x	10–	15–	20–	25–	30–	35–40
f	2	12	27	41	30	7

$\bar{x}=\dfrac{\Sigma fx}{\Sigma f}=\dfrac{3207.5}{119}=26.95$ (to 2 d.p.)

x	mid-value	f	$|x-\bar{x}|$	$f|x-\bar{x}|$
10–	12.5	2	14.45	28.90
15–	17.5	12	9.45	113.40
20–	22.5	27	4.45	120.15
25–	27.5	41	0.55	22.55
30–	32.5	30	5.55	166.50
35–40	37.5	7	10.55	73.85
		$\Sigma f=119$		$\Sigma f|x-\bar{x}|=525.35$

mean deviation $= \dfrac{525.35}{119}=4.41$ (2 d.p.)

Variance and standard deviation

Variance is denoted by σ^2 for a population and by s^2 for a sample.

The variance of the n values $x_1, x_2, \ldots, x_n$ is

$$\sigma^2=\frac{\Sigma(x-\bar{x})^2}{n} \quad \text{(definition)}$$

$$=\frac{\Sigma x^2}{n}-\bar{x}^2 \quad \begin{array}{l}\text{(computational form–}\\\text{easier for calculation)}\end{array}$$

The variance of a frequency distribution is

$$\sigma^2=\frac{\Sigma f(x-\bar{x})^2}{\Sigma f} \quad \text{(definition)}$$

$$=\frac{\Sigma fx^2}{\Sigma f}-\bar{x}^2 \quad \text{(computational form)}$$

These formulae are also used for continuous and grouped discrete frequency distributions but the class mid-value is used to represent each class.

If a is a working mean and either the class intervals are of equal width c or the deviations d_r from a have a common factor c such that $d_r=cu_r$, then

$$\sigma^2=\frac{c^2\Sigma fu^2}{\Sigma f}-b^2 \text{ where } b=\bar{x}-a.$$

Standard deviation is simply the positive square root of the variance, i.e. $\sqrt{\text{variance}}$.
It is denoted by σ for a population
and by s for a sample.

ℹ️ *Calculate the variance and standard deviation for this distribution of the test results of 200 students.*

mark	10–	20–	30–	40–	50–	60–	70–	80–90
f	18	34	58	42	24	10	6	8

class	mid-value x	f	fx	d	u $(c=10)$	u^2	fu^2
10–	15	18	270	−30	−3	9	162
20–	25	34	850	−20	−2	4	136
30–	35	58	2030	−10	−1	1	58
40–	45 $=a$	42	1890	0	0	0	0
50–	55	24	1320	10	1	1	24
60–	65	10	650	20	2	4	40
70–	75	6	450	30	3	9	54
80–90	85	8	680	40	4	16	128
			$\Sigma fx=8140$				$\Sigma fu^2=602$

$\bar{x}=\dfrac{\Sigma fx}{\Sigma f}=\dfrac{8140}{200}=40.7$

Let $a=45$, so $b=\bar{x}-a=40.7-45=-4.3$

$\text{Variance}=\dfrac{c^2\Sigma fu^2}{\Sigma f}-b^2=\dfrac{100\times602}{200}-(-4.3)^2$

$\qquad\qquad = 301-18.49$

$\qquad\qquad = 282.51$

Standard deviation $=\sqrt{282.51}=16.81$

Measures of Dispersion
Worked example and Exam questions

The table below gives the frequency distribution of the speeds of 95 cars passing a check point.

Speed (km/h)	30 –	40–	50–	55–	60–	70–
Frequency	0	20	30	35	10	0

Calculate the mean speed and the standard deviation. It is later found that the recording instrument gave a reading of 5 km/h below the true speed. Find the new mean and standard deviation.

Close the last class at 80.
Let the working mean be $a = 52\cdot5$ km/h, d the deviation of the mid-class value from a and c ($= 2\cdot5$) the scale factor for the deviation of the mid-class values.

Tabulating the results:

Class	Class mid-value	Frequency f	Deviation from a, d	$u = \dfrac{d}{c}$	fu	fu^2
30–	35	0	$-17\cdot5$	-7	0	0
40–	45	20	$-7\cdot5$	-3	-60	180
50–	$52\cdot5 = a$	30	0	0	0	0
55–	$57\cdot5$	35	5	2	70	140
60–	65	10	$12\cdot5$	5	50	250
70–	75	0	$22\cdot5$	9	0	0
80						
		$\Sigma f = 95$			$\Sigma fu = 60$	$\Sigma fu^2 = 570$

Let the mean speed be $\bar{v}$, then

$$\bar{v} = a + \frac{c\Sigma fu}{\Sigma f} = 52\cdot5 + \frac{2\cdot5 \times 60}{95} = 54\cdot1 \text{ km/h}$$

Let s_v be the standard deviation of v, then

$$s_v^2 = \frac{c^2\Sigma fu^2}{\Sigma f} - b^2, \text{ where } b = \bar{v} - a = 54\cdot1 - 52\cdot5 = 1\cdot6$$

So $s_v^2 = \dfrac{2\cdot5^2 \times 570}{95} - 1\cdot6^2 = 34\cdot94$

$\therefore s_v = 5\cdot9$ km/h
Let $V = v + 5$ (i.e. V is the true speed), then $\bar{V} = \bar{v} + 5$
$\qquad\qquad\qquad\qquad\qquad\qquad\qquad = 59\cdot1$ km/h
and $s_V = s_v = 5\cdot9$ km/h

---

1 The figures below are the yields to the nearest kilogram of a certain root crop obtained from 32 plots of equal size.

21	23	19	22	21	23	20	22
17	26	20	19	18	20	20	24
25	21	22	20	24	18	22	20
19	23	21	17	25	24	26	18

Draw up a table showing the frequencies of the various yields. Using your table and an assumed mean, estimate the mean yield and the standard deviation. *(C)*

2 A zoologist weighs 200 eggs and records the weights in the following grouped frequency table.

Weight (g)	24–	30–	36–	42–	48– 54
No. of eggs	22	45	72	43	18

Find the mean and standard deviation, correct to two decimal places.
He later discovers that his scales were incorrectly set and that each egg was underweighed by 4 g. Determine the corrected mean and standard deviation. *(S)*

3 The sum of 20 numbers is 320 and the sum of their squares is 5840. Calculate the mean of the 20 numbers and the standard deviation:
(i) Another number is added to these 20 so that the mean is unchanged. Show that the standard deviation is decreased.
(ii) Another set of 10 numbers is such that their sum is 130 and the sum of their squares is 2380. This set is combined with the original 20 numbers. Calculate the mean and standard deviation of all 30 numbers.
(C)

4 Show, from the basic definition, why the standard deviation
of a set of observations $x_1, x_2, x_3, \ldots\ldots, x_i, \ldots\ldots, x_n$ with
mean $\bar{x}$ may be found by evaluating $\sqrt{\dfrac{\Sigma x_i^2}{n} - \bar{x}^2}$
(a) Find, showing your working clearly and not using any pre-programmed function on your calculator, the standard deviation of the following frequency distribution:

x	25	26	27	28
f	2	0	15	1

(b) The average height of 20 boys is 160 cm, with a standard deviation of 4 cm. The average height of 30 girls is 155 cm, with a standard deviation of $3\cdot5$ cm. Find the standard deviation of the whole group of 50 children.
(S)

5 Suppose that the values of a random sample taken from some population are $x_1, x_2, \ldots\ldots, x_n$. Prove the formula

$$\sum_{i=1}^{n} (x_i - \bar{x})^2 = \sum_{i=1}^{n} x_i^2 - n\bar{x}^2.$$

Parplan Opinion Polls Ltd. conducted a nationwide survey into the attitudes of teenage girls. One of the questions asked was "What is the ideal age for a girl to have her first baby?" In reply, the sample of 165 girls from the Northern zone gave a mean of $23\cdot4$ years and a standard deviation of $1\cdot6$ years. Subsequently, the overall sample of 384 girls (Northern plus Southern zones) gave a mean of $24\cdot8$ years and a standard deviation of $2\cdot2$ years. Assuming that no girl was consulted twice, calculate the mean and standard deviation for the 219 girls from the Southern zone. *(A)*

6 The following table summarises the masses, measured to the nearest gram, of 200 animals of the same species.

Mass (g)	Frequency
70–79	7
80–84	30
85–89	66
90–94	57
95–99	27
100–109	13

Calculate estimates of the median and upper quartile of the distribution.
Estimate the number of animals whose actual masses are less than 81 g.
Calculate estimates of the mean and the standard deviation of the distribution. *(A)*

Index numbers

A set of data may be reduced to relative values by comparing it with a fixed (base) number. These relative values are called percentage relatives or simple index numbers. If the percentage relatives refer to prices they are called price relatives.

A percentage relative can be calculated using $\dfrac{q_n}{q_0} \times 100$

where q_0 is the quantity in a base year,
and q_n is the quantity in another year.
In situations which have many contributory factors, more complicated index numbers are found by using weighted averages of the percentage relatives of the contributory factors. If the percentage relatives are $r_1, r_2, \ldots, r_n$ with respective weights $w_1, w_2, \ldots, w_n$, then the weighted index is $\dfrac{\Sigma r_n w_n}{\Sigma w_n}$.

The expenditure of a household over a three year period is shown below.

Item	Year 1	Year 2	Year 3	Weight
House	3000(100)	2400(80)	2550(85)	110
Food	2000(100)	2200(110)	2400(120)	360
Fuel	800(100)	880(110)	950(119)	90
Travel	800(100)	900(113)	900(113)	80
Others	1100(100)	1500(136)	1600(145)	360
Total	7700(100)	7880(102.3)	8400(109.1)	

The figures in brackets are percentage relatives with year 1 as base. Verify these.

e.g. for total expenditure for Year $2 = \dfrac{7880}{7700} \times 100 = 102.3$

Index numbers for the total expenditure as weighted averages of the percentage relatives are 100, 116.3, 124.5.

Crude and standardised rates

The crude death rate for a given district is the number of deaths per thousand population,

i.e. crude death rate $= \dfrac{\text{total number of deaths}}{\text{total population}} \times 1000$

Crude death rates are poor for comparing different areas because they do not take into account the age structure of the populations under consideration. To do this a standardized death rate is used.
A standardized death rate is a weighted average in which the weights are related to the age distribution of the population of the entire country.
Crude and standardized rates can also be obtained for births, marriages and unemployment figures.

Calculation of crude and standardized death rate.

Age group	Pop.	Deaths in group	% of U.K. pop. in gp. (W)	Group deaths per 1000 pop. (D)	WD
0–4	4000	52	7	13	91
5–14	5000	4	17	0.8	13.6
15–29	4200	14	26	3.3	85.8
30–59	9000	74	38	8.2	311.6
60+	2600	104	12	40	480.0
	24800	248	100		982.0

Crude death rate $= \dfrac{248}{24800} \times 1000 = 10$ deaths per 1000.

Standardized death rate $= \dfrac{982}{100} = 9.82$ deaths per 1000.

Moving averages and their use in time series

Given a set of numbers $x_1, x_2, \ldots$ the moving average of order n is given by the following set of arithmetic means:

$$\frac{x_1 + x_2 + \ldots + x_n}{n}, \frac{x_2 + x_3 + \ldots + x_{n+1}}{n}, \frac{x_3 + x_4 + \ldots + x_{n+2}}{n},$$

If the data is monthly data, then the average is known as the n month moving average, and is usually centred on the period to which it refers.
Moving averages are used in time series analysis. In this analysis it is usual to distinguish between the following types of variation:

(a) the secular or general trend with a cycle of several decades,
(b) cyclical variations, which are fluctuations in the general trend usually of a 5–10 year period. They may be removed by subtracting suitably chosen moving averages,
(c) seasonal variations which recur annually due to the season of the year and may be removed by subtracting 4-quarterly or 12-monthly moving averages,
(d) residual variations are those fluctuations which remain when the above have been taken into account.

Given the numbers 2, 0, 1, 7, 4, 2, 2, 8, find the moving averages of order 4 and 5.

First moving average of order 4 is $\dfrac{2+0+1+7}{4} = 2.5$

Second moving average of order 4 is $\dfrac{0+1+7+4}{4} = 3$

Summarising:

	2	0	1	7	4	2	2	8
Order 4		2.5	3	3.5	3.75	4		
Order 5			2.8	2.8	3.2	4.6		

Calculation of 3-monthly moving averages as used in the analysis of time series.

Month	1	2	3
Quarter 1	9	11	12
Quarter 2	11	12	14
$\dfrac{Q_2 - Q_1}{3}$	0.67	0.33	0.67

First 3-monthly moving average $\dfrac{9+11+12}{3} = 10.67$

Adding the quantities in the final row of the table gives the sequence of moving averages:
10.67, 11.34, 11.67, 12.34

Index Numbers and Moving Averages
Worked examples and Exam questions

 (a) *Using 1980 as base year, the price index of a particular commodity in 1981 is 110. Using 1981 as base year, the price index for the same commodity in 1982 is 120. Calculate the index number for 1982 taking 1980 as base year.*
(b) *Calculate a composite index number for the following data.*

Index Number	127	118	96	112
Weight	2	1	6	3

(a) With an obvious notation

$$\frac{P_{81}}{P_{80}} \times 100 = 110 \text{ and } \frac{P_{82}}{P_{81}} \times 100 = 120$$

Multiplying gives,

$$\frac{P_{81}}{P_{80}} \times \frac{P_{82}}{P_{81}} \times 100 \times 100 = 110 \times 120$$

$$\text{i.e. } \frac{P_{82}}{P_{80}} \times 100 = \frac{120 \times 120}{100} = 132$$

∴ the index number for 1982 with 1980 as base is 132.

(b) The composite index number is

$$\frac{127 \times 2 + 118 \times 1 + 96 \times 6 + 112 \times 3}{2 + 1 + 6 + 3} = \frac{1284}{12} = 107$$

 A petrol filling station, open seven-days a week, had the following daily sales of petrol, in gallons, during a two week period.

	Mon	Tue	Wed	Thu	Fri	Sat	Sun
Week 1	700	360	590	620	680	710	350
Week 2	520	280	460	530	510	650	290

Plot these sales.
Calculate the values of the seven-day moving averages and superimpose these on the graph.
From the graph, estimate the next value of the moving average and hence estimate the sales for Monday of Week 3.

See the graph for the plot of the sales.
To calculate the moving averages the working is set out in tabular form.

	Mon	Tue	Wed	Thu	Fri	Sat	Sun
Week 1	700	360	590	620	680	710	350
Week 2	520	280	460	530	510	650	290
W2−W1	−180	−80	−130	−90	−170	−60	−60
$\frac{W2-W1}{7}$	−25·7	−11·4	−18·6	−12·9	−24·3	−8·6	−8·6

The first seven-day moving average is

$$\frac{700 + 360 + 590 + 620 + 680 + 710 + 350}{7} = 572 \cdot 9$$

Adding the quantities in the final row of the table gives the following sequence of seven-day moving averages, 572·9, 547·2, 535·8, 517·2, 504·3, 480·0, 471·4, 462·8.
See the graph for the superimposition of the values.
The estimate from the graph of the next moving average is 445.
Note, the 'best line' through the moving averages has been estimated by eye.

$$\therefore 445 = \frac{280 + 460 + 530 + 510 + 650 + 290 + M3}{7}$$

$$\therefore M3 = 7 \times 445 - 2720$$
$$= 3115 - 2720$$
$$= 395$$

∴ the estimated sales for Monday of Week 3 are 395 gallons.

Commodity	Price (£)		
	1976	1977	1978
A	4	5	7
B	10	12	15
C	8	8	10

Use the above data to complete the following table of price relatives (1976 = 100).

Commodity	Price Relatives (1976 = 100)	
	1977	1978
A	125	
B		
C		

Using weights of 5, 3 and 2 for commodities A, B and C, respectively, show that the weighted index of price relatives (1976 = 100) for 1977 is 118·5. Obtain the weighted index of price relatives for 1978.
Using the same weights, determine the weighted index of price relatives (1977 = 100) for 1978.

(S)

2 The table below refers to the mean quarterly rainfall, in cm, for the years 1975 and 1976.

	Jan.–Mar.	Apr.–June	July–Sept.	Oct.–Dec.
1975	26	14	33	a
1976	b	18	23	28

The 4-point moving averages for the above data are 29·5, 28·25, c, 26·75, d. Calculate the values of a, b, c and d. If rainfall were subject to seasonal variation only, what would you expect of the values of the moving averages? State, and explain briefly, any other types of variation that may affect the quarterly rainfall figures.

(C)

3 Explain briefly the reasons for using moving averages. A set of 4-point moving averages is calculated for seven successive values of a certain variable. The first three values of the variable are 3, 7, 5 and the first four moving averages are 6, 8, 7, 9. Calculate the last four values of the variable. Illustrate these data by drawing graphs, on the same diagram, of the values of the variable and the values of the moving averages.

(C)

143

72 Probability

Definitions, Addition rule, Conditional probability, Multiplication rule.

Definitions

An experiment has a finite number of outcomes, called the **outcome set** S.

An **event** E of an experiment is defined to be a subset of the outcome set S.

The **complement** of E, $\overline{E}$, is the subset of S where E does not occur.

Two events of the same experiment are **mutually exclusive** if they cannot occur simultaneously.

Two events are **independent** if the occurrence of one has no effect on the occurrence of the other.

If an experiment has $n(S)$ equally likely outcomes and $n(E)$ of them are the event E, then the **theoretical probability** of event E occurring is

$$P(E) = \frac{n(E)}{n(S)}$$

Note: $0 \leqslant P(E) \leqslant 1$
$P(E) = 0$ means that E is an **impossibility**.
$P(E) = 1$ means that E is a **certainty**.

If the outcome S has only n different possible events $E_1, E_2, \ldots, E_n$, then

$$P(E_1) + P(E_2) + \ldots + P(E_n) = \sum_{i=1}^{n} P(E_i) = 1$$

and $P(\overline{E}) = 1 - P(E)$

Venn diagram showing outcome set S and its subsets E and $\overline{E}$.

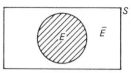

Figure 1

ℹ️ If a fair die is thrown once, then two different scores, e.g. 2 and 3, cannot occur simultaneously, so they are mutually exclusive events. If the die is thrown again, the second score is independent of the first.

ℹ️ *A fair die is thrown. List the possible outcomes. What is the probability of scoring:*
(a) a multiple of 3, (b) not a multiple of 3?

$S = \{$possible scores with a die$\} = \{1, 2, 3, 4, 5, 6\}$

(a) $E = \{$'multiple of 3' scores$\} = \{3, 6\}$

$$P(E) = \frac{n(E)}{n(S)} = \frac{2}{6} = \frac{1}{3}$$

$P(\overline{E}) = \{$'not multiple of 3' scores$\}$
$= 1 - P(E) = 1 - \frac{1}{3} = \frac{2}{3}$

Addition rule

If E_1 and E_2 are two events of the same experiment, then the **probability of E_1 or E_2 or both** occurring is given by:

Figure 2

$P(E_1 \text{ or } E_2) = P(E_1) + P(E_2) - P(E_1 \text{ and } E_2)$
or $P(E_1 \cup E_2) = P(E_1) + P(E_2) - P(E_1 \cap E_2)$

If E_1 and E_2 are mutually exclusive, the $P(E_1 \cap E_2) = 0$
and so: $P(E_1 \text{ or } E_2) = P(E_1 \cup E_2) = P(E_1) + P(E_2)$

Figure 3

What is the probability of drawing a club or an eight from a pack of cards?

$S = \{$pack of cards$\}$, $n(S) = 52$
$E_1 = \{$clubs$\}$, $n(E_1) = 13$, $P(E_1) = \dfrac{13}{52}$

$E_2 = \{$eights$\}$, $n(E_2) = 4$, $P(E_2) = \dfrac{4}{52}$

E_1 and $E_2 = \{$clubs and eights$\} = \{$eight of clubs$\}$,
$n(E_1 \text{ and } E_2) = 1$, $P(E_1 \text{ and } E_2) = \dfrac{1}{52}$

$\therefore P(E_1 \text{ or } E_2) = P(E_1) + P(E_2) = P(E_1 \text{ and } E_2)$

$$= \frac{13}{52} + \frac{4}{52} - \frac{1}{52} = \frac{16}{52} = \frac{4}{13}$$

Conditional probability

If E_1 and E_2 are two events (not necessarily from the same experiment), then the **conditional probability that E_1 will occur given that E_2 has occurred** is

$$P(E_1 | E_2) = \frac{P(E_1 \text{ and } E_2)}{P(E_2)} = \frac{n(E_1 \text{ and } E_2)}{n(E_2)}$$

If E_1 and E_2 are mutually exclusive, then $P(E_1 | E_2) = 0$.

Two events E_1 and E_2 are independent, if $P(E_1) = P(E_1 | E_2)$ and $P(E_2) = P(E_2 | E_1)$.

ℹ️ *A card is drawn from a normal pack of 52 cards. Calculate $P(club | black\ suit)$*

$$P(club | black\ suit) = \frac{n(\text{club and black suit})}{n(\text{black suit})} = \frac{13}{26} = \frac{1}{2}$$

Multiplication rule

If E_1 and E_2 are any two events, then the **probability that both E_1 and E_2 occur** is
$P(E_1 \text{ and } E_2) = P(E_1 \cap E_2) = P(E_2) \times P(E_1 | E_2)$
$= P(E_1) \times P(E_2 | E_1)$

If E_1 and E_2 are independent, then
$P(E_1 \text{ and } E_2) = P(E_1 \cap E_2) = P(E_1) \times P(E_2)$

ℹ️ *A coin is tossed and a die thrown. Find the probability that a head and a score less than 3 result.*

These are independent events.
If $E_1 = \{$head$\}$ and $E_2 = \{$score less than 3$\}$, then
$P(E_1 \text{ and } E_2) = P(E_1) \times P(E_2) = \frac{1}{2} \times \frac{2}{6} = \frac{1}{6}$

Probability
Worked examples and Exam questions

 (i) *The events A and B are such that* $P(A) = 0.43$, $P(B) = 0.48$, $P(A \cup B) = 0.78$.
Show that the events A and B are neither mutually exclusive nor independent.

(ii) *A bag contains 10 red balls, 9 blue balls and 5 white balls. Three balls are taken from the bag at random and without replacement. Find the probability that all three balls are of the same colour. Find also the probability that all three balls are of different colours.*

(i) A and B are mutually exclusive if $P(A \cup B) = P(A) + P(B)$
$P(A \cup B) = 0.78$, $P(A) + P(B) = 0.43 + 0.48 = 0.91$
$$\neq P(A \cup B)$$
$\therefore$ A and B are not mutually exclusive, in which case
$P(A \cup B) = P(A) + P(B) - P(A \cap B)$
i.e. $P(A \cap B) = P(A) + P(B) - P(A \cup B)$
$$= 0.91 - 0.78 = 0.13$$
If A and B are independent $P(A \cap B) = P(A) \times P(B)$
$P(A \cap B) = 0.13$, $P(A) \times P(B) = 0.43 \times 0.48 = 0.2064$
$$\neq P(A \cap B)$$
$\therefore$ A and B are neither mutually exclusive nor independent

(ii) With an obvious notation we have,
$10R$, $9B$, $5W$, a total of 24 balls
3 balls are drawn at random without replacement
P(3 balls of the same colour)
$= P(1R2R3R \text{ or } 1B2B3B \text{ or } 1W2W3W)$
$= P(1R2R3R) + P(1B2B3B) + P(1W2W3W)$
$= P(1R)P(2R)P(3R) + P(1B)P(2B)P(3B) + P(1W)P(2W)P(3W)$
$$= \frac{10}{24} \times \frac{9}{23} \times \frac{8}{22} + \frac{9}{24} \times \frac{8}{23} \times \frac{7}{22} + \frac{5}{24} \times \frac{4}{23} \times \frac{3}{22}$$
$$= \frac{1}{24 \times 23 \times 22} [720 + 504 + 60] = \frac{107}{1012}$$

P(3 balls all of different colour)
$= P(1R2W3B \text{ or } 1R2B3W \text{ or } 1W2R3B \text{ or } 1W2B3R \text{ or } 1B2R3W \text{ or } 1B2W3R)$
$$= \frac{10}{24} \times \frac{5}{23} \times \frac{9}{22} + \frac{10}{24} \times \frac{9}{23} \times \frac{5}{22} + \ldots$$
$$= 6 \times \frac{10}{24} \times \frac{5}{23} \times \frac{9}{22} = \frac{225}{1012}$$

WE *Mass-produced ceramic tiles are inspected for defects. The probability that a tile has air bubbles is 0.0015. If a tile has air bubbles the probability that it is also cracked is 0.55 while the probability that a tile free from air bubbles is cracked is 0.0055. What is the probability that a tile selected at random is cracked? The probability that a tile is discoloured is 0.0065. Given that discoloration is independent of the other two defects, find the probability that a tile selected at random has no defects.*

Let A be the event 'the tile has air bubbles'
Let B be the event 'the tile is cracked'
Let C be the event 'the tile is discoloured'
$P(A) = 0.0015$, $P(\bar{A}) = 0.9885$
$P(C) = P(A \text{ and } C \text{ and } \bar{A} \text{ and } C)$
$\qquad = P(A \text{ and } C) + P(\bar{A} \text{ and } C)$
$\qquad = P(A)P(C|A) + P(\bar{A})P(C|\bar{A})$
$\qquad = 0.0015 \times 0.55 + 0.9985 \times 0.0055$
$\qquad = 0.000825 + 0.0054918$
$\qquad = 0.00632$ (3 sf)
P(no defects) $= P(\bar{A} \text{ and } \bar{C} \text{ and } \bar{D})$
$\qquad = P(\bar{A}) \times P(\bar{C}) \times P(\bar{D})$
$\qquad = 0.9985 \times 0.9945 \times 0.9935$
$\qquad = 0.987$ (3 sf)

EX 1 Ten cards, bearing the letters P, R, O, P, O, R, T, I, O, N, are placed in a box. Three cards are drawn out at random without replacement. Calculate the probability that:
  (i) the first two cards bear the same letter,
  (ii) the third card bears the letter P,
  (iii) the three cards bear the letters P, O, T in that order,
  (iv) the three cards bear the letters P, O, T in any order.
  (C)

2 Four people were chosen at random from a group of 8 which comprised 4 husbands and their 4 wives. Find the probability that the sample contained:
  (a) one person from each married couple (called event A);
  (b) two males and two females (called event B).
  Find also the probability that *both* events A and B occurred.
  Deduce, or otherwise find, $Pr(B|A)$ (the probability that event B happened if it is known that event A happened). (Answers may be given as fractions in their lowest terms.)
  (O & C)

3 Four cards are to be drawn at random without replacement from a pack of ten cards numbered from 1 to 10, respectively.
  (a) Calculate the probabilities that
    (i) the largest number drawn will be 6,
    (ii) the product of the four numbers will be even,
    (iii) all four numbers drawn will be consecutive integers.
  (b) Given that at least two of the four numbers drawn were even, find the probability that every number drawn was even.
  (W)

4 The events A and B are such that $P(A) = \frac{1}{2}$, $P(A'|B) = \frac{1}{3}$, $P(A \cup B) = \frac{3}{5}$, where A' is the event 'A does not occur'. Using a Venn diagram, or otherwise, determine $P(B|A')$, $P(B \cap A)$ and $P(A|B')$. The event C is independent of A and $P(A \cap C) = \frac{1}{8}$. Determine $P(C|A')$. State, with a reason in each case, whether (i) A and B are independent, (ii) A and C are mutually exclusive.
  (C)

5 (a) A and B play a game as follows: an ordinary die is rolled and if a six is obtained then A wins and if a one is obtained then B wins. If neither a six nor a one is obtained then the die is rolled again until a decision can be made. What is the probability that A wins on (i) the first roll, (ii) the second roll, (iii) the rth roll? What is the probability that A wins?
  (b) A bag contains 4 red and 3 yellow balls and another bag contains 3 red and 4 yellow. A ball is taken from the first bag and placed in the second, the second bag shaken and a ball taken from it and placed in the first bag. If a ball is now taken from the first bag what is the probability that it is red? (You are advised to draw a tree diagram.)
  (S)

6 An unbiased die is thrown six times. Calculate the probabilities that the six scores obtained will
  (i) consist of exactly two 6's and four odd numbers,
  (ii) be 1, 2, 3, 4, 5, 6 in some order,
  (iii) have a product which is an even number,
  (iv) be such that a 6 occurs *only* on the last throw and that exactly three of the first five throws result in odd numbers.
  (J)

73 Discrete Probability Distributions
Definitions, Expectation, Variance, Two random variables.

Definitions

Suppose the outcome set S of an experiment is divided into n mutually exclusive and exhaustive events $E_1, E_2, \ldots, E_n$. A variable, X, which can assume exactly n numerical values each of which corresponds to one and only one of the events is called a **random variable**.

Two random variables are independent if any value that either may take is unaffected by any value of the other.

Let X be a discrete variable taking only the values $x_1, x_2, \ldots, x_n$ with probabilities $p_1, p_2, \ldots, p_n$ respectively.

X is called a **discrete random variable** if $\sum_{i=1}^{n} p_i = 1$.

The **probability density function** (pdf) of a discrete random variable X is a function that allocates probabilities to all the distinct values that X can take. Let $P(X=x)$ be the pdf for a random variable X defined for the discrete values of X.

Then $\sum_{\text{all}x} P(X=x) = 1$.

ℹ️ *A bag contains 6 blue and 4 red counters. 3 counters are drawn at random and not replaced. Find the probability distribution for the number of red counters drawn.*

Let the random variable X be 'the number of red counters drawn'.

$P(X=0) = P(\text{no red counters}) = P(B_1. B_2. B_3)$
$= P(B_1) \times P(B_2|B_1) \times P(B_3|B_2 \text{ and } B_1)$
$= \dfrac{6}{10} \times \dfrac{5}{9} \times \dfrac{4}{8} = \dfrac{1}{6}$

$P(X=1) = P(R_1. B_2. B_3 \text{ or } B_1. R_2. B_3 \text{ or } B_1. B_2. R_3)$
$= \left(\dfrac{4}{10} \times \dfrac{6}{9} \times \dfrac{5}{8}\right) + \left(\dfrac{6}{10} \times \dfrac{4}{9} \times \dfrac{5}{8}\right) + \left(\dfrac{6}{10} \times \dfrac{5}{9} \times \dfrac{4}{8}\right)$
$= \dfrac{1}{2}$

$P(X=2) = P(R_1. R_2. B_3 \text{ or } R_1. B_2. R_3 \text{ or } B_1. R_2. R_3)$
$= \dfrac{3}{10}$

$p(X=3) = P(R_1. R_2. R_3) = \dfrac{1}{30}$

Hence the probability distribution for X is:

x	0	1	2	3
$P(X=x)$	$\frac{1}{6}$	$\frac{1}{2}$	$\frac{3}{10}$	$\frac{1}{30}$

Note: The sum of the probabilities is 1.

Expectation

For a discrete random variable X with pdf $P(X=x)$ the **expectation** of x is $E[X] = \sum_{\text{all}x} xP(X=x)$.

$E[X]$ is interpreted as the mean value μ of X.

Properties of E
$E[a] = a$ where a is constant
$E[aX] = aE[X]$
$E[F(X) + G(X)] = E[F(X)] + E[G(X)]$
where $F(X)$ and $G(X)$ are any two functions of X
$E[G(X)] = \sum_{\text{all}x} G(X)P(X=x)$

ℹ️ For the situation above:

$E[X] = \sum_{x=0}^{3} xP(X=x)$
$= \left(0 \times \dfrac{1}{6}\right) + \left(1 \times \dfrac{1}{2}\right) + \left(2 \times \dfrac{3}{10}\right) + \left(3 \times \dfrac{1}{30}\right) = 1.2$

$E[X^2] = \sum_{x=0}^{3} x^2 P(X=x)$
$= \left(0 \times \dfrac{1}{6}\right) + \left(1 \times \dfrac{1}{2}\right) + \left(4 \times \dfrac{3}{10}\right) + \left(9 \times \dfrac{1}{30}\right) = 2$

$E[3X^2 - 2X] = 3E[X^2] - 2E[X] = 6 - 2.4 = 3.6$

Variance

The **variance** of a probability distribution associated with the random variable X is
$\text{Var}[X] = E[(X-\mu)^2]$ where $\mu = E[X]$
Computational formula:
$\text{Var}[X] = E[X^2] - (E[X])^2$

Properties of Var
If a and b are constants:
$\text{Var}[a] = 0$
$\text{Var}[aX] = a^2 \text{Var}[X]$
$\text{Var}[aX+b] = a^2 \text{Var}[X]$

ℹ️ For the situation above:

$\text{Var}[X] = E[X^2] - (E[X])^2$
$= 2 - (1.2)^2$
$= 0.56$

ℹ️ For the situation above:

$\text{Var}[3X] = 3^2 \text{Var}[X] = 9 \times 0.56 = 5.04$

Two random variables

If X and Y are any two random variables and a and b are constants, then
$$E[aX + bY] = aE[X] + bE[Y]$$
If X and Y are also independent, then
$$\text{Var}[aX + bY] = a^2 \text{Var}[X] + b^2 \text{Var}[Y]$$
From this last result
$$\text{Var}[X+Y] = \text{Var}[X-Y] = \text{Var}[X] + \text{Var}[Y]$$

ℹ️ *If X and Y are two independent random variables with $E[X] = 0.4$, $E[Y] = 0.7$, $Var[X] = 0.2$ and $Var[Y] = 0.3$, find (a) $E[2X+3Y]$, (b) $Var[X-Y]$.*

(a) $E[2X + 3Y] = 2E[X] + 3E[Y] = 2 \times 0.4 + 3 \times 0.7 = 2.9$

(b) $\text{Var}[X - Y] = \text{Var}[X] + \text{Var}[Y] = 0.2 + 0.3 = 0.5$

Discrete Probability Distributions
Worked example, Guided example and Exam questions

 The discrete random variable, X, has the following probability distribution.

x	0	1	2
$P(X=x)$	$\dfrac{1}{8}$	$\dfrac{3}{4}$	$\dfrac{1}{8}$

Find the mean and variance of X.
If two independent random variables, X_1 and X_2, have the same distribution as X, find the distribution of $X_1 - X_2$, and give its mean and variance.

By symmetry $E[X] = 1$
$\text{Var}[X] = E[X^2] - E^2[X]$

$$E[X^2] = \sum_{x=0}^{2} x^2 P(X=x) = 0 \times \frac{1}{8} + 1 \times \frac{3}{4} + 4 \times \frac{1}{8} = \frac{5}{4}$$

$$\text{Var}[X] = \frac{5}{4} - 1 = \tfrac{1}{4}$$

Let $Y = X_1 - X_2$

The diagram below shows the possible values y of Y and the associated probabilities $P(Y=y)$.

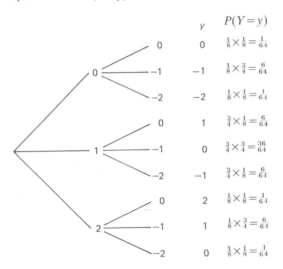

The probability distribution for Y is

y	-2	-1	0	1	2
$P(Y=y)$	$\dfrac{1}{64}$	$\dfrac{12}{64}$	$\dfrac{38}{64}$	$\dfrac{12}{64}$	$\dfrac{1}{64}$

$E[X_1] = 1 \qquad \text{Var}[X_1] = \tfrac{1}{4}$
$E[X_2] = 1 \qquad \text{Var}[X_2] = \tfrac{1}{4}$
$E[X_1 - X_2] = E[X_1] - E[X_2] = 1 - 1 = 0$
$\text{Var}[X_1 - X_2] = \text{Var}[X_1] + \text{Var}[X_2] = \tfrac{1}{4} + \tfrac{1}{4} = \tfrac{1}{2}$

 The faces of an ordinary die are re-numbered so that the faces are 1, 2, 2, 3, 3, 3. This die and an ordinary, unaltered die are thrown at the same time. The score, X, is the sum of the numbers on the uppermost faces of the two dice. Show that the probability of X being 3 is $\dfrac{1}{12}$ and of being 4 is $\dfrac{1}{6}$.

List the values that X can take and determine their respective probabilities. Hence, obtain the expected value of X, correct to three decimal places.
If the dice are thrown three times, determine the probability, correct to three significant figures, that none of the three values of X exceeds 3.

Let A be the score on the altered die and U the score on the unaltered die, then $X = A + U$ and the possible score combinations for a score of 3 are:

A	U
1	2
2	1

Thus $P(X=3) = P(A=1)P(U=2) + P(A=2)P(U=1)$
The probabilities on the RHS are easily found and hence $P(X=3)$ can be evaluated.
A similar approach can be used for $P(X=4)$.

A tree diagram shows that the range of values of X is the integral values 2 to 9 inclusive, and also assists in the calculation of the respective probabilities.
$P(X \text{ does not exceed } 3) = P(X = 2 \text{ or } 3)$, and this probability is easily found from the probability distribution of X and so $P(X \text{ does not exceed } 3 \text{ in three throws of the dice})$ can be evaluated.

---

1 Jane earns £40 for a five-day week. She works every fourth Sunday, for which she is paid double-time, and three Saturdays in every four, for each of which she is paid time and a half. Draw up a table showing the probability distribution of her wage on a day chosen at random and hence find the mean and variance of her daily wage. Find also the mean and variance of her weekly wage assuming that the Sundays on which she works fall in weeks *not* containing a 'working' Saturday. What would be the mean of her weekly wage if her 'working' Sundays fall in weeks also containing a 'working' Saturday?

(W)

2 In a game, a player rolls two balls down an inclined plane so that each ball finally settles in one of five slots and scores the number of points allotted to that slot as shown in the diagram below.

2	4	7	4	2

It is possible for both balls to settle in one slot and it may be assumed that each slot is equally likely to accept either ball. The player's score is the sum of the points scored by each ball. Draw up a table showing all the possible scores and the probability of each. If the player pays 10p for each game and receives back a number of pence equal to his score, calculate the player's expected gain or loss per 50 games.

(C)

3 In a certain gambling game a player nominates an integer x from 1 to 6 inclusive and he then throws three fair cubical dice. Calculate the probabilities that the number of x's thrown will be 0, 1, 2 and 3.
The player pays 5 pence per play of the game and he receives 48 pence if the number of x's thrown is three, 15 pence if the number of x's thrown is two, 5 pence if only one x is thrown and nothing otherwise. Calculate the player's expected gain or loss per play of the game.

(J)

4 A random variable X has the probability distribution given in Table 1, and $2Y = X - 18$. Find $E(Y)$ and $E(Y^2)$. Deduce the values of $E(X)$ and $\text{Var}(X)$.

Table 1

X	12	16	18	20	24
$P(X)$	$\frac{1}{15}$	$\frac{4}{15}$	$\frac{1}{3}$	$\frac{1}{5}$	$\frac{2}{15}$

(L)

147

74 Continuous Probability Distributions

Definitions, Expectation, Variance, Two random variables, Cumulative distribution function

Definitions

Let X be a continuous variable taking only values in the ranges x_1 up to x_2, x_2 up to x_3, . . . , x_n up to x_{n+1}, with probabilities $p_1, p_2, \ldots, p_n$ respectively.

X is called a **continuous random variable** if $\sum_{i=1}^{n} p_i = 1$.

The **probability density function** (pdf) of a continuous random variable X is a function that allocates probabilities to all of the ranges of values that X can take.

Let $f(x)$, the pdf for the random variable X, be defined over the range x_1 to x_2 only. Then

$$P(a \leqslant X \leqslant b) = \int_a^b f(x)\,dx.$$

and $\int_{x_1}^{x_2} f(x)\,dx = 1.$

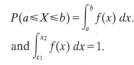

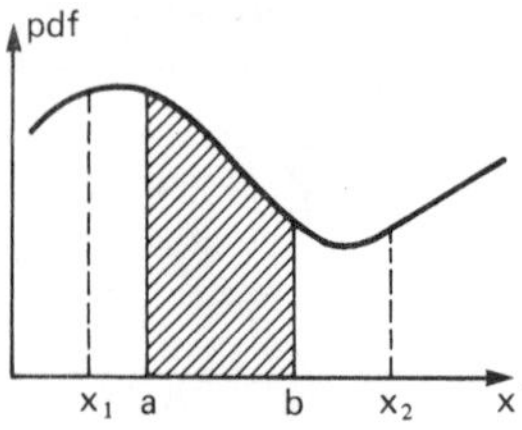

Figure 1

i *Construct the probability distribution for the ranges 0 to 1, 1 to 2, 2 to 3, for the continuous distribution defined by the pdf, $f(x) = \frac{2}{9}x(3-x)$, $0 \leqslant x \leqslant 3$.*

$$P(0 \leqslant X \leqslant 1) = \int_0^1 \frac{2}{9}x(3-x)\,dx = \frac{2}{9}\left[\frac{3x^2}{2} - \frac{x^3}{3}\right]_0^1 = \frac{7}{27}$$

$$P(1 \leqslant X \leqslant 2) = \int_1^2 \frac{2}{9}x(3-x)\,dx = \frac{2}{9}\left[\frac{3x^2}{2} - \frac{x^3}{3}\right]_1^2 = \frac{13}{27}$$

$$P(2 \leqslant X \leqslant 3) = \int_2^3 \frac{2}{9}x(3-x)\,dx = \frac{2}{9}\left[\frac{3x^2}{2} - \frac{x^3}{3}\right]_2^3 = \frac{7}{27}$$

So the probability distribution for X is

Range of X	0 up to 1	1 up to 2	2 up to 3
Probability	$\frac{7}{27}$	$\frac{13}{27}$	$\frac{7}{27}$

Note: The sum of the probabilities is 1.

Expectation

For a continuous random variable X with pdf $f(x)$, the **expectation** of X is $E[X] = \int_{\text{all}\,x} xf(x)\,dx$.

$E[X]$ is interpreted as the mean value μ of X.

Properties of E

$$E[a] = a \text{ where } a \text{ is constant}$$
$$E[aX] = aE[X]$$
$$E[F(X) + G(X)] = E[F(X)] + E[G(X)]$$

where $F(X)$ and $G(X)$ are any two functions of X.

$$E[G(X)] = \int_{\text{all}\,x} G(x)f(x)\,dx$$

i For the situation above:

$$E[X] = \int_0^3 x.\frac{2}{9}x(3-x)\,dx = \frac{2}{9}\left[x^3 - \frac{x^4}{4}\right]_0^3 = \frac{3}{2}$$

$$E[X^2] = \int_0^3 x^2.\frac{2}{9}x(3-x)\,dx = \frac{2}{9}\left[\frac{3x^4}{4} - \frac{x^5}{5}\right]_0^3 = \frac{27}{10}$$

$$E[3X^2 - 2X] = 3E[X^2] - 2E[X]$$
$$= 3(2.7) - 2(1.5) = 5.1$$

Variance

The **variance** of a probability distribution is $\text{Var}\,[X] = E[(X-\mu)^2]$ where $\mu = E[X]$.

Computational formula: $\text{Var}\,[X] = E[X^2] - (E[X])^2$.

Properties of Var

If a and b are constants: $\text{Var}\,[a] = 0$

$\text{Var}\,[aX] = a^2\,\text{Var}\,[X]\,\text{Var}\,[aX+b] = a^2\,\text{Var}\,[X]$

i For the situation above:

$$\text{Var}\,[X] = E[X^2] - (E[X])^2$$
$$= 2.7 - (1.5)^2 = 0.45$$

i For the situation above:

$\text{Var}\,[4X-8] = 4^2\,\text{Var}\,[X] = 16 \times 0.45 = 7.20$

Two random variables

If X and Y are any two random variables and a and b are constants, then

$$E[aX + bY] = aE[X] + bE[Y]$$

If X and Y are also independent, then

$$\text{Var}\,[aX + bY] = a^2\,\text{Var}\,[X] + b^2\,\text{Var}\,[Y]$$

From this last result:

$$\text{Var}\,[X+Y] = \text{Var}\,[X-Y] = \text{Var}\,[X] + \text{Var}\,[Y]$$

i *If X and Y are two independent random variables with $E[X] = 0.3$, $E[Y] = 0.4$, $Var\,[X] = 0.2$ and $Var\,[Y] = 0.3$, find: (a) $E[4X-3Y]$ (b) $Var\,[4X-3Y]$.*

(a) $E[(4X-3Y] = 4E[X] - 3E[Y]$
$= 4 \times 0.3 - 3 \times 0.4 = 0$

(b) $\text{Var}\,[4X-3Y] = 4^2\text{Var}\,[X] + 3^2\text{Var}\,[Y] = 5.9$

Cumulative distribution function

The **cumulative distribution function** is defined by

$$A = \int_{-\infty}^{a} f(x)\,dx$$

where $f(x)$ is the pdf.

As a varies so does A, i.e. $A = F(a)$

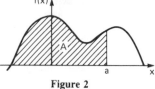

Figure 2

So $F(a) = \int_{-\infty}^{a} f(x)\,dx = P(X \leqslant a)$

$F(a)$ is the cumulative distribution function

Clearly $f(x) = F'(x)$.

If M, Q_1, Q_3 are the median, lower and upper quartiles of x, then

$$F(M) = \frac{1}{2},\ F(Q_1) = \frac{1}{4},\ F(Q_3) = \frac{3}{4}.$$

i For the random variable X with pdf

$$p(x) = \frac{1}{3} \quad 0 < x < 3$$
$$= 0 \quad \text{otherwise}$$

the cumulative distribution function is

$$F(x) = \int_{-\infty}^{x} p(x)\,dx$$

which gives $F(x) = \begin{cases} 0 & x \leqslant 0 \\ \frac{1}{3}x & 0 < x < 3 \\ 1 & x \geqslant 3 \end{cases}$

The graph of $F(x)$ against x is shown.

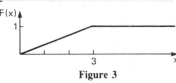

Figure 3

Continuous Probability Distributions
Worked example and Exam questions

A continuous random variable X has a probability density function defined by

$$f(x) = \frac{3x^2}{a^3}, \quad 0 \le x \le a$$
$$= 0, \text{ otherwise.}$$

(a) *Find the variance of X.*
(b) *Find the cumulative distribution function of X.*

(c) *If $Y = 1 - \dfrac{X}{a}$ find*

   (i) *$P(Y > X)$,*
   (ii) *the probability density function of Y.*

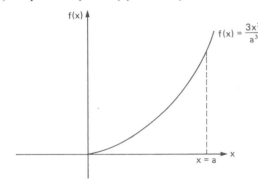

(a)
$$\text{Var}[X] = E[X^2] - E^2[X]$$

$$E[X] = \int_0^a x f(x) dx = \int_0^a \frac{3x^3}{a^3} dx = \frac{3}{4a^3} \left[x^4 \right]_0^a = \frac{3a}{4}$$

$$E[X^2] = \int_0^a x^2 f(x) dx = \int_0^a \frac{3x^4}{a^3} dx = \frac{3}{5a^3} \left[x^5 \right]_0^a = \frac{3a^2}{5}$$

$$\text{Var}[X] = \frac{3a^2}{5} - \left(\frac{3a}{4} \right)^2 = \frac{3 \times 16 - 9 \times 5}{80} a^2 = \frac{3a^2}{80}$$

(b) Let $F(x) = \int_0^x f(t) dt$ be the cumulative distribution function of X.

$$F(x) = \int_0^x \frac{3t^2}{a^3} dt = \frac{1}{a^3} \left[t^3 \right]_0^x = \frac{x^3}{a^3}$$

(c)
(i) $Y = 1 - \dfrac{X}{a}$

$$P(Y > X) = P(Y - X > 0)$$

$$Y - X = 1 - \frac{X}{a} - X = \frac{a - X - aX}{a} > 0$$

i.e. $a > X(1 + a)$

i.e. $X < \dfrac{a}{1+a}$

$$\therefore P\left(X < \frac{a}{1+a} \right) = \int_0^{\frac{a}{1+a}} f(x) dx = \int_0^{\frac{a}{1+a}} \frac{3x^2}{a^3} dx$$

$$= \frac{1}{(1+a)^3}$$

(ii) $P(Y < t) = P\left(1 - \dfrac{X}{a} < t \right) = P(X > a(1-t))$

$$= 1 - P(X < a(1-t))$$

$$P(X < a(1-t)) = \int_0^{a(1-t)} f(x) dx = \int_0^{a(1-t)} \frac{3x^2}{a^3} dx$$

$$= (1-t)^3$$

$$\therefore P(Y < t) = 1 - (1-t)^3$$

If $F_Y(x)$ is the cumulative distribution function for Y, then $F_Y(x) = P(Y < x) = 1 - (1-x)^3$ and if $f_Y(x)$ is the probability density function for Y, then $f_Y(x) = \dfrac{d}{dx} F_Y(x) = 3(1-x)^2$

---

1 A continuous random variable X can assume values only between 0 and 2 inclusive and its probability density function is given by $f(x) = k(4x - x^2)$, $0 \le x \le 2$, where k is a positive constant.
   (i) Using the fact that $P(0 \le x \le 2) = 1$, find the value of k. Using this value of k, find
   (ii) the mean value of X,
   (iii) the variance of X, and
   (iv) the probability that the value of X exceeds $1 \cdot 0$, i.e. $P(X > 1 \cdot 0)$.   *(W)*

2 A continuous random variable X can assume values only between 0 and 4 inclusive and its probability density function is given by $f(x) = \frac{1}{2} - ax$ $(0 \le x \le 4)$, where a is a constant.
   (i) Calculate, in terms of a, the value of $P(0 \le x \le 4)$.
   (ii) Using the result of (i) above, determine the value of a.
   (iii) Find the mean and variance of the random variable X.
   (iv) Find the probability that the value of X lies between 1 and 2.   *(W)*

3 The random variable X has a probability density function given by

$$P(x) = \begin{cases} kx(1-x^2) & (0 \le x \le 1), \\ 0 & \text{elsewhere,} \end{cases}$$

k being a constant. Find the value of k and find also the mean and variance of this distribution.
Find the median of the distribution.   *(O & C)*

4 The lifetime T, in hours of a certain type of electric lamp is a random variable with distribution
$$f(t) = Ae^{-t/1200}, \quad 0 \le t < \infty,$$
$$= 0, \, t < 0.$$

Find the value of A and show that the mean and standard deviation of T are both 1200 hours. To test the reliability of the production a random sample of 40 bulbs was tested and found to have a mean life of 1020 hours. Does this indicate at the 5% level of significance that the batch from which the sample was taken was sub-standard?   *(S)*

5 The random variable X has probability density function given by

$$f(x) = \begin{cases} \dfrac{1}{(b-a)} & a \le x \le b \\ 0 & \text{otherwise} \end{cases}, \text{ where } b > a.$$

Show that the mean is $(b+a)/2$, and the variance is $(b-a)^2/12$ for this distribution. Given that the mean equals 1 and the variance equals 4/3 find:
   (i) $P(X < 0)$,
   (ii) the value of z such that $P(X > z + \sigma_x) = \frac{1}{4}$, where σ_x is the standard deviation of X.   *(A)*

6 A continuous random variable, X, has probability density function $\lambda \sin x$ $(0 \le x \le \pi)$, and zero outside this range. Find a value of the consant λ, the mean, the variance, the median, the quartiles. What is the probability that a random observation lies within one standard deviation of the mean?   *(OLE)*

75 The Binomial Distribution

Definition, Binomial situations, Expectation and variance, Binomial recurrence formula, Fitting a Binomial distribution.

Definition

A discrete random variable X having a probability density function (pdf) of the form:

$$p(X=x)=\binom{n}{x}p^x(1-p)^{n-x} \text{ where } x=0, 1, 2, \ldots, n.$$

is said to have a **binomial distribution**.

We write X is Bin(n, p).
n and p are called the **parameters of the distribution**.
n is a positive integer and $0 \leq p \leq 1$.

ℹ️ Bin $(4, \frac{1}{2})$ has $n=4$ and $p=\frac{1}{2}$.

$$P(X=x)=\binom{4}{x}\left(\frac{1}{2}\right)^x\left(\frac{1}{2}\right)^{4-x}=\binom{4}{x}\left(\frac{1}{2}\right)^4$$

Verify this gives the following probability table:

x	0	1	2	3	4
$P(X=x)$	$\frac{1}{16}$	$\frac{4}{16}$	$\frac{6}{16}$	$\frac{4}{16}$	$\frac{1}{16}$

Binomial situations

Any situation having only two possible, mutually exclusive outcomes, often labelled 'success'/'failure', is called a **binomial situation**. The probability of a 'success' (or 'failure') is usually known.

ℹ️ *Seeds have a probability of germinating of 0.9. If six seeds are sown what is the probability of five or more seeds germinating?*

This is a binomial situation with $n=6$ and $P(\text{success}) = P(\text{germination}) = 0.9$

$$\therefore P(X=x)=\binom{6}{x}(0.9)^x(0.1)^{6-x}$$

$P(5 \text{ or more germinations}) = P(X=5 \text{ or } 6)$
$= P(X=5) + P(X=6)$

$$=\binom{6}{5}(0.9)^5(0.1)^1+\binom{6}{6}(0.9)^6(0.1)^0=0.88$$

Expectation and variance

If X is Bin(n, p), then:
the **expectation** of X is $E[X]=np$,
the **variance** of X is $\text{Var}[X]=np(1-p)$

ℹ️ For Bin$(6, 0.9)$:
$E[X]=6\times0.9=5.4$
$\text{Var}[X]=6\times0.9\times(1-0.9)=0.54$

Binomial recurrence formula

The **binomial recurrence formula** is

$$P(X=x+1)=\frac{n-x}{x+1}\cdot\frac{p}{(1-p)}P(X=x)$$

This enables successive probabilities to be more easily calculated once the initial probability is known.

ℹ️ For Bin$\left(4, \frac{1}{2}\right)$, $P(X=0)=\binom{4}{0}\left(\frac{1}{2}\right)^4\left(\frac{1}{2}\right)^0=\frac{1}{16}$

By the binomial recurrence formula:

$$P(X=1)=\frac{4}{1}\cdot\frac{\frac{1}{2}}{1\frac{1}{2}}\cdot P(X=0)=4\times\frac{1}{16}=\frac{4}{16}$$

$$P(X=2)=\frac{3}{2}\cdot1\cdot P(X=1)=\frac{3}{2}\times\frac{4}{16}=\frac{6}{16}$$

$$P(X=3)=\frac{2}{3}\cdot1\cdot P(X=2)=\frac{2}{3}\times\frac{6}{16}=\frac{4}{16}$$

$$P(X=4)=\frac{1}{4}\cdot1\cdot P(X=3)=\frac{1}{4}\times\frac{4}{16}=\frac{1}{16}$$

Fitting a Binomial distribution

To fit a binomial distribution to a given frequency distribution:

(a) Find the values of the parameters n and p:
n is the largest value of x,
$P \approx \frac{\bar{x}}{n}$, since $\bar{x} \approx E[X]=np$.

(b) Generate the binomial probability distribution for X as Bin(n, p).

(c) Multiply the probabilities by the total frequency of the given distribution to find the expected frequencies.

ℹ️ *Fit a binomial distribution to the following:*

x	0	1	2	3	4	Total
f	4	13	20	11	2	50

The largest value of x is 4. Take $n=4$.

$\Sigma fx=94$, $\Sigma f=50 \Rightarrow \bar{x}=1.88$. Take $p\approx\frac{1.88}{4}=0.47$.

Assume X is Bin$(4, 0.47)$.

$$P(X=x)=\binom{4}{x}(0.47)^x(0.53)^{4-x}$$

$$P(X=0)=\binom{4}{0}(0.47)^0(0.53)^4=0.0789$$

By the binomial recurrence formula:
$P(X=1)=0.2799$, $P(X=2)=0.3723$,
$P(X=3)=0.0488$, $P(X=4)=0.0488$.

$\Sigma f=50$ gives this table of expected frequencies:

x	0	1	2	3	4	Total
f_E	4	14	19	11	2	50

The Binomial Distribution
Worked examples and Exam questions

 The probability that a marksman will hit a target is $\frac{5}{6}$. He fires 9 shots. Calculate, correct to three decimal places, the probability that he will hit the target (i) at least 7 times, (ii) no more than 6 times. If he hits the target exactly 6 times, calculate the probability that the three misses are with 3 successive shots.

$P(\text{hitting target}) = P(\text{success}) = \dfrac{5}{6}$

$P(\text{not hitting target}) = P(\text{failure}) = \dfrac{1}{6}$

Taking the nine shots to be independent events the probability distribution of X, 'the number of targets hit' is

$\text{Bin}\left(9, \dfrac{5}{6}\right)$, and so $P(X=x) = \dbinom{9}{x}\left(\dfrac{5}{6}\right)^{x}\left(\dfrac{1}{6}\right)^{9-x}$

(i) $P(\text{at least 7 targets hit}) = P(X=7) + P(X=8) + P(X=9)$

$= \dfrac{9\times 8}{1\times 2}\left(\dfrac{5}{6}\right)^{7}\left(\dfrac{1}{6}\right)^{2} + 9\left(\dfrac{5}{6}\right)^{8}\left(\dfrac{1}{6}\right) + \left(\dfrac{5}{6}\right)^{9}$

$= \left(\dfrac{5}{6}\right)^{7}\left[\dfrac{36}{36} + \dfrac{45}{36} + \dfrac{25}{36}\right] = \left(\dfrac{5}{6}\right)^{7}\times\dfrac{106}{36}$

$= 0\cdot822$

(ii) $P(\text{no more than 6 targets})$
$= P(X=0) + P(X=1) + P(X=2) + \ldots + P(X=6)$
$= 1 - P(X=7) - P(X=8) - P(X=9)$
$= 1 - 0\cdot822 = 0\cdot178$

The number of ways of hitting the target exactly six times is $^{9}C_{6} = 84$. The three misses happening with three successive shots can occur in 7 ways i.e. 123, 234, 345, 456, 567, 678 or 789.

$\therefore P(\text{6 targets and 3 misses with successive shots}) = \dfrac{7}{84} = \dfrac{1}{12}$.

 Groups of 5 people are chosen at random and the number, x, of people in each group who normally wear spectacles is recorded. The results obtained for 400 groups of 5 are shown in the table.

x	0	1	2	3	4	5
f	34	106	130	90	36	4

Calculate, from the above data, the mean value of x. Assuming that the situation can be modelled by a binomial distribution having the same mean as the one calculated above, state the appropriate values for the parameters n and p. Calculate the theoretical frequencies corresponding to those in the table.

$\bar{x} = \dfrac{\Sigma fx}{\Sigma f} = 2$

$n = 5, \; p = \dfrac{\bar{x}}{n} = \dfrac{2}{5}$

The probability distribution of X, 'number in group wearing spectacles' is $\text{Bin}\left(5, \dfrac{2}{5}\right)$, and $P(X=x) = \dbinom{5}{x}\left(\dfrac{2}{5}\right)^{x}\left(\dfrac{3}{5}\right)^{5-x}$

The theoretical probability distribution is shown in the table below

x	0	1	2	3	4	5
$P(X=x)$	0·078	0·259	0·346	0·230	0·077	0·010

The theoretical frequencies are found by multiplying each probability by 400, giving

x	0	1	2	3	4	5
f_T	31·2	103·6	138·4	92·0	30·8	4·0

EX

1 Previous experience indicates that, of the students entering upon a particular diploma course, 90% will successfully complete it. One year, 15 students commence the course. Calculate, correct to 3 decimal places, the probability that:
  (i) all 15 successfully complete the course;
  (ii) only 1 student fails;
  (iii) no more than 2 students fail;
  (iv) at least 2 students fail.   *(C)*

2 (i) There are 8 red apples and 2 green apples in a bag. If 2 apples are taken out at random, determine the probability that just one of them will be red.
(ii) It is found by experience that 20% of the electric toasters made by a certain manufacturer are faulty. The toasters are packed in boxes, each box containing 10 toasters. Calculate the probability that in a box selected at random (a) just 3 will be faulty; (b) just 2 will be faulty. Determine the most likely number of faulty toasters to be found in a box chosen at random.   *(OLE)*

3 State the conditions necessary for a Binomial Distribution to result when a series of events occurs. Illustrate your answer by means of an example.
Four ordinary dice are thrown. Find:
  (a) the probability that at least one of the dice shows a six;
  (b) the probability that the highest score showing is a two.
   (S)

4 A company has ten telephone lines. At any instant the probability that any particular line is engaged is 1/5. State the expected number of free telephone lines. Calculate for any instant, correct to two significant figures, the probability that
  (i) all the lines are engaged,
  (ii) at least one line is free,
  (iii) exactly two lines are free.   *(A)*

5 (a) Berg and Korner have a long rivalry in tennis: in the last 25 games, Berg has won 15 times. They start a new series of 8 games. Assuming that the binomial model may be applied, what is the probability that Berg will win at least 6 of these?
(b) Prove that the mean of the binomial distribution is np. A set of 100 pods, each containing 4 peas, was examined to see how many of the peas were good. the following were the results:

No. of good peas in pod	0	1	2	3	4
No of pods	7	20	35	30	8

  (i) What is the probability of getting a good pea?
  (ii) Calculate the theoretical frequencies of 0, 1, 2, 3, 4 good peas, using the associated theoretical binomial distribution.   *(S)*

6 (i) In a certain large population the numbers of red and blue elements are in the ratio 9:1. Find, to three significant figures, the probability that in a random selection of ten there are exactly 5 of each.
(ii) In another large population there are red, blue and green elements in the ratio 9:1:2. Find the probability that, in a random selection of 15, there are exactly 5 green elements.
Find the probability that, in a random selection of 15, there are 5 of each colour. Find also the most likely distribution of colours in a random selection of 15.   *(A)*

7 Prove that the mean number of successes in a series of n independent trials, each of which has a probability p of success, is np. Show that the standard deviation is $\sqrt{\{np(1-p)\}}$.   *(O & C)*

76 The Poisson Distribution
Definitions, Uses, Poisson or binomial, Additive property.

Definitions

A discrete random variable X having a probability density function (pdf) of the form

$$P(X=x)=e^{-\mu}\cdot\frac{\mu^x}{x!}\text{ where }x=0,1,2,\ldots$$

is said to have a **Poisson distribution.**
We write X is Po(μ)
$\mu(>0)$ is the **parameter of the distribution.**

Note: There is no upper limit to the value of x. This is usually determined by practical considerations.

If X is $P(\mu)$, then $E[X]=\mu$ and Var$[X]=\mu$.

The **Poisson recurrence formula** is

$$P(X=x+1)=\frac{\mu}{x+1}P(X=x)\text{ for }x=0,1,\ldots$$

This enables successive probabilities to be more easily calculated once the initial probability is known.

ℹ️ For Po(1.5), $\mu=1.5$

$$P(X=x)=e^{-1.5}\cdot\frac{(1.5)^x}{x!}\text{ where }x=0,1,2,\ldots$$

This gives the following probability table:

x	0	1	2	3	4	5 or more
$P(X=x)$	0.223	0.335	0.251	0.126	0.047	0.018

ℹ️ For Po(1.5), $E[X]=1.5$ and Var$[X]=1.5$.

ℹ️ For Po(1.5), $P(0)=0.223$
By the Poisson recurrence formula:

$$P(1)=\frac{1.5}{0+1}P(0)=1.5\times0.223=0.335$$

etc.

Uses

There are two main uses of the Poisson distribution.

1. Estimation of probabilities of random events which have a small probability of occurrence. Typical applications of this are telephone calls arriving at a switchboard, insurance claims, accident rates, flaws in manufactured material. Usually the mean rate of occurrence per unit time will be given, although this may be scaled accordingly.

2. Approximation to a binomial distribution with the same mean, i.e. $\mu=np$, and usually $n>50$ and $p<0.1$.

ℹ️ *Telephone calls arrive at a switchboard at the rate of* 50 *per hour. Find the probabilities of* 0, 1 *or* 2 *calls arriving in any* 5 *minute period.*

The average rate of calls per 5 minute period
$=50\div12=4.17$ calls
If the random variable X is 'the number of calls in any 5 minute period', then X is Po(4.17).

$$P(X=x)=e^{-4.17}\cdot\frac{(4.17)^x}{x!}$$

$P(X=0)=0.02$, $P(X=1)=0.06$, $P(X=2)=0.13$

ℹ️ *A large population has* 0.5% *defective. A sample of* 200 *is taken at random. Using a Poisson approximation find the probabilities of* 0, 1 *or* 2 *defectives in the sample.*

This is a binomial situation with $n=200$, $p=0.005$.
So $\mu=np=200\times0.005=1.00$
If X is 'the number of defectives', then X is Po(1).

$$P(X=x)=e^{-1}\cdot\frac{(1)^x}{x!}$$

$P(X=0)=0.37$, $P(X=1)=0.37$, $P(X=2)=0.18$

Poisson or binomial

Given a frequency distribution of the binomial/Poisson type, the decision on whether to fit a binomial or Poisson distribution is often made by considering the mean and variance of the distribution.
The closer these two are in value the more likely a Poisson distribution is applicable.

ℹ️ The expected frequencies for a theoretical binomial and Poisson fit are compared with a given distribution below.

x	0	1	2	3	4	total
f	21	18	7	3	1	50
f_B	18	21	9	2	0	50
f_P	20	18	8	2	1	50

For the given distribution $\bar{x}=0.9$ and $s^2=0.97$.
The Poisson fit is more appropriate.

Additive property

The **additive property of the Poisson distribution:**
if X is Po(x) and Y is Po(y),
then $X+Y$ is Po$(x+y)$.

ℹ️ An item is made of two manufactured parts A and B. Flaws in A occur randomly and with a mean rate of 7.8 per 1000, while flaws in B also occur randomly (and independently of those in A) with a mean rate of 5.6 per 1000. Assuming Poisson distributions for the flaws in A and B, the item composed of A and B will have flaws distributed Po(7.8+5.6)=Po(13.4).

The Poisson Distribution
Worked example, Guided example and Exam questions

 Telephone calls reach a switchboard independently and at random, external ones at a mean rate of 1 in any 5 minute period, and internal ones at a mean rate of 2 in any 5 minute period.
Calculate the probability that there will be more than 2 calls in any period of 2 minutes.

Let the random variable E be 'the number of external calls per 2 minute period', E is $Po(0\cdot4)$.
Let the random variable I be 'the number of internal calls per 2 minute period', I is $Po(0\cdot8)$.
Using the additive property of Poisson distributions,
$E+I$ is $Po(0\cdot4+0\cdot8)$ i.e. $Po(1\cdot2)$.

$P(E+I>2) = 1 - P(E+I=0) - P(E+I=1) - Po(E+I=2)$
$P(E+I=0) = e^{-1\cdot2} = 0\cdot301$

$P(E+I=1) = e^{-1\cdot2} \times \dfrac{1\cdot2}{1} = 0\cdot361$

$P(E+I=2) = e^{-1\cdot2} \times \dfrac{(1\cdot2)^2}{2!} = 0\cdot271$

$\therefore P(E+I>2) = 1 - 0\cdot301 - 0\cdot361 - 0\cdot217$
$\qquad = 0\cdot121$

$\therefore P(\text{more than 2 calls in any 2 minute period}) = 0\cdot121$

 The frequency distribution of the number of accidents in each week in a period of 104 weeks in a factory is given in the table.

Number of accidents(x)	0	1	2	3	4	5
Frequency(f)	30	45	20	6	2	1

Fit a Poisson distribution to these data, calculating the frequencies it predicts.

Calculate $\bar{x}$ the mean number of accidents per week. Then the random variable X 'the number of accidents per week' is $Po(\bar{x})$.
Calculate $P(X=0) = e^{-\bar{x}}$ and then use the Poisson recurrence formula to find $P(X=1)$ to $P(X=5)$.
Multiplying each of these probabilities by 104 will give the predicted frequencies using a Poisson model.

 1 Sketch the frequency polygon of a Poisson distribution with mean 2. Telephone calls arrive at a switchboard at random intervals at an average rate of 24 calls per hour. Find the probability of receiving:
(a) no calls in 5 minutes;
(b) more than 4 calls in 5 minutes;
(c) Estimate the probability of receiving more than 50 calls in $1\frac{1}{2}$ hours. *(O & C)

2 The frequency distribution of the number of accidents in each week in a period of 2 years (104 weeks) in a factory is given in the table.

Number of accidents	0	1	2	3	4	5
Frequency	33	42	19	8	1	1

Fit a Poisson distribution to these data, calculating the frequencies it predicts. State why the binomial distribution is not appropriate in this case. *(A)

3 Show that the variance of a Poisson distribution is equal to the mean. The number of bacteria in 1 ml of inoculum has a Poisson distribution with mean 2.0. If at least 3 bacteria are needed for a dose to be infective, find the probability that a dose of 1 ml will cause infection. Find approximate limits, symmetrical about the mean, between which lies 95% of the distribution of the number of bacteria in 100 ml of inoculum. *(O & C)*

4 Define the Poisson distribution and derive its mean and variance. In the first year of the life of a certain type of machine, the number of times a maintenance engineer is required has a Poisson distribution with mean four. Find the probability that more than four calls are necessary. The first call is free of charge and subsequent calls cost £20 each. Find the mean cost of maintenance in the first year.
(J)

5 A footballer finds that the number of goals he scores in a match has a Poisson distribution with mean $\frac{1}{4}$. What is the distribution of the number of goals he scores in n matches? How many matches must he play in order to be 95% sure of scoring at least 20 goals? *(OLE)*

6 A random variable X has a Poisson distribution given by $Pr(X=r) = p_r = e^{-\lambda}\lambda^r/r!$, $r=0, 1, 2, \ldots$. Prove that the mean of X is λ. Give two examples (other than that suggested below) of situations where you would expect a Poisson distribution to occur.
The number of white corpuscles on a slide has a Poisson distribution with mean $3\cdot2$. By considering the values of r for which $p_{r+1}/p_r > 1$ find the most likely number of white corpuscles on a slide. Calculate correct to 3 decimal places the probability of obtaining this number. If two such slides are prepared what is the probability, correct to 3 decimal places, of obtaining at least two white corpuscles in total on the two slides? *(S)*

7 Derive the mean and the variance of a Poisson distribution. Two types of flaw, A and B, may occur in a manufactured cloth. The numbers of flaws of type A and of type B occurring per metre length of the cloth are independent random variables having Poisson distributions with means $0\cdot5$ and 1, respectively.
(a) Find the probabilities, to three significant figures, that a length of 1 metre of the cloth will have:
(i) 2 or fewer flaws of type A,
(ii) no flaw of either type.
(b) Show that the probability of a length of 1 metre of the cloth containing 1 flaw only is exactly three times that of it containing 1 flaw of each type.
(c) Removing a type A flaw from the cloth costs 8 pence and removing a type B flaw costs 2 pence. Find the mean and the standard deviation of the cost of removing flaws per 1 metre length of cloth. *(W)*

8 Explain under what conditions it is appropriate to apply a Poisson model to describe a distribution. Referring to your project work if you wish, give brief details of two situations in which you used this model. It is known that $0\cdot1\%$ of people having an influenza injection of type A suffer an adverse reaction. If 2250 people are to receive the injection, what is the probability that
(a) exactly two people will suffer an adverse reaction;
(b) more than 3 people will suffer an adverse reaction?
Given that 2000 people receive an injection of type B and no one suffers an adverse reaction to this injection, is this sufficient evidence to suggest that there is a smaller probability of an adverse reaction to type B than to type A? Give reasons for your answer. *(L)*

9 A large number of screwdrivers from a trial production run is inspected. It is found that the cellulose acetate handles are defective on 1% and that the chrome steel blades are defective on $1\frac{1}{2}\%$ of the screwdrivers, the defects occurring independently.
(a) What is the probability that a sample of 80 contains more than two defective screwdrivers?
(b) What is the probability that a sample of 80 contains at least one screwdriver with both a defective handle and a defective blade? *(O & C)*

77 The Normal Distribution

Definition, Standard Normal distribution, Use of tables, Using the standard variable.

Definition

A continuous random variable X having a probability density function (pdf) of the form

$$f(x)=\frac{1}{\sigma\sqrt{2\pi}}\exp[-\frac{(x-\mu)^2}{2\sigma^2}],\ -\infty<x<\infty$$

is said to have a **Normal distribution**.

We write X is $N(\mu, \sigma^2)$.

$\mu(-\infty<\mu<\infty)$ and $\sigma(>0)$ are the **parameters of the distribution**.

$E[X]=\mu$ and $Var[X]=\sigma^2$

The normal distribution curve is a characteristic 'bell shape' symmetrical about $x=\mu$ (the mean).

Figure 1

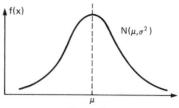

The area under the Normal curve is unity (1).

Standard Normal distribution

The Normal distribution with $\mu=0$, $\sigma=1$ is called a **standard Normal distribution**. The corresponding variable is usually denoted by Z, so Z is $N(0, 1)$.

The pdf for Z is $\phi(x)=\frac{1}{\sqrt{2\pi}}\exp\left(\frac{-x^2}{2}\right)$

By definition

$$P(Z<x)=\int_{-\infty}^{x}\phi(x)\,dx=\int_{-\infty}^{x}\frac{1}{\sqrt{2\pi}}\exp\left(\frac{-x^2}{2}\right)dx$$

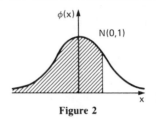

Figure 2

$P(Z<x)$ is the shaded area in the diagram. It is usually referred to as $\Phi(x)$. The integral which gives this area under the curve is difficult to evaluate but it is given in Normal distribution tables.

Use of tables

Most Normal distribution function tables are given only for $x\geqslant 0$. Other probabilities are derived by suitable transformations. When evaluating probabilities using these tables, draw a sketch to ensure that the correct quantity is being evaluated.

For a given $a\geqslant 0$,
$P(Z<a)=\Phi(a)$
This is simply the value in the table at $x=a$

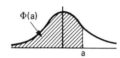

$P(Z>a)=1-\Phi(a)$
since the total area under the curve is equal to 1.

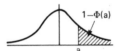

$P(Z>-a)=\Phi(a)$
$P(Z<-a)=1-\Phi(a)$
by the symmetry of the normal curve

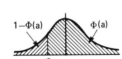

For $b<c$,
$P(b<Z<c)=\Phi(c)-\Phi(b)$

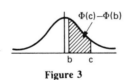

Figure 3

The tables may also be used 'in reverse' to find z if $\Phi(z)$ is known.

i *Find the values of:*
(*a*) $\Phi(1.85)$ (*b*) $\Phi(-0.63)$ (*c*) $P(Z>-1.5)$
(*d*) $P(0.5<Z<1.8)$ (*e*) $P(-2.1<Z<1.6)$

(a) $\Phi(1.84)=0.9671$
direct from tables

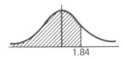

(b) $\Phi(-0.63)=1-\Phi(0.63)$
$=1-0.7357$
$=0.2643$

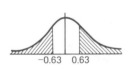

(c) $P(Z>-1.5)=1-\Phi(-1.5)$
$=1-(1-\Phi(1.5))$
$=\Phi(1.5)$
$=0.9332$

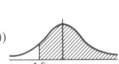

(d) $P(0.5<Z<1.8)$
$=\Phi(1.8)-\Phi(0.5)$
$=0.9641-0.6915$
$=0.2726$

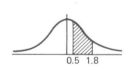

(e) $P(-2.1<Z<1.6)$
$=\Phi(2.1)-(1-\Phi(1.6))$
$=0.9821-(1-0.9452)$
$=0.9273$

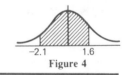

Figure 4

Using the standard variable

Probabilities associated with any given normal distribution may be found by using the standard variable as follows.

1. Transform the given random variable X, which is $N(\mu, \sigma^2)$, into the standard variable Z, which is $N(0, 1)$ using:

$$Z=\frac{X-\mu}{\sigma}$$

2. Use the standard Normal distribution function tables to find the probabilities of the transformed values as shown above.

i *An industrial process mass produces an item whose weights are normally distributed with mean* 18.5 kg *and standard deviation* 1.5 kg. *What is the probability that an item chosen at random weighs more than* 21.5 kg?

The weights are N(18.5, 2.25).

The standard variable is $Z=\frac{W-18.5}{1.5}$.

$\therefore P(W>21.5)=P\left(Z>\frac{21.5-18.5}{1.5}\right)$

$=P(Z>2)=1-\Phi(2)$
$=0.0227$

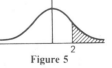

Figure 5

The Normal Distribution
Worked example, Guided example and Exam questions

WE *An industrial process mass produces items which are normally distributed. 11·55% of them weigh over 20 kg and 5·89% weigh under 10 kg. Calculate the mean weight and standard deviation for this distribution.*

Let $Z = \dfrac{W-\mu}{\sigma}$ be standardized variable, where μ and σ are the mean and standard deviation for the distribution.

When $W = 20$, $\Phi(z_{20}) = 1 - 0·1155 = 0·8845$
 i.e. $z_{20} = 1·198$ (from tables)

 hence $1·198 = \dfrac{20-\mu}{\sigma}$ [1]

When $W = 10$, $\Phi(z_{10}) = 0·0589$
 i.e. $\Phi(-z_{10}) = 1 - 0·0589 = 0·9411$
 so $-z_{10} = 1·564$ (from tables)

 hence $1·564 = \dfrac{-(10-\mu)}{\sigma}$ [2]

Solving [1] and [2] for μ and σ gives $\mu = 15·66$ and $\sigma = 3·62$.

GE *Eggs are classified by weight according to the following table:*

Class	2	3	4	5	6
Weight (grams)	65–70	60–65	55–60	50–55	45–50

100 hens of breed A are found to lay eggs at the rate of 180 per day, the eggs being of mean weight 63 g with standard deviation 5 g. 100 hens of breed B lay 210 eggs per day of mean weight 52 g and standard deviation 6 g. What ratio of hens of breed A to hens of breed B should be kept in order that equal numbers of eggs of class 3 and class 5 should be produced daily?
Assume that the weights of eggs from each breed of hen are normally distributed.

Summarising diagrammatically the information for breed A, and breed B gives:

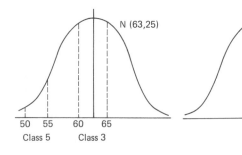

$$Z_A = \frac{w_A - 63}{5} \qquad\qquad Z_B = \frac{w_B - 52}{6}$$

Calculating the probability of obtaining a class 3 egg gives
$P(-0·6 \leqslant Z_A \leqslant 0·4) + P(1·33 \leqslant Z_B \leqslant 2·17)$
$\Phi(0·4) - (1 - \Phi(0·6)) + \Phi(2·17) - \Phi(1·33)$
$0·6554 - 1 + 0·7257 + 0·9850 - 0·9082$
$0·3811 + 0·0768$
∴ the number of class 3 eggs will be
     $0·3811 \times 180 + 0·0768 \times 210 = 68·6 + 16·1$
per 200 hens, 100 of each breed.

A similar calculation shows that the number of class 5 eggs will be
     $0·3208 \times 210 + 0·0491 \times 180 = 67·4 + 8·8$
per 200 hens, 100 of each variety.

For the same number of class 3 eggs and class 5 eggs we require
$68·6A + 16·1B = 67·4B + 8·8A$

---

where A = number of breed A in units of 100
and B = number of breed B in units of 100.
From this equation we obtain the required ratio, $A : B = 0·86 : 1$

EX

1 Observation of a very large number of cars at a certain point on a motorway establishes that the speeds are normally distributed. 90% of cars have speeds less than 75·7 mph and only 5% have speeds less than 60 mph. Determine the mean speed μ and the standard deviation σ. Give your answers correct to 2 decimal places. Because of a fuel economy drive the mean speed of motorists is reduced. Assuming that the standard deviation σ has remained unchanged and that 2% now exceed 75 mph, find the new mean speed. What percentage of motorists now exceed the 70 mph speed limit?

                            (W)

2 In a given manufacturing process, components are rejected if they have a particular dimension greater than 60·4 mm or less than 59·7 mm. It is found that 3% are rejected as being too large and 5% are rejected as being too small. Assuming the dimension is Normally distributed, find the mean and standard deviation of the distribution of the dimension, correct to 1 decimal place. Use the mean and standard deviation you have calculated to estimate the percentage of rejects if the limits for *acceptance* are changed to 59·6 mm and 60·3 mm.

                            (W)

3 The length of an engine part must be between 4·81 cm and 5·20 cm. In mass production it is found that 0·8% are too short and 3% are too long. If these lengths are normally distributed about mean μ with standard deviation σ, find two equations of the form $\mu + A\sigma = B$. Solve these equations to find the mean and standard deviation. Each part costs £4 to produce; those that turn out to be too long are shortened, at an extra cost of £2; those that turn out to be too short have to be scrapped. Find the expected total cost of producing 100 parts that meet the specification.

                        (O & C)

4 A sample of 100 apples is taken from a load. The apples have the following distribution of sizes.

Diameter to nearest cm	6	7	8	9	10
Frequency	11	21	38	17	13

Determine the mean and standard deviation of these diameters.
Assuming that the distribution is approximately normal with this mean and this standard deviation find the range of size of apples for packing, if 5% are to be rejected as too small and 5% are to be rejected as too large.

                        (O & C)

5 A machine packs flour into bags which nominally contain 1 kg but there is a variation in the actual weight (kg), which is described by a normal random variable of mean μ and variance σ^2. Previous investigations indicate that $\sigma = 0·03$ kg and that the probability that a bag is underweight is $0·02$. Find the value of μ at which the machine is operating. An attempt is made to improve the machine with the hope that, while it operates with the same value of μ, σ will be reduced. Find the value of σ which is required to ensure that the probability that a bag is underweight is $0·001$. Assuming this improved value of σ to have been achieved, show that the new probability that a random bag will weigh more than 1·1 kg is just less than $0·03$.

                        (L)

78 Uses of the Normal Distribution

Fitting a Normal distribution, Normal approximation to binomial, Normal approximation to Poisson.

Fitting a Normal distribution

There are two cases to consider when fitting a Normal distribution.

1. **Given a frequency distribution**
   (a) Calculate $\bar{x}$ and s^2 from the given data. Use them as estimates of μ and σ^2.
   (b) Note the upper class bound for each class.
   (c) Standardise the upper class bounds using
   $$z = \frac{x - \bar{x}}{s}.$$
   (d) Find $\Phi(z)$ for each standardised upper class bound. This gives a set of cumulative probabilities.
   (e) Obtain class probabilities by subtracting successive Φs.
   (f) Calculate expected frequencies by multiplying each probability by the total given frequency.

2. **Given μ and σ^2**
   (a) Determine a practical range, i.e. $\mu - 3\sigma$ to $\mu + 3\sigma$.
   (b) Divide this range into about 10 classes.
   (c) Proceed as in 1.

ℹ️ *Fit a Normal distribution to this data:*

class	up to 15	15 up to 20	20 up to 25	25 up to 30	30 up to 35	35 and over
f	5	12	26	34	16	7

Take the first class as 10 up to 15 and the last class as 35 up to 40.
Using mid-class values:
$\Sigma fx = 2575$, $\Sigma f = 100 \Rightarrow \bar{x} = 25.75$
$\Sigma fx^2 = 70075 \Rightarrow s^2 = 37.7$ and $s = 6.14$.
The upper class bounds are:
$$15, 20, 25, 30, 35, \infty$$
The standardised upper class bounds are:
$$-1.74, -0.94, -0.12, 0.89, 1.51, \infty$$
Φ for each standardised upper class bound:
$$0.0409, 0.1736, 0.4522, 0.7549, 0.9345, 1$$
The probabilities for each class are:
$$0.0409, 0.1327, 0.2786, 0.3027, 0.1796, 0.0655$$
$\Sigma f = 100$, so the expected frequencies are:
$$4, 13, 28, 30, 18, 7 \text{ (total 100)}$$

Normal approximation to binomial

The Normal distribution may be used to approximate the binomial distribution when n is large ($n > 50$) and p is not too big or small ($0.2 \leqslant p \leqslant 0.8$). If n is very large, then the approximation is good even if p is near to 0 or 1.

If X is Bin(n, p), then X is approximately N$(np, np(1-p))$ since $E[X] = np$ and Var$[X] = np(1-p)$.

To compensate for the change from a discrete distribution (the binomial) to a continuous distribution (the Normal), a continuity correction is made. The discrete integer value a in the binomial distribution becomes the class interval $[(a-0.5)$ up to $(a+0.5)]$ in the Normal distribution.

So the discrete variable 3 becomes the class interval 2.5 up to 3.5, and the discrete value '>5' becomes '>5.5'.

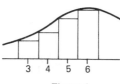

Figure 1

ℹ️ *A machine manufacturing nails makes approximately 15% that are outside set tolerance limits. If a random sample of 200 is taken, find the probability that more than 20 will be outside the tolerance limits.*

If the random variable X is 'number of nails outside limits', then X is Bin (200, 0.15).
$E[X] = 200 \times 0.15 = 30$ and Var$[X] = 30 \times 0.85 = 25.5$.
So X is approximately N(30, 25.5).
We require $P(X > 20.5)$ using the continuity correction.
Standard variable is $Z = \dfrac{X - 30}{5.05}$

$$\therefore P(X > 20.5) = P\left(Z > \frac{20.5 - 30}{5.05}\right) = P(Z > -1.88)$$
$$= P(Z < 1.88) = \Phi(1.88)$$
$$= 0.9699$$

i.e. It is almost certain that there will be more than 20 faulty nails in the sample of 200.

Normal approximation to Poisson

The Normal distribution may be used to approximate the Poisson distribution when μ is large ($\mu > 20$).
If X is Po(μ), then X is approximately N(μ, μ) since $E[X] = \mu$ and Var$[X] = \mu$.
The continuity correction is also required because the Poisson distribution is discrete.

ℹ️ *An accident 'black-spot' averages 2 per week. Find the probability that there are 24 or more accidents in a 12 week period.*

Over a 12 week period the mean number of accidents would be $2 \times 12 = 24$.
If X is the random variable 'number of accidents in a 12 week period', then X is Po(24).
Use the Normal approximation N(24, 24).

Standard variable is $Z = \dfrac{X - 24}{4.90}$

We require $P(X \geqslant 23.5)$ using the continuity correction.

$$P(X \geqslant 23.5) = P\left(Z > \frac{23.5 - 24}{4.90}\right) = P(Z > -0.102)$$
$$= \Phi(0.102) = 0.541.$$

i.e. The required probability is 0.541.

Uses of the Normal Distribution
Worked example and Exam questions

 (a) *Estimate the probability that a fair coin comes down heads more than 290 in 500 tosses.*

(b) *Assuming that a biased coin comes down heads with probability $\dfrac{290}{500}$, estimate the least integer value of r such that the probability of getting more than r heads in 500 tosses is less than $\dfrac{1}{1000}$.*

(a) $P(\text{success}) = P(\text{a head}) = \frac{1}{2}$
Let the random variable X be 'the number of heads in 500 tosses', then X is Bin$(500, \frac{1}{2})$
$np = 250$ and $np(1-p) = 125$
$\therefore$ since n is large and p is not too large or too small we take a normal approximation and let X be N(250, 125) approximately.

The standard variable is $Z = \dfrac{X-\mu}{\sigma} = \dfrac{X-250}{\sqrt{125}}$

$\therefore P(X>290) = P\left(Z > \dfrac{290 \cdot 5 - 250}{\sqrt{125}}\right)$ Note, the continuity correction
$= P(Z > 3 \cdot 622)$
$= 1 - \Phi(3 \cdot 622)$
$= 0 \cdot 00022$

(b) $P(\text{success}) = P(\text{a head}) = \dfrac{290}{500}$

Let the random variable X be 'the number of heads in 500 tosses', then X is Bin$\left(500, \dfrac{29}{50}\right)$

$np = 290$, $np(1-p) = 121 \cdot 8$
$\therefore$ since n is large and p is not too large or too small we take a normal approximation and let X be N(290, 121·8) approximately.

The standard variable is $Z = \dfrac{X-\mu}{\sigma} = \dfrac{X-290}{\sqrt{121 \cdot 8}}$

$P(X>r) = P\left(Z > \dfrac{r+0 \cdot 5 - 290}{\sqrt{121 \cdot 8}}\right) = 1 - \Phi\left(\dfrac{r+0 \cdot 5 - 290}{\sqrt{121 \cdot 8}}\right)$
$< 0 \cdot 001$

i.e. $\Phi\left(\dfrac{r+0 \cdot 5 - 290}{\sqrt{121 \cdot 8}}\right) > 0 \cdot 999$

i.e. $\dfrac{r - 289 \cdot 5}{\sqrt{121 \cdot 8}} = 3 \cdot 1$, and $r = 323 \cdot 71$

Hence the least integer value of r is 324.

---

 1 The probability of success in each of 5 independent trials is p. Show that the mean number of successes is $5p$. (If you quote a formula for the mean of a binomial distribution, you should prove that it is correct.) A farmer has 5 cows in calf. Assuming these calves are equally likely to be bulls or heifers, find the probability that there will be at least 2 heifers. He estimates bulls to be worth £200 each and heifers to be worth £300 each. Find the expected value of the 5 calves. A second farmer has 50 cows in calf. Use the normal distribution to estimate the probability that they will produce at least 35 heifers.

(O & C)

2 One of two dice is loaded so that there is a probability of $0 \cdot 2$ of throwing a six with it, nothing being known about the other scores. The other die is fair. A person is given one of these dice (which is just as likely to be the fair as the biased one), together with the above information and is asked to discover which die it is. He decides to throw the die 10 times; if there are two or more sixes he will assert that the die is biased, otherwise he will assert that it is fair.

Calculate the probability of his asserting that the die is (i) biased when it is, in fact, fair; (ii) fair when it is, in fact, biased. What is the probability that his choice will be incorrect?
If, instead, he decided to throw the die 240 times and will assert that the die is biased if there are N or more sixes, use the Normal approximation to the Binomial distribution to estimate N if the probability of his asserting that it is fair when it is biased is to be $0 \cdot 2$.

(S)

3 (a) Estimate the probability that a fair coin would come down heads 270 times or more in 500 tosses.
(b) Assuming that a biased coin comes down heads with probability 270/500, estimate the least integer value of r such that the probability of getting r heads or more in 500 tosses is less than 1/1000.

(O & C)

4 Calculate the normal approximation to the probability of obtaining exactly 30 successes in 90 independent trials in each of which the probability of success is $\frac{1}{3}$. In a sequence of 100 independent identical trials the number of successes is 36. Estimate the probability of success at a given trial and the variance of the proportion of successes in 100 such trials. Is the observed proportion of successes significantly different from $\frac{1}{3}$?

(OLE)

5 A machine manufactures glass bottles of which, on average, 1 in 20 are found to be defective. If a random sample of five bottles is taken from the production line, calculate the probability that
(i) there is at least one defective bottle,
(ii) there are at most two defective bottles.
If a random sample of 1000 bottles is taken from the production line, calculate the probability that there are at least 35, but no more than 55, defective bottles.

(A)

6 Explain, briefly, the circumstances under which a Binomial distribution may be approximated by:
(a) a Normal distribution, (b) a Poisson distribution.
Give examples, from projects where possible, of the use of these approximations.
Neutrophils and basophils are two distinct types of white cells. The proportion of white cells which are neutrophils is $0 \cdot 55$, and the proportion which are basophils is $0 \cdot 005$. Calculate the probability of there being more than sixty in all of the two types (neutrophils and basophils) in a random sample of 100 white cells. Calculate also the probability of there being three or fewer basophils in a random sample of 200 white cells.

(L)

7 In a certain factory 70% of the microprocessors produced are found to be imperfect and have to be discarded. Find the probability that, in a randomly chosen sample of 5 microprocessors produced, the number of perfect specimens will be: (i) 0, (ii) 1, (iii) 2, (iv) 3, (v) 4, (vi) 5. Verify that the probability of more than 3 perfect specimens is less than $\dfrac{1}{30}$. If a sample of 200 is taken from the same set of microprocessors, find the probability that there are at least 50 perfect specimens.

(A)

79 Sampling

Random sampling, Sample statistics, Sampling distributions, Sums and differences, Finite population sampled without replacement.

Random sampling

A **sample** is a subset of a population.

A **random sample** is one in which each member of the population has an equal chance of being selected.

Tables of random numbers are often used to construct random samples.

ⓘ A random sample of 5 is required from 50 observations labelled 0 to 49.

From a table of random digits we have
20, 74, 94, 22, 93, 45, 44, 16
The numbers outside the allowed range are ignored.

Sample statistics

A **sample statistic** is any quantity which depends only on the data of the sample.

If a large number of random samples, of the same size, are taken from the same population, then the same statistic calculated for all of the samples will form a distribution, called the **sampling distribution** of the statistic.

The standard deviation of a sampling distribution is called the **standard error** of the sample statistic.

ⓘ A die was thrown 4 times and the mean score calculated. This was repeated 50 times and gave rise to the sampling distribution below.

$\bar{x}$	1–	1.5–	2–	2.5–	3–	3.5–	4–	4.5–	5–	5.5–
f	0	0	5	5	12	14	5	7	2	0

For this case $\sigma_{\bar{x}}^2 = 0.6256$
So the standard error is $\sigma_{\bar{x}} = 0.79$.

Sampling distributions

Mean: $\bar{x}$ is approximately $N\left(\mu, \dfrac{\sigma^2}{n}\right)$. This is the **Central Limit Theorem**. If the parent population is normal this is an exact result. The standard variable $Z = \dfrac{\bar{x} - \mu}{\sqrt{\dfrac{\sigma^2}{n}}}$ is $N(0, 1)$ approximately.

Proportion: P is approximately $N\left(\Pi, \dfrac{\Pi(1-\Pi)}{n}\right)$ for $n \geqslant 30$ and Π is the population proportion.

The standard variable $Z = \dfrac{P - \Pi}{\sqrt{\dfrac{\pi(1-\pi)}{n}}}$ is $N(0, 1)$ approximately.

ⓘ A sample of size 16 is drawn from a population with mean 52 and variance 64. Find the probability that the sample mean is greater than 55.

The sampling distribution of sample means is approximately $N\left(52, \dfrac{64}{16}\right)$, i.e. $N(52, 4)$.

$P(\bar{x} > 55) = P\left(Z > \dfrac{55 - 52}{2}\right)$
$= 1 - \Phi(1.5)$
$= 1 - 0.9332$
$= 0.0668$

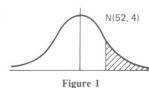

N(52, 4)

Figure 1

Sums and differences

If X and Y are two independent random variables which are $N(\mu_1, \sigma_1^2)$ and $N(\mu_2, \sigma_2^2)$ respectively, then $X + Y$ is $N(\mu_1 + \mu_2, \sigma_1^2 + \sigma_2^2)$ and $X - Y$ is $N(\mu_1 - \mu_2, \sigma_1^2 + \sigma_2^2)$.

This result readily extends to the X_i $(i = 1, \ldots, n)$ random variables which are $N(\mu_i, \sigma_i^2)$ respectively, then ΣX_i is $N(\Sigma \mu_i, \Sigma \sigma_i^2)$.

Difference of means; $\bar{x} - \bar{y}$ is approximately $N\left(\mu_1 - \mu_2, \dfrac{\sigma_1^2}{n_1} + \dfrac{\sigma_2^2}{n_2}\right)$ for $n \geqslant 30$.

The standard variable $Z = \dfrac{(x - y) - (\mu_1 - \mu_2)}{\sqrt{\dfrac{\sigma_1^2}{n_1} + \dfrac{\sigma_2^2}{n_2}}}$ is $N(0, 1)$ approximately.

ⓘ A girl travels to college by walking part of the way and travelling the rest by train. Over a period of time she estimates that the walking time and train time are approximately $N(12, 3)$ and $N(20, 6)$ minutes respectively. Find the probability that if she leaves home 40 minutes before a lecture starts that she will be late.

Assume the two components are independent. The distribution of the total journey time is $N(12 + 20, 3 + 6) = N(32, 9)$

$P(T > 40) = P\left(Z > \dfrac{40 - 32}{3}\right)$
$= 1 - \Phi(2.33)$
$= 0.0099$

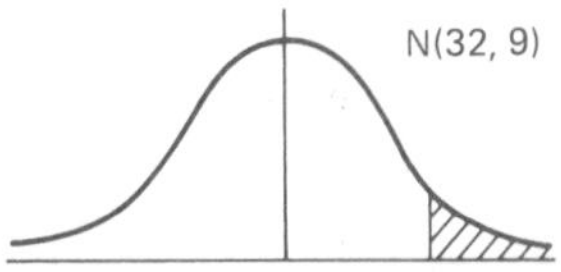

N(32, 9)

Figure 2

Finite population sampled without replacement

If a sample of size n is taken from a finite population of size N and the **sampling** is **without replacement**, then for the sample mean $\bar{x}$,

$$E[\bar{x}] = \mu \quad \text{and} \quad \text{Var}[\bar{x}] = \dfrac{\sigma^2}{n}\left(\dfrac{N - n}{N - 1}\right)$$

where μ and σ^2 are the population mean and variance.

ⓘ A random sample of 20 is taken from a population of size 80 without replacement. Find the expectation and variance of the sample if the population mean is 2.85 and standard deviation 0.07.

For the sample mean $\bar{x}$:
$E[\bar{x}] = \mu = 2.85$

$$\text{Var}[\bar{x}] = \dfrac{\sigma^2}{n}\left(\dfrac{N - n}{N - 1}\right) = \dfrac{(0.07)^2}{20}\left(\dfrac{80 - 20}{80 - 1}\right)$$

$$= 0.0002$$

Sampling
Worked example, Guided example and Exam questions

 The inside diameters of bearings supplied by a factory have a mean of 21·04 mm and a standard deviation of 0·03 mm. The diameters of axles supplied by a second factory have a mean value of 20·92 mm and a standard deviation of 0·05 mm. What is the mean and standard deviation of the random variable defined to be the diameter of a bearing less the diameter of an axle? Assuming that both dimensions are normally distributed, what percentage of axles and bearings taken at random will not fit?

Let D_B be the random variable 'diameter of a bearing',
$\mu_B = 21\cdot4$ and $\sigma_B = 0\cdot03$
Let D_A be the random variable 'diameter of an axle',
$\mu_A = 20\cdot92$ and $\sigma_A = 0\cdot05$
Then the random variable $D_B - D_A$ is 'the diameter of a bearing less the diameter of an axle' and
$$E[D_B - D_A] = E[D_B] - E[D_A] = \mu_B - \mu_A = 21\cdot04 - 20\cdot92$$
$$= 0\cdot12 \text{ mm}$$
$$\text{Var}[D_B - D_A] = \text{Var}[D_B] + \text{Var}[D_A] = \sigma_B^2 + \sigma_A^2;$$
$$= 0\cdot03^2 + 0\cdot05^2 = 0\cdot0034 \text{ mm}^2$$
Assuming D_B is N(21·04, 0·03²) and D_A is N(20·92, 0·05²)
then $(D_B - D_A)$ is N(0·12, 0·0034) with standard variable
$$Z = \frac{(D_B - D_A) - 0\cdot12}{0\cdot0583}.$$
If axle and bearing do not fit $D_B - D_A < 0$, i.e. the axle is bigger than the bearing.
$$P(D_B - D_A < 0) = P\left(Z < \frac{-0\cdot12}{0\cdot0583}\right)$$
$$= P(Z < -2\cdot058)$$
$$= 1 - \Phi(2\cdot058)$$
$$= 1 - 0\cdot9802$$
$$= 0\cdot0198$$
∴ 1·98% of bearings and axles will not fit.

GE *The random number 482 is obtained from a table of random digits. Use it to obtain a random observation from each of the following distributions, quoting as many significant figures as are justified in each answer.*
(a) Poisson with mean 1·6.
(b) Normal with mean 4 and variance 4.

(a) We assign the random digits 000 to 999 to the distribution Po(1·6) in proportion to the probabilities corresponding to the values of X, the random variable which is Po(1·6).
The working is set out in tabular form below, with the first two rows only shown.

Probability	Cumulative probability (to 3dp)	Allocation of random numbers
$P(X=0) = e^{-1\cdot6}$ $= 0\cdot2019$	0·202	001–202
$P(X=1) = e^{-1\cdot6} \times \dfrac{1\cdot6}{1!}$ $= 3\cdot3230$	0·525	203–525

The random number 482 is in the range 203–525 corresponding to the value 1 of the random variable X which is Po(1·6).
(b) The distribution to be considered is N(4, 4).
Take 4 ± 6 (approximately $\mu \pm 3\sigma$) i.e. −2 to 10 as an effective range. Divide this range into 8 classes. Find the upper class bounds, the standardised upper class bounds and the cumulative probability corresponding to each of the

standardised upper class bounds (as in the method of Unit 78, Uses of the Normal Distribution). The random numbers are then allocated as above. It will be easy to see the interval in which the random observation lies, simple proportion in this interval gives an estimate of the random observation.

---

1 The number 1437 is obtained from a table of random digits. Use it to select an observation at random from each of the following distributions. Your method should be clearly indicated by your written working, and your solution should contain as many significant figures as the given random number permits.
(a) Rectangular distribution with range 13 to 15.
(b) Binomial distribution with parameters 4, ¾.
(c) Poisson distribution with mean 2·5.
(d) Normal distribution with mean 11, variance 4.

(OLE)

2 A discrete random variable X has probability distribution given by $P(X=0) = 0\cdot1$, $P(X=0 \text{ or } 1) = 0\cdot3$; $P(X=0 \text{ or } 1 \text{ or } 2) = 0\cdot7$, $P(X=0 \text{ or } 1 \text{ or } 2 \text{ or } 3) = 1\cdot0$. Obtain the expectation and variance of X. Describe how you would use a table of single-figure random numbers to generate a random sample of size n from the above distribution.
Use your procedure, for the case $n = 10$, with the random numbers given below, to obtain unbiased estimates of the expectation and variance of X.
Random numbers: 0, 9, 3, 3, 3, 7, 4, 7, 5, 6.

(C)

3 In a packaging factory, the empty containers for a certain product have a mean weight of 400 g with a standard deviation of 10 g. The mean weight of the contents of a full container is 800 g with a standard deviation of 15 g. Find the expected total weight of 10 full containers and the standard deviation of this weight, assuming that the weights of containers and contents are independent.
Assuming further that these weights are normally distributed random variables, find the proportion of batches of 10 full containers which weigh more than 12·1 kg.
If 1% of the containers are found to be holding weights of product which are less than the guaranteed minimum amount, deduce this minimum weight.

(O & C)

4 A population of size N has mean μ and variance σ^2. Random samples of size n are taken without replacement. Write down the expectation and the variance of the means of these samples.
In the game of bridge hands of size 13 are dealt to each of four players in such a way that each hand can be considered to be a random sample without replacement from a standard pack of 52 cards.
Each player has to decide upon a 'bid' for his hand, and to help him to this one particular player decides to allot points to his cards. An ace receives 4 points, a king 3 points, a queen 2 points and a jack 1 point; all others receiving 0 points. He thus arrives at a total score for this hand. Calculate the expectation and variance of the mean score and deduce the expectation and variance of the total.
Assuming that the distribution of this total is normal, calculate the probability of a total score greater than 18.

(A)

Point estimation, Interval estimation.

Point estimation	**Point estimation** involves using a statistic from a random sample to find an estimator for the corresponding population parameter.	

Point estimation involves using a statistic from a random sample to find an estimator for the corresponding population parameter.

Note: 'Estimator' is used for a statistic; 'estimate' is used for the numerical value of that statistic.

An **unbiased estimator** is a sample statistic whose expectation is equal to the population parameter.

The **best** or **most efficient estimator** is the unbiased estimator which has the smallest variance.

i Some important results are:

	sample statistic	best estimator for population
mean	$\bar{x}$	$\bar{x}$ for μ
variance	s^2	$\dfrac{ns^2}{n-1}$ for σ^2
proportion	r	r for Π

Interval estimation

Interval estimation involves using the data from a random sample to find an interval within which an unknown population parameter is expected to lie with a given degree of confidence (probability).

The interval is called a **confidence interval** and the two extreme values are called the **confidence limits**.

If $\bar{x}$ is the mean of a random sample of size n from $N(\mu, \sigma^2)$, where σ^2 is known, then a symmetric $B\%$ confidence interval for μ

is given by $\bar{x} \pm z . \dfrac{\sigma}{\sqrt{n}}$

i.e. $\bar{x} - z . \dfrac{\sigma}{\sqrt{n}} < \mu < x + z . \dfrac{\sigma}{\sqrt{n}}$

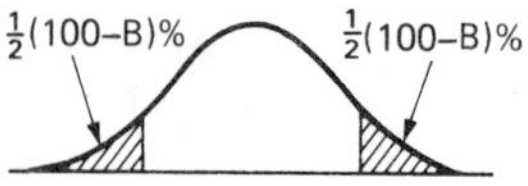

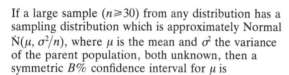

$\frac{1}{2}(100-B)\%$ $\frac{1}{2}(100-B)\%$

Figure 1

where z is the $\frac{1}{2}(100-B)\%$ point of $N(0, 1)$.

If a large sample $(n \geqslant 30)$ from any distribution has a sampling distribution which is approximately Normal $\dot{N}(\mu, \sigma^2/n)$, where μ is the mean and σ^2 the variance of the parent population, both unknown, then a symmetric $B\%$ confidence interval for μ is

$$\bar{x} - z . \frac{s}{\sqrt{n}} < \mu < \bar{x} + z . \frac{s}{\sqrt{n}}$$

where $\bar{x}$, s^2 are the mean and variance of the sample and z is the $\frac{1}{2}(100-B)\%$ point of $N(0, 1)$.

If r is the proportion of a random sample of size n from a population that has a particular property, then an approximate $B\%$ confidence interval for the population proportion Π having the property is

$$r - z\sqrt{\frac{r(1-r)}{n}} < \Pi < r + z\sqrt{\frac{r(1-r)}{n}}$$

where z is the $\frac{1}{2}(100-B)\%$ point of $N(0, 1)$.

i The weight of each of ten specimens of a certain type of beetle were found to be (in grams): 14.3, 13.8, 13.6, 14.6, 15.4, 14.8, 13.1, 14.2, 16.8, 15.1

Given that the weights are approximately Normally distributed with variance 1.44, construct a 95% confidence interval for μ, the mean of the population weights.

$$\bar{x} = \frac{\Sigma x}{n} = \frac{145.7}{10} = 14.57$$

$\sigma^2 = 1.44$, so $\sigma = 1.2$

For a 95% confidence interval, z is the $\frac{1}{2}(100-95)\%$ $=2.5\%$ point of $N(0, 1)$.
$\Phi(z) = 1 - 0.025 = 0.975 \Rightarrow z = 1.96$
$\therefore$ a 95% confidence interval for μ is given by:

$$14.57 - 1.96 \times \frac{1.2}{\sqrt{10}} < \mu < 14.57 + 1.96 \times \frac{1.2}{\sqrt{10}}$$

i.e. $13.83 < \mu < 15.31$

i *Before a by-election, for which there are two candidates A and B, a survey was made of 350 voters, chosen at random, and it was found that 198 of them intend to vote for A. Give 95% confidence limits for the percentage of voters favourable to A at the time of the survey.*

Using proportions the property to be considered is 'is an A voter'.

Assuming that the total population votes for either A or B, the sample proportion is given by

$$r = \frac{198}{350} = 0.566$$

For a 95% confidence level, z is the $\frac{1}{2}(100-95)\%$ $=2.5\%$ point of $N(0, 1)$.
$\Phi(z) = 1 - 0.025 = 0.975 \Rightarrow z = 1.96$
$\therefore$ a 95% confidence interval for Π is given by

$$0.566 \pm 1.96 \sqrt{\frac{0.566(1-0.566)}{350}}$$

i.e. $0.514 < \Pi < 0.618$
$\therefore$ between 51.4% and 61.8% of voters would vote for A.

Estimation
Worked examples and Exam questions

 A bag contains 10 balls of which 3 are blue and 7 are yellow. A random sample of 3 balls is taken, without replacement, and $\hat{p}$ denotes the proportion of blue balls in the sample.
Tabulate the probability distribution of $\hat{p}$, and hence verify that $\hat{p}$ is an unbiased estimate of the population proportion.

P(drawing a blue ball) $= 0 \cdot 3$
∴ if the random variable X is 'number of blue balls in a sample of 3' then X is Bin(3, $0 \cdot 3$).
$P(X = 0) = (0 \cdot 7)^3 = 0 \cdot 343$
$P(X = 1) = 3(0 \cdot 7)^2(0 \cdot 3) = 0 \cdot 441$
$P(X = 2) = 3(0 \cdot 7)(0 \cdot 3)^2 = 0 \cdot 189$
$P(X = 3) = (0 \cdot 3)^3 = 0 \cdot 027$
∴ the probability distribution of $\hat{p}$ is

x	0	$\frac{1}{3}$	$\frac{2}{3}$	1
$P(\hat{p} = x)$	$0 \cdot 343$	$0 \cdot 441$	$0 \cdot 189$	$0 \cdot 027$

$E[\hat{p}] = \Sigma x P(\hat{p} = x)$
$\qquad = 0 \times 0 \cdot 343 + \frac{1}{3} \times 0 \cdot 441 + \frac{2}{3} \times 0 \cdot 189 + 1 \times 0 \cdot 027$
$\qquad = 0 \cdot 3$, the population proportion.
∴ $\hat{p}$ is an unbiased estimator for the population proportion.

 A certain city has about 1 million adult inhabitants of whom an unknown proportion p have never spent a holiday in a foreign country. A random sample of 1000 of the adult inhabitants is taken, and 735 people in the sample are found never to have spent a holiday in a foreign country. Find a 95% confidence interval for p.

The attribute of interest is 'has never spent a holiday in a foreign country'.

The sample proportion is $\frac{735}{1000} = 0 \cdot 735$.

The 95% confidence interval for p is

$p_s - 1 \cdot 96 \sqrt{\dfrac{p_s(1-p_s)}{n}} < p < p_s + 1 \cdot 96 \sqrt{\dfrac{p_s(1-p_s)}{n}}$

where $p_s = 0 \cdot 735$
$\qquad n = 1000$

Substituting and evaluating leads to
$0 \cdot 708 < p < 0 \cdot 762$ as the 95% confidence interval for the population proportion p.

---

 1 Explain briefly what the standard error of the mean is used for. Among the first 150 customers at a new snack bar 90 order coffee. Assuming that this is a random sample from the population of future customers, estimate 95% confidence limits for the proportion of future customers who will order coffee. If the proportion of future customers who order coffee is exactly 60%, find the probability that 2, 3 or 4 of the next 5 customers will order coffee.
(O & C)

2 X is a random variable. Define var (X), the variance of X, in terms of $E(X)$ and $E(X^2)$. Deduce that var $(aX + b) = a^2$ var (X), where a and b are constants. The weights in grams of the contents of 200 packets of soap powder are summarised in the table below.

Weight (Centre of Interval)	899	900	901	902
Frequency	40	90	58	12

Find the mean and the standard deviation of these measurements. Assuming they are a random sample from a normal distribution, find 99% confidence limits for the population mean, correct to the nearest $0 \cdot 02$ g.
(O & C)

3 A point whose co-ordinates are (X, Y) with respect to rectangular axes is chosen at random where $0 < X < 1$ and $0 < Y < 1$. What is the probability that the point lies inside the circle whose equation is $x^2 + y^2 = 1$? In a computer simulation 1000 such points were generated and 784 of them lay inside the circle. Obtain an estimate for π and give an approximate 90% confidence interval for your estimate. Show that about 290 000 points need to be selected in order to be 90% certain of obtaining a value for π which will be in error by less than $0 \cdot 005$.
(S)

4 In the production of an item of furniture, part A fits into part B. For part A, the relevant outer dimension is x, and for B the corresponding inner dimension is y. Both x and y are normally distributed, having means μ_x, μ_y and standard deviations σ_x, σ_y respectively. State the mean and variance of $y - x$.
It is given that $\mu_x = 2 \cdot 05$ cm, $\mu_y = 2 \cdot 10$ cm, $\sigma_x = 0 \cdot 03$ cm, $\sigma_y = 0 \cdot 04$ cm. In assembly, a part A is selected at random and an attempt is made to fit it into a part B, also selected at random. Find the percentage of pairs so selected which must be rejected because part A is too large to fit into part B.
In order to reduce this percentage, the setting of the machine which produces part A is adjusted so that the mean μ_x is altered, the variance remaining unchanged. A sample of 50 of part A has a mean value for x of $2 \cdot 01$ cm. Write down a symmetrical two-sided 99% confidence interval for the new value of μ_x, giving the limits to two decimal places. Calculate the corresponding range of values of the percentage of pairs for which part A is too large to fit into part B.
(J)

5 There are n_0 fish in a lake. A random sample of m of these fish is taken. The fish in this sample are tagged and released unharmed back into the lake. After a suitable interval, a second random sample of size n is taken. The random variable R is the number of fish in this second sample that are found to have been tagged. Assuming that the probability that a fish is captured is independent of whether it has been tagged or not, and that n_0 is sufficiently large for a binomial approximation to be used, obtain the expectation of R in terms of m, n and n_0. Suppose that $m = 100$, $n = 4000$ and that the observed value of R is 20. Obtain an approximate symmetric 98% confidence interval for the proportion of fish in the lake which are tagged. Deduce an approximate 98% confidence interval for n_0.
(C)

6 Distinguish between the expressions $\dfrac{\Sigma(x - \bar{x})^2}{n}$ and $\dfrac{\Sigma(x - \bar{x})^2}{n-1}$, both of which are used in connection with variance for a set of observations. A random sample of 100 observations is taken from a distribution. The sum of the observations is 1000 and the sum of their squares is 19 900.
(a) Explain how you would estimate the mean and the variance of the distribution from which the random sample was taken, and give the values of these estimates.
(b) Estimate the mean and the variance of the distribution of the mean of random samples of size 100 from the original distribution.
(c) Construct a 95% confidence interval for the mean of the distribution, and use it to test whether this mean could be 9.
(OLE)

81 Hypothesis Testing
Statistical hypotheses, Critical regions, Types of error.

Statistical hypotheses

A **statistical hypothesis** is an assumption about the value of a statistic of a distribution.

A **null hypothesis,** H_0, is a statistical hypothesis which can be tested in some way.

An **alternative hypothesis,** H_1, is the one which is accepted if the null hypothesis is rejected.

A **test of a null hypothesis,** or **significance test,** is a rule, based on the results of a random sample, whereby acceptance or rejection of H_0 is decided.

$\boxed{i}$ The sample distribution of the mean is $N(\mu, \sigma^2/n)$ with standardised variable $z = \dfrac{\bar{x} - \mu}{\sigma/\sqrt{n}}$

Typically H_0 will be an assumption about μ, the population mean, for instance $H_0: \mu = \mu_0$. The value $\bar{x}$, obtained from a sample, is tested on this assumption.
Possible alternative hypotheses in this case are:
$H_1: \mu \neq \mu_0$; $H_1: \mu > \mu_0$; $H_1: \mu < \mu_0$.

Critical regions

The **critical region** corresponding to every test must be found.

If the sample value falls in the critical region, then H_0 is rejected, otherwise it is accepted.

A test statistic is said to be **significant** if it falls in the critical region, otherwise it is **non-significant.**

The critical region depends upon:
(a) the significance levels of the test,
(b) the nature of H_1, the alternative hypothesis.

The **significance level** of the test gives the probability assigned to rejecting H_0.

There are two types of alternative hypothesis, **one-tailed** and **two-tailed,** which give rise to two types of test. A one-tailed test considers only an increase or only a decrease in the parameter and there is only one critical region whose area is equal to the level of significance. A two-tailed test considers any change in the parameter and there are two equal critical regions whose area sum is equal to the level of significance.

Consider the following tests based on H_0; $\mu = \mu_0$.

(i) $H_0: \mu = \mu_0$
   $H_1: \mu \neq \mu_0$
Significance level $\alpha\%$
A two-tailed test since any change in μ is considered.

(ii) $H_0: \mu = \mu_0$
   $H_1: \mu > \mu_0$
Significance level $\alpha\%$
A one-tailed test since only an increase in μ is considered.

(iii) $H_0: \mu = \mu_0$
   $H_1: \mu < \mu_0$
Significance level $\alpha\%$
A one-tailed test since only a decrease in μ is considered.

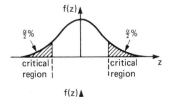
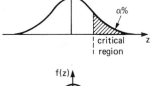
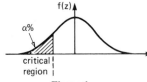

Figure 1

$\boxed{i}$ *Nails produced by a machine have a mean value of 1.50 in. A random sample of 100 nails has a mean length of 1.51 in with a standard deviation of 0.05 in. Do these results indicate that the mean length of nails produced has changed at the 5% significance level?*

The sampling distribution of the mean $\bar{x}$ is $N(1.50, \sigma^2/n)$.

Since σ^2 is unknown, estimate $\hat{\sigma}^2 = \dfrac{ns^2}{(n-1)}$

where s is the sample standard deviation.

So $\dfrac{\hat{\sigma}^2}{n} = \dfrac{(0.05)^2}{99}$ $\therefore \dfrac{\hat{\sigma}}{\sqrt{n}} = \dfrac{0.05}{\sqrt{99}} = 0.005$

Take $\bar{x} \sim N(1.50, 0.005)$

with standardised variable $Z = \dfrac{\bar{x} - 1.50}{0.005}$.

The null hypothesis is $H_0: \mu = 1.50$.
The alternative hypothesis is $H_1: \mu \neq 1.50$.
Significance level is 5%.
This is a two-tail test since any change in μ is to be considered.

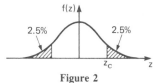

Figure 2

The sample mean $\bar{x} = 1.51$ in.

$z_{\text{test}} = \dfrac{1.51 - 1.50}{0.005} = 4.12 \, (2\,\text{d.p.})$

$\Phi(z_c) = 1 - 0.025 = 0.975$
   $z_c = 1.96$

$\therefore -1.96 < z_{\text{test}} < 1.96$
   $z_{\text{test}} = 4.12 > 1.96$, which is significant
       so H_0 is rejected.

Types of error

A **Type I error** is made when the null hypothesis H_0 is rejected when it should have been accepted. The probability of making this type of error is the level of significance of the test.
We write: P (Type I error) $= \alpha$.

A **Type II error** is made when the null hypothesis H_0 is accepted when it should have been rejected. The probability of making this type of error is not usually easy to calculate.
We write: P (Type II error) $= \beta$.

The **power function** of a test of some statistic is the value of $(1 - \beta)$ and indicates the **power** of the test to reject a wrong hypothesis.

Hypothesis Testing
Worked examples and Exam questions

WE *The O-level results in mathematics from a large school, considered over a number of years, showed an average of 57% passes. In 1975 from a group of 100 students taking the examination the number of passes was 64. Test, at the 1% level, the hypothesis that this was a significantly high number of passes.*

Let the O-level mathematics candidates over the period of years be the population from which each year's candidates are considered to be drawn.
Let R be the proportion of pass candidates in this population.
We have $H_0:R=0\cdot57$, $H_1:R>0\cdot57$ (the test is one-tail).

If r is the sample proportion, the test statistic

$$Z=\frac{r-R}{\sqrt{\dfrac{R(1-R)}{n}}} \text{ is } N(0,1) \text{ where } n=100$$

$$z_{\text{test}}=\frac{0\cdot64-0\cdot57}{\sqrt{\dfrac{0\cdot57\times0\cdot43}{100}}}=1\cdot414, \; z_{1\%}=2\cdot33$$

Since $z_{\text{test}}=1\cdot414<2\cdot33$ the result is not significant and we do not reject H_0. Thus it cannot be concluded that the 1975 students are significantly better than normal.

WE *The top forms of two junior schools both took the same examination on transfer to secondary school. In school X there were 32 pupils and the mean mark obtained was 51. In school Y there were 37 pupils and the mean mark was 46. The standard deviation for marks in this examination, calculated from a large number of primary school candidates, was $\sigma=11\cdot5$.*
Test the hypothesis that the pupils from school X were better than those from school Y, using a 5% significance level.

Let μ_X and μ_Y be the population mean marks for the examination candidates from the two schools.
We have $H_0:\mu_X=\mu_Y$, $H_1:\mu_X>\mu_Y$ (the test is one-tailed)

The test statistic is $Z=\dfrac{(\bar{x}_X-\bar{x}_Y)-(\mu_X-\mu_Y)}{\sqrt{\dfrac{\sigma_X^2}{n_X}+\dfrac{\sigma_Y^2}{n_Y}}}$

where $\bar{x}_X=51$ and $n_X=32$
$\qquad \bar{x}_Y=46$ and $n_Y=37$
$\qquad \sigma_X^2=\sigma_Y^2=11\cdot5$
and $\mu_X-\mu_Y=0$ if H_0 is true.

$$z_{\text{test}}=\frac{51-46}{11\cdot5\sqrt{\dfrac{1}{32}+\dfrac{1}{37}}}=1\cdot801 \text{ and } z_{5\%}=1\cdot65$$

Since $z_{\text{test}}=1\cdot801>1\cdot65$ the result is significant at the 5% level and we reject H_0 in favour of H_1.
Thus is can be considered that the pupils from school X performed significantly better than those from school Y.

EX
1 The mass of jam in a jar is x grams, the mass of the jar is y grams and the mass of the lid is z grams. x, y and z are independent and normally distributed with means 502, 90 and 10 and standard deviations $0\cdot60$, $0\cdot24$ and $0\cdot07$ respectively. Deduce the mean and standard deviation of the total mass of the jar full of jam, complete with lid. These full jars are packed in boxes of 25. Find the mean and standard deviation of the mass of the total contents of a box.
The average mass of jam per jar in a box is found to be $502\cdot2$ grams. Test, at the 5% level, whether or not this is
  (i) significantly different from the expected value,
  (ii) significantly greater than the expected value.   (J)

2 Two hypotheses concerning the probability density function of a random variable are

$$H_0:f(x)=\begin{cases}1 & 1<x<2,\\0 & \text{otherwise};\end{cases}$$

$$H_1:f(x)=\begin{cases}3x^2/7 & 1<x<2,\\0 & \text{otherwise}.\end{cases}$$

Give a sketch of the probability density function for each case. The following test procedure is decided upon: A single observation of X is made and if X is less than a particular value a, where $1<a<2$, then H_0 is accepted; otherwise H_1 is accepted. Find a such that, when H_0 is true, the test procedure leads, with probability $0\cdot1$, to the acceptance of H_1. With this value of a, find the probability that, when H_1 is true, the test procedure leads to the acceptance of H_0.

$\hspace{12cm}(C)$

3 When you are testing a hypothesis, explain the conditions under which you would use (a) a one-tailed test and (b) a two-tailed test. Give an example of a typical problem for each of (a) and (b), referring to your projects if you wish. The breaking strengths of a particular brand of thread are known to have a normal distribution with mean μ and standard deviation $1\cdot4$ units. A random sample of 36 newly produced pieces of thread are found to have a mean breaking strength of $9\cdot3$ units. Test, at the 5% level of significance, the null hypothesis that $\mu=9\cdot7$ units against the alterna'ive hypothesis that $\mu<9\cdot7$ units. We are given that x is a typical breaking strength for the above random sample and that $\Sigma(x^2)=3240$. Assuming that neither the mean nor the standard deviation of the population of breaking strengths had in fact been known, find estimates of the population mean and variance.

$\hspace{12cm}(L)$

4 Assuming that the mean and variance of a random variable X having a Binomial distribution with parameters n and p are np and $np(1-p)$ respectively, prove that the mean and variance of a proportion based on a sample of size n are p and $p(1-p)/n$ respectively, where p is the true proportion. Of a random sample of 50 shoppers in a certain city store 13 stated that they lived more than 10 miles from the city centre. Of a random sample of shoppers from another store in the same city 9 lived more than 10 miles from the city centre. Stating your null and alternative hypotheses and using a significance level of 5%
  (i) test that the true proportion in both stores could be $0\cdot15$;
  (ii) show that the two samples do not offer evidence of a difference in proportions between the two stores.

$\hspace{12cm}(S)$

5 A person claims that he has an almost foolproof system for identifying the suit of a playing card placed face downwards. To test whether the person is merely guessing he is asked to identify the suits of 4 playing cards placed face downwards. It is agreed to accept his claim only if he correctly identifies at least 3 of the 4 cards. Viewing this as a hypothesis testing problem, write down an appropriate null hypothesis and an appropriate alternative hypothesis. Calculate the significance level of the test and the probability that the person will be regarded as merely guessing when in fact his system has a probability of $0\cdot8$ of correctly identifying the suit of a card.

$\hspace{12cm}(J)$

82 Linear Regression

Observations and errors, Method of least squares, Equations of the lines of regression.

Observations and errors

A **linear relationship** between two variables x and y can be represented mathematically as $y = \alpha + \beta x$. In many practical situations x, the independent variable, can be measured with little or no error, while y, the dependent variable, is subject to **random experimental error**. These random experimental errors may be due to limitations in any experimental apparatus used and/or random fluctuations in experimental conditions.

If an experiment is performed for n values of x, x_i for $i = 1, 2, \ldots, n$, then there will be n corresponding equations for y, given by $y_i = \alpha + \beta x_i + \varepsilon_i$.

The ε_i are error terms which take into account random experimental errors, and are usually considered to be independent (the error of one experiment has no effect on the error of another experiment) and to be distributed normally with zero mean and known variance.

Methods of least squares

If the n pairs (x_i, y_i) for an experiment are plotted on a graph, then the points are scattered about a straight line.

The **method of least squares** 'fits' the 'best' line AB to the points by making $\Sigma(P_iQ_1)^2$ a minimum (i.e. minimising the sum of the squared deviations of each point from the 'best' line). This is called the **line of regression of y on x**.

If $\Sigma(P_iR_i)^2$ is minimised, then the corresponding line of best fit is called the **line of regression of x on y**.

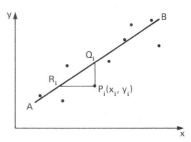

Figure 1

Equations of the lines of regression

The line of regression of y on x (used for predicting y-values given x-values) is

$$y - \bar{y} = \frac{C_{xy}}{C_{xx}}(x - \bar{x})$$

where $\bar{y} = \dfrac{\Sigma y_i}{n}$, $\bar{x} = \dfrac{\Sigma x_i}{n}$,

$$C_{xy} = \Sigma xy - n\bar{x}\bar{y},$$
$$C_{xx} = \Sigma x^2 - n\bar{x}^2.$$

Note: Since $C_{xx} = ns_x^2$, i.e. $s_x^2 = \dfrac{C_{xx}}{n}$ is the variance of x,

$\dfrac{C_{xy}}{n} = s_{xy}$ is called the **covariance of x and y**.

The **line of regression of x on y** (used for predicting x-values given y-values) is

$$x - \bar{x} = \frac{C_{xy}}{C_{yy}}(y - \bar{y})$$

where $C_{yy} = \Sigma y^2 - n\bar{y}^2$.

Since $(\bar{x}, \bar{y})$ satisfies both equations for the lines of regression, it follows that both lines pass through the point $(\bar{x}, \bar{y})$.

$b = \dfrac{C_{xy}}{C_{xx}}$ and $a = \bar{y} - b\bar{x}$ are often referred to as the **least squares estimates** for β and α in the equation $y_i = \alpha + \beta x_i + \varepsilon_i$.

Suppose there is a large number N of independent sets of n experimental pairs (x_i, y_i) and from each set determine estimates a and b for α and β. The set of values of a so formed will have a sampling distribution, as will the set of values of b.

It can be shown that with the four assumptions made above about the errors ε_i, then

$$\frac{b - \beta}{\sigma / \sqrt{C_{xx}}} \sim N(0, 1), \quad \frac{a - \alpha}{\sigma\sqrt{\dfrac{1}{n} + \dfrac{\bar{x}}{C_{xx}}}} \sim N(0, 1)$$

ℹ *For the following set of bivariate data.*

x	4	15	22	27
y	3.2	6.7	11.3	12.5

(a) *plot the points on a graph,*
(b) *calculate the two least squares regression lines,*
(c) *draw these regression lines on the graph,*
(d) *calculate an estimate for x when $y = 7$.*

(a) and (c)

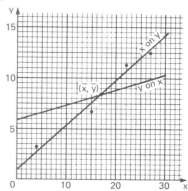

Figure 2

(b) $\Sigma x = 68$, so $\bar{x} = 17$. $\Sigma y = 33.7$, so 8.425.
$\Sigma x^2 = 1454$, so $C_{xx} = 876$.
$\Sigma y^2 = 339.1$, so $C_{yy} = 55.2$.
$\Sigma xy = 699.4$, so $C_{xy} = 126.5$.

The line of regression of y on x is given by

$$(y - 8.425) = \frac{126.5}{876}(x - 17)$$

i.e. $y = 0.144x + 5.97$

The line of regression of x on y is given by

$$(x - 17) = \frac{126.5}{55.2}(y - 8.425)$$

i.e. $x = 2.29y - 2.33$

(d) Use the line of regression of x on y.
When $y = 7$, $x = 2.29(7) - 2.33 = 13.7$.

Linear Regression
Guided example and Exam questions

In a heathland region there are a large number of silver birch trees where the ground is dry but very few where the ground is marshy. The number x of silver birch trees and the ground moisture content y are found in each of 10 equal areas (which have been chosen to cover the range of x in all such areas). The following is a summary of the results of the survey:

$\Sigma x = 495$, $\Sigma y = 425$, $\Sigma x^2 = 31475$, $\Sigma xy = 17300$, $\Sigma y^2 = 20125$.

Find the equation of the regression line of y on x. Estimate the ground moisture content in an area equal to one of the chosen areas which contains 60 silver birch trees.

The equation of the line of regression of y on x is

$$y - \bar{y} = \frac{C_{xy}}{C_{xx}}(x - \bar{x})$$

where $C_{xy} = \Sigma xy - n\bar{x}\bar{y}$ and $C_{xx} = \Sigma x^2 - n\bar{x}^2$

Using the given information C_{xy} and C_{xx} can be calculated and hence the required regression line found.
Substituting $x = 60$ in the regression line gives the corresponding value of y, the ground moisture content.

1 The principle of least squares is used to find the regression line of y on x. Illustrate the distances the sum of whose squares is minimised on a rough sketch showing the x- and y-axes with a few points and their regression line. From 20 pairs of values of x and y the following were calculated:
$\Sigma x = 50$, $\Sigma x^2 = 140$, $\Sigma xy = 27$, $\Sigma y = 24$.
(a) Find the mean values of x and y.
(b) Find the line of regression of y on x in the form $y = a + bx$.
(c) Draw a graph showing this regression line for the range $0 \leqslant x \leqslant 5$.   *(O & C)*

2 Eight candidates sat examinations in Mathematics and Physics. Their corresponding marks were:

Mathematics (x)	63	72	41	56	44	89	70	45
Physics (y)	48	71	50	46	35	92	42	48

(i) Plot these points on a scatter diagram.
(ii) Calculate the equation of the regression line of y on x by the method of least squares.
(iii) Calculate the co-ordinates of the points at which this regression line intersects the lines $x = 10$ and $x = 90$.
(iv) Hence plot the regression line on your diagram.   *(A)*

3 The maximum value, in parts per million (PPM), of atmospheric carbon dioxide in Hawaii for each given year is shown in the table below.

Year (x)	PPM of carbon dioxide (y)
1958	318
1961	320
1964	322
1967	325
1970	329
1973	333
1976	335

Plot a scatter diagram, and obtain the equation of the regression line of y on x by the method of least squares. Show this line on your diagram. Use your equation to estimate the mean annual increase in atmospheric carbon dioxide and the year in which the proportion of carbon dioxide may be expected to reach 350 PPM, if present trends continue.   * *(A)*

4 The values of the dependent variable y corresponding to values of the independent variable x are shown in the following table:

x	0	1	2	3	4
y	2	3	5	4	6

Find the equation of the line of regression of y on x and hence estimate the value of y when $x = 3 \cdot 5$. Explain in what sense the line of regression is a line of best fit to the data.   *(O & C)*

5 (X_i, Y_i), $i = 1, 2, \ldots, n$ is a sample from a bivariate population. The least-square regression lines of Y on X and X on Y are calculated. Why would you not expect the two lines to coincide? Under what circumstances would they coincide? In the table, Y_i is the mass (in grammes) of potassium bromide which will dissolve in 100 grammes of water at a temperature of $X_i°C$.

X	10	20	30	40	50
Y	61	64	70	73	78

Find the equation of the regression line of Y on X. Find, also, the product-moment correlation coefficient between X and Y.   *(S)*

6 In an investigation into prediction using the stars and planets, a celebrated astrologist Horace Cope predicted the ages at which thirteen young people would first marry. The complete data, of predicted and actual ages at first marriage, are now available and are summarised in the following table:

Person	Predicted age x (years)	Actual age y (years)
A	24	23
B	30	31
C	28	28
D	36	35
E	20	20
F	22	25
G	31	45
H	28	30
I	21	22
J	29	27
K	40	40
L	25	27
M	27	26

(i) Draw a scatter diagram of these data.
(ii) Calculate the equation of the regression line of y on x and draw this line on the scatter diagram.
(iii) Comment upon the results obtained, particularly in view of the data for person G. What further action would you suggest?   *(A)*

7 To each value of x, there corresponds a value y of a random variable Y. Forty observations (x, y) are summarised thus:
$\Sigma x = 96$, $\Sigma y = 26$, $\Sigma x^2 = 270$, $\Sigma xy = 58$, $\Sigma y^2 = 18$.
(a) Find the line of regression of Y on x in the form $y = a - bx$.
(b) Draw a graph for $0 \leqslant x \leqslant 6$ showing the regression line; show also on your graph the point $(\bar{x}, \bar{y})$. Indicate roughly on your graph a region in which you would expect almost all the observations (x, y) to lie.
(c) For what value of x would the mean value of the corresponding Y be zero?   *(O & C)*

83 Correlation

Bivariate distributions, Correlation coefficients, Fisher's transformation, *t*-test.

Bivariate distributions

A population with two variables gives rise to **bivariate distributions.**

In such distributions it is often necessary to know any interdependence or correlation.
If the variables are plotted in the *xy*-coordinate plane, the result is a **scatter diagram**.

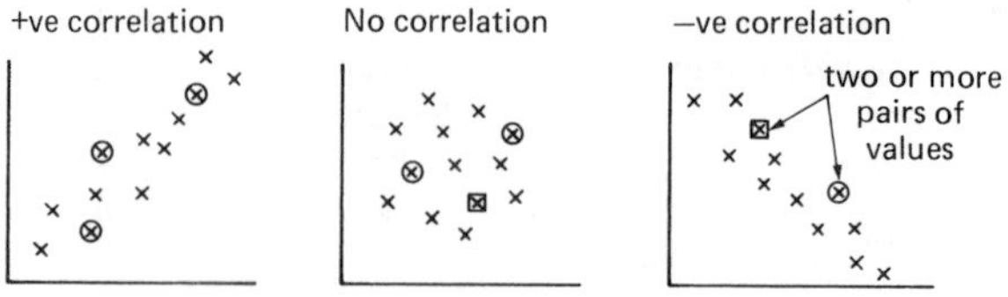

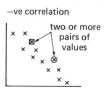

Figure 1

Correlation coefficients

The **product moment correlation coefficient** is given by

$$r = \frac{C_{xy}}{\sqrt{C_{xx}C_{yy}}}$$

where $C_{xy} = \Sigma xy - n\bar{x}\bar{y}$,
$C_{xx} = \Sigma x^2 - n\bar{x}^2$,
$C_{yy} = \Sigma y^2 - n\bar{y}^2$.
r takes the sign of C_{xy}.
It can be shown that $-1 \leqslant r \leqslant 1$.
$r = 1$ for perfect positive correlation.
$r = 0$ for no correlation.
$r = -1$ for perfect negative correlation.

Spearman's rank correlation coefficient is given by

$$r_s = 1 - \frac{6\Sigma d^2}{n(n^2-1)}$$

where d is the rank difference for each pair of values. For tied ranks it is conventional to give both places the average rank of the equal values.

 For the data given below, calculate:
(a) The product moment correlation coefficient,
(b) Spearman's rank correlation coefficient.

x	1	2	3	4	5
y	2	1	3	3	6

(a) From the above data:
$$\Sigma x = 15,\ \Sigma y = 15,\ \Sigma x^2 = 55,\ \Sigma y^2 = 59,\ \Sigma xy = 55.$$
From their definitions:
$$\bar{x} = 3,\ \bar{y} = 3,\ C_{xx} = 10,\ C_{yy} = 14,\ C_{xy} = 10.$$
So $r = \dfrac{10}{\sqrt{10 \times 14}} = 0.845$ (+ve because C_{xy} is +ve)

(b) s and t are the ranks of x and y and $d = (s - t)$

s	5	4	3	2	1
t	4	5	2.5	2.5	1
d^2	1	1	0.25	0.25	0

$$r = 1 - \frac{6(2.5)}{5(24)}$$
$$= 0.875$$

Fisher's transformation

Fisher's transformation, z, of r is given by

$$z = F_i(r) = \frac{1}{2}\ln\frac{(1+r)}{(1-r)} = \tanh^{-1}r$$

and z is approximately $N(\tanh^{-1}\rho, 1/(n-3))$.
To test hypotheses about the population correlation coefficient, ρ:
(a) Transform r and ρ to obtain z_r and z_ρ.

(b) Calculate $z_{\text{test}} = \dfrac{z_r - z_\rho}{\sqrt{1/(n-3)}}$.

(c) Test z_{test} against $N(0, 1)$.

If r_1 and r_2 are the correlation coefficients of two independent random samples, then
$z_1 = F_i(r_1)$ and $z_2 = F_i(r_2)$ are approximately
$N(\tanh^{-1}\rho_1, 1/(n_1-3))$ and $N(\tanh^{-1}\rho_2, 1/(n_2-3))$
and $z_1 - z_2$ is $N\left(\tanh^{-1}(\rho_1 - \rho_2), \dfrac{1}{(n_1-3)} + \dfrac{1}{(n_2-3)}\right)$
approximately.
To test the hypothesis that two samples come from the same population:
(a) Transform r_1 and r_2 to obtain z_1 and z_2

(b) Calculate $z_{\text{test}} = \dfrac{z_1 - z_2}{\sqrt{\dfrac{1}{(n_1-3)} + \dfrac{1}{(n_2-3)}}}$.

(c) Test z_{test} against $N(0, 1)$.

 A random sample of size 45 is taken from a bivariate normal distribution and has a value of
$r = 0.78$.
Test H_0: $\rho = 0.65$ *against* H_1: $\rho > 0.65$.

$$z_r = F_i(r) = F_i(0.78) = 1.045$$
$$z_\rho = F_i(\rho) = F_i(0.65) = 0.7675$$
$$z_{\text{test}} = \frac{1.045 - 0.7675}{\sqrt{\dfrac{1}{42}}} = 1.798$$

From tables, $\Phi(1.64) = 0.95$.
$\therefore z_{\text{test}} = 1.798 > 1.64$
Reject H_0.

one-tail test

Figure 2

The correlation coefficients from two independent samples of sizes $n_1 = 50$ *and* $n_2 = 45$ *are* $r_1 = 0.65$ *and* $r_2 = 0.45$. *Test* H_0: $\rho_1 = \rho_2$ *against* H_1: $\rho_1 \neq \rho_2$.

$z_1 = F_i(r_1) = 0.7675$ and $z_2 = F_i(r_2) = 0.485$.

$$z_{\text{test}} = \frac{0.7675 - 0.485}{\sqrt{\dfrac{1}{47} + \dfrac{1}{42}}} = 1.33$$

From tables, $\Phi(1.96) = 0.975$
$\therefore z_{\text{test}} = 1.33 < 1.96$.
Accept H_0.

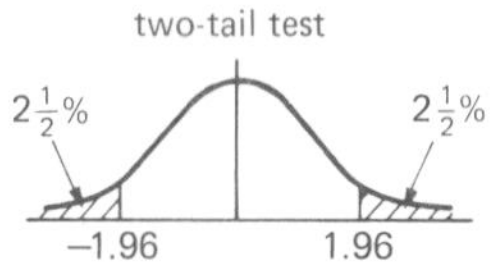

two-tail test

Figure 3

---

t-test

The **t-test** is used to test the hypothesis that the true population correlation coefficient is zero ($\rho = 0$).

(a) Calculate $t = \sqrt{\dfrac{r^2(n-2)}{(1-r^2)}}$, n is the sample size.

(b) Test against $t_{\alpha\%}(n-2)$ where $\alpha\%$ is the level of significance, and is the α percentage point of the *t*-distribution.

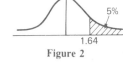 *For the data of the first* *above test the null hypothesis* H_0: $\rho = 0$ *at the 1% level.*

$$t = \sqrt{\frac{0.845^2(5-2)}{(1-0.845^2)}} = 7.49,\ t_{1\%}(3) = 5.84$$

Since $7.49 > 5.84$ reject H_0: $\rho = 0$ at the 1% level.

Correlation
Worked example and Exam questions

x	1·1	1·9	3·0	4·2	5·1	5·8	7·0	8·3	9·3	10·0	10·9	12·1
y	5	7	6	6	8	9	11	12	14	20	43	60

(*i*) *Calculate the product moment correlation coefficient for these data.*

(*ii*) *Assuming that the above data comprise a random sample from a population with a bivariate normal distribution, test the hypothesis $H_0: \varrho = 0·95$, where ϱ is the true population correlation coefficient. Use a 5% level of significance.*

(*iii*) *Calculate Spearman's rank correlation coefficient.*

(i) $r^2 = \dfrac{C_{xy}}{C_{xx}C_{yy}}$ where $C_{xy} = \Sigma xy - n\bar{x}\bar{y}$
$$C_{xx} = \Sigma x^2 - n\bar{x}^2$$
$$C_{yy} = \Sigma y^2 - n\bar{y}^2$$

$\bar{x} = \dfrac{\Sigma x}{12} = 6·56$, $\bar{y} = \dfrac{\Sigma y}{12} = 16·75$

$\Sigma xy = 1856·5$, $\Sigma x^2 = 660·71$, $\Sigma y^2 = 6601$

$C_{xy} = 1856·5 - 12 \times 6·56 \times 16·75 = 537·94$

$C_{xx} = 660·71 - 12 \times 6·56^2 = 144·31$

$C_{yy} = 6601 - 12 \times 16·75^2 = 3234·25$

$r^2 = \dfrac{537·94^2}{144·31 \times 3234·25} = 0·62$, $r = 0·79$ (positive $\because C_{xy}$ is positive)

(ii) Let $z = Fi(r)$; $H_0: \varrho = 0·95$, $H_1: \varrho \neq 0·95$ (two tail test)
(z is the Fisher transformation of r; see Unit 81, Hypothesis Testing.)

$z_r = Fi(0·79) = 1·072$, from tables

$z_\rho = Fi(0·95) = 1·83$, from tables

z_r is $N\left(\tan h^{-1}, \dfrac{1}{n-3}\right)$ with standardised variable

$\dfrac{z_r - z_\rho}{\sqrt{\dfrac{1}{n-3}}}$

$z_{\text{test}} = \dfrac{1·072 - 1·83}{\sqrt{\dfrac{1}{9}}} = -2·297$

The critical value for a two tail test with a 5% significance level is $\pm 1·96$. Since $z_{\text{test}} = -2·297 < -1·96$ the result is significant so we reject H_0 that the sample comes from a population with correlation coefficient $\varrho = 0·95$.

(iii)

x	1·1	1·9	3·0	4·2	5·1	5·8	7·0	8·3	9·3	10·0	10·9	12·1
r	12	11	10	9	8	7	6	5	4	3	2	1
y	5	7	6	6	8	9	11	12	14	20	43	60
s	12	11	$9\frac{1}{2}$	$9\frac{1}{2}$	8	7	6	5	4	3	2	1
d	0	0	$-\frac{1}{2}$	$\frac{1}{2}$	0	0	0	0	0	0	0	0

Note: $d = r - s$

$R = 1 - \dfrac{6\Sigma d^2}{n(n^2-1)} = 1 - \dfrac{6 \times \frac{1}{2}}{12 \times 143} = 0·999$

EX

1 The moisture contents in per cent and the crushing loads in tonnes of 10 test specimens are given in the following table:

Moisture	Crushing load
0·6	1·5
7·2	4·1
3·1	4·3
5·4	4·6
9·6	2·1
1·2	1·8
2·5	2·9
8·4	2·3
7·7	3·0
5·9	3·8

Convert these data to ranks and find Spearman's rank correlation coefficient. Plot a scatter diagram of the original data. Comment on your results. Is there a relationship between the crushing load and the moisture content? *(O & C)

2 In the two papers of an A-level Higher Mathematics examination, ten candidates gained the marks shown in the table below.

Candidate	A	B	C	D	E	F	G	H	I	J
Paper I	95	83	74	92	84	89	36	71	49	71
Paper II	78	92	72	84	81	93	63	63	66	73

Calculate: (a) the product-moment correlation coefficient; (b) Spearman's coefficient of correlation by ranks.*(A)

3 (a) The marks of eight candidates in English and Mathematics are:

Candidate	1	2	3	4	5	6	7	8
English (x)	50	58	35	86	76	43	40	60
Mathematics (y)	65	72	54	82	32	74	40	53

Rank the results and hence find a rank correlation coefficient between the two sets of marks.

(b) Using the data in part (a), obtain the product-moment correlation coefficient. To assist in the lengthy calculation, you may use the information $S_x = 16·67$. *(S)*

4 The heights h, in cm, and weights W, in kg, of 10 people are measured. It is found that $\Sigma h = 1710$, $\Sigma W = 760$, $\Sigma h^2 = 293\,162$, $\Sigma hW = 130\,628$ and $\Sigma W^2 = 59\,390$. Calculate the correlation coefficient between the values of h and W. What is the equation of the regression line of W on h?

(O & C)

5 The items in a sample each have associated with them the variables X and Y. Explaining any symbols that you use, write down a formula for the sample product-moment correlation coefficient between X and Y and the equation of the regression line of Y on X.

State the conclusions that you would draw if you obtained values for the correlation coefficient of (i) 0, (ii) -1.

For a sample of 100 such items, the following data are known.

$\Sigma x = 36$, $\Sigma y = 25$, $\Sigma xy = 21$,
$\Sigma x^2 = 1012·96$, $\Sigma y^2 = 366·25$.

(x, y are actual values taken by the variables respectively.) Representing the mean values of X and Y for the data by $\bar{x}$ and $\bar{y}$ respectively, calculate $\Sigma(x-\bar{x})(y-\bar{y})$, $\Sigma(x-\bar{x})^2$, $\Sigma(y-\bar{y})^2$ and the sample product-moment correlation coefficient between X and Y.

Given that the equation of the regression line of Y on X for the data is $y = a + bx$, calculate a and b. *(J)*

84 χ^2

χ^2 distribution, Tables, Goodness of fit.

χ^2 distribution

In many statistical situations **observed frequences**, O, are compared with **expected frequencies, E.**

In such cases it is possible to calculate the statistic $\chi^2 = \sum \dfrac{(O-E)^2}{E}$.

The χ^2 **distribution** is a function of ν, the number of **degrees of freedom**. ν is an **integral valued parameter**. For a particular value of ν, the appropriate χ^2 distribution is denoted by $\chi^2(\nu)$.

Tables

The χ^2 distribution is tabulated as percentage points.

A **percentage point** of a χ^2 distribution is that value of χ^2 which has a specified percentage of the distribution lying to its right.

The $p\%$ point of $\chi^2(\nu)$ is written $\chi^2_{p\%}(\nu)$ and is that value of $\chi^2(\nu)$ which has $p\%$ of the distribution lying to its right (see diagram).

The tables for the χ^2 distribution are usually given for those selected values of ν and p which are found to be adequate for most practical purposes.

To calculate $\chi^2_{5\%}(\nu)$ for $\nu > 100$, use the result
$$\chi^2_{5\%}(\nu) = \tfrac{1}{2}(1.645 + \sqrt{2\nu - 1})^2$$

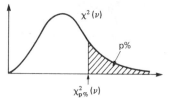

Figure 1

$\boxed{i}$ *From suitable tables verify (a) $\chi^2_{5\%}(4) = 9.49$,*
(b) $\chi^2_{1\%}(4) = 11.34$ (c) $\chi^2_{10\%}(1) = 2.71$.

$\boxed{i}$ $\chi^2_{5\%}(110) = \tfrac{1}{2}(1.645 + \sqrt{220 - 1})^2$
$\qquad\qquad = 135.20$ (2 d.p.)

Goodness of fit

The χ^2 distribution is used to test the **goodness of fit** of a given table of observed frequencies to a theoretical model. It is often used to test whether or not a given distribution is binomial, Poisson or Normal.

To apply the test it is usual to have a total frequency of at least 50 and a minimum class frequency of 5.

If class frequencies fall below this minimum level, then two or more adjacent classes should be combined.

For a given distribution, which is thought to be binomial, Poisson or Normal, proceed as follows.

(a) Calculate the expected frequencies, E, under H_0, the null hypothesis that the distribution is binomial, Poisson or Normal.

(b) Combine any adjacent classes so that no expected frequency is less than 5. If this has to be done combine the corresponding classes of the observed frequencies.

(c) Calculate $\dfrac{(O-E)^2}{E}$ for each class.

(d) Calculate the statistic $\chi^2_{\text{test}} = \sum \dfrac{(O-E)^2}{E}$, where the sum is over all classes.

(e) Determine ν. In general $\nu =$ number of classes (n) $-$ number of restrictions.
For a binomial distribution:
(i) if p is known (by hypothesis), then $\nu = n-1$,
(ii) if p has to be estimated (using $\bar{x} = np$) from the observed frequencies, then $\nu = n-2$.
For a Poisson distribution:
(i) if λ is known, then $\nu = n-1$,
(ii) if λ has to be estimated (using $\bar{x} = \lambda$) from the observed frequencies, then $\nu = n-2$.
For a Normal distribution:
(i) if μ and σ are known, then $\nu = n-1$,
(ii) if μ and σ have to be estimated from the observed frequencies, then $\nu = n-3$.

(f) Find $\chi^2_{a\%}(\nu)$ from tables, where $a\%$ is the significance level assigned to the test.

(g) Compare χ^2_{test} with $\chi^2_{a\%}(\nu)$.
If $\chi^2_{\text{test}} > \chi^2_{a\%}(\nu)$, then reject H_0, otherwise accept it.

$\boxed{i}$ *Four identical coins were tossed 160 times and the observed frequencies of the number of heads per toss is shown in the table.*

Number of heads	0	1	2	3	4
Observed frequency (O)	10	46	54	37	13

Test at the 5% level if the coins are biased.

Let H_0: $P(H) = \tfrac{1}{2}$ i.e. the coins are unbiased.
The random variable X 'the number of heads per toss' is Bin$(4, \tfrac{1}{2})$. This probability distribution is calculated (see Binomial Distribution p. 150) and given below.

x	0	1	2	3	4
$P(X=x)$	0.0625	0.25	0.375	0.25	0.0625

Multiplying each of these probabilities by 160 gives the following table of expected frequencies.

Number of heads	0	1	2	3	4
Expected frequency (E)	10	40	60	40	10

The calculation of χ^2 is set out below.

x	0	1	2	3	4
O	12	46	54	37	13
E	10	40	60	40	10
$(O-E)$	2	6	-6	-3	3
$\dfrac{(O-E)^2}{E}$	0.4	0.9	0.6	0.225	0.9

$$\chi^2_{\text{test}} = \sum \frac{(O-E)^2}{E} = 3.025$$

$\nu = 4$ since there are 5 classes with one restriction, the total frequency.
So test $\chi^2_{\text{test}} = 3.025$ against $\chi^2_{5\%}(4) = 9.49$.

Since $\chi^2_{\text{test}} = 3.025 < 9.49$ this is not significant so do not reject H_0, i.e. accept it.
There is no evidence that the coins are biased.

χ^2
Worked example and Exam questions

 For a period of three months 100 similar gerbils were given a new type of food. The table below shows the recorded changes in mass.

Change in mass (g) x	Observed frequency f
$-\infty < x \leqslant -15$	2
$-15 < x \leqslant -10$	3
$-10 < x \leqslant -5$	8
$-5 < x \leqslant 0$	14
$0 < x \leqslant 5$	16
$5 < x \leqslant 10$	24
$10 < x \leqslant 15$	15
$15 < x \leqslant 20$	9
$20 < x \leqslant 25$	6
$25 < x < \infty$	3

It is thought that these data follow a normal distribution, with mean 5 and standard deviation 10. Use the χ^2 distribution at the 5% level of significance to test this hypothesis.

Describe how the test would be modified if the mean and standard deviation were unknown.

Let the random variable X be 'change in mass over three months', $H_0 : X$ is $N(5, 10^2)$ and $H_1 : X$ is not $N(5, 10^2)$. Assuming H_0 the expected frequencies for the given class intervals can be calculated. This calculation is set out in tabular form below.
If X is $N(5, 10^2)$ then the standardised variable is

$$Z = \frac{X-5}{10}$$

Class	Observed frequency	Upper class bound.	Standard upper class bound.	$\Phi(z)$	Class prob.	Expected class frequency
$-\infty < x \leqslant -15$	2	-15	$-2 \cdot 0$	$0 \cdot 0202$	$0 \cdot 0202$	$2 \cdot 0$
$-15 < x \leqslant -10$	3	-10	$-1 \cdot 5$	$0 \cdot 0668$	$0 \cdot 0466$	$4 \cdot 7$
$-10 < x \leqslant -5$	8	-5	$-1 \cdot 0$	$0 \cdot 1587$	$0 \cdot 0919$	$9 \cdot 2$
$-5 < x \leqslant 0$	14	0	$-0 \cdot 5$	$0 \cdot 3085$	$0 \cdot 1498$	$15 \cdot 0$
$0 < x \leqslant 5$	16	5	$0 \cdot 0$	$0 \cdot 5000$	$0 \cdot 1915$	$19 \cdot 2$
$5 < x \leqslant 10$	24	10	$0 \cdot 5$	$0 \cdot 6915$	$0 \cdot 1915$	$19 \cdot 2$
$10 < x \leqslant 15$	15	15	$1 \cdot 0$	$0 \cdot 8413$	$0 \cdot 1498$	$15 \cdot 0$
$15 < x \leqslant 20$	9	20	$1 \cdot 5$	$0 \cdot 9332$	$0 \cdot 0919$	$9 \cdot 2$
$20 < x \leqslant 25$	6	25	$2 \cdot 0$	$0 \cdot 9773$	$0 \cdot 0441$	$4 \cdot 4$
$25 < x < \infty$	3	∞	∞	$1 \cdot 0000$	$0 \cdot 0227$	$2 \cdot 3$

Combining adjacent cells where necessary so that no cell has a frequency less than 5 gives rise to the following table of observed and expected frequencies.

O	5	8	14	16	24	15	9	9
E	$6 \cdot 7$	$9 \cdot 2$	$15 \cdot 0$	$19 \cdot 2$	$19 \cdot 2$	$15 \cdot 0$	$9 \cdot 2$	$6 \cdot 7$
$O-E$	$-1 \cdot 7$	$-1 \cdot 2$	$-1 \cdot 0$	$-3 \cdot 2$	$4 \cdot 8$	0	$-0 \cdot 2$	$2 \cdot 3$
$\dfrac{(O-E)^2}{E}$	$0 \cdot 431$	$0 \cdot 157$	$0 \cdot 067$	$0 \cdot 533$	$1 \cdot 200$	0	$0 \cdot 004$	$0 \cdot 790$

$$\chi^2_{\text{test}} = \sum \frac{(O-E)^2}{E} = 3 \cdot 182$$

We test this against $\chi^2_{5\%}(7) = 14 \cdot 07$
Note: there are 7 degrees of freedom, 8 cells less 1 restriction.
$\chi^2_{\text{test}} = 3 \cdot 182 < 14 \cdot 07$, $\therefore$ we do not reject H_0 that the data follow a normal distribution with mean 5 and standard deviation 10.
If the mean and standard deviation were unknown they would be estimated from the given data and used to recalculate the expected frequencies; the test would be as above except that there would be two degrees of freedom less.

EX

1 (a) A gambler threw 15 ones and 45 larger numbers in 60 throws of a 6-sided die. Use χ^2 and a 5% significance level to test his assertion that this die is biased.
(b) Whatever the result of your test, assume that the die is biased and find approximate 95% confidence limits for the probability of obtaining a one in a single throw.
*(O & C)

2 Over a period of 50 weeks the numbers of road accidents reported to a police station are shown in the table below.

No. of accidents	0	1	2	3
No. of weeks	23	13	10	4

Find the mean number of accidents per week.
Use this mean, a 5% level of significance, and your table of χ^2 to test the hypothesis that these data are a random sample from a population with a Poisson distribution.

(O & C)

3 A manufactured article can be made by three different methods A, B, C. The numbers of defective articles found in random samples taken from trials of each method are shown in the table:

Method	A	B	C	Total
Defective	14	11	5	30
Total	30	40	30	100

Use χ^2 and a 5% significance level to determine whether there is evidence that the percentage of defectives is not the same for all three methods.

(O & C)

4 For a period of six months 100 similar hamsters were given a new type of feedstuff. The gains in mass are recorded in the table below:

Gain in mass (g) x	Observed frequency f	Gain in mass (g) x	Observed frequency f
$-\infty < x \leqslant -10$	3	$10 < x \leqslant 15$	16
$-10 < x \leqslant -5$	6	$15 < x \leqslant 20$	14
$-5 < x \leqslant 0$	9	$20 < x \leqslant 25$	8
$0 < x \leqslant 5$	15	$25 < x \leqslant 30$	3
$5 < x \leqslant 10$	24	$30 < x \leqslant \infty$	2

It is thought that these data follow a normal distribution, with mean 10 and variance 100. Use the χ^2 distribution at the 5% level of significance to test this hypothesis. Describe briefly how you would modify this test if the mean and variance were unknown.

(A)

5 Analysis of the goals scored per match by a certain football team gave the following results:

No. of goals per match:	0	1	2	3	4	5	6	7
No. of matches:	14	18	29	18	10	7	3	1

Calculate the mean of the above distribution and the frequencies (each correct to 1 decimal place) associated with a Poisson distribution having the same mean. Perform a χ^2 goodness of fit test to determine whether or not the above distribution can be reasonably modelled by this Poisson distribution. (S)

169

85 Contingency Tables
Definitions, Testing for independence.

Definitions

A **contingency table** is an array which displays data relating to two factors.
A table with m rows and n columns is called an $m \times n$ contingency table.

factor B

		b_1	b_2	.	.	.	b_n
	a_1	f_{11}	f_{12}	.	.	.	f_{1n}
factor	a_2	f_{21}	f_{22}	.	.	.	f_{2n}
A	.	.	.				.
	.	.	.				.
	.	.	.				.
	a_m	f_{m1}	f_{m2}	.	.	.	f_{mn}

Two schools enter their pupils for a mathematics test with the results shown in the table

School

		X	Y
Test result	Pass	75	63
	Fail	21	26

This is a 2×2 contingency table.

One factor, which relates to the schools, is shown as the column headings. The other factor, which relates to the test result is shown as the row headings.

Testing for independence

To test if the two factors are independent in a **2×2 contingency table.**

(a) State the null hypothesis H_0: the two factors are independent.

(b) Calculate the row totals, column totals and the grand total (=sum of the row totals=sum of the column totals).

(c) Calculate the expected frequency E for each cell of the table using
$$E = \frac{\text{cell row total} \times \text{cell column total}}{\text{grand total}}$$

(d) Calculate $(O-E)$ for each cell of the table where O is the observed frequency.

(e) Calculate the value of χ^2_{test} using
$$\chi^2_{\text{test}} = \Sigma \frac{(O-E)^2}{E}$$
(see χ^2 p. 168).

(f) State the number of degrees of freedom ν. For a 2×2 contingency table $\nu = 1$. This is because, although there are 4 variables (the expected frequencies), there are 3 restrictions (3 of the four row and column totals must be given), so $\nu = 4 - 3 = 1$.

(g) Compare the χ^2_{test} value with the $\chi^2(1)$ distribution (since $\nu = 1$) at the appropriate significance level.

(h) Apply **Yates' continuity correction** if necessary. When the χ^2 test is applied to a situation with only one degree of freedom, i.e. $\nu = 1$, Yates' continuity correction should be applied. It gives
$$\chi^2_c = \Sigma \frac{(|O-E|-0.5)^2}{E}$$

Since χ^2_c is always less than χ^2 it is not necessary to apply the correction if H_0 is accepted. If an uncorrected χ^2 would reject H_0 while χ^2_c would accept H_0, then it usually indicates that a larger sample should be taken.

To test if the two factors are independent in a **$m \times n$ contingency table.**
Proceed as with the 2×2 contingency table but the number of degrees of freedom is given by
$$\nu = (m-1)(n-1)$$
Yates' continuity correction is not needed here since $\nu \neq 1$.

For the above 2×2 contingency table, test if the factors 'School' and 'Test result' are independent at the 5 per cent level of significance.

Null hypothesis H_0: the two factors 'School' and 'Test result' are independent.

The totals are given in the table below.

	X	Y	
Pass	75	63	138
Fail	21	26	47
	96	89	185

The expected frequencies are given in the table below.

	X	Y	
Pass	$\frac{96 \times 138}{185} = 72$	$\frac{89 \times 138}{185} = 66$	138
Fail	$\frac{96 \times 47}{185} = 24$	$\frac{89 \times 47}{185} = 23$	47
	96	89	185

$(O-E)$ values are given in the table below.

	X	Y
Pass	$75 - 72 = 3$	$63 - 66 = -3$
Fail	$21 - 24 = -3$	$26 - 23 = 3$

Using $\chi^2_{\text{test}} = \Sigma \frac{(O-E)^2}{E}$

$$= \frac{3^2}{72} + \frac{(-3)^2}{66} + \frac{(-3)^2}{24} + \frac{3^2}{23}$$

$$= 1.03 \ (2 \text{ d.p.})$$

Since it is a 2×2 contingency table $\nu = 1$

$\chi^2_{5\%}(1) = 3.81$. So $\chi^2_{\text{test}} = 1.03 < 3.81$.

$\therefore$ Accept H_0, i.e. the two factors 'School' and 'Test result' are independent at 5% significance level. Yates' continuity correction is not necessary because H_0 is accepted.

Contingency Tables
Worked example and Exam questions

Four machines are used to manufacture items which are then graded into three categories. A summary of the production for a given period is shown in the table below.

Grade	Machine A	B	C	D
Top	16	29	23	42
Ordinary	11	6	7	26
Reject	3	15	10	12

Find the expected frequencies on the hypothesis that there is no difference in the quality of the product from each machine.
Use the χ^2 distribution and a 5% level of significance to test the above hypothesis.

Let H_0: grade and machine are independent.
In the table of expected frequencies shown below the row and column totals are first computed, then the total of the row totals (which is also the total of the column totals) is also recorded.
Each expected frequency is then calculated using the standard result

$$\frac{\text{cell row total} \times \text{cell column total}}{\text{grand total}}$$

where grand total means total of row totals.
So, for example, the expected frequency for top grade items from machine A is $\frac{110 \times 30}{200} = 16 \cdot 5$.

Expected Frequencies	Machine A	B	C	D	Totals
Top	16·5	27·5	22·0	44·0	110
Ordinary	7·5	12·5	10·0	20·0	50
Reject	6·0	10·0	8·0	16·0	40
Totals	30	50	40	80	200

The following table records observed frequency minus expected frequency for each cell.

$O - E$	Machine A	B	C	D
Top	−0·5	1·5	1·0	−2·0
Ordinary	3·5	−6·5	−3·0	6·0
Reject	−3·0	5·0	2·0	−4·0

Thus from the tables above we can calculate

$$\chi^2_{\text{test}} = \sum \frac{(O-E)^2}{E} = 13 \cdot 45$$

This value is tested against $\chi^2_{5\%}(6) = 12 \cdot 59$

Note: there are 6 degrees of freedom, a 3×4 contingency table has $(3-1) \times (4-1)$ degree of freedom.

$\chi^2_{\text{test}} = 13 \cdot 45 > 12 \cdot 59$

∴ there is some evidence to suggest that we should reject H_0, i.e. that there is some association between grade and machine.

1 Fifty people were chosen at random in each of two towns and asked whether they had watched a certain TV programme. Their replies are summarized in the table below.

	Watched Programme	Did Not Watch	Totals
Town A	24	26	50
Town B	16	34	50
Totals	40	60	100

If the numbers of people who watched this programme bear the same proportion to the total population in each of these towns, show that the number in the town A sample who would be expected to have watched this programme is 20. Hence or otherwise use χ^2 to show that it is reasonable to assume that the proportion of people who watched this programme was the same in both towns. Find approximate 95% confidence limits for this proportion.

*(O & C)

2 A random sample of 100 housewives were asked by a market research team whether or not they used Sudsey Soap. 58 said yes and 42 said no. In a second random sample of 80 housewives, 62 said yes and 18 said no. By considering a suitable 2×2 contingency table, test whether these two samples are consistent with each other.

(O & C)

3 Children from five schools are entered for an examination in which four grades are awarded. The table gives the frequency distribution of the results.

School	Ackney	Beetham	Cramham	Dotheboys	Egonham
Grade A	55	45	55	40	55
Grade B	45	60	50	40	55
Grade C	50	90	55	10	45
Grade D	50	105	40	10	45

Test whether the proportion of entrants obtaining the various grades varies significantly among the schools. If the schools are of equal size and can be assumed to enter their best pupils, explain why your test does not give a fair comparison of academic standards in the schools. Do you think the above data provide any information about relative standards? If you do, conduct a test and state your conclusions.

(OLE)

4 Inflatia, in common with many other Western countries, is in the middle of an economic recession. As part of a nationwide enquiry into which economic measures will be most acceptable to the general public, a survey was undertaken in the town of Tucville. The responses to the question 'Would you support an incomes policy based on a flat rate increase of £500 per annum for every worker?' are summarised below, together with the employment status of the respondents.

	Employment Status			
Response	Skilled and Union Member	Skilled and non Union Member	Unskilled and Union Member	Unskilled and non Union Member
Yes	7	7	9	12
No	24	21	9	11
Don't Know	29	27	17	27

Use the χ^2 distribution and a 5% level of significance to test the hypothesis that there is no association between response to the above question and employment status. Form a new 2×2 contingency table from the above data by omitting all the "Don't know" responses and then pooling the remaining responses to obtain one column for "Skilled" and one column for "Unskilled". Use a 5% level of significance to test the hypothesis of no association between the factors in this new table.

(A)

Logarithmic graph paper

Logarithmic graph paper uses scales which are graduated logarithmically. By plotting original data on a logarithmic scale, logarithms are automatically taken and plotted in a single process.

A logarithmic scale may have one or more cycles. Each cycle is graduated from 1 to 10 and may be used to represent 10^n to 10^{n+1}, where n is an integer.

The number of cycles used and the values to be marked on them are determined by the range of the data.

There are two basic types of logarithmic graph paper.

1. **Full logarithmic paper** (or **log–log paper**)
On this paper both axes are marked with logarithmic scales. A relation of the form $y = ax^n$ gives a straight line if x is plotted against y on log–log paper since $\log y = \log a + n \log x$. The gradient is n and the intercept on the y-axis is a.

Note: Use an ordinary ruler to measure the distances used to calculate the gradient. Do not use the numbers marked on the logarithmic scales.

2. **Semi-logarithmic paper** (or **log–linear paper**)
On this paper one scale is linear and the other is logarithmic. A relation of the form $y = ab^x$ gives a straight line if x is plotted on the linear scale and y on the logarithmic scale since $\log y = \log a + x \log b$. The gradient is $\log b$ and the intercept on y-axis is a.

Note: Use an ordinary ruler to find the gradient as before. The length of one cycle is taken as one unit of length on the logarithmic scale.

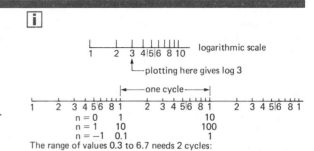

The range of values 0.3 to 6.7 needs 2 cycles:

Figure 2

ⓘ *The given data are thought to behave as $y = a^x$. Use log paper to estimate a.*

x	y
0	1
1	2.1
2	5.29
3	13.82
4	19.45

From the graph:
4.1 cm ≈ 1 log cycle
5.8 cm ≈ 1.4 cycles
∴ $\log a \approx \dfrac{1.4}{4} = 0.35$
$\Rightarrow a \approx 2.24$

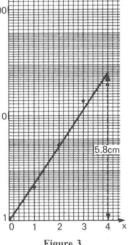

Figure 3

Poisson probability chart

The **Poisson probability chart** is used to find $P(X \geq c)$ for a distribution which is $Po(a)$. It gives the probabilities for different values of a and c. The chart is used in the following way.

(a) Determine a (using $a = np$). Find its position on the 'a scale'.

(b) Find where the ordinate at a cuts the curve marked c.

(c) read the corresponding value of p on the 'p scale'. This gives the probability for $x \geq c$.

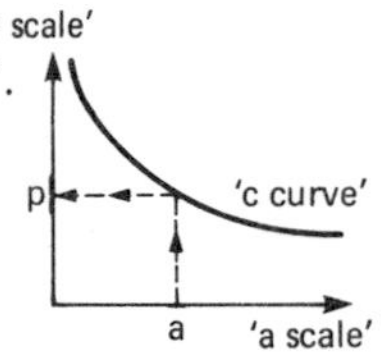

Figure 1

ⓘ *A company making microprocessor parts has 1% of its production faulty. What is the probability of getting at least 5 defectives in a box of 200?*

$n = 200$, $p = 0.01$, so $a = np = 2$. Verify on a Poisson chart that for $a = 2$ and $c = 5$, $p = 0.015$. So the probability of getting at least 5 defectives in a box of 200 is 0.015.

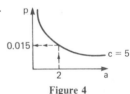

Figure 4

Arithmetic probability graph paper.

Arithmetic probability graph paper has one linear and one non-linear scale. It is designed so that when the cumulative frequency of a Normal distribution is plotted (on the non-linear scale) against a variable the result is a straight line. This gives a method of testing if a sample is from a normal population.

Note: If the distribution has been transformed, then the transformed variable must be used.

Skewed distributions result in curved lines as shown.

The mean μ can be estimated since it is the value of the variable corresponding to 50% cumulative frequency.

The standard deviation can be estimated using
$$4\sigma = x(97.72) - x(2.28)$$
$$\text{or} \quad 3\sigma = x(93.32) - x(6.68)$$
where $x(97.72)$ is the value of the variable corresponding to 97.72% cumulative frequency, etc.

The range of values of the variable determines which of these results is used but the first is preferred.

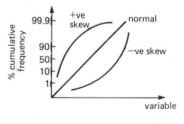

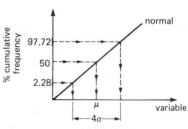

Figure 5

Special Graph Papers
Worked example and Exam questions

The following is a random sample of size 20, which it is believed may be from a normal population
11·4, 18·0, 12·4, 22·5, 17·0, 0·1, 31·1, 20·4, 34·6, 18·5, 26·8, 2·6, 1·1, 7·0, 21·0, 7·6, 19·6, 36·4, 36·3, 21·9.
Plot this sample on arithmetic probability paper.
Estimate from this graph the mean and variance of the sample.

Inspecting the given set of values indicates 0 to 40 as a suitable range, this range is divided into eight equal classes each of width 5. A frequency, cumulative frequency and percentage cumulative frequency table is constructed as shown below.

Class	0–	5–	10–	15–	20–	25–	30–	35–40
Frequency	2	3	2	4	4	1	2	2
Cumulative frequency	2	5	7	11	15	16	18	20
Percentage cumulative frequency	10	25	35	55	75	80	90	100

The percentage cumulative frequency is plotted on arithmetic probability paper as shown below. On the assumption that the sample is from a normal population an estimate for the mean of the sample is obtained by reading the value of the variable corresponding to 50% cumulative frequency, and this is estimated to be 18·2 from the graph.
We use
$$s = \frac{x(93·32) - x(6·68)}{3}$$
to estimate s, the standard deviation of the sample, where $x(93·32)$ is the value of the variable corresponding to 93·32% percentage cumulative frequency, and similarly for $x(6·68)$.

From the graph $s = \frac{34·5 - 1·5}{3} = 11$, so the variance of the sample is estimated to be 121.
∴ the sample has estimated mean 18·2 and estimated variance 121. (See graph overleaf)

---

1 The frequency distribution of 200 masses, each recorded to the nearest gram, is given in the table.

Mass (g)	11	12	13	14	15
Frequency	16	28	118	22	16

Use arithmetic probability paper to fit a normal distribution to the data, stating its mean and standard deviation. Find the frequencies predicted by the fitted distribution for the groups in the table.
*(A)

2 The number of cosmic particles per minute arriving at an apparatus was recorded over a period of 1000 minutes and these observations are tabulated below.

No. of particles	0	1	2	3	4	5	6
No. of minutes	300	372	201	97	25	4	1

Use Poisson probability paper to estimate the mean rate of arrival of cosmic particles. State the variance and find the probability that, in any given interval of one minute, the rate of arrival has a value which is more than five standard deviations from the mean. Also determine the probability that, in any given interval of three minutes:
 (i) no particles arrive,
 (ii) no more than two particles arrive. *(A)*

3 (a) Nine men of similar build were given x standard units of alcohol, allowed to rest for one hour, and then asked to complete a simple task. The time y, taken to complete this task was measured, in seconds, for each man. The results are given below.

Amount of alcohol x	Time y
1·1	1·4
1·8	4·5
3·3	11·2
3·9	19·0
5·2	31·4
6·5	45·9
6·7	55·6
8·0	76·8
9·1	97·1

It is thought that the model for these data is of the form $y = ax^b$. Verify that this is a reasonable assumption by plotting the above data on log-log paper. Estimate the parameters a and b from your graph. Indicate briefly how the method of least squares might have been used to find estimates of a and b.
(b) The table below gives the masses, in kilograms, of 150 working coalminers of similar height.

Mass (kilograms)	Frequency
50–54	4
55–59	8
60–64	25
65–69	33
70–74	35
75–79	27
80–84	9
85–89	5
90–94	4

Using arithmetical probability paper, verify that it is reasonable to assume that these data came from an underlying normal population. Estimate from your graph the mean and standard deviation of this distribution. *(A)*

4 (a) All schools in the county of Kentwall were involved in a "Sponsored Pumpkin Grow" during the summer of 1980. Each pupil was given two pumpkin seeds and asked to come back at the end of the summer at which time the mass of their largest pumpkin was recorded in kilograms. The table below gives the results for two random samples of pupils, one from the Much Wopping School and one from Markum School.

Much Wopping Primary School

6·07, 8·32, 9·40, 6·68, 4·24, 8·45, 10·00, 7·79, 5·79

Markum High School

5·86, 2·78, 6·61, 4·57, 6·50, 7·72, 2·30, 8·45, 4·41, 5·86, 3·35

Using the *same* sheet of arithmetical probability paper:
 (i) plot the sample from Much Wopping Primary School,
 (ii) plot the sample from Markum High School. Compare the distributions.

(b) The number of misprints on 200 randomly selected pages from the 1981 editions of the Daily Planet, a quality newspaper, were recorded. The table below summarises these results.

Number of misprints per page

x	0	1	2	3	4	5	6	7	8	more than 8
Frequency										
f	5	12	31	40	38	29	22	14	5	4

Use Poisson probability paper to verify that the Poisson distribution with mean 4 is a reasonable model for these data. Determine graphically an estimate for the probability of more than 11 misprints on a page. *(A)*

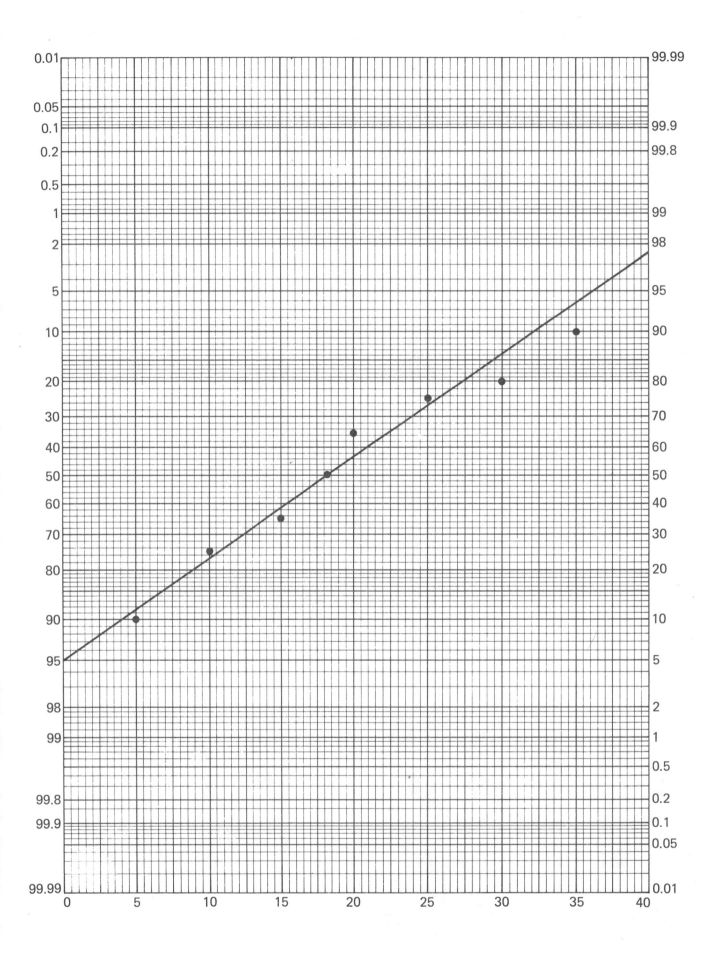

Part III
Answers

Guided Examples Answers

1 Polynomials

$P(x) = 2x^2 + 3x - 2$

$x = \dfrac{1}{2}$ or -2

3 Quadratics

$\alpha\beta = 5$

(i) $x^2 - 4x + 8 = 0$

(ii) $(x-5)(x^2 + 4x + 8) = 0$

4 Sequences and series

$r = -\dfrac{1}{2}$

$S_\infty = \dfrac{2a}{3}$

6 Permutations and combinations

$\dfrac{6!}{2!3!} = 60$

$\dfrac{5!}{2!2!} + \dfrac{5!}{3!} + \dfrac{5!}{2!3!} = 60$

8 Inequations

$x \leqslant 1$ or $x \geqslant 4$

$x \leqslant 1$ or $x \geqslant 4$ or $2 \leqslant x \leqslant 3$

10 Exponential and logarithmic functions

$2x - 4x^2 + \dfrac{17}{3}x^3 - 8x^4; \ |x| < \dfrac{1}{2}$

13 The circle

C_1 centre $(0, 8)$, radius $= 4\sqrt{2}$

C_2 centre $(9, -1)$, radius $= 5\sqrt{2}$

$167°$

23 Vectors

(a) $\overrightarrow{OU} = (\mu + \lambda)(\mathbf{i} + \mathbf{j})$

$\overrightarrow{ST} = (\mu - \lambda)(\mathbf{i} - \mathbf{j})$

(b) O is the centre of mass of triangle PQR.

24 Vectors and geometry

$\lambda : \mu = 1 : 6$

$\mathbf{r} = \mathbf{i} - 2\mathbf{j} + \mathbf{k} + t(13\mathbf{i} + 4\mathbf{j} - 5\mathbf{k})$

$\mathbf{p} = -12\mathbf{i} - 6\mathbf{j} + 6\mathbf{k}$

25 Complex numbers

(i) $1 - i$ or $1 + i$; $1 + 2i$ or $2 + i$; $1 - 2i$ or $2 - i$.

26 Complex numbers and graphs

$z = 2\left(\cos\dfrac{\pi}{4} + i\sin\dfrac{\pi}{4}\right),$

$w = 6\left(\cos\left(\dfrac{2\pi}{3}\right) + i\sin\left(\dfrac{2\pi}{3}\right)\right);$

(i) $\dfrac{1}{2}\left(\cos\left(\dfrac{-\pi}{4}\right) + i\sin\left(\dfrac{-\pi}{4}\right)\right);$

(ii) $12\left(\cos\left(\dfrac{11\pi}{12}\right) + i\sin\left(\dfrac{11\pi}{12}\right)\right);$

(iii) $\dfrac{1}{3}\left(\cos\left(\dfrac{-5\pi}{12}\right) + i\sin\left(\dfrac{-5\pi}{12}\right)\right).$

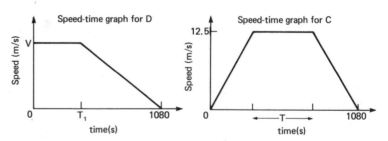

27 Differentiation

(a) (i) $e^x(\cos x + \sin x)$,

(ii) $\dfrac{3x^2(1 + \sec x) - x^3 \sec x \tan x}{(1 + \sec x)^2}$

(b) $x^4(1 + 5\ln x) - \sin x$

29 Applications of differentiation

(a) 1s, 3s; (b) 4 m, 0 m; (c) -6 ms^{-2}, 6 ms^{-2};

(d) -3 ms^{-1}.

30 Changes

4%.

31 Special points

Stationary points: Inflexion at $(0, 0)$, Minimum at $(3, -27)$;

Inflexion at $(2, -16)$.

32 Curve sketching

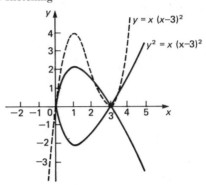

33 Integration

(a) $5\frac{2}{3}$, (b) $\dfrac{\pi}{4}$, (c) $\dfrac{\pi}{8} - \dfrac{1}{4}$, (d) $\dfrac{1}{2}\ln 5$.

34 Methods of integration

(a) $\dfrac{e^{4x}}{32}(8x^2 - 4x + 1) + c$; (b) $x - \dfrac{1}{2}\ln\left|\dfrac{x-1}{x+1}\right| + c$;

(c) $\ln\left|x + \sqrt{(1+x^2)}\right| + c$.

35 Applications of integration

(a) 64π cubic units; (b) 64π cubic units.

36 Differential equations

$y = 2(x + 1)$

40 Matrics

$k = a^2 + bc$

43 Graphs in kinematics

(a) 4500 m, (b) 13·3 m/s, (c) 59·2 m/s^2.

44 Relative motion

N31·4°E, 12.41 pm, 9·5 nautical miles.

45 1-D particle dynamics

(a) 4930 N, (b) 680 N.

47 Work and energy

(i) 20 Nm, (ii) 48 Nm, (iii) 28 Nm, (iv) $2\frac{1}{3}$N; 5m.

49 Impulse and momentum

790 kN; $\dfrac{1}{25}$ s.

53 Vertical circular motion

$\sqrt{[2ag(1-\cos\theta)]}$; $mg(3\cos\theta-2)$.

54 Variable forces

(ii) $v^2 = u^2(1-e^{-2kx})$.

55 Simple harmonic motion

$2\pi\sqrt{\dfrac{2a}{g}}$; a; $\dfrac{a}{2}$ below O.

57 Coplanar concurrent forces

2 N along GP.

58 Moments

(i) 5 kg; (ii) 2 kg.

59 Equilibrium

Force from wall, $\dfrac{75\sqrt{3}}{4}$ N; force from ground, $168\frac{3}{4}$ N;

force from rod, $37\frac{1}{2}$ N.

60 Three force problems

Tension, 30 N; Reaction, 18N.

61 Friction

(a) $H = W\tan 2\alpha$; (b) $\theta = \alpha$.

63 Equivalent systems of forces

(a) R = 5 N, S = 8 N; (b) R = 5 N, S = 2 N; $11\sqrt{3}$ Nm.

64 Centre of mass

$\bar{x} = \dfrac{5r}{3}$, $\bar{y} = 0$.

65 Suspending and toppling

(a) 1 m; (b) 3 m from AB, 3.8 m from AE; (c) 38.3°.

73 Discrete probability distributions

x	2	3	4	5	6	7	8	9
$P(X=x)$	$\dfrac{1}{36}$	$\dfrac{3}{36}$	$\dfrac{6}{36}$	$\dfrac{6}{36}$	$\dfrac{6}{36}$	$\dfrac{6}{36}$	$\dfrac{5}{36}$	$\dfrac{3}{36}$

$E[X] = 5.833$
$P(X=2$ or 3 for 3 throws$) = 0.00137$

76 The Poisson distribution

Predicted frequencies, 34.1, 38.1, 21.2, 7.9, 2.2, 0.5.

79 Sampling

For the 8 classes; less than -2, -2 up to 0, 0 up to 2, 2 up to 4, 4 up to 6, 6 up to 8, 8 up to 10, greater than 10, the allocation of the random numbers 001 to 000 to the classes is 001, 002–023, 024–159, 160–500, 501–841, 842–977, 978–999, 000.
The random observation is in the class 160–500 and simple proportion gives the random observation to be 3.894.

82 Linear regression

$y = -0.536x + 69.033$
For $x = 60$, $y = 36.873$

Answers to Examination Questions

1 Polynomials

1 $a = -9$, $b = 7$; $(x-2)(x-3)(2x+1)$.
2 $a = -1$, $b = 6$.
3 -7.
4 (i) -12; (ii) -60; (iii) 0; $2x+1$ is factor; $(x-3)(x+1)$.
5 $a = -13$, $b = 26$; $3x-26$.
6 $a = -18$, $b = 8$; $(x-2)(x+4)(2x-1)$.
7 $a = -10$.
8 (a) $a = -21$, $b = 8$. (b) $P = 6$, $Q = -8$, $c = 4$.
9 $(2x-1)(2x+5)(2x-3)$.
10 $a = -9$, $b = 2$, $c = 8$.

11 (a) $k = 1$, $-\dfrac{1}{2} \pm \sqrt{\dfrac{3}{4}}$. (b) $2x+3$.

12 $3x-2$; $x-1$.
13 $(3-x)(1-x)^2$; $x = 1$ or $x \geqslant 3$.
14 (i) $a = b = -1$, $c = 0$. (ii) $x^3 + x^2 - 6x + 4$.

2 Rational functions

1 $A = 3/2$, $B = -\dfrac{1}{2}$.

2 $2/(2+x) - 1/(1+x^2)$.
3 $2/(x-1) + 3/(x+2) - 5/(x+2)^2$.
4 $1/(1+x) + 2/(1-2x^2)$.

5 $1/(1-2x) - 8/(2-x)$.
6 $2/(2+x) + 1/(1-2x)$.
7 $4/(2+x) + 1/(1-2x) + 2/(1-2x)^2$.
8 $3/(1-3x) - 2/(1-2x)$.
9 $4/(1+2x) + 2/(1-x) + 3/(1-x)^2$.
10 $1/(x+7) - 1/(x+9)$.
11 $1/(x+2) + 1/(x+3)$.
12 $4/(x-4) + 1/(x+2)$.
13 $16/(x-2) - 14/(x-1)$.
14 $3/(x+1) - 2/(x+1)^2$.
15 $-4/9(x+1) + 4/9(x-2) + 2/3(x-2)^2$.
16 $-1/3(1+2x) + 5/3(x+2) - 4/(x+2)^2$.
17 $-1/(x+1) + x/(x^2+1)$.
18 $1/2(x-1) + (x+1)/2(x^2+1)$.
19 $1 + 3/(x+4) + 5/(x-2)$.
20 $(x-3) - 1/(x+1) + 8/(x+2)$.

3 Quadratics

1 $-4 \leqslant x \leqslant 4$; $q = 3$ or 4.
2 $qx^2 + px + 1 = 0$.
3 (a) $2 < k < 6$. (b) -7, 11. (c) $11x^2 - 27x + 11 = 0$.
4 $\begin{cases}\text{(i) 2. (ii) 5. (iii) } 4x^2 - 21x + 1 = 0. \\ 5/2; \ m = -5/2n = \dfrac{1}{2}.\end{cases}$

5 ± 2.
6 (i) $-11/4$. (ii) $4x^2 + 11x + 9 = 0$.
7 (a) $4 < x < 5$. (b) $-6 < k < 6$.
8 $n > 3$ or $n < -3$.

4 Sequences and series

1 $c = -8$, $r = \dfrac{1}{2}$.

2 6273.
3 (a) $d=2$, $a=3$. (b) 1,1,1 or 4,6,9.
4 (i) £1656.03. (ii) $n>500\,000$.
5 $-\cos 2\alpha$.

6 $x+1$ valid if $x>-\dfrac{1}{2}$

7 $\dfrac{\pi}{4}<|\theta|<\dfrac{3\pi}{4}$

8 ——————

9 $d=\dfrac{1}{2}$ $a=2\frac{1}{2}$; 205

10 100; 2046; 2146.
11 (i) $7(p+1)$ (ii) $7(p+1)(10p+11)$.

5 Summation of series

1 (a) (i) $\dfrac{1}{6}n(n+1)(4n-1)$. (ii) $2x(1-(2x)^n)/(1-2x)$.

(b) (i) $2^{n+1}-2-n$; (ii) 6.
2 73710.
3 ——————
4 ——————
5 ——————
6 $8n^2+8n+1$.
7 ——————

6 Permutations and combinations

1 (i) 40320; (ii) 6720; (iii) 907200; (iv) 1152
2 (a) 5040. (b) 1440. (c) 144.
3 840; 96.
4 (a) 120; (b) 60.
5 3240.
6 50.
7 10080; 30.
8 (i) 12; (ii) 12.
9 362880; (i) 3024; (ii) 2880.

7 Binomial theorem

1 $32-40x+20x^2-5x^3+\dfrac{5}{8}x^4-\dfrac{1}{31}x^5$; $32\cdot208$.

2 (a) $1-2x-2x^2-4x^3$; $-\dfrac{1}{4}<x<\dfrac{1}{4}$. (b) ± 8.

3 $-192x^3-432x$.
4 $1-6x+24x^2-80x^3$.

5 $2+\dfrac{2x^2}{9}+\dfrac{2x^4}{81}$.

6 245/64.
7 253.
8 $32\cdot808$.

9 $1+\dfrac{1}{2}x+\dfrac{3}{8}x^2+\dfrac{5}{16}x^3$.

10 $(a+bn+cn(n-1))/n!$; $a=2$, $b=4$, $c=1$.

11 $1+\dfrac{1}{2}x-\dfrac{1}{8}x^2$; $1+\dfrac{1}{2}x+\dfrac{3}{8}x^2$; $3\cdot315$.

12 (i) $1+x+x^2+x^3+x^4$; (ii) $1+2x+3x^2+4x^3+5x^4$.

8 Inequations

1 (a) $\frac{1}{2}<x<2\frac{1}{2}$. (b) $-1<x<1$ or $x>2$.
2 (i) $x\geqslant-2$; (ii) $-2\leqslant x\leqslant-1$ or $x\geqslant1$.
3 $x=0$ or -2; $x>0$ or $x<-2$
4 (i) $x<-1$ or $x>2/3$; (ii) $x<-1$ or $0<x<2/3$;
   (iii) $-1<x<0$ or $x>2/3$.
5 $-4<x<4$.

6 $-\dfrac{3}{4}<x<3$.

7 $5/6+\ln 4$.

9 Indices and logarithms

1 (a) (i) $z=3$; (ii) $z=\dfrac{1}{4}$; (iii) $x=\dfrac{1}{16}$; (iv) $y=-1$.

(b) $z^2-4z+3=0$, $y=0$ or 1.
(c) 1.

2 (a) $\dfrac{x+1}{x(x-1)}$.

(b) (i) $\dfrac{1}{2}$; (ii) $-\dfrac{3}{2}$.

3 (i) $3\frac{1}{4}$; (ii) $\dfrac{1}{125}$.

4 (a) 2. (b) $v=x^{\frac{1}{2}(p-q)}$.
5 (a) $x=625$ or $(625)^{-1}$. (b) $y=2$
6 (i) $x=-2\cdot06$; (ii) $y=8$.

7 (a) $xy=3^{2p+\frac{q}{2}}$, $\dfrac{x^2}{y}=3^{4p-\frac{q}{2}}$.

(b) $x=\ln 2$.
8 (a) (i) 0; (ii) 3.

(b) $x=\ln\left(\dfrac{5}{2}\right)$, $y=-\ln 6$.

9 $(x,y)=(3,9)$ or $(9,3)$.
10 (i) $y=1\cdot983$; (ii) $0\cdot69897$.

11 (i) $k=2\cdot303$, $\dfrac{dy}{dx}=10^x\ln 10$;

(ii) $g(x)=\ln x$, $\dfrac{dy}{dx}=x^x(1+\ln x)$.

12 $(x,y)=(8,2)$ or $(2,8)$.

10 Exponential and logarithmic functions

1 $1-\dfrac{28}{3}x^3$; $|x|<\dfrac{1}{2}$.

2 $1+\dfrac{x}{2!}+\dfrac{x^2}{4!}+\dfrac{x^3}{6!}+...+\dfrac{x^n}{(2n)!}+...$; $\dfrac{x^{\frac{n}{2}}}{(2n)!}$.

3 (a) $1+x+\dfrac{x^2}{2!}+\dfrac{x^3}{3!}+...+\dfrac{x^n}{n!}+...$.

(b) $1-\dfrac{3}{2}x+\dfrac{3}{4}x^2$.

4 (i) $-2x-2x^2-\dfrac{8}{3}x^3$; (ii) $-3x-\dfrac{9}{2}x^2-9x^3$;

$3x^2+10x^3$; $\dfrac{1}{n}(2(3)^n-3(2^n))$; $|x|<\dfrac{1}{3}$.

5 (a) $a=-\dfrac{1}{4}$. (b) $0\cdot099\,504$.

6 $4x+\dfrac{16}{3}x^3+\dfrac{64}{5}x^5$; $0\cdot510\,82$.

11 Co-ordinates and graphs

1 (ii) Area $=25$; (iii) $\sqrt{5}$.
2 (i) $t=1$; (ii) Area $=10$.
3 $x^3-4x^2+4=0$.
4 $X(10,0)$ $Y(0,5)$; Area $=25$.
5 $(1,1)$ $y=18-2x$.
6 (a) $P(1,7,0)$ $Q(5/3,5,-8/3)$. (b) 1:3. (c) $(4,6,-18)$.
7 $D(2,-4)$.
8 (i) $B(2,3)$; (ii) $1/\sqrt{50}$; (iii) area $=6$.

12 The straight line

1 (i) $y=3x-6$; (ii) $(1,-3)$.
2 $k=1,-2$.
3 $k=3/2$.
4 $a=2\sqrt{5}$, $b=4\sqrt{5}$.
5 $3y=2x+11$.
6 $x^2+y^2-30x-15y+225=0$.

7 (i) $\dfrac{3}{4}$; (ii) 3.

13 The circle

1 $x^2+y^2-6x+8y=0$; $P(3,-3)$ $Q(2,-4)$; $\sqrt{2}$.
2 10 units; $x^2+y^2-6x-2y-15=0$.
3 Pt of contact (11,8).
4 $2y=x+a$; $3y=16-x$; $x^2-2x+y^2-10y+1=0$; $x^2-7x+y^2=\frac{1}{4}$.
5 $x^2-4x+y^2-12y+32=0$; $x^2-12x+y^2-4y+32=0$. Tangents: $y=x$, $y=4-x$, $y=12-x$.
6 (i) $x^2-8x+y^2-6y=0$;
    (ii) $x^2-4x+y^2-4y+4=0$.

14 Conic sections

1 $y=2hx-h^2$. (i) $Q\left(\frac{1}{4}h,-\frac{1}{2}h^2\right)$. (ii) $y=-8x^2$.
    (iii) $(-2,-32)$.
2 $A(9,6)$; tgt $3y=x+9$; normal $y+3x=33$; $B(121/9,-22/3)$; $DE=20$ $AE=\sqrt{40}$ $AD=6\sqrt{10}$.
3 ——————
4 $V((a^2-b^2)\cos t/a,\ (a^2-b^2)\sin t/b)$; $a^2x^2+b^2y^2=(a^2-b^2)^2$.
5 $pqy+x=c(p+q)$; $p^2y+x=2pc$.
6 $y^2=x-3$.

15 Loci

1 ——————
2 (iii) $P\left(\sqrt{\frac{3}{4}},-\sqrt{\frac{3}{4}}\right)$; Area $=1\cdot5$.
3 (i) $y=-3x-3$ or $y=-12x$. (ii) $\left(\frac{1}{3},-4\right)$.
4 (i) $2y=x+4$; (ii) $3y=14-4x$; $(11x-16)^2+(11y-30)^2=650$.
5 $108\cdot8$.
6 $y^2=ax-a^2$.
7 $2y=3tx-at^3$; $(at^2/4,-at^3/8)$ Exception if $t=0$.

16 Polar co-ordinates

1 $r=\frac{3}{4}\sec\theta$ $r=3\sqrt{3}\ \text{cosec}\ \theta/4$.
2 (i) $r=a$; (ii) $\theta=\alpha$; (iii) $r=c\sin\theta$;
    (iv) $r=d\sin(\beta-\varphi)/\sin(\beta-\varphi)$.
3 $r=\frac{3}{4}a$.

17 Experimental laws

1 $a=2\cdot5$, $k=5\cdot56$.
2 $n=1\cdot4$, $a=0\cdot7$, $y=1\cdot6$.
3 $b=1\cdot56$, $a=0\cdot78$.
4 $\log y=\log a+k\log x$; $k=1\cdot5$, $a=2\cdot6$. (a) $2\cdot45$. (b) $14\cdot6$.
5 $a=30$, $b=-4$, $x\leqslant7\cdot5$.
6 $a=0\cdot25$, $b=2\cdot3$.
7 (a) $2y+3x=4$. (b) $\ln5+\ln x=y\ln6$.
8 $a=7$, $b=5$.
9 $b=0\cdot35$, $a=140$.
10 (i) $\log y=\log a+(x+1)\log b$; (ii) $px=1-q\left(\frac{y}{x}\right)$.

18 Trigonometrical functions

1 ——————
2 (a) $0\cdot499074$. (b) $0\cdot728580$. (c) $1\cdot493944$.
3 (a) $3°45'21''$. (b) $60°42'22''$. (c) $125°47'7''$.
4 (a) $\pi/4$. (b) $1\cdot288$. (c) $3\cdot067$.
5 (a) $28°38'52''$. (b) $45°$. (c) $89°10'50''$.
6 $17°11'30''$.
7 $114°35'30''$.
8 $75°$.
9 $1\cdot318$ m.
10 $18\cdot75$ cm.
11 $7\cdot4$ cm.
12 $8\cdot12$ cm.
13 $\sin=\tan=0\cdot0015$, $\cos=1\cdot0000$.
14 $\sin=\tan=0\cdot0058$, $\cos=1\cdot0000$.
15 (a) 3/5, 4/5. (b) 3/4, 3/5. (c) 7/25, 24/7.
16 (a) $0\cdot0095$ rad. (b) $0\cdot0099$ rad. (c) $0\cdot1$ rad. (d) $0\cdot14$ rad.
17 $0\cdot04$ rad.
18 $0\cdot07$ rad.
19 $48\cdot6°$.
20 339 cm^2.

19 Trigonometrical graphs

1 $\theta=0\cdot34$ rad.
2 ——————
3 $\theta=45°$, $165°$.
4 $x=33$, 327.
5 ——————
6 x co-ords 0, $1\cdot8\pi$, $4\cdot3\pi$, $5\cdot6\pi$; $\sin\frac{1}{2}x=\frac{x}{10\pi}$.
7 (a) $-12/13$. (b) $-12/5$. (c) $-7/17$. (d) $13/12$.
8 ——————
9 $\theta=\pi6$, $\pm\pi/2$, $5\pi/6$; $-\pi\leqslant\theta\leqslant-\pi/2$, $\pi/6\leqslant\theta\leqslant\pi/2$, $5\pi/6\leqslant\theta\leqslant\pi$.

20 Trigonometrical identities

1 (a) $\cos\theta=\frac{1-t^2}{1+t^2}$; $122\cdot4°$, $20\cdot8°$.
2 $A=54\cdot7°$, $125\cdot3°$, $234\cdot7°$, $305\cdot3°$; $k<1$.
3 (b) (i) $-7/25$; (ii) $\frac{1}{2}$.
4 (i) 60, 120, 240, 300; (ii) 31, 211;
    (iii) 30, 150, 210, 330; (iv) 0, 109, 5, 250·5, 360.
5 ——————
6 (i) 45, 125, 165, 245, 285; (ii) 19·5, 160·5, 210, 330.
7 $a=3$, $r=\sqrt{13}$, $\alpha=33\cdot7°$; $3+\sqrt{13}$.
8 $\sqrt{5}\sin(x-63\cdot4)$; (i) $[-\sqrt{5},\sqrt{5}]$; (ii) $90°$, $216\cdot8°$.
9 $48\cdot6°$, $131\cdot4°$, $150°$, $330°$.
10 (i) $45°$, $105°$, $165°$, $225°$, $285°$; (ii) $210°$, $330°$.
11 $45°$, $90°$, $135°$, $225°$, $270°$, $315°$.
12 $7t^2-10t+3=0$; $46\cdot4°$.

21 Plane triangles

1 (a) $53\cdot1°$; $126\cdot9°$.
2 (i) $2\sqrt{117}$; (ii) $16\cdot1°$.
3 (i) $A=81\cdot8°$ $B=38\cdot2°$; (ii) $30°$, $270°$.
4 $58\cdot1°$, $121\cdot9°$; $13\cdot8$ cm, $3\cdot2$ cm.
5 $\pi/6$, $5\pi/6$.
6 (b) $\pi/2$.
7 $1\cdot58$ m.
8 ——————
9 (i) $241\cdot6$ m; (ii) 295 m.
10 $124°$, $28°$, $28°$.
11 $AP^2=a^2+c^2-2ac\cos(\theta+\pi/3)$.

22 3-D Figures

1 $420\cdot5$ m.
2 62 m.
3 (i) $20/\sqrt{3}$ m; (ii) $73\cdot9°$.
4 $69\cdot8$ m.
5 $219\cdot8$ m; $138\cdot6$ m; $326\cdot6$ m; Area 7616 m^2.
6 $51\cdot2°$; $7\cdot79$ cm; $70\cdot5°$.
7 (i) $\sqrt{13/3}\ a^2$; (ii) $2a^3/\sqrt{3}$; (iii) $6a/\sqrt{13}$; (iv) 11/19.
8 $\theta=\tan^{-1}((s-q)/r\cos\alpha)$.
9 ——————
10 $93\cdot6°$.

23 Vectors

1 $|\mathbf{p}+\mathbf{q}|=11\cdot5$ units.
2 $\mathbf{m}=\frac{1}{2}(\mathbf{a}+\mathbf{b})$; $\mathbf{t}=\frac{2}{3}(\mathbf{a}+\mathbf{b})$; $OM:OT=3:4$.
3 $\mathbf{AB}=\mathbf{b}-\mathbf{a}$, $\mathbf{PQ}=\frac{1}{10}(5\mathbf{b}-3\mathbf{a})$;
    (i) $\frac{n}{10}(5\mathbf{b}-3\mathbf{a})$; (ii) $\frac{1}{2}(1+k)\mathbf{b}-\frac{1}{2}\mathbf{a}$; $n=\frac{5}{3}$, $k=\frac{1}{3}$.
4 $AB:BC=1:2$; $\vec{OD}=6\mathbf{q}-4\mathbf{p}$; $\vec{OE}=-3\mathbf{q}+5\mathbf{p}$.
5 (a) $\mathbf{x}=\frac{1}{2}(3\mathbf{b}-\mathbf{a})$; (b) $\mathbf{y}=\frac{1}{4}(\mathbf{c}+3\mathbf{b})$; (d) $XY:YZ=1:1$.
6 (a) $\vec{OM}=\frac{1}{2}\mathbf{b}$; (b) $\vec{OT}=\frac{1}{3}(\mathbf{a}+2\mathbf{b})$; (c) $\vec{OX}=\frac{1}{6}(8\mathbf{a}+\mathbf{b})$;
    (d) $\vec{OY}=\frac{1}{9}(8\mathbf{a}+\mathbf{b})$.
7 $PQ:QR=1:4$.

24 Vectors and geometry

1 (i) $\frac{1}{5}\binom{3}{4}$; (ii) $m=5$, $n=1$; $\frac{-2\sqrt{5}}{25}$.

2 $\binom{4}{2}$; $s\binom{1}{-2}$; $t\binom{2}{2}$; $\frac{-\sqrt{10}}{5}$.

3 (a) $AC:CB=5:3$.

4 $17\cdot3°$; $\overrightarrow{OP}=\mathbf{i}+3\mathbf{j}-2\mathbf{k}$; $\mathbf{r}.(4\mathbf{i}+3\mathbf{j}-3\mathbf{k})=19$.

5 (i) $30°$; $\overrightarrow{PQ}=\begin{pmatrix}2+q-2p\\-3+p\\5+q-p\end{pmatrix}$; $p=1$, $q=-2$.

6 (a) $3\mathbf{j}+4\mathbf{k}$; (b) $\mathbf{r}.(3\mathbf{j}+4\mathbf{k})=0$; (c) $\left(\frac{\sqrt{26}}{26}\right)$.

25 Complex numbers

1 $32+47i$; $32-47i$; $(6-5i)(6+5i)(7+2i)(7-2i)$.

2 0, $\pm\sqrt{3}$, $\pm\frac{\sqrt{3}}{3}$.

3 $z=2-i$, $w=2+3i$.

4 (a) $-1-i\sqrt{3}$. (b) $p=6$, $q=25$.

5 $a=\pm\frac{\sqrt{2}}{2}$, $b=\pm\frac{\sqrt{2}}{2}$; $z=\frac{\sqrt{2}}{2}-1+i\frac{\sqrt{2}}{2}$ or $-\frac{\sqrt{2}}{2}-1-i\frac{\sqrt{2}}{2}$.

6 $\frac{x}{x^2+y^2}-i\frac{y}{x^2+y^2}$.

7 (a) $z^2=x^2+y^2+i\,2xy$; $\sqrt{i}=\pm\frac{\sqrt{2}}{2}(1+i)$.

(b) $\frac{7+24i}{5}$; $5x^2-14x+125=0$.

8 $p=-5-4i$, $q=1+7i$; $a=-3$, $b=-1$.

9 z^2+2z+5.

10 x^2-2x+2; $(1+i)$ or $(1-i)$ or $(-1+2i)$ or $(-1-2i)$.

11 $p=3$, $q=5$; $2-i$.

12 $\pm(3+2i)$ and $\pm(3-2i)$; $z=5+3i$ or $2+i$; $5-3i$ or $2-i$.

13 $p=\pm3$, $q=\pm1$.

26 Complex numbers and graphs

1 (a) $a=\pm\frac{1}{2}$, $b=\pm\frac{1}{2}$; $z_1=2\left(\cos\frac{\pi}{2}+i\sin\frac{\pi}{2}\right)$,

$z_2=\frac{1}{2}\left(\cos\frac{\pi}{4}+i\sin\frac{\pi}{4}\right)$, $z_3=\frac{1}{2}\left[\cos\left(\frac{-3\pi}{4}\right)+i\sin\left(\frac{-3\pi}{4}\right)\right]$.

(b) $2+i2\sqrt{3}$; $z_4=4\left[\cos\frac{\pi}{3}+i\sin\frac{\pi}{3}\right]$;

$z_5=4\left[\cos\left(\frac{-2\pi}{3}\right)+i\sin\left(\frac{-2\pi}{3}\right)\right]$.

2 $\frac{1}{2}\left[\cos\left(\frac{-\pi}{3}\right)+i\sin\left(\frac{-\pi}{3}\right)\right]$.

3 (a) $4\left(\cos\frac{\pi}{3}+i\sin\frac{\pi}{3}\right)$. (b) $\frac{1}{2}\left[\cos\left(\frac{-\pi}{6}\right)+i\sin\left(\frac{-\pi}{6}\right)\right]$.

4 (a) $1,\theta$; $2\cos\frac{\theta}{2}$, $\frac{\theta}{2}$; $2\sin\frac{\theta}{2}$, $\frac{\pi}{2}+\frac{\theta}{2}$.

5 (i) disc, centre $(0,0)$, radius 1;
(ii) disc, centre $(1,1)$, radius 2.

6 (a) $127°$. (b) (i) circle, centre $(1,1)$, radius 1;
(ii) perpendicular bisector of line joining $(1,0)$ to $(0,-1)$;
(iii) half line, $x=1$ ($y>0$).

7 (i) 1; (ii) $2+2i$.

8 (a) $-2+4i$. (b) $\frac{2-i}{30}$; $P\hat{O}Q=90°$.

9 $x^2+y^2+10x+16=0$; $P\left(\frac{-16}{5},\frac{-12}{5}\right)$.

10 1, $\tan^{-1}\left(\frac{3}{4}\right)-\pi$; radius 1, centre $(0,0)$.

27 Differentiation

1 (i) $-\frac{2}{x^2}-\frac{1}{2x^{\frac{3}{2}}}$; (ii) $\frac{x^2+2x-1}{(x+1)^2}$.

2 (i) $12x+1$; (ii) $4x^3-2-\frac{2}{x^3}$; (iii) $\frac{6x+3}{4x^{\frac{3}{4}}}$.

3 ————

4 (i) $-\frac{3}{2x^{\frac{5}{2}}}$; (ii) $2x(3x+2)$; (iii) $3+\frac{2}{x^3}$.

5 (a) (i) $\frac{3}{2}x^{\frac{1}{2}}+\frac{1}{2x^{\frac{1}{2}}}$; (ii) $1-\frac{2}{x^3}+\frac{8}{x^5}$.

(b) $y=x^{\frac{1}{3}}$, $\frac{dy}{dx}=\frac{1}{3}$.

6 ————

7 (a) $x^2(3\ln x+1)$. (b) $\frac{-x\sin x-1-\cos x}{x^2}$.

8 $\frac{dy}{dx}=\frac{6x}{(2x^2+1)^2}>0$ for $x>0$; $-1\leqslant y\leqslant0$.

9 $\frac{dy}{dx}=\frac{2(x^2+4)}{(x^2-4)^2}$; $\frac{1}{3}\leqslant y\leqslant\frac{5}{3}$.

10 (a) ————

(b) $\left(\frac{\pi}{6},-3\sqrt{3}\right)$, $\left(\frac{5\pi}{6},3\sqrt{3}\right)$.

28 Methods of differentiaton

1 (i) $x(x^3+4)(22x^3-9x+16)$; (ii) $\frac{7x^4+8x^3-1}{2(x+1)^{\frac{3}{2}}}$;

(iii) $-1\left(x+\frac{1}{x}\right)^{-2}\left(1-\frac{1}{x^2}\right)$;

(iv) $\frac{1}{2}\cos\frac{x}{2}\cos^3 x-3\sin\frac{x}{2}\cos^2 x\sin x$;

(v) $4\tan^3 2x.\,2\sec^2 2x$.

2 (i) (a) $6x(x^2+1)^2$. (b) $12\sin^3 3x\cos 3x$. (c) $\frac{3x+1}{(2x+1)^{\frac{1}{2}}}$.

3 (a) $\frac{3}{2}x^{\frac{1}{2}}-\frac{1}{2x^{\frac{1}{2}}}$. (b) $2\sin 2x$.

4 (i) $2x\cos 3x-3x^2\sin 3x$; (ii) $e^x\left(\ln x+\frac{1}{x}\right)$;

(iii) $6x(x^2+2)^2$; (iv) $\frac{3+x}{(x^2+1)^{\frac{3}{2}}}$.

5 (i) $8\tan^3 2x\sec^2 2x$; (ii) $-\frac{1}{(2x-3)^2}$; (iii) $x(2\ln x+1)$.

6 $\frac{y-x^2}{y^2-x}$.

8 $\frac{dy}{dx}=\cot\left(\frac{t}{2}\right)$.

29 Applications of differentiation

1 (a) $1\cdot2$ s. (b) $10\frac{2}{3}$ m.

2 (a) $14\cdot5$ m/s. (b) $15\cdot52$ m.

3 (i) -28 m/s²; (ii) 0 and 12 s; (iii) 18 s.

4 $9y=27x-16$; $9y+2=0$.

5 (a) $18x + 9x^{\frac{1}{2}} + 1$. (b) $\left(-\frac{1}{2}, \frac{3}{2}\right)$, $\left(\frac{1}{2}, -\frac{3}{2}\right)$.

6 ———————

7 $\dfrac{\sqrt{3}}{2} - \dfrac{\pi}{2}$.

8 $y = x - \dfrac{\pi}{2} + 2$; $y = -x + \dfrac{\pi}{2}$.

9 (i) $6 \sin 3x \sec^3 3x$; (ii) $\dfrac{t(t+2)}{(t+1)^2}$; (iii) $-\dfrac{5}{4}$.

10 (a) $y + x = \dfrac{4}{3}$. (b) $(0, 1)$, $\left(\dfrac{2}{3}, -\dfrac{1}{3}\right)$.

(c) max $y = 1$, min $y = -\dfrac{1}{3}$. (d) $(1, 0)$, $\left(-\dfrac{1}{3}, \dfrac{2}{3}\right)$.

30 Changes

1 100π cm^3 s^{-1}.
2 (a) $r = 4$ cm. (b) 8π cm^2/s.
3 860 m^2/s.
4 (a) $1 \cdot 2\pi$. (b) $0 \cdot 05$ cm.
5 4 cm^2 s^{-1}.

6 (i) 10π m^2 s^{-1}; (ii) $\dfrac{x}{y} = \dfrac{3}{2}$; (iii) $1 \cdot 5$ s.

7 $\dfrac{dv}{dt} = -\dfrac{v^2}{u^2}\dfrac{du}{dt}$; $\dfrac{1}{8}$ cm s^{-1}.

8 $16\frac{1}{2}$ s; $-\dfrac{a}{27}$ cm/s.

9 $0 \cdot 02$.
10 $1 \cdot 39T$.

31 Special points

1 (i) $p = 2$, $q = 1$; (ii) 6.
2 $a = -2$, $b = 4$.
3 Depth: $32x^{-2}$ cm; width $3 \cdot 42$ cm, length $13 \cdot 68$ cm, depth $2 \cdot 74$ cm.

4 $x = \dfrac{1}{3}$, maximum.

5 (a) $x = 0$, maximum; $x = -4$, minimum.

6 $\dfrac{dy}{dx} = 2t(1+t)\,\mathrm{e}^t$; $\dfrac{d^2y}{dx^2} = 2(1+t)(1+2t+2t^2+2t^3)\,\mathrm{e}^t$; minimum at $(0, 1)$.

7 $\left(-1, \dfrac{1}{3}\right)$.

8 (i) min $(a, 0)$, max $\left(x = \dfrac{a+2}{3}\right)$;

(ii) max $(a, 0)$, min $\left(x = \dfrac{a+2}{3}\right)$.

9 ———————

10 $0 < V < \dfrac{32}{81}\pi a^3$.

32 Curve sketching

1 ———————
2 $x = 0$ or -2; minimum at $(0, 0)$, maximum at $(-2, -4)$.

3 (i) Inflexion at $(0, 0)$, maximum at $\left(\dfrac{3}{4}, \dfrac{27}{256}\right)$;

(ii) asymptote $x = -2$, maximum at $\left(0, \dfrac{1}{4}\right)$.

4 (i) $y \to 1$ as $x \to \pm\infty$; (ii) $\dfrac{2}{3} < y < 2$;

(iv) maximum at $(-1, 2)$, minimum at $\left(1, \dfrac{2}{3}\right)$.

5 (a) minimum at $(0, 0)$, maximum at $\left(\dfrac{2}{3}, \dfrac{4}{27}\right)$.

(b) $-\infty < y < 1$. (c) asymptotes $x = 0$ and $y = \pm 1$.

6 Minimum $\left(-\dfrac{\pi}{4}, -\dfrac{1}{\sqrt{2}}\,\mathrm{e}^{-\frac{\pi}{4}}\right)$; maximum $\left(\dfrac{3\pi}{4}, \dfrac{1}{\sqrt{2}}\,\mathrm{e}^{\frac{3\pi}{4}}\right)$.

7 $y \to 0$ as $x \to \pm\infty$, $y \to \pm\infty$ as $x \to \pm 1$; maximum at $\left(-\dfrac{1}{2}, -4\right)$, minimum at $(-2, -1)$.

8 $\dfrac{dy}{dx} = 8x^3 - 2x$; maximum at $(0, 1)$, minima at $\left(-\dfrac{1}{2}, \dfrac{7}{8}\right)$ and $\left(\dfrac{1}{2}, \dfrac{7}{8}\right)$.

9 (ii) $\dfrac{dy}{dx} = -\dfrac{1}{x^2(1-x^2)^{\frac{1}{2}}} < 0$ for all x; (iii) 2.

33 Integration

1 $\dfrac{9}{8}$.

2 (a) $4\left(2x - \dfrac{1}{x^2}\right)\left(1 + \dfrac{1}{x^3}\right)$. (b) $13\frac{1}{2}$.

3 (a) (i) $4x - \dfrac{4x^3}{3} + \dfrac{x^5}{5} + c$; (ii) $\dfrac{2}{3}x^{\frac{3}{2}} + 2x^{\frac{1}{2}} + c$.

(b) $-\dfrac{8}{3}$.

4 (i) (a) $-\dfrac{1}{x} - \dfrac{3}{x^2} - \dfrac{3}{x^3}$. (b) $-\dfrac{2}{9}(1-3x)^{\frac{3}{2}} + c$.

(c) $\dfrac{x}{2} - \dfrac{\sin 3x}{6} + c$; (ii) (a) $\dfrac{1}{30}$; (b) 1; (c) $\dfrac{\sqrt{3}}{2}$.

5 (i) $\dfrac{\mathrm{e}^{3x}}{3} - \dfrac{1}{x} + c$; (ii) $\dfrac{8 + \mathrm{e}^3}{3}$; (iii) $\dfrac{\pi}{4} + \dfrac{1}{2}$.

6 (a) $13\frac{1}{2}$; (ii) $6(\sqrt{3} - 1)$; (iii) $-\dfrac{2}{3\pi}$.

7 (i) $\dfrac{(x^4 - 3)^6}{24} + c$; (ii) $\ln \sec x + c$; (iii) $\dfrac{\mathrm{e}^{x^2}}{2} + c$.

8 $-\cos x - \dfrac{\cos^3 x}{3} + c$.

9 $p = 2$, $q = 1$; $\dfrac{\pi}{4}$; $\ln 2$; $\ln 2 + \pi$.

34 Methods of integration

1 $\dfrac{1}{8}$.

2 $0 \cdot 16$.

3 (a) $33\frac{2}{5}$. (b) $\dfrac{2}{3}$. (c) $\ln\left(\dfrac{7}{5}\right)$.

4 (i) $\dfrac{1}{x-2} + \dfrac{3}{5-x}$; (ii) $4 \ln 2$.

5 (a) (i) $\dfrac{1}{2} \ln 2$; (ii) $\dfrac{(2\mathrm{e}^3 + 1)}{9}$. (b) $2\mathrm{e}^{-\sqrt{x}} + c$.

6 $\dfrac{\pi}{2}$; $\dfrac{1}{4} - \dfrac{\sqrt{3}}{8}$.

7 (i) $\sqrt{(2x+1)} + c$; (ii) $\dfrac{x}{2} - \dfrac{1}{4}\ln|2x+1| + c$.

8 (a) $\dfrac{\pi}{12}$. (b) $x\,\mathrm{e}^{2x} - \dfrac{\mathrm{e}^{2x}}{2} + c$. (c) $\dfrac{1}{x-1} + \ln\left|\dfrac{x-1}{2x+1}\right| + c$.

35 Applications of integration

1 8.

2 (a) (i) $(1, 0)$, $(2, 0)$, $(3, 0)$; (ii) $\frac{1}{4}$.

3 (a) Inflexion at $(0, 0)$, minimum at $\left(-\frac{3}{4}, -\frac{27}{128}\right)$.

(b) $(-1, 0)$, $(0, 0)$; (c) $\frac{1}{10}$.

4 $\frac{63\pi}{2}$ cubic units.

5 (i) $(1, -4)$ and $(25, 20)$; (ii) $193 \cdot 6$ square units.

6 $A(6, 8)$, $B(-6, 8)$; $112 \cdot 34$ square units; $\frac{544\pi}{3}$ cubic units.

7 (i) $\frac{14}{3}$; (ii) $\frac{188\pi}{15}$.

8 $\frac{\pi}{12} + \frac{\sqrt{3}}{8}$ square units.

9 $2 \cdot 15$ cubic units.

36 Differential equations

1 $y = \frac{4x}{10 - 4x}$.

2 $\ln y = \frac{x^2}{2} \ln x - \frac{x^2}{4} + \frac{1}{4}$.

3 $2y = \ln (x^2 + x)$.

4 $\frac{e^{-2y}}{2} + e^{-y} = \frac{3}{2} - \frac{x}{2} - \frac{\sin 2x}{4}$.

5 $\tan y = \ln (1 + x) + 1$.

6 $y = \frac{e^x}{1 + e^x}$.

7 $\ln \left(\frac{y}{20}\right) = \frac{1}{(1 + t)^2} - 1$; as $t \to \infty$, $y \to \frac{20}{e}$.

8 $t = \frac{1}{\beta V} \ln (3a)$, $m = \frac{V e^{\beta V t}}{a(3 + e^{\beta V t})}$.

37 Numerical solution of equations

1 $1 \cdot 03$.
2 2 roots; $3 \cdot 98$; $-1 \cdot 06$.
3 $2 \cdot 02$.
4 $1 \cdot 179747$.
5 $0 \cdot 53$.
6 $18 \cdot 5$.
7 $0 \cdot 627$; $e^2 - e - 2$.

38 Numerical integration

1 $10/9$.
2 $0 \cdot 467$; $0 \cdot 475$; $0 \cdot 476$; $3 \cdot 116$.
3 $-0 \cdot 026$.
4 $57 \cdot 3$; $57 \cdot 29$.
5 $0 \cdot 867$.
6 $3 \cdot 69$; $6 \cdot 71$.
7 $0 \cdot 815$; $0 \cdot 815$.
8 $0 \cdot 160$.
9 $2 \cdot 4948$.

39 Functions

1 (a) Max $= 5$. (b) $(-\infty, 5)$.
2 (i) fg; (ii) g^{-1}; (iii) g^2; (iv) g^2f; (v) gf^{-1}.

3 (a) $f: x \to 2 - \frac{1}{x} x \neq 0$. (b) (i) $\frac{1}{3}$; (ii) $5/3$.
4 —————
5 (a) 0. (b) $9^{2n+2} - 5^{2n+2}$.
6 $(-\infty, \infty)$ $g \neq 0$; $f^{-1}: x \to a^x$ $(x \in R_+, a > 1)$; g^{-1}:
$x \to \frac{1}{x}$ $(x \in R_+)$.

7 $x = 1$; $1/5$.
8 Periods π, π, π Range $[-1, 1]$.
9 —————
10 $b > -9$; $x = 11$, -5; Non-commutative.
11 $x \to \sin x$.

40 Matrices

1 (i) $M^2 = \begin{pmatrix} 3 & -5 \\ 5 & 8 \end{pmatrix}$, $M^3 = \begin{pmatrix} 1 & -18 \\ 18 & 19 \end{pmatrix}$, $M^{-1} = \frac{1}{7}\begin{pmatrix} 3 & 1 \\ -1 & 2 \end{pmatrix}$;

$x = 2$, $y = 1$;

(ii) $\begin{pmatrix} 0 & -2 \\ -2 & 0 \end{pmatrix}$, $(-1, -3)$.

2 (i) $(13, 19)$, (ii) $\frac{1}{10}\begin{pmatrix} 3 & -1 \\ -2 & 4 \end{pmatrix}$, (iii) $a = 2$, $b = 3$,

(iv) $(5a, 5a)$, (v) $y = x$.

3 (a) I. (b) $\begin{pmatrix} 1 & -1 & -1 \\ -1 & 3 & -1 \\ 0 & 1 & 3 \end{pmatrix}$.

4 (a) $m_2 = \frac{m_1}{1 + m_1}$, $y = 0$.

5 $4x + 5y = 0$; $x + y = 0$, $2x - 3y = 0$.

41 Force diagrams

1

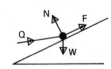

W – weight
N, R – normal reactions
F – friction

2 (a)

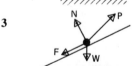

W – weight of ladder
N, R – normal reactions
F – friction
P – normal reaction
between man and
ladder.

(b)

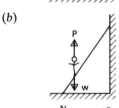

w – weight of man
P – normal reaction

3

P – pull
W – weight
N – normal reaction
F – friction

4

Q – push
W – weight
N – normal reaction
F – friction

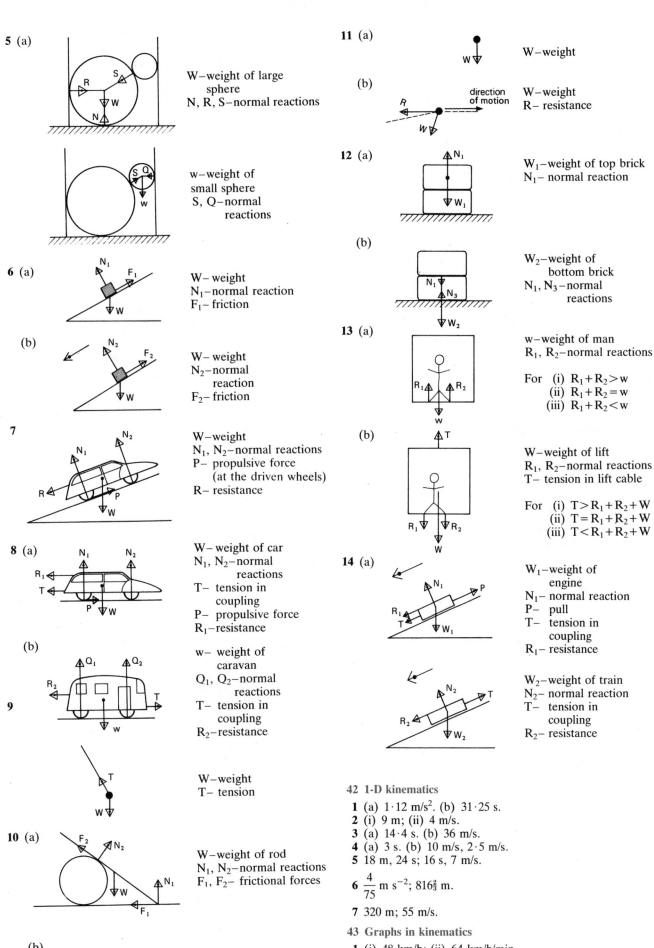

5 (a) W–weight of large sphere
N, R, S–normal reactions

w–weight of small sphere
S, Q–normal reactions

6 (a) W– weight
N_1–normal reaction
F_1– friction

(b) W– weight
N_2–normal reaction
F_2– friction

7 W–weight
N_1, N_2–normal reactions
P– propulsive force (at the driven wheels)
R– resistance

8 (a) W– weight of car
N_1, N_2–normal reactions
T– tension in coupling
P– propulsive force
R_1–resistance

(b) w– weight of caravan
Q_1, Q_2–normal reactions
T– tension in coupling
R_2–resistance

9 W–weight
T– tension

10 (a) W–weight of rod
N_1, N_2–normal reactions
F_1, F_2– frictional forces

(b) w–weight of cylinder
N_2, N_3–normal reactions
F_2–friction

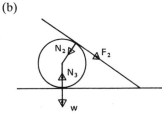

11 (a) W–weight

(b) direction of motion
W–weight
R– resistance

12 (a) W_1–weight of top brick
N_1– normal reaction

(b) W_2–weight of bottom brick
N_1, N_3–normal reactions

13 (a) w–weight of man
R_1, R_2–normal reactions

For (i) $R_1+R_2>w$
(ii) $R_1+R_2=w$
(iii) $R_1+R_2<w$

(b) W–weight of lift
R_1, R_2–normal reactions
T– tension in lift cable

For (i) $T>R_1+R_2+W$
(ii) $T=R_1+R_2+W$
(iii) $T<R_1+R_2+W$

14 (a) W_1–weight of engine
N_1– normal reaction
P– pull
T– tension in coupling
R_1– resistance

W_2–weight of train
N_2– normal reaction
T– tension in coupling
R_2– resistance

42 1-D kinematics

1 (a) $1 \cdot 12$ m/s^2. (b) $31 \cdot 25$ s.
2 (i) 9 m; (ii) 4 m/s.
3 (a) $14 \cdot 4$ s. (b) 36 m/s.
4 (a) 3 s. (b) 10 m/s, $2 \cdot 5$ m/s.
5 18 m, 24 s; 16 s, 7 m/s.

6 $\dfrac{4}{75}$ m s^{-2}; $816\frac{2}{3}$ m.

7 320 m; 55 m/s.

43 Graphs in kinematics

1 (i) 48 km/h; (ii) 64 km/h/min.
2 (ii) 3 s; (iii) 15 m/s; (iv) $1\frac{2}{3}$ m/s^2.
3 (i) 18 m s^{-2}, 3 m s^{-2}; (ii) 1035 m; (iii) $3 \cdot 88$ km h^{-1}.

4 (ii) (a) $\dfrac{2V(a+r)}{3ar}$. (b) $\dfrac{7V(a+r)}{6ar}$. (iii) $\dfrac{4}{7}V$.

5 (i) $\dfrac{4T}{3}$; (ii) $\dfrac{2VT}{3}$.

44 Relative motion

1 6 m s^{-1} from S60°E; from S41°E.
2 11·1 knots on bearing 339°; 340·6°.
3 2·7 h, 2·6 h, 2·4 h.
4 $P(30t - 10, 0)$, $Q(0, 10 - 40t)$, $PQ = 10\sqrt{(25t^2 - 14t + 2)}$, approximately 17 minutes past noon, 50 km per hour at S36·9°W.

45 1-D particle dynamics

1 5·6 N.
2 (i) $9\frac{3}{8}$ m/s^2; (ii) 16·25 N; (iii) 95 m.
3 (i) 167·5 N; (ii) 2·58 m s^{-2}.
4 $T = 1200$ N; (i) $4\frac{1}{6}$ s; (ii) $10^5/_{12}$ m.
5 1·26 kg, 6·315 m/s^2.
6 0·4, 2·066 m s^{-2}.

46 Connected particles

1 (i) $13\frac{1}{3}$ N; (ii) $\dfrac{3\sqrt{10}}{10}$ s.

2 (a) 2 m s^{-2}. (b) 3 m s^{-1}. (c) 3·6 m.
3 (a) 1·4 m/s^2. (b) 2·52 N. (c) 3 s. (d) 4·2 m/s.

4 (i) $\dfrac{g}{2}$; (ii) $\dfrac{3mg}{2}$; (iii) $\dfrac{3}{2}mgh$; (iv) $\dfrac{5h}{2}$, $3\sqrt{\dfrac{h}{g}}$; (v) $\dfrac{\sqrt{hg}}{4}$.

5 3·92 ms^{-2}, 42 N; $\dfrac{4}{7}$ s.

47 Work and energy

1 52 J; (i) $25d$ J; (ii) $40d$ J; $d = 0·8$ m.
2 (i) 10·08 kJ (ii) 24 kJ; 82·8 N

3 (i) mgd; (ii) $\sqrt{2gd}$; (iii) $\dfrac{8}{5}mgd$; (iv) $3mgd$.

4 $x = \dfrac{2E}{mg}$, $\lambda = \dfrac{m^2g^2l}{2E}$.

48 Power

1 400 N.
2 (i) 8400 N, 240 N; (ii) 3590 N, 990 N; (iii) 12·6 kW.
3 (i) 1500 N; (ii) (a) 0·05 m s^{-2}; (b) 1650 N, (c) 24·75 kW; (iii) $33\frac{1}{3}$ kW.
4 500 N, 15 kW.
5 10 m s^{-1}, 0·2 m s^{-2}.

6 1·8 kW; $\dfrac{1}{15}$ m s^{-2}; 25 minutes.

49 Impulse and momentum

1 (i) 6 m/s; (ii) 24 Ns; (iii) 5832 J.
2 (a) 2120 Ns. (b) 10·6 kN.

3 $x = \dfrac{u}{6}$, $y = \dfrac{u}{4}$.

4 u, $2u$.
5 (i) 15·25 m s^{-1}, 900 m s^{-1}; (ii) 6 m s^{-1}; (iii) 0·07 m s^{-1}; (iv) 96·7 kJ, 8900 Ns.

50 Impact

1 1·5 m s^{-1}; (i) 25 s; (ii) $6\frac{7}{8}$ m.

2 $\dfrac{1}{2}$.

3 $e^{-\frac{1}{2}}$.

4 (i) $\dfrac{1}{2}$.

5 $4V$, $45\,mV^2$; $-V$, $14V$.

6 $\dfrac{2}{7}$; $\dfrac{3u}{1+k}$; $k > \dfrac{10}{11}$.

51 Projectiles

1 (i) 39·3 m; (ii) 60 m; (iii) $4\sqrt{3}$ s; (iv) 138·6 m.
2 (a) 27 m. (b) 24 m. (c) 48°.
3 73·7 m; 29·6 m s^{-1}.

4 $2\sqrt{\dfrac{a}{g}}$.

5 $\tan\alpha + \tan\beta = \dfrac{5}{x}$, $\tan\alpha\tan\beta = \dfrac{5y + x^2}{x^2}$.

52 Motion in a Horizontal Circle

1 $\dfrac{1}{2}$ kg; 81·8 revs/min.

2 (a) $\sqrt{\dfrac{3ag}{2}}$ (b) ——————

3 $m(g\cos\alpha + l\omega^2\sin^2\alpha)$, $\omega^2 = \dfrac{g}{l\cos\alpha}$.

4 $\dfrac{5mu^2}{9a} + \dfrac{5}{8}mg$; $\dfrac{5mu^2}{9a} - \dfrac{5}{8}mg$.

5 $2\pi\sqrt{\dfrac{ma}{F}} \approx 2·55$ h.

53 Motion in a Vertical Circle

1 60° to upward vertical; $h = \dfrac{27a}{16}$.

2 (ii) $u^2 = \dfrac{5ag}{2}$, $v = \sqrt{\dfrac{ag}{2}}$.

3 $\sqrt{\dfrac{2ag}{3}}$.

4

5 (b) $\theta = 60$, $v = \sqrt{3ag}$; $\sqrt{ag}$.
6 0·994 m; clock gains; new length 1·005 m.

54 Variable forces

1 $\left(\dfrac{g}{k} + \dfrac{e^{-2kx}}{k}(kw^2 - g)\right)^{\frac{1}{2}}$.

2 (i) 48·2 m; (ii) 24 s.

3 (i) $v = Ue^{k(a-x)}$; $v = \dfrac{U}{1 + Ukt}$; (iii) $x = a + \dfrac{1}{k}\ln(1 + Ukt)$.

4 $\dfrac{dv}{dt} = -g\left(1 + \dfrac{v^2}{V^2}\right)$.

5 (i) 2·77; (ii) $v = \dfrac{e^{\frac{t}{2}}}{4 + e^{\frac{t}{2}}}$; $x_{0·8} = 4\ln 2$, $x_{0·6} = 2\ln 2$.

6 $\left(1 + \dfrac{k}{g}u\right)e^{-kt} - \dfrac{g}{k}$.

55 Simple harmonic motion

1 (i) (a) $\sqrt{10}$ m; (b) π s; (c) $2\sqrt{10}$ m/s; (d) $\dfrac{\pi}{4}$ s; (ii) (a) 8 m/s^2. (b) 64 N; (iii) $x = \sqrt{10}\sin 2t$ m,
$t_1 = \sin^{-1}\left(\dfrac{\sqrt{10}}{10}\right)$ s, $t_2 = \sin^{-1}\left(\dfrac{\sqrt{10}}{5}\right)$ s,
$\sin^{-1}\left(\dfrac{\sqrt{10}}{5}\right) - \sin^{-1}\left(\dfrac{\sqrt{10}}{10}\right)$ s.

2 (i) $x = 3$, $v = -12$; (ii) $x = 5$.
3 ——————
4 (i) 0·08 m; (ii) 0·6 m s^{-1}; 3 N.
5 $\ddot{x} = -5$ m s^{-2}; 7 cm.

56 Vectors in dynamics

1 6 s.
2 $2\sqrt{5}$ N; $8\mathbf{i} + 16\mathbf{j}$; $2\sqrt{13}$ m s^{-1}.

3 (i) $3\mathbf{i} + 4\mathbf{j}$, $\dfrac{4\mathbf{i} - 4\lambda\mathbf{j}}{\sqrt{(1+\lambda^2)}}$; (ii) $(3t-1)\mathbf{i} + 4t\mathbf{j}$,
$\dfrac{4t\mathbf{i} - (4\lambda t + 8\sqrt{(1+\lambda^2)})\mathbf{j}}{\sqrt{(1+\lambda^2)}}$;
$\dfrac{t(4 - 3\sqrt{(1+\lambda^2)})}{\sqrt{(1+\lambda^2)}}\mathbf{i} - \dfrac{4t(\sqrt{(1+\lambda^2)} + \lambda)}{\sqrt{(1+\lambda^2)}}\mathbf{j}$.

4 (i) ―――――

(ii) 1 hour, 4 km; (iii) 3 hours;

(iv) 10 km/h at $\tan^{-1}\left(\dfrac{4}{3}\right)$ to x-axis.

5 ―――――

6 $m\left(-\dfrac{v}{2}\mathbf{i}+\dfrac{v\sqrt3}{6}\mathbf{j}\right)$; $\dfrac{v\sqrt3}{9}$ at 30° clockwise to x-axis.

7 (i) $\mathbf{F}.\mathbf{v}$; (ii) $\displaystyle\int_0^T \mathbf{F}.\mathbf{v}\,dt$; $-4a\sin 2t\,\mathbf{i}+2a\cos 2t\,\mathbf{j}$;

$-4m\mathbf{r}$; $6a^2m$; $\dfrac{\pi}{4}$.

57 Coplanar concurrent forces

1 $P=24$ N, $Q=12\sqrt3$ N.
2 8·83 N at 316° 34′; $P=8·82$ N, $Q=0·25$ N.
3 $P=10$ N at 090°; 223° 51′; 18·03 N.
4 (i) 302·4 N at 41° 24′; (ii) $Q=3P$.
5 $\mathbf{F}_1=9\mathbf{i}+12\mathbf{j}$, $\mathbf{F}_2=3\mathbf{i}-3\mathbf{j}$, $\mathbf{F}_3=8\mathbf{i}+4\mathbf{j}$,
$\mathbf{F}_R=20\mathbf{i}+13\mathbf{j}$.
6 (a) (i) $\overrightarrow{AC}=\mathbf{a}+\mathbf{b}$; (ii) $\overrightarrow{AD}=2\mathbf{b}$; (iii) $\overrightarrow{AE}=2\mathbf{b}-\mathbf{a}$;
(iv) $\overrightarrow{AF}=\mathbf{b}-\mathbf{a}$. (b) $7\mathbf{i}-\mathbf{j}$; $\overrightarrow{OA}=4\mathbf{i}+3\mathbf{j}$;
$\overrightarrow{OB}=14\mathbf{i}-2\mathbf{j}$, $\overrightarrow{CO}=-6\mathbf{i}+8\mathbf{j}$; 15 N.

58 Moments and couples

1 $\mathbf{F}_1=3\mathbf{i}+6\mathbf{j}$, $\mathbf{F}_2=\mathbf{i}-2\mathbf{j}$, $\mathbf{F}_R=4\mathbf{i}+4\mathbf{j}$.
2 $X=25$ g, 65 g, 45 g.
3 5 kg, 10 kg; (i) 30 N; (ii) 3·75 m.
4 2·75 m from A; 20·5 kg, 39·5 kg.
5 (a) 14²⁄₇ N. (b) 171³⁄₇ N.
6 $P=\dfrac{w(6-y)}{12-x-y}$, $Q=\dfrac{w(6-x)}{12-x-y}$; (ii) 3 cm.
7 (a) 2·4; (b) 56 N and 88 N.
8 (a) 10 N. (b) $\tan^{-1}\left(\dfrac{1}{3\sqrt3}\right)=10°\,54′$. (c) $10\sqrt7$ N;

90 Nm anticlockwise.

9 $\dfrac{3W}{4}$, c of g at $\dfrac{7a}{3}$ from A; $\dfrac{W}{2}$ and $\dfrac{W}{4}$.

59 Equilibrium

1 (i) $15\sqrt6$ N, 30 N (twice), $30\sqrt3$ N;
(ii) $1·5(\sqrt3-1)$ kg, 3 kg.

2 (a) $\dfrac{W\sqrt3}{6}$ (b) $\sqrt{\dfrac{13}{12}}\,W$ at 73° 53′ to horizontal.

(c) $\dfrac{W}{3}$. (d) 71° to horizontal.

3 $\dfrac{1180\sqrt3}{3}$ N; $X=\dfrac{590\sqrt3}{3}$ N; $Y=290$ N (down).

60 Three force problems

1 (i) 64·5 N; (ii) 60° 15′.
2 (i) 22°; (ii) 21 N and 41·5 N.
3 (i) 40·9°; (ii) $50\sqrt7$ N.
5 $\dfrac{38}{65}$.

61 Friction

1 $\dfrac{8}{27}$.

2 7·5 N; 5 m s^{-2}.

3 $\dfrac{W\sqrt3}{6}$; W.

4 (i) $\dfrac{P}{mg}$. (ii) ―――――

62 Bodies in contact

1 Equilibrium broken at C.

2 (i) C; (ii) $\dfrac{W}{2}\sqrt{(1+9\mu^2)}$ at $\tan^{-1}(3\mu)$ to vertical;

(iii) $\tan^{-1}\left(\dfrac{2}{3\mu}\right)$.

3 $\dfrac{1}{\sqrt3}$.

63 Equivalent systems of forces

1 7 N at 38·2° to AC; 4·5 m.
2 (a) $2\sqrt{10}$ N at $\tan^{-1}3$ to x-axis. (b) 71·6°.
(c) $a=8\frac{2}{8}$ m, $y=3x-26$.
3 48 Nm in sense $ABCD$.

4 65 N at $\tan^{-1}8$ to OA; line of action $\dfrac{3}{8}$ m from O in direction
OA and 3 m from O in direction CO; 3 N in direction BA.

5 $k=\dfrac{35}{6}$; 43·2 cm.

6 (i) 10 P; (ii) 53·1°; (iii) $3a$; (iv)
(v) $30a$ in sense $ABCD$. (vi) 6 N, $3\frac{1}{2}$ N, $4\frac{1}{2}$ N.

64 Centres of mass

1 (a) $3\frac{1}{2}$ cm. (b) $2\frac{1}{3}$ cm.

2 (a) $\dfrac{40}{59}$ mm from O. (b) $\bar{x}=10$, $\bar{y}=6$.

3 7·88 cm.

4 $\left(\dfrac{7a}{18},\dfrac{4a}{9}\right)$.

65 Suspending and toppling

1 $\dfrac{2}{3}$.

2 $\dfrac{14b}{15}$ from AB, $\dfrac{19a}{15}$ from AD; 45°.

3 $\sin\beta=\dfrac{3+k}{8(k-8)}$.

66 Pictorial representation

1 309°, 18°, 15°, 12°, 6°.
2 ―――――
3 ―――――
4 5·48 cm 6 cm 37·5°.
5 ―――――

68 Mode and means

1 5·65 cm.
2 51·6.
3 1011·3 millibars.
4 (a) (i) 3·5; (ii) 2·994.
5 ―――――

6 (i) $\bar{a}+\bar{b}$; (ii) $\dfrac{1}{2}(\bar{a}+\bar{b})$; (iii) $100+10\bar{a}+\bar{b}$; (iv) AB;
(v) A^2B^3.
7 (a) (i) 5.

69 Median and quantiles

1 (i) 46·25; (ii) 29·80, 9·125%.
2 (b) 233 thou. (d) 2·165 tonnes. (e) 516 250 tonnes.
3 (ii) 35 years; (iii) Mean $=38·75$.
4 6·6, 4, 9·25.
5 437, 412·5, 435 kNm^{-2}.

70 Measures of dispersion

1 Mean 21·25 kg, s.d. 2·54 kg.
2 38·7 g, 6·69 g; 42·7 g, 6·69 g.
3 Mean 16, s.d. 37·9; (ii) mean 15, s.d. 50·7.
4 (a) 0·799. (b) 4·42.
5 mean 25·85 years, s.d. 1·99 years.
6 median 89·27 g, Q_3 93·62 g, 16 animals, mean 89·725 g, s.d. 6·52 g.

71 Index numbers and moving averages

1 125, 120, 100; 175, 150, 125; 157·5, 132·5.
2 $a = 45$, $b = 21$, $c = 29·25$, $d = 22·5$.
3 9, 11, 3, 13.

72 Probability

1 (i) 1/9; (ii) 1/5; (iii) 1/120; (iv) 1/20.
2 (a) 8/35. (b) 18/35; 3/35, 3/8.
3 (a) (i) 1/21; (ii) 41/42; (iii) 1/30. (b) 1/31.

4 1/5, 1/5, 3/7, 3/4. (i) not independent,

(ii) not mutually exclusive.

5 (a) (i) 1/6; (ii) 1/9; (iii) $(2/3)^{r-1}$ 1/6; 1/2. (b) 31/56.

6 (i) 5/192; (ii) 5/324; (iii) 1/64; (iv) 5/216.

73 Discrete probability distributions

1 Mean £7·57 Variance 12·96; £53, 3·00; £53, 99·0.
2 4, 4/25; 6, 8/25; 8, 4/25; 9, 4/25; 11, 4/25; 14, 1/25; 7·6 pence; £1·20 loss.
3 125/216, 75/216, 15/216, 1/216; 2 pence loss.
4 2/15, 34/15, 68/15, 2024/225.

74 Continuous probability distributions

1 (i) 3/16; (ii) 5/4; (iii) 19/80; (iv) 11/16.

2 (i) $2 - 8a$; (ii) $\frac{1}{8}$; (iii) mean $\frac{4}{3}$, variance $\frac{8}{9}$; (iv) $\frac{5}{16}$.

3 $k = 4$; mean 8/15; variance 11/225; median 0·541.
4 $A = 1/1200$, result not significant.

5 (i) $\frac{1}{4}$; (ii) $z = \frac{2}{3}(3 - \sqrt{3})$.

6 $\lambda = \frac{1}{2}$, mean $\frac{\pi}{2}$, variance $\frac{\pi^2}{4} - 2$, median $\frac{\pi}{2}$, quartiles $\frac{\pi}{3}, \frac{2\pi}{3}$; $\cos\left(\frac{\pi^2}{4} - 2\right)$.

75 The Binomial Distribution

1 (i) 0·206; (ii) 0·343; (iii) 0·816; (iv) 0·451.
2 (i) 16/45; (ii) (a) 0·201. (b) 0·302; 2.
3 (a) 671/1296. (b) 15/1296.
4 (i) 0·00000010; (ii) 1·00; (iii) 0·000074.
5 (a) 0·315. (b) (i) 0·53 Fit 0·049, 0·22, 0·37, 0·28, 0·079.
6 (i) 0·00149; (ii) 0·0624, 0·0000930.
7 ─────

76 The Poisson distribution

1 (a) 0·135. (b) 0·0527. (c) 0·00783.
2 Frequencies 35, 38, 21, 8, 2, 0.
3 0·143; limits 172·3, 227·7.
4 0·371, £78.53.
5 29 matches.
6 3, 0·223, 0·988.
7 (a) (i) 0·986; (ii) 0·223. (c) 6 pence per metre.
8 (a) 0·267. (b) 0·191 Insufficient evidence.
9 (a) 0·323. (b) 0·0119.

77 The normal distribution

1 Mean 68·82 mph, s.d. 5·36; new mean 63·98 mph, 13·1%.
2 Mean 60·0 mm, s.d. 0·2 mm, 10·0%.
3 Mean 5·03 cm, s.d. 0·09 cm, £409·27.
4 Mean 8 cm, s.d. 1·16 cm, Range 6·09 to 9·91 cm.
5 1·06, 0·020.

78 Uses of the normal distributions

1 £1250, 0·0036.
2 (i) 0·515; (ii) 0·376; 0·445 $N = 43$.
3 (a) 0·0404. (b) $r = 305$.
4 0·089; not significant.
5 (i) 0·226; (ii) 0·9988; 0·7752.
6 0·0221, 0·99389.
7 Probs 0·168, 0·360, 0·309, 0·132, 0·0284, 0·00243; 0·9474.

79 Sampling

1 (a) 13·2874. (b) 0. (c) 1. (d) 8·872 (other answers possible).
2 Mean 1·9, Variance 0·89; mean 1·9, variance 0·54.
3 Mean 12000 g, s.d. 57·01 g; 0·4304; 765·1 g.
4 Mean score 0·769, 0·101. Total score 10, 17·07; 0·0198.

80 Estimation

1 limits 0·521, 0·678; 0·835.
2 900·21, 0·83 limits 900·06, 900·36 g.
3 $\pi/4$, $\pi = 3·136$, limits 3·05, 3·22.
4 0·1587 C.I. 2·00, 2·02 Range 2·28 to 5·48.
5 Interval for proportion 0·00241, 0·00759; for number 13168, 41569.
6 (a) 10, 100. (b) 10, 1. (c) Interval 8·04, 11·96 Mean could be 9.

81 Hypothesis testing

1 602 g, 0·65 g; box 15050 g, 3·25 g. (i) Not significant; (ii) significant.
2 $a = 1·9$; 0·837.
3 Mean < 9·7; Mean 9·3 units, Variance 3·61.
4 (i) 1st significant, 2nd not significant.

5 $H_0\ p = \frac{1}{4}$, $H_1\ p > \frac{1}{4}$ sig level 5·08% 0·1808.

82 Linear regression

1 (a) 2·5, 1·2, $y = 6·7 - 2·2\,x$.
2 (ii) $y = 0·855\,x + 2·668$. (iii) 11·22, 79·62.
3 $y = x - 1641$; incorrect use of information.
4 $y = 0·9\,x + 2·2$; $y = 5·35$;
5 $Y = 0·43\,X + 56·3$; $r = 0·9949$.
6 (ii) $y = 1·03\,x + 0·533$.
7 (a) $y = 0·917 - 0·11\,x$.

83 Correlation

1 0·188.
2 (a) 0·753. (b) 0·767.
3 (a) 0·143. (b) 0·152.
4 0·603. $W = 0·888\,h - 75·9$.
5 $r = 0·02$; $a = 0·246$; $b = 0·012$.

84 χ^2

1 (a) 3, not biased. (b) 95% CI 0·140, 0·360.
2 Mean 0·9; 4·61, Poisson population.
3 6·63% defective is not the same.
4 3·24. Accept hypothesis.
5 frequencies 10·0, 23·1, 26·5, 20·3, 11·7, 5·4, 2·1, 0·7. Accept Poisson hypothesis.

85 Contingency tables

1 2·67; Not significant, proportion the same 95% CI 0·304, 0·496.
2 Significant at 1%. Not consistent.
3 Different proportions, highly significant.
4 8·45. Not significant. No association; 8·03 Significant.

86 Special graph papers

1 Mean = 12·97 s = 0·95. Frequencies 12, 48, 80, 48, 12.
2 Mean 1·19, Variance 1·91; 0·00003; (i) 0·028 (ii) 0·308.
3 (a) $a = 1·16$, $b = 2$; (b) Mean 70·4, s.d. 8·46.
4 (b) 0·00009.

Index

188